Handbook of Experimental Pharmacology

Volume 203

For further volumes:
http://www.springer.com/series/164

Mathias Schwanstecher
Editor

Diabetes - Perspectives in Drug Therapy

 Springer

Editor
Prof.Dr.Dr. Mathias Schwanstecher
Fakultät Lebenswissenschaften,
Molekulare Pharmakologie und
Toxikologie
Technische Universität Braunschweig
Beethovenstraße 55
38106 Braunschweig
Germany
M.Schwanstecher@tu-braunschweig.de

ISSN 0171-2004   e-ISSN 1865-0325
ISBN 978-3-642-17213-7   e-ISBN 978-3-642-17214-4
DOI 10.1007/978-3-642-17214-4
Springer Heidelberg Dordrecht London New York

Library of Congress Control Number: 2011923422

Cover design: SPi Publisher Services

Printed on acid-free paper

Springer is part of Springer Science+Business Media (www.springer.com)

Preface

Weight loss – in many cases, as little as 4 kg – with its pleiotropic benefits and optimal safety profile appears as the most effective means of managing type 2 diabetes (T2DM) (Harris 1991; Pories et al. 1995; Sjostrom et al. 2004; Pontiroli et al. 2005; Ratner et al. 2005; Pi-Sunyer et al. 2007; Dixon et al. 2008). In addition, the pharmacotherapeutic armamentarium seems well equipped with ten different classes of antidiabetic drugs; providing potent tools to achieve predefined HbA1c goals (i.e. insulin, sulphonylureas, metformin, thiazolidinediones, alpha-glucosidase inhibitors, glinides, GLP1-analogues, DPP-4 inhibitors, pramlintide, and colesevalam) (Rodbard et al. 2009).

What then are we struggling for? First, in most of the T2DM patients, present clinical praxis fails to attain sustained weight loss and glycemic control (Nathan et al. 2009). Second, even if optimal management of HbA1c, lipid profile and blood pressure could hypothetically be supplied, increased morbidity and mortality rates would still leave much room for improvement (Mourad and Le Jeune 2008). Third, unravelling the molecular pathophysiology of nutrient excess should allow to target the thrifty genotype roots of obesity and T2DM directly and should thus facilitate the development of highly efficient novel therapies (Neel 1999). Respectively, distinct encouragement evolves from potential mechanisms underlying treatments through metformin and bariatric surgery (Cummings et al. 2004; Foretz et al. 2010).

The chapters of this book report cutting-edge research on molecular events in adiposity and T2DM, thus opening the way for innovative drug-based therapeutic strategies. Beyond that, profound insights and exciting ideas are unveiled. Please, go ahead and explore!

Braunschweig
October 2010

M. Schwanstecher

References

Cummings DE, Overduin J, Foster-Schubert KE (2004) Gastric bypass for obesity: mechanisms of weight loss and diabetes resolution. J Clin Endocrinol Metab 89:2608–2615

Dixon JB, O'Brien PE, Playfair J, Chapman L, Schachter LM, Skinner S, Proietto J, Bailey M, Anderson M (2008) Adjustable gastric banding and conventional therapy for type 2 diabetes: a randomized controlled trial. JAMA 299:316–323

Foretz M, Hebrard S, Leclerc J, Zarrinpashneh E, Soty M, Mithieux G, Sakamoto K, Andreelli F, Viollet B (2010) Metformin inhibits hepatic gluconeogenesis in mice independently of the LKB1/AMPK pathway via a decrease in hepatic energy state. J Clin Invest 120:2355–2369

Harris MI (1991) Epidemiological correlates of NIDDM in Hispanics, whites, and blacks in the U.S. population. Diabetes Care 14:639–648

Mourad JJ, Le Jeune S (2008) Blood pressure control, risk factors and cardiovascular prognosis in patients with diabetes: 30 years of progress. J Hypertens Suppl 26:S7–S13

Nathan DM, Buse JB, Davidson MB, Ferrannini E, Holman RR, Sherwin R, Zinman B (2009) Medical management of hyperglycaemia in type 2 diabetes mellitus: a consensus algorithm for the initiation and adjustment of therapy: a consensus statement from the American Diabetes Association and the European Association for the Study of Diabetes. Diabetologia 52:17–30

Neel JV (1999) The "thrifty genotype" in 1998. Nutr Rev 57:S2–S9

Pi-Sunyer X, Blackburn G, Brancati FL, Bray GA, Bright R, Clark JM, Curtis JM, Espeland MA, Foreyt JP, Graves K, Haffner SM, Harrison B, Hill JO, Horton ES, Jakicic J, Jeffery RW, Johnson KC, Kahn S, Kelley DE, Kitabchi AE, Knowler WC, Lewis CE, Maschak-Carey BJ, Montgomery B, Nathan DM, Patricio J, Peters A, Redmon JB, Reeves RS, Ryan DH, Safford M, Van Dorsten B, Wadden TA, Wagenknecht L, Wesche-Thobaben J, Wing RR, Yanovski SZ (2007) Reduction in weight and cardiovascular disease risk factors in individuals with type 2 diabetes: one-year results of the look AHEAD trial. Diabetes Care 30:1374–1383

Pontiroli AE, Folli F, Paganelli M, Micheletto G, Pizzocri P, Vedani P, Luisi F, Perego L, Morabito A, Bressani Doldi S (2005) Laparoscopic gastric banding prevents type 2 diabetes and arterial hypertension and induces their remission in morbid obesity: a 4-year case-controlled study. Diabetes Care 28:2703–2709

Pories WJ, Swanson MS, MacDonald KG, Long SB, Morris PG, Brown BM, Barakat HA, deRamon RA, Israel G, Dolezal JM, et al. (1995) Who would have thought it? An operation proves to be the most effective therapy for adult-onset diabetes mellitus. Ann Surg 222:339–350, discussion 350–332

Ratner R, Goldberg R, Haffner S, Marcovina S, Orchard T, Fowler S, Temprosa M (2005) Impact of intensive lifestyle and metformin therapy on cardiovascular disease risk factors in the diabetes prevention program. Diabetes Care 28:888–894

Rodbard HW, Jellinger PS, Davidson JA, Einhorn D, Garber AJ, Grunberger G, Handelsman Y, Horton ES, Lebovitz H, Levy P, Moghissi ES, Schwartz SS (2009) Statement by an American Association of Clinical Endocrinologists/ American College of Endocrinology consensus panel on type 2 diabetes mellitus: an algorithm for glycemic control. Endocr Pract 15:540–559

Sjostrom L, Lindroos AK, Peltonen M, Torgerson J, Bouchard C, Carlsson B, Dahlgren S, Larsson B, Narbro K, Sjostrom CD, Sullivan M, Wedel H (2004) Lifestyle, diabetes, and cardiovascular risk factors 10 years after bariatric surgery. N Engl J Med 351:2683–2693

Contents

Contributors

Tamara L. Allen Cellular and Molecular Metabolism Laboratory, Baker IDI Heart and Diabetes Institute, P.O Box 6492, St Kilda Road Central, 8008 Victoria, Australia

Fabrizio Andreelli Département d'Endocrinologie Métabolisme et Cancer, Inserm U1016, Institut Cochin, Université Paris Descartes, CNRS (UMR 8104), 24, rue du Faubourg Saint Jacques, 75014 Paris, France

Jonathan R.S. Arch Clore Laboratory, University of Buckingham, Buckingham, MK18 1EG, UK

Francisco Castaneda Max-Planck-Institute of Molecular Physiology Olto-Hahn-Str. 11, 44227 Dortmund, Germany, rolf.kinne@mpi-dortmund.mpg.de

Michael A. Cawthorne Clore Laboratory, University of Buckingham, Buckingham, UK

Gitanjali Dharmadhikari Centre for Biomolecular Interactions Bremen, University of Bremen, Leobener Straße NW2, Room B2080, mailbox 330440, 28334, 28359 Bremen, Germany

Vincenzo Di Marzo Endocannabinoid Research Group, Institute of Biomolecular Chemistry, National Research Council, Via Campi Flegrei 34, Comprensorio Olivetti, 80078 Pozzuoli (NA), Italy

Mark D. Erion Merck & Co. Inc., 126 E Lincoln Ave, Rahway, NJ 07065, USA, mark-erion@merck.com

Mark A. Febbraio Cellular and Molecular Metabolism Laboratory, Baker IDI Heart and Diabetes Institute, P.O Box 6492, St Kilda Road Central, 8008 Victoria, Australia, mark.febbraio@bakeridi.edu.au

Baptist Gallwitz Medizinische Klinik IV, Otfried-Müller-Str. 10, 72076 Tübingen, Germany, baptist.gallwitz@med.uni-tuebingen.de

Antje Garten University of Leipzig, Hospital for Children and Adolescents, Liebigstr. 20a, 04103 Leipzig, Germany

Monique Heald Clore Laboratory, University of Buckingham, Buckingham, UK

Jin-ichi Inokuchi Division of Glycopathology and CREST, Japan Science and Technology Agency, Institute of Molecular Biomembranes and Glycobiology, Tohoku Pharmaceutical University, 4-4-1, komatsushima, Aoba-ku, Sendai 981-8558, Miyagi Japan, jin@tohoku-pharm.ac.jp

Wieland Kiess University of Leipzig, Hospital for Children and Adolescents, Liebigstr. 20a, 04103 Leipzig, Germany, wieland.kiess@medizin.uni-leipzig.de

Rolf K.H. Kinne Max-Planck-Institute of Molecular Physiology, Otto-Hahn-Str. 11, 44227 Dortmund, Germany

Antje Körner University of Leipzig, Hospital for Children and Adolescents, Liebigstr. 20a, 04103 Leipzig, Germany

Jürgen Kratzsch University of Leipzig, Institute for Laboratory Medicine, Clinical Chemistry and Molecular Diagnostics, Paul-List-Str.13a, 04103 Leipzig, Germany

Kathrin Maedler Centre for Biomolecular Interactions Bremen, University of Bremen, Leobener Straße NW2, Room B2080, mailbox 330440, 28334, 28359 Bremen, Germany, kmaedler@uni-bremen.de

Vance B. Matthews Cellular and Molecular Metabolism Laboratory, Baker IDI Heart and Diabetes Institute, P.O Box 6492, St Kilda Road Central, 8008 Victoria, Australia

Raphael Mechoulam Medicinal Chemistry and Natural Products Department, Medical Faculty, Hebrew University of Jerusalem, Ein Kerem, Jerusalem 91120, Israel

Paula I. Moreira Medicine and Center for Neuroscience and Cell Biology, University of Coimbra, 3004-517 Coimbra, Portugal, pismoreira@gmail.com

Catarina R. Oliveira Medicine and Center for Neuroscience and Cell Biology, University of Coimbra, 3004-517 Coimbra, Portugal, catarina.n.oliveira@gmail.com

Stefanie Petzold University of Leipzig, Hospital for Children and Adolescents, Liebigstr. 20a, 04103 Leipzig, Germany

Fabiana Piscitelli Endocannabinoid Research Group, Institute of Biomolecular Chemistry, National Research Council, Via Campi Flegrei 34, Comprensorio Olivetti, 80078 Pozzuoli (NA), Italy

Scott C. Potter Lilly AMB, 10300 Campus Point Drive, San diego, CA 92121, USA, pottersc@lilly.com

Desiree M. Schumann Boehringer-Ingelheim, Cardiometabolic Diseases Research, Biberach, Germany

Susanne Schuster University of Leipzig, Hospital for Children and Adolescents, Liebigstr. 20a, 04103 Leipzig, Germany

Christina Schwanstecher Molekulare Pharmakologie und Toxikologie, Technische Universität Braunschweig, Beethovenstraße 55, 38106 Braunschweig, Germany

Mathias Schwanstecher Molekulare Pharmakologie und Toxikologie, Technische Universität Braunschweig, Beethovenstraße 55, 38106 Braunschweig, Germany, M.Schwanstecher@tu-braunschweig.de

Joachim Størling Hagedorn Research Institute, Gentofte, Denmark

Paul D. van Poelje Pfizer Inc., Eastern Point Road, Groton, CT 06340, USA, paul.vanpoelje@Pfizer.com

Benoit Viollet Département d'Endocrinologie Métabolisme et Cancer, Inserm U1016, Institut Cochin, Université Paris Descartes, CNRS (UMR 8104), 24, rue du Faubourg Saint Jacques, 75014 Paris, France, benoit.viollet@inserm.fr.

Minghan Wang Department of Metabolic Disorders, Amgen Inc., One Amgen Center Drive, Mail Stop 29-1-A, Thousand Oaks, CA 91320, USA, mwang@amgen.com

Targeting Type 2 Diabetes

Christina Schwanstecher and Mathias Schwanstecher

Contents

Abstract The evolving concept of how nutrient excess and inflammation modulate metabolism provides new opportunities for strategies to correct the detrimental health consequences of obesity. In this review, we focus on the complex interplay among lipid overload, immune response, proinflammatory pathways and organelle dysfunction through which excess adiposity might lead to type 2 diabetes. We then consider evidence linking dysregulated CNS circuits to insulin resistance and results on nutrient-sensing pathways emerging from studies with calorie restriction. Subsequently, recent recommendations for the management of type 2 diabetes are discussed with emphasis on prevailing current therapeutic classes of biguanides, thiazolidinediones and incretin-based approaches.

Keywords Type 2 diabetes · Insulin resistance · Metformin · Pioglitazone · DPP-4

C. Schwanstecher and M. Schwanstecher (✉)
Molekulare Pharmakologie und Toxikologie, Technische Universität Braunschweig,
Beethovenstraße 55, 38106 Braunschweig, Germany
e-mail: M.Schwanstecher@tu-braunschweig.de

M. Schwanstecher (ed.), *Diabetes - Perspectives in Drug Therapy*,
Handbook of Experimental Pharmacology 203,
DOI 10.1007/978-3-642-17214-4_1, © Springer-Verlag Berlin Heidelberg 2011

1   Inflamed About Obesity: Adipose Tissue and Insulin Resistance

Within the last decade, increasing rates of obesity drove adipose tissue into the focus of scientific interest. This wave gained additional vigorous support from recognition of the tissue's critical role in a broad array of homeostatic processes, particularly its link to the immune response in metabolically triggered inflammation (Hotamisligil 2006; Kahn et al. 2006b; Schenk et al. 2008; Shoelson et al. 2006). Promoting development of insulin resistance and type 2 diabetes (T2DM), this chronic inflammation was unequivocally shown to be initiated within adipose tissue, thus rendering adipocytes to represent the major interface connecting metabolism to the immune system (Hotamisligil 2006; Kahn et al. 2006b; Schenk et al. 2008; Shoelson et al. 2006).

Interestingly, this conclusion was convincingly reinforced by evolutionary aspects (Hotamisligil 2006): (1) favouring the ability to withstand starvation and survive epidemics of infectious disease, evolution was likely to give rise to systems highly capable of both storing energy and eliciting a powerful immune response; (2) within ancestral structures, key metabolic and immune functions were initially combined. This is – for example – indicated by the fly's fat body, a structure recognized to represent the mammalian homologue of adipose tissue, liver and haematopoietic/immune systems (Leclerc and Reichhart 2004; Sondergaard 1993) (Fig. 1). In the fly this locus is crucial for sensing nutrient availability and coordinating energy status with metabolism and survival/immune response (Sondergaard 1993). Thus, in terms of an evolutionary approach, shared developmental heritage between the adipose tissue and the immune system points towards overlapping pathways controlling both metabolic and immune functions through common key regulatory molecules and signalling cascades (Hotamisligil 2006). This might provide the basis, enabling nutrients to act through pathogen-sensing mechanisms as Toll-like receptors and hence – in case of overload – to elicit metabolically induced inflammation (Sondergaard 1993; Beutler 2004; Shi et al. 2006; Song et al. 2006).

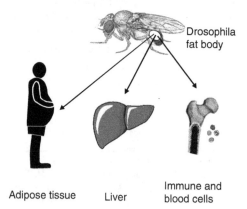

Fig. 1 Developmental divergence of adipose tissue, liver and the haematopoietic system

Successive enlargement of predominantly visceral adipocytes in progredient weight gain is believed to generate areas of microhypoxia and in turn, activation of the proinflammatory JNK/activator protein 1 (AP1) and IKK/NF-kappaB signalling pathways as well as endoplasmic reticulum (ER) stress (Hosogai et al. 2007; Wang et al. 2007; Ye et al. 2007; Schenk et al. 2008) (Fig. 2). Consequently, cytokine release is triggered with recruitment of macrophages, which – gathered around apoptotic adipocytes – are primarily prone to removing cell debris (Cinti et al. 2005; Strissel et al. 2007; Ye et al. 2007). In addition, however, and presumably aggravated by microhypoxia, these macrophages liberate large amounts of proinflammatory cytokines. Through paracrine activation of JNK/IKK signalling within neighbouring adipocytes, this is thought to represent the critical step towards induction of insulin resistance (Schenk et al. 2008) (Fig. 2).

Besides hypoxia, ER stress is triggered through overloading cells with nutrients, particularly with fatty acids (Gregor and Hotamisligil 2007; Ron and Walter 2007; Eizirik et al. 2008). Once activated, ER stress results in stimulation of the unfolded

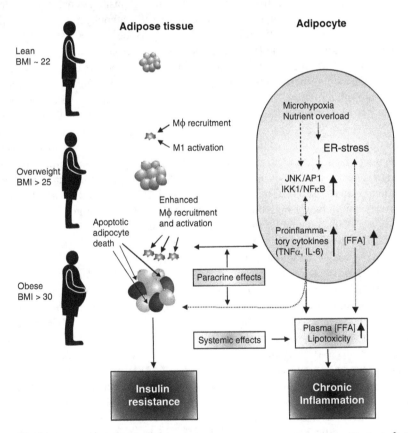

Fig. 2 Adipose tissue, inflammation and insulin resistance. BMI, body mass index in [kg/m^2]; ER, endoplasmic reticulum; FFA, free fatty acids; Mɸ, bone marrow-derived macrophages; M1, M1-type inflammatory macrophages

protein response (UPR) (Gregor and Hotamisligil 2007; Ron and Walter 2007; Eizirik et al. 2008). The UPR in turn induces a set of genes involved in protein processing, thus seeking to halt transcription and promote folding (Gregor and Hotamisligil 2007; Ron and Walter 2007; Eizirik et al. 2008). Activation of the UPR, however, aggravates insulin resistance by stimulating expression and release of proinflammatory cytokines (e.g., TNFα, IL-6 and Cd2) (Li et al. 2005b; Gregor and Hotamisligil 2007; Ron and Walter 2007; Eizirik et al. 2008). These in turn incite JNK/IKKbeta pathways, which leads to serine phosphorylation of IRS-1 and consecutive disruption of insulin signalling (see also below and Fig. 4) (Ozcan et al. 2004). In line with a critical role of ER stress in development of insulin resistance, knockout of a transcription factor boosting the expression of UPR-relieving chaperones (Xbp1, see also below and Fig. 6) decreased insulin sensitivity in HFD-fed mice (Ozcan et al. 2004), with orally active chemical chaperones (4-phenyl butyrate or tauroursodeoxycholic acid) antagonizing this effect (Ozcan et al. 2006).

Selective knockout of JNK1, IKKbeta or TNFalpha within myeloid progenitor cells protected obese mice from HFD-induced loss of insulin sensitivity, further affirming the essential role of inflammation in the pathophysiology of insulin resistance (Arkan et al. 2005; Solinas et al. 2007). Bone marrow-derived macrophages within adipose tissue track with the degree of obesity and were shown to comprise 40% of the tissues' total cell content in obese rodents or humans versus 10% in lean counterparts (Weisberg et al. 2003) (Fig. 3). During induction of obesity proinflammatory M1-type macrophages are attracted, while prior HFD feeding the anti-inflammatory M2 type is resident (Lumeng et al. 2007). M2 macrophages are specified by releasing high levels of anti-inflammatory factors such as IL-4 and IL-10 (Mantovani et al. 2004). In contrast, M1 macrophages are characterized by increased inflammatory gene expression, proinflammatory cytokine release and reactivity to fatty acids and lipopolysaccharides (LPS) (Mantovani et al. 2004). The surface markers F4/80 and CD11b are found in both types, but only M1 macrophages are positive for CD11c (Nguyen et al. 2007; Strissel et al. 2007). Notably, PPARγ activation polarizes towards the anti-inflammatory M2 type suggesting the PPARγ agonistic thiazolidinediones (TZDs, e.g., pioglitazone, rosiglitazone) to act partially through this mechanism (Bouhlel et al. 2007; Hevener et al. 2007; Odegaard et al. 2007). M1-type macrophages were demonstrated to express enhanced levels of pattern recognition receptors TLR2/TLR4, suggesting a critical role in fatty acid triggered proinflammatory activation of adipocytes and leukocytes (Nguyen et al. 2007). Consistently, HFD-induced insulin resistance was prevented through TLR4 knockout (Shi et al. 2006; Tsukumo et al. 2007). Similar protection was observed in mice deficient of the surface receptor for gut-derived bacterial LPS (CD14) and the chemoattractants CCL2 and osteopontin, which are released by stimulated adipocytes and appear critical in monocyte recruitment (Kanda et al. 2006; Weisberg et al. 2006; Cani et al. 2007; Nomiyama et al. 2007).

Myeloid specific knockout of JNK1 fully reversed insulin resistance of the liver, suggesting (1) analogous to the situation within adipose tissue, activation of inflammatory pathways within hepatocytes to account for defective insulin

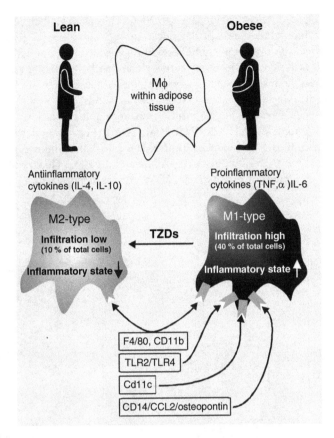

Fig. 3 Body weight-dependent invasion of M1-type macrophages into adipose tissue. Mφ , bone marrow-derived macrophages; M1, M1-type inflammatory macrophages; M2, M2-type anti-inflammatory macrophages

signalling and (2) Kupffer cells (i.e., bone-derived macrophages within the liver) to represent essential paracrine triggers of proinflammatory pathways within hepatocytes (Solinas et al. 2007). Consistently, insulin resistance was aggravated by selective overexpression of IKKβ within hepatocytes and alleviated through inhibition of NF-kappaB and liver-specific knockout of IKKβ (Arkan et al. 2005; Cai et al. 2005). Notably, myeloid specific knockout of JNK1 did not protect from hepatic steatosis, indicating that the lipid-rich milieu of non-alcoholic fatty liver (NAFLD) disease might induce chronic inflammation and not vice versa (Solinas et al. 2007).

In adipose tissue, liver and skeletal muscle insulin signalling involves tyrosine phosphorylation of insulin receptor substrate 1 (IRS-1) with downstream activation of phosphatidylinositol 3-kinase (PI3K), PKB/Akt and PKC-λ/ζ (Taniguchi et al. 2006; Thirone et al. 2006; Solinas et al. 2007) (Fig. 4). On the level of adipose tissue and skeletal muscle, this cascade results in translocation of GLUT4 and stimulation of cellular glucose uptake (Taniguchi et al. 2006; Thirone et al. 2006; Solinas et al. 2007). In the liver, glucose release is reduced through inhibition of glycogenolysis and gluconeogenesis (Wahren and Ekberg 2007). Serine phosphorylation of IRS-1 (e.g., ser307, 661, 731, 1101) impedes this signalling by hindering association of the substrate to the insulin receptor (Aguirre et al. 2002; Gual et al. 2005). Proinflammatory cytokines, fatty acids, certain amino acids and multiple additional factors have been shown to induce insulin resistance at least partially through this mechanism (Aguirre et al. 2002; Gual et al. 2005) (Fig. 4). Importantly, serine phosphorylation of IRS-1 is induced by JNK and IKKβ, thus linking inflammation to insulin resistance (Karin et al. 2001; Aguirre et al. 2002; Gual et al. 2005). In turn, numerous models of obesity and T2DM were associated with both increased

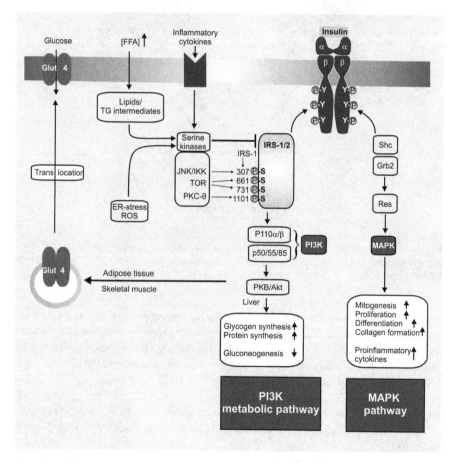

Fig. 4 Serine phosphorylation of IRS1/2 and insulin resistance. ER-stress, endoplasmic reticulum stress; [FFA], extracellular concentration of free fatty acids; Glut4, glucose transporter subtype 4; IRS-1/2, insulin receptor substrate subtype 1 or 2; MAPK, mitogen activated protein kinase; ROS, reactive oxygen species; TG intermediates, triglyceride intermediates. Inflammatory signalling does not affect the MAPK pathway

JNK/IKKβ signalling and IRS-1 serine phosphorylation (Yuan et al. 2001; Hirosumi et al. 2002; Itani et al. 2002; Arkan et al. 2005; Bandyopadhyay et al. 2005; Cai et al. 2005).

Marked improvement of insulin resistance in dietary restriction significantly precedes weight loss, implying cellular pathways capable of sensing deficient nutritional supply (Kelley et al. 1993; Assali et al. 2001; Cai et al. 2005). These include regulatory subunits of PI3K (i.e., p50, p55, p85), the serine/threonine protein kinase mTOR and the protein deacetylase sirtuin 1 (SIRT1) (Um et al. 2004; McCurdy et al. 2005; Cornier et al. 2006; Tzatsos and Kandror 2006; Civitarese et al. 2007; Krebs et al. 2007; Sun et al. 2007). Initiation of calorie restriction reduced expression of p50/55, which was paralleled by enhanced PI3K activity and regain of insulin sensitivity (Fig. 4). In humans, similar results were reported for the mTOR/p70S6 kinase pathway and expression of SIRT1.

2   The Lipotoxicity Concept

In the course of developing T2DM, insulin resistance significantly precedes overt hyperglycaemia (DeFronzo 1988; Abdul-Ghani and DeFronzo 2010). Thus, obese with normal glucose tolerance and without family history of T2DM revealed 35–50% loss of whole body insulin-induced glucose disposal, with 80% of this loss attributed to skeletal muscle (Bogardus et al. 1984; DeFronzo et al. 1985; Kahn et al. 2006b). Similar levels of insulin resistance were reported for many disease states (e.g., essential hypertension, ischemic heart disease, dyslipidemia, i.e., increased plasma triglyceride/decreased HDL cholesterol, polycystic ovary syndrome, chronic kidney failure, myotonic dystrophy, lipodystrophy, acute injury or sepsis), secondary to drug therapy (e.g., protease inhibitors in HIV therapy, glucocorticoids and beta-blockers) and in association with the normal aging process (Abdul-Ghani and DeFronzo 2010).

Elevated plasma levels in free fatty acids (FFA) typically trait with obesity and there is strong evidence causally linking this hallmark of obesity to insulin resistance, thus establishing the "lipotoxicity" concept (Unger 1995). For example, reduction of insulin sensitivity through FFA was shown to be dose dependent and both acute (4 h) as well as chronic (4 days) elevation were effective (DeFronzo et al. 1978; Boden and Chen 1995; Dresner et al. 1999; Griffin et al. 2000; Itani et al. 2002; Yu et al. 2002; Belfort et al. 2005; Richardson et al. 2005). Further support of the lipotoxicity concept emerges from association of insulin resistance with intramyocellular (IMCL) fat content (~1% of whole skeletal muscle fat content): (1) NMR biopsy studies indicated IMCL versus extramyocellular (EMCL) fat to strictly correlate with development of insulin resistance independent of total body fat mass and glucose tolerance (Boden and Chen 1995; Krssak et al. 1999; Perseghin et al. 1999); (2) acipimox-induced reduction of FFA, dose-dependently paralleled decreased IMCL content of fatty acid CoenzymA (FACoA) (Bajaj et al. 2005); (3) gradual augmentation of IMCL fat content was coupled to successive loss of insulin sensitivity (Boden et al. 2001); (4) adiponectin-induced amelioration of insulin resistance associates with enhanced IMCL fat oxidation and reduced fat content (Yamauchi et al. 2001); (5) muscle-specific overexpression of lipoprotein lipase traits with increased IMCL fat content and severe loss of insulin-induced muscle glucose uptake, but does not affect insulin sensitivity of adipose tissue and liver (Kim et al. 2001); (6) bariatric surgery sequels normalization of insulin sensitivity paralleled by reduced IMCL fat content, despite persisting severe obesity (BMI 39 vs. 49 kg/m^2) (Greco et al. 2002).

Transport of FFA into muscle cells appears driven by a combination of lipophilic diffusion plus facilitated transport with three proteins putatively involved: (1) the plasma-membrane fatty acid binding protein (FABP), (2) fatty acid transport protein (FATP) and (3) fatty acid translocase (CD36) (Abumrad et al. 1999; Bonen et al. 2003; Pownall and Hamilton 2003) (Fig. 5). CD36 deficient mice displayed reduced FFA transport rates suggesting CD36 to be particularly critical (Febbraio et al. 1999). Consistently, it might be involved in IMCL fat accumulation, since in obese and T2DM individuals, CD36-mediated transport is increased with concomitant reduction of fat oxidation (Bonen et al. 2003, 2004; Kiens et al. 1999; Roepstorff et al. 2004).

Once having reached myocellular (MCL) cytoplasm, FFA are either inclined to mitochondrial β-oxidation or triglyceride synthesis (intramuscular triglyceride, IMTG) (Coleman and Lee 2004) (Fig. 5). While IMTG itself appears inert in terms of insulin signalling, IMTG intermediates (1) lysophosphatidic acid (LPA), (2) phosphatidic acid (PA) and (3) diacylglycerol (DAG) are thought to play a key role in development of insulin resistance (see Fig. 5) (Pan et al. 1997; Bandyopadhyay et al. 2006; Liu et al. 2007). An additional product of MCL FFA metabolism presumably critical in insulin action is ceramide, the synthesis of which is driven through high concentrations of saturated C15-17 FFA, particularly palmitate (Summers 2006; Holland et al. 2007). In obesity and models with lipid/heparin infusion, enhanced MCL levels of fatty acyl-CoA, IMTG intermediates and ceramide paralleled with increased plasma [FFA] (Bandyopadhyay et al. 2006; Goodpaster et al. 2001; Holland et al. 2007; Liu et al. 2007; Pan et al. 1997). Consistently, inhibition of lipolysis or stopping of infusions was associated with reduced MCL concentrations of these metabolites (Yu et al. 2002; Bajaj et al. 2005). Yet, how do these derivates of lipid metabolism link to insulin signalling? Importantly, LPA, PA, DAG and ceramide can activate the mTOR/p70S6K, JNK and IKK pathways with subsequent suppression of insulin action through serine phosphorylation of IRS1 (Sathyanarayana et al. 2002; Jean-Baptiste et al. 2005; Sampson and Cooper 2006; Wang et al. 2006) (Fig. 5). Ceramide also directly interferes with PKB/AKT (Stratford et al. 2001). Accordingly, pharmacologic or genetic suppression of these pathways or inhibition of ceramide production rescued from insulin resistance through nutrient/lipid overload or obesity in humans and various animal models (Hundal et al. 2002; Um et al. 2004; Arkan et al. 2005; Cai et al. 2005; Nguyen et al. 2005; Tzatsos and Kandror 2006; Holland et al. 2007; Krebs et al. 2007). Another pre-eminent link to IRS-1 inactivation (i.e., serine phosphorylation) is provided by activation of the protein kinase C family of serine-threonine kinases

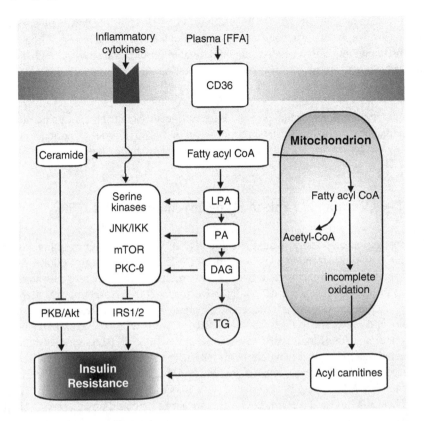

Fig. 5 Insulin resistance in skeletal muscle and liver: the "lipotoxicity concept". DAG, diacyl-glycerol; [FFA], extracellular concentration of free fatty acids; IRS-1/2, insulin receptor substrate subtype 1 or 2; LPA, lysophosphatidic acid; PA, phosphatidic acid

with mice deficient in PKC-θ being protected against fat-induced insulin resistance and insulin insensitive obese humans displaying MCL stimulation of PKC-β2, δ and θ (Schmitz-Peiffer et al. 1997; Griffin et al. 1999; Itani et al. 2001; Aguirre et al. 2002; Itani et al. 2002; Yu et al. 2002; Kim et al. 2004; Samuel et al. 2010) (Figs. 4 and 5).

Reduced fatty acid oxidative capacity and impaired mitochondrial function of skeletal muscle are thought to further contribute to IMCL accumulation of fatty acid metabolites and thus aggravate loss of insulin action in obesity and T2DM (Mootha et al. 2003; Patti et al. 2003; Morino et al. 2006; Houmard 2008). Another line of evidence suggests deterioration of insulin signalling by enhanced mitochondrial release of incompletely oxidized products (i.e., acylcarnitines) (Koves et al. 2008) (Fig. 5).

Notably, expression of nitric oxide synthase is enhanced through IRS-1/PI3-K signalling and thus restraint of this pathway in states of insulin resistance might contribute to endothelial dysfunction/accelerated atherosclerosis in T2DM and obesity (Breen and Giacca 2010). Inactivation of IRS1/IRS2, however, does not affect insulin receptor signalling through the adapter proteins Shc/Grb2 with downstream stimulation of the mitogen-activated protein kinase (MAPK) pathway and subsequent promotion of cellular proliferation and differentiation (Cusi et al. 2000; Krook et al. 2000) (Fig. 4). Thus, hyperinsulinaemia will induce proliferation of vascular smooth muscle, collagen formation and production/release of proin-flammatory cytokines, which is thought to significantly contribute to the pernicious consequences of insulin resistance per se in terms of arterial hypertension and cardiovascular disease (CVD) (Hsueh and Law 1999; Jiang et al. 1999).

3   β-Cell Failure: Link to Hyperglycaemia and T2DM

In most cases of obesity, pancreatic β-cells are capable of outweighing loss of insulin sensitivity through increased insulin release (Kahn et al. 2006b). Deve-lopment of hyperglycaemia and overt T2DM indicate breakdown of this compen-sation in consequence of β-cell failure (Defronzo 2009). Susceptibility to β-cell failure is thought to critically depend on genetic background (Kahn et al. 2006b; Muoio and Newgard 2008). Obese BTBR/leptinob mice, for example, display defective islet proliferation and severe diabetes, while C57BL6/leptinob mice are protected against diabetes (Stoehr et al. 2000). In humans, maturity onset diabetes of the young (MODY) results from mutations in β-cell key proteins (e.g., MODY1: HNF4alpha; MODY2: glucokinase; MODY4: PDX1) (Winter et al. 1999). Typi-cal, obesity-associated T2DM, however, appears to be based on an array of single nucleotide polymorphisms (SNPs) resulting in increased vulnerability of pancre-atic β-cells to overnutrition and chronic inflammation (Barroso 2005; Muoio and Newgard 2008). Interestingly, one of these SNPs (E23K within KCNJ11) appears to reside within the pore forming subunit of the ATP-sensitive potassium channel (K_{ATP}, see also below), and was demonstrated to discretely modify the channels' open probability in the heterozygous state (Schwanstecher et al. 2002; Schwanstecher and Schwanstecher 2002; Florez et al. 2004; Li et al. 2005a; O'Rahilly 2009).

Within pancreatic β-cells, overnutrition plus increased plasma [FFA] might induce CPT1 plus enzymes of FA β-oxidation leading to enhanced intramitochon-drial [acetyl-CoA], allosteric activation of pyruvate carboxylase and enhanced pyruvate cycling (Chen et al. 1994; Poitout and Robertson 2002; Prentki et al. 2002; Khaldi et al. 2004; Muoio and Newgard 2008) (Fig. 6). This in turn might result in the defect of glucose-stimulated insulin release (GSIS) characteristic of prediabetic and diabetic states: constitutive basal insulin hypersecretion with loss of the first phase of glucose-induced insulin release, which is paralleled by lack of the glucose-stimulated rapid increment in pyruvate flux (Poitout and Robertson 2002; Prentki et al. 2002; Muoio and Newgard 2008).

Genetic inactivation of PERK/eIF2a resulted in β-cell dysfunction and diabetes suggesting that ER stress might be of key importance for β-cell failure (Harding et al. 2001; Muoio and Newgard 2008) (Fig. 6). This pathway couples the UPR

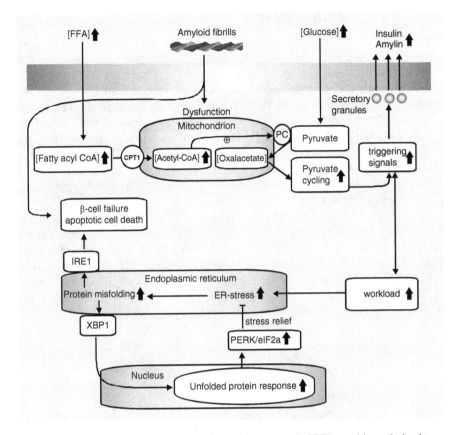

Fig. 6 β-cell failure in T2DM: the "glucolipotoxicity concept". CPT1, carnitine palmitoyltransferase 1; eIF2a, eukaryotic translation initiation factor-2a; ER-stress, endoplasmic reticulum stress; [FFA], extracellular concentration of free fatty acids; IRE1, inositol-requiring kinase 1; PC, pyruvate carboxylase; PERK, protein kinase RNA(PKR)-like endoplasmic reticulum associated kinase; XBP1, X-box binding protein 1

to inhibition of protein translation. In obesity, compensatory enhancement of insulin release might exceed ER capacity with downstream activation of the UPR/PERK/eIF2a pathways. Chronic conditions of insulin resistance plus over-nutrition might then result in successive desensitization of these pathways and finally end up in cumulative cellular damage and T2DM-typical apoptotic loss of β-cell mass. Consistently, heterozygous eIF2a mutant HFD-fed mice displayed distension of β-cell ER, β-cell failure and diabetes (Scheuner et al. 2005).

Moreover, deposits of islet amyloid polypeptide (IAPP) might contribute to β-cell failure (Fig. 6). IAPP is secreted from pancreatic β- and delta-cells and toxic fibrils were demonstrated in islets from humans with T2DM (Cooper et al. 1987; Westermark et al. 1987). Consistent with this concept, mice overexpressing human IAPP developed increased rates of β-cell apoptosis, defective GSIS, glucose intolerance and diabetes (Matveyenko and Butler 2006).

4 Role of the CNS in Glucose Homeostasis

The CNS represents a key player in glucose metabolism, transposing neural, hormonal and nutrient signals in response to the ingestion of food into direct regulatory control of glucagon/insulin secretion, peripheral insulin sensitivity and hepatic glucose output (Obici et al. 2001; Obici et al. 2002a, b; Obici et al. 2002c; Lam et al. 2005a; Lam et al. 2005b; Lam et al. 2005c; Pocai et al. 2005a; Pocai et al. 2005b; Sandoval et al. 2008);. Impairment of this function in overnutrition and obesity suggests an essential role in the development of T2DM (Coppari et al. 2005; Morrison et al. 2005; Pocai et al. 2006; Ono et al. 2008).

In rodents, insulin was demonstrated to control hepatic glucose production (HGP) through neurons within the medio-basal hypothalamus (Obici et al. 2002b; Pocai et al. 2005) (Fig. 7). Consistently, HGP was reduced by insulin infusions into this region, while selective deletion of the insulin receptor within neurons expressing Agouti-related peptide (AGRP) yielded liver insulin resistance (Obici et al. 2002a, b; Pocai et al. 2005; Konner et al. 2007). Effects on HGP were blunted by PI3K inhibition and glibenclamide, suggesting downstream signalling through PI3K to finally result in activation of the ATP-sensitive potassium channel (K_{ATP}; see also below) (Obici et al. 2002a, b; Pocai et al. 2005).

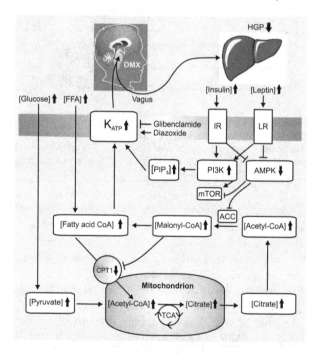

Fig. 7 Hypothalamic nutrient-sensing and glucose homeostasis. ACC, acetyl-CoA carboxylase; CPT1, carnitine palmitoyltransferase 1; DMX, motor nucleus of the vagus in the brainstem; [FFA], extracellular concentration of free fatty acids; HGP, hepatic glucose production; IR, insulin receptor; K_{ATP}, ATP-sensitive potassium channel; LR, leptin receptor; TCA, tricarboxylic acid cycle

Leptin resembles insulin in inducing satiety as well as controlling HGP through hypothalamic nuclei (Liu et al. 1998; Schwartz et al. 2000; Gutierrez-Juarez et al. 2004) (Fig. 7). Receptor reactivation within the nucleus arcuatus (ARC) of leptin receptor deficient mice suggests the effect on HGP to be independent of the peptide's effect on food intake and body weight (Coppari et al. 2005). Downstream signalling seems to include JAK2/STAT3, PI3K, AMPK and mTOR (Dennis et al. 2001; Bates et al. 2003; Andersson et al. 2004; Morton et al. 2005; Buettner et al. 2006; Cota et al. 2006; Plum et al. 2006). Similar to pancreatic β-cells (see also above), ER stress might be involved in overnutrition/obesity-induced central leptin resistance through interference with STAT3 (Zhang et al. 2008; Ozcan et al. 2009). Accordingly, chemical chaperones were shown to rescue leptin sensitivity in models with resistance (Zhang et al. 2008; Ozcan et al. 2009).

Distinct from its direct peripheral effects, GLP1 was shown to affect central glucose control (Perrin et al. 2004; Sandoval et al. 2008). Instillation into ARC and portal infusions both reduced HGP, with parasympathetic afferent signalling presumably explaining the peripheral effect (Dardevet et al. 2005; Ionut et al. 2005; Sandoval et al. 2008). Akin to direct action of GLP1 on islet β-cells, central administration also enhanced pancreatic glucose-induced insulin release (Knauf et al. 2005; Sandoval et al. 2008, 2009).

CNS [glucose] was demonstrated to affect peripheral glucose homeostasis (Lam et al. 2005a; Sandoval et al. 2009). In line, HGP was suppressed by hypothalamic glucose instillation (Lam et al. 2005a; Sandoval et al. 2009). Attenuation of this effect through local inhibition of LDH (lactate dehydrogenase) and stimulation of pyruvate dehydrogenase suggests – as in pancreatic β-cells – mitochondrial oxidation to exert an essential role in neuronal glucose signalling (Fig. 7). Further, alike the hepatic GLP-1 effect, direct intestinal application of glucose induces – besides inhibition of food intake – improved glucose tolerance through stimulation of a vagal afferent response (Woltman and Reidelberger 1995; Zhou et al. 2008).

Similarly, hypothalamic [FFA] has been postulated to be involved in control of HGP (Obici et al. 2002a, 2003; Lam et al. 2005b; Pocai et al. 2006) (Fig. 7). Hypothalamic [FFA] sensing was suggested to critically depend on an increase of cytosolic [malonyl CoA] which in turn induces enhanced levels of FA intermediates (Obici et al. 2003). These are thought to result in activation of K_{ATP} channels with consecutive cellular hyperpolarization triggering downstream neuronal circuits (see also below) (Obici et al. 2003; Lam et al. 2005b). Consistently, elevated cerebral [malonyl CoA] incited satiety, anorexia and weight loss and hypothalamic application of FA-CoA/block of CPT1 decreased HGP through activation of vagal efferent nerves to the liver (Loftus et al. 2000; Obici et al. 2002b; Shimokawa et al. 2002; Hu et al. 2005; Lam et al. 2005b; He et al. 2006; Pocai et al. 2006; Wolfgang and Lane 2006; Proulx et al. 2008) (Fig. 7).

Defective central AMPK stimulation was associated with the inadequate counter regulatory response of diabetic individuals to repeated episodes of hypoglycaemia (Alquier et al. 2007). Concurrently, hypothalamic AMPK activity increased under conditions of calorie restriction and decreased under CNS administration of glucose or leptin (Andersson et al. 2004; Minokoshi et al. 2004). Also, enhanced hypothalamic AMPK activity was paralleled by increased HGP and muscle glyco-gen synthesis (Perrin et al. 2004). Besides AMPK, the mTOR pathway was related to central nutrient sensing, with loss of leptin sensitivity under conditions of chronic overfeeding coupled to reduced mTOR signalling in key neurons (see also above) (Cota et al. 2006; Ono et al. 2008) (Fig. 7).

Analogous to their role in pancreatic β-cells, K_{ATP} channels were supposed to be of key importance in central nutrient sensing/homeostatic control (Obici et al. 2002b; Lam et al. 2005a; Pocai et al. 2005; Sandoval et al. 2008) (Fig. 7). According to this concept, central suppression of HGP is mediated by activation of these channels through increased levels of cytosolic FA intermediates with consecutive triggering of vagal output to the liver (Obici et al. 2002b; Lam et al. 2005a; Pocai et al. 2005; Sandoval et al. 2008). Consistently, the effects of locally applied insulin, glucose, oleic acid and GLP1 on HGP were eliminated by the channel blocker glibenclamide and reproduced by the opener diazoxide. Also in rodents, glucose intolerance was associated with reduced hypothalamic expression or selec-tive inactivation of K_{ATP} (Gyte et al. 2007; Parton et al. 2007).

Dopamine agonists (e.g., bromocriptinand rotigotin) are well known to decrease food intake, reduce body weight and ameliorate insulin resistance, while dopamine antagonists, particularly with selectivity for receptor subtype D4 (e.g., clozapine and olanzapine), are associated with the opposite effects (Cincotta et al. 1997; Meguid et al. 2000; Ader et al. 2005; Lieberman et al. 2005; Houseknecht et al. 2007; Chintoh et al. 2008). Within the serotoninergic system, selective knockout of receptor subtype 5HT2C produced obesity, insulin resistance and glucose intolerance with reversal through specific reactivation in POMC neurons (Mirshamsi et al. 2004; Xu et al. 2008b). In addition, administration of a selective 5HT2C agonist corrected insulin sensitivity and glucose tolerance in HFD-fed rodents (Zhou et al. 2007).

Dysregulation of the endocannabinoid (EC) system has been proposed to play a pre-eminent role in the pathophysiology of obesity, insulin resistance and T2DM with strong evidence in support of this concept arising from clinical studies on therapy with the CB1 receptor inverse agonist rimonabant (see also below) (Di Marzo 2008).

5 Nutrient-Sensing Pathways in Calorie Restriction

Dietary restriction in adult rhesus monkeys (30%, 20-year follow-up) prevented insulin resistance, glucose intolerance and diabetes paralleled by a 50% decrement in neoplasia and CVD (Colman et al. 2009). Inflammation markers, immune senescence, sarcopenia and brain atrophy were markedly attenuated (Colman et al. 2009). In humans, significant amelioration was demonstrated in terms of obesity, insulin resistance, glucose intolerance, inflammation and cardiac function (Fontana and Klein 2007). Insulin, triiodothyronine, testosterone and cardiovascu-lar risk factors (LDL cholesterol, C-reactive protein, blood pressure, intima-media thickness of the carotid arteries) were lowered while adiponectin was increased (Fontana et al. 2004; Fontana and Klein 2007). Based on studies in other species, these changes were partly attributed to reduced growth hormone (GH)/IGF-1 signalling (Fontana et al. 2010). Consistently, GH-deficient individuals revealed reduced incidence of diabetes, cancer and CVD, albeit obesity and hyperlipidaemia were aggravated (Shevah and Laron 2007). Also, specific single nucleotide polymorphisms (SNPs) within the IGF-1 receptor or downstream factor FOXO were shown to trait with human longevity (Kuningas et al. 2007).

6 Current Therapy

6.1 Published Algorithms

Based on the complex pathophysiology of T2DM it was recently concluded, that therapy should (1) target established defects instead of plasma [HbA1c] and (2) start as early as possible to prevent progressive β-cell failure (Defronzo 2009). Currently, these postulates are best met by the treatment recommendations of the AACE and ACE (American Association of Clinical Endocrinologists and American College of Endocrinology) (Fig. 8), which differ from corresponding recommendations of the ADA/EASD (American Diabetes Association and European Association for the Study of Diabetes) with respect to (Nathan et al. 2009; Rodbard et al. 2009; Rodbard and Jellinger 2010) (1) placing increased emphasis on incretin-based therapies (i.e., GLP-1 agonists and DPP4-inhibitors); (2) lower priorities on thiazolidinediones due to frequent adverse events of weight gain/fluid retention and increased risk of congestive heart failure and bone fractures; (3) much lower priority on the use of sulfonylureas due to (a) significant risk of hypoglycaemia and weight gain; (b) effectiveness for only a short period of time because of lacking β-cell protection; (4) adjustment of HbA1c goal for patients with hypoglycaemia/ hypoglycaemia unawareness (see also above) or long duration diabetes and/or established coronary heart disease; (5) stratification by HbA1c at the time point of presentation for therapy:

(a) In case of an initial HbA1c < 7.5%, lifestyle modification alone might be sufficient to achieve the goal of 6.5%. If this fails, then monotherapy is recommended with metformin representing the preferred agent.
(b) In case of an initial HbA1c between 7.6 and 9.0%, pharmacotherapy should be started with a dual approach, because monotherapy would hardly be sufficient to attain the goal of 6.5% and appear inadequate to address the underlying pathophysiology (i.e., insulin resistance plus advanced β-cell failure plus inflammation plus lipotoxicity). In addition to metformin, GLP-1 agonists/ DPP4-antagonists are recommended as first choice with optional substitution through TZDs in case of metabolic syndrome and/or NAFLD.
(c) If the initial HbA1c is above 9.0%, therapy should start with either a dual or triple approach. Triple therapy should include TZDs in addition to metformin plus GLP-1 agonist/DPP4 antagonist. In case of failure to reach the target of

Fig. 8 T2DM treatment
algorithm, simplified
according to the AACE/ACE
consensus statement
(Rodbard et al. 2009). A1c,
HbA1c; CVD, cardiovascular
disease; FPG, fasting plasma
glucose concentration;
NAFLD, non-alcoholic fatty
liver disease; PPG,
postprandial plasma glucose
concentration; T2DM, type
2 diabetes mellitus

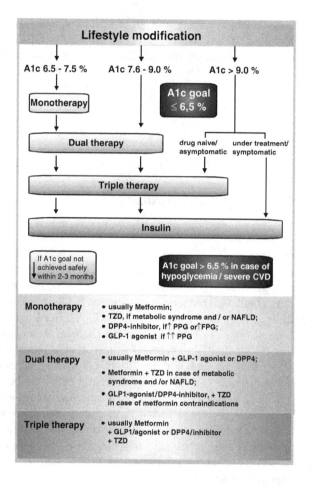

6.5% on a previous regime or symptomatic hyperglycaemia, the algorithm recommends moving directly to insulin therapy.

In all regimes, patients should be closely monitored at 2–3 months intervals with appropriate adjustment (i.e., either increasing dosage or moving to the next level) if the target HbA1c is not met.

6.2 *Metformin*

Current guidelines for management of type 2 DM unison recommend lifestyle change followed by rapid or even concomitant introduction of metformin as the drug of first choice (e.g., Nathan et al. 2009; Rodbard et al. 2009). Metformin inhibits complex I of the respiratory chain, which is thought to result in a disruption of liver cell energy metabolism with consecutive fall in cytosolic [ATP]/[ADP] and subsequent inhibition of [ATP] dependent steps in gluconeogenesis (El-Mir et al. 2000; Owen et al. 2000). Based on recent evidence and in contrast with a widely accepted concept, this action does not seem to involve LKB1-AMPK/SIK1/CRTC2 signalling (Foretz et al. 2010). The LKB1/AMPK pathway might, however, be involved in other effects of the drug.

Metformin's clinical efficacy is mainly based on enhancing liver glucose uptake, implicating decreased HGP through suppression of gluconeogensis and glycogenolysis (Goodarzi and Bryer-Ash 2005). Moreover, it may discretely augment glucose disposal within the gastrointestinal tract, increase fatty acid oxidation, enhance insulin sensitivity in skeletal muscle and adipose tissue and elevate the postprandial GLP-1 response (Mannucci et al. 2001; Lenhard et al. 2004; Bailey 2005). Metformin might also moderately reduce inflammatory markers (e.g., CRP), but did not lower the number of adipose tissue macrophages (Chu et al. 2002; Di Gregorio et al. 2005; Bodles et al. 2006). In human leukocytes, it was shown to decrease ROS levels and in a rat model of T2DM to ameliorate mitochondrial oxidative stress (Bonnefont-Rousselot et al. 2003; Rosen and Wiernsperger 2006). Within the UKPDS substudy (UKPDS 1998) of overweight patients, metformin was the only drug (i.e., in comparison with sulfonylureas and insulin) displaying positive effects on CVD and reduced all cause mortality. Metformin monotherapy decreased fasting plasma glucose within the first year by ~65 mg/dl and HbA1c by 1.5–2%, but only 25% of patients achieved the goal of HbA1c <7.0% (DeFronzo and Goodman 1995; DeFronzo 1999). Secondary to initiation, however, glycaemic control progressively deteriorated due to progressive decline in insulin release (UKPDS 1998) indicating that metformin does not protect β-cell function.

6.3 Thiazolidinediones

The thiazolidinediones (TZDs, pioglitazone and rosiglitazone), are thought to exert their therapeutic effects through PPARγ agonism (Spiegelman 1998). Downstream to activation, PPAR-γ dimerizes with the retinoid X receptor and induces complex transcriptional control through interaction with PPAR-response elements (Kliewer et al. 1997; Mootha et al. 2003). PPAR-γ is predominantly expressed in adipocytes, thus regulating adipogenesis and glucose/lipid metabolism (Olefsky 2000; Willson et al. 2000; McGuire and Inzucchi 2008). However, PPAR-γ also resides within hepatocytes, macrophages, skeletal muscle, cardiac muscle and vascular endothelium (Olefsky 2000; Willson et al. 2000; McGuire and Inzucchi 2008).

TZDs induce adipocyte differentiation and thus increase body weight (Okuno et al. 1998). Mean adipocyte size, however, is reduced and fat redistributed from visceral to subcutaneous depots (Adams et al. 1997; Okuno et al. 1998). Plasma levels of proinflammatory cytokines (e.g., TNF-alpha, IL-6) are decreased and those of adiponectin elevated (Adams et al. 1997; Shimizu et al. 2006). Importantly, [CD36] is upregulated within skeletal muscle, thus presumably increasing peripheral FFA uptake and contributing to decreased plasma [FFA], with potentially positive implications for β-cell function, liver metabolism, endothelial function, myocardial viability and general state of chronic inflammation (Oliver and Opie 1994; Martin et al. 1997; Spiegelman 1998; McGarry and Dobbins 1999; Tripathy et al. 2003). Consistently, recruitment of M1-type macrophages within adipose tissue was found to be reduced (Patsouris et al. 2009).

Similar to metformin and sulfonylureas, TZD monotherapy reduces HbA1c by 1–2% (Inzucchi 2002). This was demonstrated to be mainly due to significant amelioration of insulin sensitivity with the effect on the liver being less pronounced than that on skeletal muscle (Petersen et al. 2000). Importantly however, and in contrast to metformin, the improvement of glycaemic control was durable, providing strong evidence in support of a β-cell protective action (Diani et al. 2004; Gerstein et al. 2006; Kahn et al. 2006a). Beneficial outcomes in terms of CVD were yet demonstrated for pioglitazone only, which might be due to an additional PPAR-alpha agonistic action not observed for rosiglitazone (Smith 2001; Dormandy et al. 2005).

Although generally well tolerated, there are some serious concerns: (1) Risk of heart failure appears to be increased, presumably due to weight gain and fluid retention induced by PPARγ agonism (see also above) (McGuire and Inzucchi 2008; Barnett 2009). (2) Incidence of bone fractures seems elevated (Vestergaard 2009). (3) Increased risk of cardiac ischemia was reported for rosiglitazone (but not for pioglitazone) (McGuire and Inzucchi 2008; Barnett 2009).

6.4 GLP-1 Agonists and DPP-4 Inhibitors

Intestinal augmentation of insulin secretion (i.e., the incretin effect) is generally attributed to GLP-1 and GIP. This effect appears to be diminished in T2DM, with reduced plasma [GLP-1] and normal/elevated [GIP] (Nauck et al. 1986; Nauck et al. 1993). Cellular responsiveness to GLP-1, however, is conserved, while that to GIP tends to be decreased (Nauck et al. 1993).

Both GLP-1 and GIP are degraded by the enzyme DPP-4, a cell-surface exopeptidase that preferentially cleaves peptides with a proline or alanine residue in the second aminoterminal position (Deacon et al. 1995a). DPP-4 is not specific for GLP-1 and GIP but known to cleave numerous additional peptides (e.g., Substance P, Neuropeptide Y, Peptide YY, interferons, Macrophage-derived chemokines) and play a role in the immune system through interaction with various molecules, including cytokines and chemokines (Drucker and Nauck 2006; Stulc and Sedo 2010).

DPP-4 is ubiquitously expressed with particularly high concentrations in intestinal mucosa, suggesting that the majority of GLP-1 and GIP is inactivated prior to entry into the systemic circulation (Hansen et al. 1999). Plasma half-lives for intravenously applied exogenous GLP-1 and GIP were 1–2 min and 5–7 min, respectively (Hansen et al. 1999; Mentlein 1999).

Effects of GLP-1 and GIP are mediated through interaction with their specific plasma membrane receptors, which both belong to the seven-transmembrane domain receptor family of G-protein coupled receptors (GPCRs) (Baggio and Drucker 2007). Downstream signalling involves activation of Gs, adenylate cyclase and PKA and in case of pancreatic β-cells subsequent increase of cytosolic $[Ca^{2+}]$ and insulin release (Mentlein 1999; Baggio and Drucker 2007). GLP-1 and GIP receptors are expressed in a multitude of tissues, including hypothalamic neurons (see also above) (Fehmann et al. 1995; Baggio and Drucker 2007).

In patients with T2DM, GLP-1 was demonstrated to enhance glucose-dependent insulin secretion, normalize glucagon release, slow gastric emptying, diminish food intake and induce weight loss, thus suggesting significant therapeutic potential (Deacon et al. 1995b; Flint et al. 1998; Zander et al. 2002; Nauck and Meier 2005). Use of GLP-1, however, is hampered by its short half-life within circulation, leading to the development of two incretin based approaches (Mentlein et al. 1993; Deacon et al. 1995b): (1) DPP-4 resistant GLP-1 receptor agonists (e.g., exenatide and liraglutide) with half-lives in the range of hours. (2) DPP-4 inhibitors (e.g., sitagliptin, vildagliptin and saxagliptin) that act through augmentation of endogenous [GLP-1] and [GIP] by reducing their rate of degradation.

In clinical trials, GLP-1 agonists improved HbA1c similar to insulin regimes (−0.97% vs. placebo) with weight loss of 1.4 kg and 4.8 kg versus placebo or insulin, respectively (Davidson 2009). Slightly weaker but comparable glycaemic control (HbA1c reduction 0.74 %) was demonstrated for DPP-4 inhibitors which, however, in contrast to the GLP-1 agonist class proved to be weight neutral (Davidson 2009). In addition, clinical evidence suggests either GLP-1 agonists as well as DPP-4 inhibitors to exert significant β-cell protective effects (Mari et al. 2006, 2007, 2008; Xu et al. 2008a; Bunck et al. 2009; Davidson 2009). Adverse events in both classes were generally mild, including nausea, vomiting, mild hypoglycaemia and nasopharyngitis (Davidson 2009).

7   Conclusions

Complexity of T2DM pathophysiology may force novel therapeutic strategies to focus on root causes of the disease – overnutrition, energy imbalance and cellular metabolic overload, implying calorie restriction as one promising strategy. Recent evidence, however, suggests this approach to be particularly delicate. While the CB-1 receptor antagonist rimonabant induced durable effects in terms of weight loss, insulin sensitivity and glycaemic control, drug action appeared to be inherently coupled to increased rates of depressed mood (Di Marzo 2008; Leite et al. 2009). Further progress in unravelling the neuronal circuits involved in feeding control might help to solve this problem. The example, however, highlights intrinsic obstacles associated with targets involved in the enigmatic pathways of metabolic control (Fig. 9).

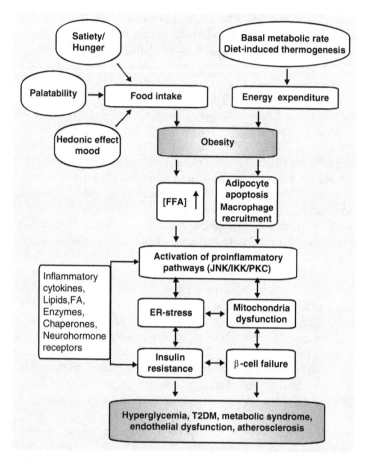

Fig. 9 Therapeutic targets at the interface between obesity and detrimental metabolic sequelae. ER-stress, endoplasmic reticulum stress; [FFA], extracellular concentration of free fatty acids; T2DM, type 2 diabetes mellitus

References

Abdul-Ghani MA, DeFronzo RA (2010) Pathogenesis of insulin resistance in skeletal muscle. J Biomed Biotechnol 2010:476279

Abumrad N, Coburn C, Ibrahimi A (1999) Membrane proteins implicated in long-chain fatty acid uptake by mammalian cells: CD36, FATP and FABPm. Biochim Biophys Acta 1441:4–13

Adams M, Montague CT, Prins JB, Holder JC, Smith SA, Sanders L, Digby JE, Sewter CP, Lazar MA, Chatterjee VK, O'Rahilly S (1997) Activators of peroxisome proliferator-activated receptor gamma have depot-specific effects on human preadipocyte differentiation. J Clin Invest 100:3149–3153

Ader M, Kim SP, Catalano KJ, Ionut V, Hucking K, Richey JM, Kabir M, Bergman RN (2005) Metabolic dysregulation with atypical antipsychotics occurs in the absence of underlying disease: a placebo-controlled study of olanzapine and risperidone in dogs. Diabetes 54: 862–871

Aguirre V, Werner ED, Giraud J, Lee YH, Shoelson SE, White MF (2002) Phosphorylation of Ser307 in insulin receptor substrate-1 blocks interactions with the insulin receptor and inhibits insulin action. J Biol Chem 277:1531–1537

Alquier T, Kawashima J, Tsuji Y, Kahn BB (2007) Role of hypothalamic adenosine 5'-monophosphate-activated protein kinase in the impaired counterregulatory response induced by repetitive neuroglucopenia. Endocrinology 148:1367–1375

Andersson U, Filipsson K, Abbott CR, Woods A, Smith K, Bloom SR, Carling D, Small CJ (2004) AMP-activated protein kinase plays a role in the control of food intake. J Biol Chem 279:12005–12008

Arkan MC, Hevener AL, Greten FR, Maeda S, Li ZW, Long JM, Wynshaw-Boris A, Poli G, Olefsky J, Karin M (2005) IKK-beta links inflammation to obesity-induced insulin resistance. Nat Med 11:191–198

Assali AR, Ganor A, Beigel Y, Shafer Z, Hershcovici T, Fainaru M (2001) Insulin resistance in obesity: body-weight or energy balance? J Endocrinol 171:293–298

Baggio LL, Drucker DJ (2007) Biology of incretins: GLP-1 and GIP. Gastroenterology 132: 2131–2157

Bailey CJ (2005) Treating insulin resistance in type 2 diabetes with metformin and thiazolidinediones. Diabetes Obes Metab 7:675–691

Bajaj M, Suraamornkul S, Romanelli A, Cline GW, Mandarino LJ, Shulman GI, DeFronzo RA (2005) Effect of a sustained reduction in plasma free fatty acid concentration on intramuscular long-chain fatty Acyl-CoAs and insulin action in type 2 diabetic patients. Diabetes 54: 3148–3153

Bandyopadhyay GK, Yu JG, Ofrecio J, Olefsky JM (2005) Increased p85/55/50 expression and decreased phosphotidylinositol 3-kinase activity in insulin-resistant human skeletal muscle. Diabetes 54:2351–2359

Bandyopadhyay GK, Yu JG, Ofrecio J, Olefsky JM (2006) Increased malonyl-CoA levels in muscle from obese and type 2 diabetic subjects lead to decreased fatty acid oxidation and increased lipogenesis; thiazolidinedione treatment reverses these defects. Diabetes 55: 2277–2285

Barnett AH (2009) Redefining the role of thiazolidinediones in the management of type 2 diabetes. Vasc Health Risk Manag 5:141–151

Barroso I (2005) Genetics of Type 2 diabetes. Diabet Med 22:517–535

Bates SH, Stearns WH, Dundon TA, Schubert M, Tso AW, Wang Y, Banks AS, Lavery HJ, Haq AK, Maratos-Flier E, Neel BG, Schwartz MW, Myers MG Jr (2003) STAT3 signalling is required for leptin regulation of energy balance but not reproduction. Nature 421:856–859

Belfort R, Mandarino L, Kashyap S, Wirfel K, Pratipanawatr T, Berria R, Defronzo RA, Cusi K (2005) Dose–response effect of elevated plasma free fatty acid on insulin signaling. Diabetes 54:1640–1648

Beutler B (2004) Innate immunity: an overview. Mol Immunol 40:845–859

Boden G, Chen X (1995) Effects of fat on glucose uptake and utilization in patients with non-insulin-dependent diabetes. J Clin Invest 96:1261–1268

Boden G, Lebed B, Schatz M, Homko C, Lemieux S (2001) Effects of acute changes of plasma free fatty acids on intramyocellular fat content and insulin resistance in healthy subjects. Diabetes 50:1612–1617

Bodles AM, Varma V, Yao-Borengasser A, Phanavanh B, Peterson CA, McGehee RE Jr, Rasouli N, Wabitsch M, Kern PA (2006) Pioglitazone induces apoptosis of macrophages in human adipose tissue. J Lipid Res 47:2080–2088

Bogardus C, Lillioja S, Mott D, Reaven GR, Kashiwagi A, Foley JE (1984) Relationship between obesity and maximal insulin-stimulated glucose uptake in vivo and in vitro in Pima Indians. J Clin Invest 73:800–805

Bonen A, Benton CR, Campbell SE, Chabowski A, Clarke DC, Han XX, Glatz JF, Luiken JJ (2003) Plasmalemmal fatty acid transport is regulated in heart and skeletal muscle by contraction, insulin and leptin, and in obesity and diabetes. Acta Physiol Scand 178:347–356

Bonen A, Parolin ML, Steinberg GR, Calles-Escandon J, Tandon NN, Glatz JF, Luiken JJ, Heigenhauser GJ, Dyck DJ (2004) Triacylglycerol accumulation in human obesity and type 2 diabetes is associated with increased rates of skeletal muscle fatty acid transport and increased sarcolemmal FAT/CD36. FASEB J 18:1144–1146

Bonnefont-Rousselot D, Raji B, Walrand S, Gardes-Albert M, Jore D, Legrand A, Peynet J, Vasson MP (2003) An intracellular modulation of free radical production could contribute to the beneficial effects of metformin towards oxidative stress. Metabolism 52:586–589

Bouhlel MA, Derudas B, Rigamonti E, Dievart R, Brozek J, Haulon S, Zawadzki C, Jude B, Torpier G, Marx N, Staels B, Chinetti-Gbaguidi G (2007) PPARgamma activation primes human monocytes into alternative M2 macrophages with anti-inflammatory properties. Cell Metab 6:137–143

Breen DM, Giacca A (2010) Effects of insulin on the vasculature. Curr Vasc Pharmacol. Sep 1 [Epub ahead of print]

Buettner C, Pocai A, Muse ED, Etgen AM, Myers MG Jr, Rossetti L (2006) Critical role of STAT3 in leptin's metabolic actions. Cell Metab 4:49–60

Bunck MC, Diamant M, Corner A, Eliasson B, Malloy JL, Shaginian RM, Deng W, Kendall DM, Taskinen MR, Smith U, Yki-Jarvinen H, Heine RJ (2009) One-year treatment with exenatide improves beta-cell function, compared with insulin glargine, in metformin-treated type 2 diabetic patients: a randomized, controlled trial. Diab Care 32:762–768

Cai D, Yuan M, Frantz DF, Melendez PA, Hansen L, Lee J, Shoelson SE (2005) Local and systemic insulin resistance resulting from hepatic activation of IKK-beta and NF-kappaB. Nat Med 11:183–190

Cani PD, Amar J, Iglesias MA, Poggi M, Knauf C, Bastelica D, Neyrinck AM, Fava F, Tuohy KM, Chabo C, Waget A, Delmee E, Cousin B, Sulpice T, Chamontin B, Ferrieres J, Tanti JF, Gibson GR, Casteilla L, Delzenne NM, Alessi MC, Burcelin R (2007) Metabolic endotoxemia initiates obesity and insulin resistance. Diabetes 56:1761–1772

Chen C, Hosokawa H, Bumbalo LM, Leahy JL (1994) Regulatory effects of glucose on the catalytic activity and cellular content of glucokinase in the pancreatic beta cell. Study using cultured rat islets. J Clin Invest 94:1616–1620

Chintoh AF, Mann SW, Lam L, Lam C, Cohn TA, Fletcher PJ, Nobrega JN, Giacca A, Remington G (2008) Insulin resistance and decreased glucose-stimulated insulin secretion after acute olanzapine administration. J Clin Psychopharmacol 28:494–499

Chu NV, Kong AP, Kim DD, Armstrong D, Baxi S, Deutsch R, Caulfield M, Mudaliar SR, Reitz R, Henry RR, Reaven PD (2002) Differential effects of metformin and troglitazone on cardiovascular risk factors in patients with type 2 diabetes. Diab Care 25:542–549

Cincotta AH, Tozzo E, Scislowski PW (1997) Bromocriptine/SKF38393 treatment ameliorates obesity and associated metabolic dysfunctions in obese (ob/ob) mice. Life Sci 61:951–956

Cinti S, Mitchell G, Barbatelli G, Murano I, Ceresi E, Faloia E, Wang S, Fortier M, Greenberg AS, Obin MS (2005) Adipocyte death defines macrophage localization and function in adipose tissue of obese mice and humans. J Lipid Res 46:2347–2355

Civitarese AE, Carling S, Heilbronn LK, Hulver MH, Ukropcova B, Deutsch WA, Smith SR, Ravussin E (2007) Calorie restriction increases muscle mitochondrial biogenesis in healthy humans. PLoS Med 4:e76

Coleman RA, Lee DP (2004) Enzymes of triacylglycerol synthesis and their regulation. Prog Lipid Res 43:134–176

Colman RJ, Anderson RM, Johnson SC, Kastman EK, Kosmatka KJ, Beasley TM, Allison DB, Cruzen C, Simmons HA, Kemnitz JW, Weindruch R (2009) Caloric restriction delays disease onset and mortality in rhesus monkeys. Science 325:201–204

Cooper GJ, Willis AC, Clark A, Turner RC, Sim RB, Reid KB (1987) Purification and characterization of a peptide from amyloid-rich pancreases of type 2 diabetic patients. Proc Natl Acad Sci USA 84:8628–8632

Coppari R, Ichinose M, Lee CE, Pullen AE, Kenny CD, McGovern RA, Tang V, Liu SM, Ludwig T, Chua SC Jr, Lowell BB, Elmquist JK (2005) The hypothalamic arcuate nucleus: a key site for mediating leptin's effects on glucose homeostasis and locomotor activity. Cell Metab 1: 63–72

Cornier MA, Bessesen DH, Gurevich I, Leitner JW, Draznin B (2006) Nutritional upregulation of p85alpha expression is an early molecular manifestation of insulin resistance. Diabetologia 49:748–754

Cota D, Proulx K, Smith KA, Kozma SC, Thomas G, Woods SC, Seeley RJ (2006) Hypothalamic mTOR signaling regulates food intake. Science 312:927–930

Cusi K, Maezono K, Osman A, Pendergrass M, Patti ME, Pratipanawatr T, DeFronzo RA, Kahn CR, Mandarino LJ (2000) Insulin resistance differentially affects the PI 3-kinase- and MAP kinase-mediated signaling in human muscle. J Clin Invest 105:311–320

Dardevet D, Moore MC, DiCostanzo CA, Farmer B, Neal DW, Snead W, Lautz M, Cherrington AD (2005) Insulin secretion-independent effects of GLP-1 on canine liver glucose metabolism do not involve portal vein GLP-1 receptors. Am J Physiol Gastrointest Liver Physiol 289: G806–G814

Davidson JA (2009) Advances in therapy for type 2 diabetes: GLP-1 receptor agonists and DPP-4 inhibitors. Cleve Clin J Med 76(Suppl 5):S28–S38

Deacon CF, Johnsen AH, Holst JJ (1995a) Degradation of glucagon-like peptide-1 by human plasma in vitro yields an N-terminally truncated peptide that is a major endogenous metabolite in vivo. J Clin Endocrinol Metab 80:952–957

Deacon CF, Nauck MA, Toft-Nielsen M, Pridal L, Willms B, Holst JJ (1995b) Both subcutaneously and intravenously administered glucagon-like peptide I are rapidly degraded from the NH2-terminus in type II diabetic patients and in healthy subjects. Diabetes 44:1126–1131

DeFronzo RA (1988) Lilly lecture 1987. The triumvirate: beta-cell, muscle, liver. A collusion responsible for NIDDM. Diabetes 37:667–687

DeFronzo RA (1999) Pharmacologic therapy for type 2 diabetes mellitus. Ann Intern Med 131:281–303

Defronzo RA (2009) Banting Lecture. From the triumvirate to the ominous octet: a new paradigm for the treatment of type 2 diabetes mellitus. Diabetes 58:773–795

DeFronzo RA, Goodman AM (1995) Efficacy of metformin in patients with non-insulin-dependent diabetes mellitus. The Multicenter Metformin Study Group. N Engl J Med 333:541–549

DeFronzo RA, Soman V, Sherwin RS, Hendler R, Felig P (1978) Insulin binding to monocytes and insulin action in human obesity, starvation, and refeeding. J Clin Invest 62:204–213

DeFronzo RA, Gunnarsson R, Bjorkman O, Olsson M, Wahren J (1985) Effects of insulin on peripheral and splanchnic glucose metabolism in noninsulin-dependent (type II) diabetes mellitus. J Clin Invest 76:149–155

Dennis PB, Jaeschke A, Saitoh M, Fowler B, Kozma SC, Thomas G (2001) Mammalian TOR: a homeostatic ATP sensor. Science 294:1102–1105

Di Gregorio GB, Yao-Borengasser A, Rasouli N, Varma V, Lu T, Miles LM, Ranganathan G, Peterson CA, McGehee RE, Kern PA (2005) Expression of CD68 and macrophage chemoattractant protein-1 genes in human adipose and muscle tissues: association with cytokine expression, insulin resistance, and reduction by pioglitazone. Diabetes 54:2305–2313

Di Marzo V (2008) The endocannabinoid system in obesity and type 2 diabetes. Diabetologia 51:1356–1367

Diani AR, Sawada G, Wyse B, Murray FT, Khan M (2004) Pioglitazone preserves pancreatic islet structure and insulin secretory function in three murine models of type 2 diabetes. Am J Physiol Endocrinol Metab 286:E116–E122

Dormandy JA, Charbonnel B, Eckland DJ, Erdmann E, Massi-Benedetti M, Moules IK, Skene AM, Tan MH, Lefebvre PJ, Murray GD, Standl E, Wilcox RG, Wilhelmsen L, Betteridge J, Birkeland K, Golay A, Heine RJ, Koranyi L, Laakso M, Mokan M, Norkus A, Pirags V, Podar T, Scheen A, Scherbaum W, Schernthaner G, Schmitz O, Skrha J, Smith U, Taton J (2005) Secondary prevention of macrovascular events in patients with type 2 diabetes in the PROactive Study (PROspective pioglitAzone Clinical Trial In macroVascular Events): a randomised controlled trial. Lancet 366:1279–1289

Dresner A, Laurent D, Marcucci M, Griffin ME, Dufour S, Cline GW, Slezak LA, Andersen DK, Hundal RS, Rothman DL, Petersen KF, Shulman GI (1999) Effects of free fatty acids on glucose transport and IRS-1-associated phosphatidylinositol 3-kinase activity. J Clin Invest 103:253–259

Drucker DJ, Nauck MA (2006) The incretin system: glucagon-like peptide-1 receptor agonists and dipeptidyl peptidase-4 inhibitors in type 2 diabetes. Lancet 368:1696–1705

Eizirik DL, Cardozo AK, Cnop M (2008) The role for endoplasmic reticulum stress in diabetes mellitus. Endocr Rev 29:42–61

El-Mir MY, Nogueira V, Fontaine E, Averet N, Rigoulet M, Leverve X (2000) Dimethylbiguanide inhibits cell respiration via an indirect effect targeted on the respiratory chain complex I. J Biol Chem 275:223–228

Febbraio M, Abumrad NA, Hajjar DP, Sharma K, Cheng W, Pearce SF, Silverstein RL (1999) A null mutation in murine CD36 reveals an important role in fatty acid and lipoprotein metabolism. J Biol Chem 274:19055–19062

Fehmann HC, Goke R, Goke B (1995) Cell and molecular biology of the incretin hormones glucagon-like peptide-I and glucose-dependent insulin releasing polypeptide. Endocr Rev 16:390–410

Flint A, Raben A, Astrup A, Holst JJ (1998) Glucagon-like peptide 1 promotes satiety and suppresses energy intake in humans. J Clin Invest 101:515–520

Florez JC, Burtt N, de Bakker PI, Almgren P, Tuomi T, Holmkvist J, Gaudet D, Hudson TJ, Schaffner SF, Daly MJ, Hirschhorn JN, Groop L, Altshuler D (2004) Haplotype structure and genotype-phenotype correlations of the sulfonylurea receptor and the islet ATP-sensitive potassium channel gene region. Diabetes 53:1360–1368

Fontana L, Klein S (2007) Aging, adiposity, and calorie restriction. JAMA 297:986–994

Fontana L, Meyer TE, Klein S, Holloszy JO (2004) Long-term calorie restriction is highly effective in reducing the risk for atherosclerosis in humans. Proc Natl Acad Sci USA 101:6659–6663

Fontana L, Partridge L, Longo VD (2010) Extending healthy life span – from yeast to humans. Science 328:321–326

Foretz M, Hebrard S, Leclerc J, Zarrinpashneh E, Soty M, Mithieux G, Sakamoto K, Andreelli F, Viollet B (2010) Metformin inhibits hepatic gluconeogenesis in mice independently of the LKB1/AMPK pathway via a decrease in hepatic energy state. J Clin Invest 120:2355–2369

Gerstein HC, Yusuf S, Bosch J, Pogue J, Sheridan P, Dinccag N, Hanefeld M, Hoogwerf B, Laakso M, Mohan V, Shaw J, Zinman B, Holman RR (2006) Effect of rosiglitazone on the frequency of diabetes in patients with impaired glucose tolerance or impaired fasting glucose: a randomised controlled trial. Lancet 368:1096–1105

Goodarzi MO, Bryer-Ash M (2005) Metformin revisited: re-evaluation of its properties and role in the pharmacopoeia of modern antidiabetic agents. Diabetes Obes Metab 7:654–665

Goodpaster BH, He J, Watkins S, Kelley DE (2001) Skeletal muscle lipid content and insulin resistance: evidence for a paradox in endurance-trained athletes. J Clin Endocrinol Metab 86:5755–5761

Greco AV, Mingrone G, Giancaterini A, Manco M, Morroni M, Cinti S, Granzotto M, Vettor R, Camastra S, Ferrannini E (2002) Insulin resistance in morbid obesity: reversal with intramyocellular fat depletion. Diabetes 51:144–151

Gregor MF, Hotamisligil GS (2007) Thematic review series: adipocyte biology. Adipocyte stress: the endoplasmic reticulum and metabolic disease. J Lipid Res 48:1905–1914

Griffin ME, Marcucci MJ, Cline GW, Bell K, Barucci N, Lee D, Goodyear LJ, Kraegen EW, White MF, Shulman GI (1999) Free fatty acid-induced insulin resistance is associated with activation of protein kinase C theta and alterations in the insulin signaling cascade. Diabetes 48:1270–1274

Griffin JM, Gilbert KM, Pumford NR (2000) Inhibition of CYP2E1 reverses CD4+ T-cell alterations in trichloroethylene-treated MRL+/+ mice. Toxicol Sci 54:384–389

Gual P, Le Marchand-Brustel Y, Tanti JF (2005) Positive and negative regulation of insulin signaling through IRS-1 phosphorylation. Biochimie 87:99–109

Gutierrez-Juarez R, Obici S, Rossetti L (2004) Melanocortin-independent effects of leptin on hepatic glucose fluxes. J Biol Chem 279:49704–49715

Gyte A, Pritchard LE, Jones HB, Brennand JC, White A (2007) Reduced expression of the KATP channel subunit, Kir6.2, is associated with decreased expression of neuropeptide Y and agouti-related protein in the hypothalami of Zucker diabetic fatty rats. J Neuroendocrinol 19:941–951

Hansen L, Deacon CF, Orskov C, Holst JJ (1999) Glucagon-like peptide-1-(7-36)amide is transformed to glucagon-like peptide-1-(9-36)amide by dipeptidyl peptidase IV in the capillaries supplying the L cells of the porcine intestine. Endocrinology 140:5356–5363

Harding HP, Zeng H, Zhang Y, Jungries R, Chung P, Plesken H, Sabatini DD, Ron D (2001) Diabetes mellitus and exocrine pancreatic dysfunction in perk-/- mice reveals a role for translational control in secretory cell survival. Mol Cell 7:1153–1163

He W, Lam TK, Obici S, Rossetti L (2006) Molecular disruption of hypothalamic nutrient sensing induces obesity. Nat Neurosci 9:227–233

Hevener AL, Olefsky JM, Reichart D, Nguyen MT, Bandyopadhyay G, Leung HY, Watt MJ, Benner C, Febbraio MA, Nguyen AK, Folian B, Subramaniam S, Gonzalez FJ, Glass CK, Ricote M (2007) Macrophage PPAR gamma is required for normal skeletal muscle and hepatic insulin sensitivity and full antidiabetic effects of thiazolidinediones. J Clin Invest 117:1658–1669

Hirosumi J, Tuncman G, Chang L, Gorgun CZ, Uysal KT, Maeda K, Karin M, Hotamisligil GS (2002) A central role for JNK in obesity and insulin resistance. Nature 420:333–336

Holland WL, Brozinick JT, Wang LP, Hawkins ED, Sargent KM, Liu Y, Narra K, Hoehn KL, Knotts TA, Siesky A, Nelson DH, Karathanasis SK, Fontenot GK, Birnbaum MJ, Summers SA (2007) Inhibition of ceramide synthesis ameliorates glucocorticoid-, saturated-fat-, and obesity-induced insulin resistance. Cell Metab 5:167–179

Hosogai N, Fukuhara A, Oshima K, Miyata Y, Tanaka S, Segawa K, Furukawa S, Tochino Y, Komuro R, Matsuda M, Shimomura I (2007) Adipose tissue hypoxia in obesity and its impact on adipocytokine dysregulation. Diabetes 56:901–911

Hotamisligil GS (2006) Inflammation and metabolic disorders. Nature 444:860–867

Houmard JA (2008) Intramuscular lipid oxidation and obesity. Am J Physiol Regul Integr Comp Physiol 294:R1111–R1116

Houseknecht KL, Robertson AS, Zavadoski W, Gibbs EM, Johnson DE, Rollema H (2007) Acute effects of atypical antipsychotics on whole-body insulin resistance in rats: implications for adverse metabolic effects. Neuropsychopharmacology 32:289–297

Hsueh WA, Law RE (1999) Insulin signaling in the arterial wall. Am J Cardiol 84:21J–24J

Hu Z, Cha SH, van Haasteren G, Wang J, Lane MD (2005) Effect of centrally administered C75, a fatty acid synthase inhibitor, on ghrelin secretion and its downstream effects. Proc Natl Acad Sci USA 102:3972–3977

Hundal RS, Petersen KF, Mayerson AB, Randhawa PS, Inzucchi S, Shoelson SE, Shulman GI (2002) Mechanism by which high-dose aspirin improves glucose metabolism in type 2 diabetes. J Clin Invest 109:1321–1326

Inzucchi SE (2002) Oral antihyperglycemic therapy for type 2 diabetes: scientific review. JAMA 287:360–372

Ionut V, Hucking K, Liberty IF, Bergman RN (2005) Synergistic effect of portal glucose and glucagon-like peptide-1 to lower systemic glucose and stimulate counter-regulatory hormones. Diabetologia 48:967–975

Itani SI, Pories WJ, Macdonald KG, Dohm GL (2001) Increased protein kinase C theta in skeletal muscle of diabetic patients. Metabolism 50:553–557

Itani SI, Ruderman NB, Schmieder F, Boden G (2002) Lipid-induced insulin resistance in human muscle is associated with changes in diacylglycerol, protein kinase C, and IkappaB-alpha. Diabetes 51:2005–2011

26   C. Schwanstecher and M. Schwanstecher

Jean-Baptiste G, Yang Z, Khoury C, Greenwood MT (2005) Lysophosphatidic acid mediates pleiotropic responses in skeletal muscle cells. Biochem Biophys Res Commun 335:1155–1162

Jiang ZY, Lin YW, Clemont A, Feener EP, Hein KD, Igarashi M, Yamauchi T, White MF, King GL (1999) Characterization of selective resistance to insulin signaling in the vasculature of obese Zucker (fa/fa) rats. J Clin Invest 104:447–457

Kahn SE, Haffner SM, Heise MA, Herman WH, Holman RR, Jones NP, Kravitz BG, Lachin JM, O'Neill MC, Zinman B, Viberti G (2006a) Glycemic durability of rosiglitazone, metformin, or glyburide monotherapy. N Engl J Med 355:2427–2443

Kahn SE, Hull RL, Utzschneider KM (2006b) Mechanisms linking obesity to insulin resistance and type 2 diabetes. Nature 444:840–846

Kanda H, Tateya S, Tamori Y, Kotani K, Hiasa K, Kitazawa R, Kitazawa S, Miyachi H, Maeda S, Egashira K, Kasuga M (2006) MCP-1 contributes to macrophage infiltration into adipose tissue, insulin resistance, and hepatic steatosis in obesity. J Clin Invest 116:1494–1505

Karin M, Takahashi T, Kapahi P, Delhase M, Chen Y, Makris C, Rothwarf D, Baud V, Natoli G, Guido F, Li N (2001) Oxidative stress and gene expression: the AP-1 and NF-kappaB connections. Biofactors 15:87–89

Kelley DE, Wing R, Buonocore C, Sturis J, Polonsky K, Fitzsimmons M (1993) Relative effects of calorie restriction and weight loss in noninsulin-dependent diabetes mellitus. J Clin Endocrinol Metab 77:1287–1293

Khaldi MZ, Guiot Y, Gilon P, Henquin JC, Jonas JC (2004) Increased glucose sensitivity of both triggering and amplifying pathways of insulin secretion in rat islets cultured for 1 wk in high glucose. Am J Physiol Endocrinol Metab 287:E207–E217

Kiens B, Roemen TH, van der Vusse GJ (1999) Muscular long-chain fatty acid content during graded exercise in humans. Am J Physiol 276:E352–E357

Kim JK, Fillmore JJ, Chen Y, Yu C, Moore IK, Pypaert M, Lutz EP, Kako Y, Velez-Carrasco W, Goldberg IJ, Breslow JL, Shulman GI (2001) Tissue-specific overexpression of lipoprotein lipase causes tissue-specific insulin resistance. Proc Natl Acad Sci USA 98:7522–7527

Kim JK, Fillmore JJ, Sunshine MJ, Albrecht B, Higashimori T, Kim DW, Liu ZX, Soos TJ, Cline GW, O'Brien WR, Littman DR, Shulman GI (2004) PKC-theta knockout mice are protected from fat-induced insulin resistance. J Clin Invest 114:823–827

Kliewer SA, Sundseth SS, Jones SA, Brown PJ, Wisely GB, Koble CS, Devchand P, Wahli W, Willson TM, Lenhard JM, Lehmann JM (1997) Fatty acids and eicosanoids regulate gene expression through direct interactions with peroxisome proliferator-activated receptors alpha and gamma. Proc Natl Acad Sci USA 94:4318–4323

Knauf C, Cani PD, Perrin C, Iglesias MA, Maury JF, Bernard E, Benhamed F, Gremeaux T, Drucker DJ, Kahn CR, Girard J, Tanti JF, Delzenne NM, Postic C, Burcelin R (2005) Brain glucagon-like peptide-1 increases insulin secretion and muscle insulin resistance to favor hepatic glycogen storage. J Clin Invest 115:3554–3563

Konner AC, Janoschek R, Plum L, Jordan SD, Rother E, Ma X, Xu C, Enriori P, Hampel B, Barsh GS, Kahn CR, Cowley MA, Ashcroft FM, Bruning JC (2007) Insulin action in AgRP-expressing neurons is required for suppression of hepatic glucose production. Cell Metab 5:438–449

Koves TR, Ussher JR, Noland RC, Slentz D, Mosedale M, Ilkayeva O, Bain J, Stevens R, Dyck JR, Newgard CB, Lopaschuk GD, Muoio DM (2008) Mitochondrial overload and incomplete fatty acid oxidation contribute to skeletal muscle insulin resistance. Cell Metab 7:45–56

Krebs M, Brunmair B, Brehm A, Artwohl M, Szendroedi J, Nowotny P, Roth E, Furnsinn C, Promintzer M, Anderwald C, Bischof M, Roden M (2007) The Mammalian target of rapamycin pathway regulates nutrient-sensitive glucose uptake in man. Diabetes 56:1600–1607

Krook A, Bjornholm M, Galuska D, Jiang XJ, Fahlman R, Myers MG Jr, Wallberg-Henriksson H, Zierath JR (2000) Characterization of signal transduction and glucose transport in skeletal muscle from type 2 diabetic patients. Diabetes 49:284–292

Krssak M, Falk Petersen K, Dresner A, DiPietro L, Vogel SM, Rothman DL, Roden M, Shulman GI (1999) Intramyocellular lipid concentrations are correlated with insulin sensitivity in humans: a 1H NMR spectroscopy study. Diabetologia 42:113–116

Kuningas M, Magi R, Westendorp RG, Slagboom PE, Remm M, van Heemst D (2007) Haplotypes in the human Foxo1a and Foxo3a genes; impact on disease and mortality at old age. Eur J Hum Genet 15:294–301

Lam TK, Gutierrez-Juarez R, Pocai A, Rossetti L (2005a) Regulation of blood glucose by hypothalamic pyruvate metabolism. Science 309:943–947

Lam TK, Pocai A, Gutierrez-Juarez R, Obici S, Bryan J, Aguilar-Bryan L, Schwartz GJ, Rossetti L (2005b) Hypothalamic sensing of circulating fatty acids is required for glucose homeostasis. Nat Med 11:320–327

Leclerc V, Reichhart JM (2004) The immune response of Drosophila melanogaster. Immunol Rev 198:59–71

Leite CE, Mocelin CA, Petersen GO, Leal MB, Thiesen FV (2009) Rimonabant: an antagonist drug of the endocannabinoid system for the treatment of obesity. Pharmacol Rep 61:217–224

Lenhard JM, Croom DK, Minnick DT (2004) Reduced serum dipeptidyl peptidase-IV after metformin and pioglitazone treatments. Biochem Biophys Res Commun 324:92–97

Li L, Shi Y, Wang X, Shi W, Jiang C (2005a) Single nucleotide polymorphisms in K(ATP) channels: muscular impact on type 2 diabetes. Diabetes 54:1592–1597

Li Y, Schwabe RF, DeVries-Seimon T, Yao PM, Gerbod-Giannone MC, Tall AR, Davis RJ, Flavell R, Brenner DA, Tabas I (2005b) Free cholesterol-loaded macrophages are an abundant source of tumor necrosis factor-alpha and interleukin-6: model of NF-kappaB- and map kinase-dependent inflammation in advanced atherosclerosis. J Biol Chem 280: 21763–21772

Lieberman JA, Stroup TS, McEvoy JP, Swartz MS, Rosenheck RA, Perkins DO, Keefe RS, Davis SM, Davis CE, Lebowitz BD, Severe J, Hsiao JK (2005) Effectiveness of antipsychotic drugs in patients with chronic schizophrenia. N Engl J Med 353:1209–1223

Liu L, Karkanias GB, Morales JC, Hawkins M, Barzilai N, Wang J, Rossetti L (1998) Intracerebroventricular leptin regulates hepatic but not peripheral glucose fluxes. J Biol Chem 273:31160–31167

Liu L, Zhang Y, Chen N, Shi X, Tsang B, Yu YH (2007) Upregulation of myocellular DGAT1 augments triglyceride synthesis in skeletal muscle and protects against fat-induced insulin resistance. J Clin Invest 117: 1679–1689

Loftus TM, Jaworsky DE, Frehywot GL, Townsend CA, Ronnett GV, Lane MD, Kuhajda FP (2000) Reduced food intake and body weight in mice treated with fatty acid synthase inhibitors. Science 288:2379–2381

Lumeng CN, Deyoung SM, Bodzin JL, Saltiel AR (2007) Increased inflammatory properties of adipose tissue macrophages recruited during diet-induced obesity. Diabetes 56:16–23

Mannucci E, Ognibene A, Cremasco F, Bardini G, Mencucci A, Pierazzuoli E, Ciani S, Messeri G, Rotella CM (2001) Effect of metformin on glucagon-like peptide 1 (GLP-1) and leptin levels in obese nondiabetic subjects. Diab Care 24:489–494

Mantovani A, Sica A, Sozzani S, Allavena P, Vecchi A, Locati M (2004) The chemokine system in diverse forms of macrophage activation and polarization. Trends Immunol 25:677–686

Mari A, Nielsen LL, Nanayakkara N, DeFronzo RA, Ferrannini E, Halseth A (2006) Mathematical modeling shows exenatide improved beta-cell function in patients with type 2 diabetes treated with metformin or metformin and a sulfonylurea. Horm Metab Res 38:838–844

Mari A, Degn K, Brock B, Rungby J, Ferrannini E, Schmitz O (2007) Effects of the long-acting human glucagon-like peptide-1 analog liraglutide on beta-cell function in normal living conditions. Diab Care 30:2032–2033

Mari A, Scherbaum WA, Nilsson PM, Lalanne G, Schweizer A, Dunning BE, Jauffret S, Foley JE (2008) Characterization of the influence of vildagliptin on model-assessed -cell function in patients with type 2 diabetes and mild hyperglycemia. J Clin Endocrinol Metab 93:103–109

Martin G, Schoonjans K, Lefebvre AM, Staels B, Auwerx J (1997) Coordinate regulation of the expression of the fatty acid transport protein and acyl-CoA synthetase genes by PPARalpha and PPARgamma activators. J Biol Chem 272:28210–28217

Matveyenko AV, Butler PC (2006) Beta-cell deficit due to increased apoptosis in the human islet amyloid polypeptide transgenic (HIP) rat recapitulates the metabolic defects present in type 2 diabetes. Diabetes 55:2106–2114

McCurdy CE, Davidson RT, Cartee GD (2005) Calorie restriction increases the ratio of phosphatidylinositol 3-kinase catalytic to regulatory subunits in rat skeletal muscle. Am J Physiol Endocrinol Metab 288:E996–E1001

McGarry JD, Dobbins RL (1999) Fatty acids, lipotoxicity and insulin secretion. Diabetologia 42:128–138

McGuire DK, Inzucchi SE (2008) New drugs for the treatment of diabetes mellitus: part I: Thiazolidinediones and their evolving cardiovascular implications. Circulation 117:440–449

Meguid MM, Fetissov SO, Blaha V, Yang ZJ (2000) Dopamine and serotonin VMN release is related to feeding status in obese and lean Zucker rats. NeuroReport 11:2069–2072

Mentlein R (1999) Dipeptidyl-peptidase IV (CD26) – role in the inactivation of regulatory peptides. Regul Pept 85:9–24

Mentlein R, Gallwitz B, Schmidt WE (1993) Dipeptidyl-peptidase IV hydrolyses gastric inhibitory polypeptide, glucagon-like peptide-1(7-36)amide, peptide histidine methionine and is responsible for their degradation in human serum. Eur J Biochem 214:829–835

Minokoshi Y, Alquier T, Furukawa N, Kim YB, Lee A, Xue B, Mu J, Foufelle F, Ferre P, Birnbaum MJ, Stuck BJ, Kahn BB (2004) AMP-kinase regulates food intake by responding to hormonal and nutrient signals in the hypothalamus. Nature 428:569–574

Mirshamsi S, Laidlaw HA, Ning K, Anderson E, Burgess LA, Gray A, Sutherland C, Ashford ML (2004) Leptin and insulin stimulation of signalling pathways in arcuate nucleus neurones: PI3K dependent actin reorganization and KATP channel activation. BMC Neurosci 5:54

Mootha VK, Lindgren CM, Eriksson KF, Subramanian A, Sihag S, Lehar J, Puigserver P, Carlsson E, Ridderstrale M, Laurila E, Houstis N, Daly MJ, Patterson N, Mesirov JP, Golub TR, Tamayo P, Spiegelman B, Lander ES, Hirschhorn JN, Altshuler D, Groop LC (2003) PGC-1alpha-responsive genes involved in oxidative phosphorylation are coordinately downregulated in human diabetes. Nat Genet 34:267–273

Morino K, Petersen KF, Shulman GI (2006) Molecular mechanisms of insulin resistance in humans and their potential links with mitochondrial dysfunction. Diabetes 55(Suppl 2): S9–S15

Morrison CD, Morton GJ, Niswender KD, Gelling RW, Schwartz MW (2005) Leptin inhibits hypothalamic Npy and Agrp gene expression via a mechanism that requires phosphatidylinositol 3-OH-kinase signaling. Am J Physiol Endocrinol Metab 289:E1051–E1057

Morton GJ, Gelling RW, Niswender KD, Morrison CD, Rhodes CJ, Schwartz MW (2005) Leptin regulates insulin sensitivity via phosphatidylinositol-3-OH kinase signaling in mediobasal hypothalamic neurons. Cell Metab 2:11–20

Muoio DM, Newgard CB (2008) Mechanisms of disease: molecular and metabolic mechanisms of insulin resistance and beta-cell failure in type 2 diabetes. Nat Rev Mol Cell Biol 9: 193–205

Nathan DM, Buse JB, Davidson MB, Ferrannini E, Holman RR, Sherwin R, Zinman B (2009) Medical management of hyperglycaemia in type 2 diabetes mellitus: a consensus algorithm for the initiation and adjustment of therapy: a consensus statement from the American Diabetes Association and the European Association for the Study of Diabetes. Diabetologia 52:17–30

Nauck MA, Meier JJ (2005) Glucagon-like peptide 1 and its derivatives in the treatment of diabetes. Regul Pept 128:135–148

Nauck M, Stockmann F, Ebert R, Creutzfeldt W (1986) Reduced incretin effect in type 2 (non-insulin-dependent) diabetes. Diabetologia 29:46–52

Nauck MA, Heimesaat MM, Orskov C, Holst JJ, Ebert R, Creutzfeldt W (1993) Preserved incretin activity of glucagon-like peptide 1 [7-36 amide] but not of synthetic human gastric inhibitory polypeptide in patients with type-2 diabetes mellitus. J Clin Invest 91:301–307

Nguyen MT, Satoh H, Favelyukis S, Babendure JL, Imamura T, Sbodio JI, Zalevsky J, Dahiyat BI, Chi NW, Olefsky JM (2005) JNK and tumor necrosis factor-alpha mediate free fatty acid-induced insulin resistance in 3T3-L1 adipocytes. J Biol Chem 280:35361–35371

Nguyen MT, Favelyukis S, Nguyen AK, Reichart D, Scott PA, Jenn A, Liu-Bryan R, Glass CK, Neels JG, Olefsky JM (2007) A subpopulation of macrophages infiltrates hypertrophic adipose tissue and is activated by free fatty acids via Toll-like receptors 2 and 4 and JNK-dependent pathways. J Biol Chem 282:35279–35292

Nomiyama T, Perez-Tilve D, Ogawa D, Gizard F, Zhao Y, Heywood EB, Jones KL, Kawamori R, Cassis LA, Tschop MH, Bruemmer D (2007) Osteopontin mediates obesity-induced adipose tissue macrophage infiltration and insulin resistance in mice. J Clin Invest 117:2877–2888

Obici S, Feng Z, Tan J, Liu L, Karkanias G, Rossetti L (2001) Central melanocortin receptors regulate insulin action. J Clin Invest 108:1079–1085

Obici S, Feng Z, Morgan K, Stein D, Karkanias G, Rossetti L (2002a) Central administration of oleic acid inhibits glucose production and food intake. Diabetes 51:271–275

Obici S, Zhang BB, Karkanias G, Rossetti L (2002b) Hypothalamic insulin signaling is required for inhibition of glucose production. Nat Med 8:1376–1382

Obici S, Feng Z, Karkanias G, Baskin DG, Rossetti L (2002c) Decreasing hypothalamic insulin receptors causes hyperphagia and insulin resistance in rats. Nat Neurosci 5:566–572

Obici S, Feng Z, Arduini A, Conti R, Rossetti L (2003) Inhibition of hypothalamic carnitine palmitoyltransferase-1 decreases food intake and glucose production. Nat Med 9:756–761

Odegaard JI, Ricardo-Gonzalez RR, Goforth MH, Morel CR, Subramanian V, Mukundan L, Red Eagle A, Vats D, Brombacher F, Ferrante AW, Chawla A (2007) Macrophage-specific PPAR-gamma controls alternative activation and improves insulin resistance. Nature 447:1116–1120

Okuno A, Tamemoto H, Tobe K, Ueki K, Mori Y, Iwamoto K, Umesono K, Akanuma Y, Fujiwara T, Horikoshi H, Yazaki Y, Kadowaki T (1998) Troglitazone increases the number of small adipocytes without the change of white adipose tissue mass in obese Zucker rats. J Clin Invest 101:1354–1361

Olefsky JM (2000) Treatment of insulin resistance with peroxisome proliferator-activated receptor gamma agonists. J Clin Invest 106:467–472

Oliver MF, Opie LH (1994) Effects of glucose and fatty acids on myocardial ischaemia and arrhythmias. Lancet 343:155–158

Ono H, Pocai A, Wang Y, Sakoda H, Asano T, Backer JM, Schwartz GJ, Rossetti L (2008) Activation of hypothalamic S6 kinase mediates diet-induced hepatic insulin resistance in rats. J Clin Invest 118:2959–2968

O'Rahilly S (2009) Human genetics illuminates the paths to metabolic disease. Nature 462:307–314

Owen MR, Doran E, Halestrap AP (2000) Evidence that metformin exerts its anti-diabetic effects through inhibition of complex 1 of the mitochondrial respiratory chain. Biochem J 348(Pt 3):607–614

Ozcan U, Cao Q, Yilmaz E, Lee AH, Iwakoshi NN, Ozdelen E, Tuncman G, Gorgun C, Glimcher LH, Hotamisligil GS (2004) Endoplasmic reticulum stress links obesity, insulin action, and type 2 diabetes. Science 306:457–461

Ozcan U, Yilmaz E, Ozcan L, Furuhashi M, Vaillancourt E, Smith RO, Gorgun CZ, Hotamisligil GS (2006) Chemical chaperones reduce ER stress and restore glucose homeostasis in a mouse model of type 2 diabetes. Science 313:1137–1140

Ozcan L, Ergin AS, Lu A, Chung J, Sarkar S, Nie D, Myers MG Jr, Ozcan U (2009) Endoplasmic reticulum stress plays a central role in development of leptin resistance. Cell Metab 9:35–51

Parton LE, Ye CP, Coppari R, Enriori PJ, Choi B, Zhang CY, Xu C, Vianna CR, Balthasar N, Lee CE, Elmquist JK, Cowley MA, Lowell BB (2007) Glucose sensing by POMC neurons regulates glucose homeostasis and is impaired in obesity. Nature 449:228–232

Patsouris D, Neels JG, Fan W, Li PP, Nguyen MT, Olefsky JM (2009) Glucocorticoids and thiazolidinediones interfere with adipocyte-mediated macrophage chemotaxis and recruitment. J Biol Chem 284:31223–31235

Patti ME, Butte AJ, Crunkhorn S, Cusi K, Berria R, Kashyap S, Miyazaki Y, Kohane I, Costello M, Saccone R, Landaker EJ, Goldfine AB, Mun E, DeFronzo R, Finlayson J, Kahn CR, Mandarino LJ (2003) Coordinated reduction of genes of oxidative metabolism in humans with insulin resistance and diabetes: Potential role of PGC1 and NRF1. Proc Natl Acad Sci USA 100: 8466–8471

Pan DA, Lillioja S, Kriketos AD, Milner MR, Baur LA, Bogardus C, Jenkins AB, Storlien LH (1997) Skeletal muscle triglyceride levels are inversely related to insulin action. Diabetes 46: 983–988

Perrin C, Knauf C, Burcelin R (2004) Intracerebroventricular infusion of glucose, insulin, and the adenosine monophosphate-activated kinase activator, 5-aminoimidazole-4-carboxamide-1-beta-D-ribofuranoside, controls muscle glycogen synthesis. Endocrinology 145:4025–4033

Perseghin G, Scifo P, De Cobelli F, Pagliato E, Battezzati A, Arcelloni C, Vanzulli A, Testolin G, Pozza G, Del Maschio A, Luzi L (1999) Intramyocellular triglyceride content is a determinant of in vivo insulin resistance in humans: a 1H-13C nuclear magnetic resonance spectroscopy assessment in offspring of type 2 diabetic parents. Diabetes 48:1600–1606

Petersen KF, Krssak M, Inzucchi S, Cline GW, Dufour S, Shulman GI (2000) Mechanism of troglitazone action in type 2 diabetes . Diabetes 49:827–831

Plum L, Ma X, Hampel B, Balthasar N, Coppari R, Munzberg H, Shanabrough M, Burdakov D, Rother E, Janoschek R, Alber J, Belgardt BF, Koch L, Seibler J, Schwenk F, Fekete C, Suzuki A, Mak TW, Krone W, Horvath TL, Ashcroft FM, Bruning JC (2006) Enhanced PIP3 signaling in POMC neurons causes KATP channel activation and leads to diet-sensitive obesity. J Clin Invest 116:1886–1901

Pocai A, Lam TK, Gutierrez-Juarez R, Obici S, Schwartz GJ, Bryan J, Aguilar-Bryan L, Rossetti L (2005) Hypothalamic K(ATP) channels control hepatic glucose production. Nature 434:1026–1031

Pocai A, Lam TK, Obici S, Gutierrez-Juarez R, Muse ED, Arduini A, Rossetti L (2006) Restoration of hypothalamic lipid sensing normalizes energy and glucose homeostasis. J Clin Invest 116:1081–1091

Poitout V, Robertson RP (2002) Minireview: Secondary beta-cell failure in type 2 diabetes – a convergence of glucotoxicity and lipotoxicity. Endocrinology 143:339–342

Pownall HJ, Hamilton JA (2003) Energy translocation across cell membranes and membrane models. Acta Physiol Scand 178:357–365

Prentki M, Joly E, El-Assaad W, Roduit R (2002) Malonyl-CoA signaling, lipid partitioning, and glucolipotoxicity: role in beta-cell adaptation and failure in the etiology of diabetes. Diabetes 51(Suppl 3):S405–S413

Proulx K, Cota D, Woods SC, Seeley RJ (2008) Fatty acid synthase inhibitors modulate energy balance via mammalian target of rapamycin complex 1 signaling in the central nervous system. Diabetes 57:3231–3238

Richardson DK, Kashyap S, Bajaj M, Cusi K, Mandarino SJ, Finlayson J, DeFronzo RA, Jenkinson CP, Mandarino LJ (2005) Lipid infusion decreases the expression of nuclear encoded mitochondrial genes and increases the expression of extracellular matrix genes in human skeletal muscle. J Biol Chem 280:10290–10297

Roepstorff C, Vistisen B, Roepstorff K, Kiens B (2004) Regulation of plasma long-chain fatty acid oxidation in relation to uptake in human skeletal muscle during exercise. Am J Physiol Endocrinol Metab 287:E696–E705

Rodbard HW, Jellinger PS (2010) The American Association of Clinical Endocrinologists/American College of Endocrinology (AACE/ACE) algorithm for managing glycaemia in patients with type 2 diabetes mellitus: comparison with the ADA/EASD algorithm. Diabetologia 53 (11):2458–2460

Rodbard HW, Jellinger PS, Davidson JA, Einhorn D, Garber AJ, Grunberger G, Handelsman Y, Horton ES, Lebovitz H, Levy P, Moghissi ES, Schwartz SS (2009) Statement by an American Association of Clinical Endocrinologists/American College of Endocrinology consensus panel on type 2 diabetes mellitus: an algorithm for glycemic control. Endocr Pract 15:540–559

Ron D, Walter P (2007) Signal integration in the endoplasmic reticulum unfolded protein response. Nat Rev Mol Cell Biol 8:519–529

Rosen P, Wiernsperger NF (2006) Metformin delays the manifestation of diabetes and vascular dysfunction in Goto-Kakizaki rats by reduction of mitochondrial oxidative stress. Diabetes Metab Res Rev 22:323–330

Sampson SR, Cooper DR (2006) Specific protein kinase C isoforms as transducers and modulators of insulin signaling. Mol Genet Metab 89:32–47

Samuel VT, Petersen KF, Shulman GI (2010) Lipid-induced insulin resistance: unravelling the mechanism. Lancet 375:2267–2277

Sandoval DA, Bagnol D, Woods SC, D'Alessio DA, Seeley RJ (2008) Arcuate glucagon-like peptide 1 receptors regulate glucose homeostasis but not food intake. Diabetes 57: 2046–2054

Sandoval DA, Obici S, Seeley RJ (2009) Targeting the CNS to treat type 2 diabetes. Nat Rev Drug Discov 8:386–398

Sathyanarayana P, Barthwal MK, Kundu CN, Lane ME, Bergmann A, Tzivion G, Rana A (2002) Activation of the Drosophila MLK by ceramide reveals TNF-alpha and ceramide as agonists of mammalian MLK3. Mol Cell 10:1527–1533

Schenk S, Saberi M, Olefsky JM (2008) Insulin sensitivity: modulation by nutrients and inflammation. J Clin Invest 118:2992–3002

Scheuner D, Vander Mierde D, Song B, Flamez D, Creemers JW, Tsukamoto K, Ribick M, Schuit FC, Kaufman RJ (2005) Control of mRNA translation preserves endoplasmic reticulum function in beta cells and maintains glucose homeostasis. Nat Med 11:757–764

Schmitz-Peiffer C, Browne CL, Oakes ND, Watkinson A, Chisholm DJ, Kraegen EW, Biden TJ (1997) Alterations in the expression and cellular localization of protein kinase C isozymes epsilon and theta are associated with insulin resistance in skeletal muscle of the high-fat-fed rat. Diabetes 46:169–178

Schwanstecher C, Schwanstecher M (2002) Nucleotide sensitivity of pancreatic ATP-sensitive potassium channels and type 2 diabetes. Diabetes 51(Suppl 3):S358–S362

Schwanstecher C, Meyer U, Schwanstecher M (2002) K(IR)6.2 polymorphism predisposes to type 2 diabetes by inducing overactivity of pancreatic beta-cell ATP-sensitive K(+) channels. Diabetes 51:875–879

Schwartz MW, Woods SC, Porte D Jr, Seeley RJ, Baskin DG (2000) Central nervous system control of food intake. Nature 404:661–671

Shevah O, Laron Z (2007) Patients with congenital deficiency of IGF-I seem protected from the development of malignancies: a preliminary report. Growth Horm IGF Res 17:54–57

Shi H, Kokoeva MV, Inouye K, Tzameli I, Yin H, Flier JS (2006) TLR4 links innate immunity and fatty acid-induced insulin resistance. J Clin Invest 116:3015–3025

Shimizu H, Oh IS, Tsuchiya T, Ohtani KI, Okada S, Mori M (2006) Pioglitazone increases circulating adiponectin levels and subsequently reduces TNF-alpha levels in Type 2 diabetic patients: a randomized study. Diabet Med 23:253–257

Shimokawa T, Kumar MV, Lane MD (2002) Effect of a fatty acid synthase inhibitor on food intake and expression of hypothalamic neuropeptides. Proc Natl Acad Sci USA 99:66–71

Shoelson SE, Lee J, Goldfine AB (2006) Inflammation and insulin resistance. J Clin Invest 116: 1793–1801

Smith U (2001) Pioglitazone: mechanism of action. Int J Clin Pract Suppl 121:13–18

Solinas G, Vilcu C, Neels JG, Bandyopadhyay GK, Luo JL, Naugler W, Grivennikov S, Wynshaw-Boris A, Scadeng M, Olefsky JM, Karin M (2007) JNK1 in hematopoietically derived cells contributes to diet-induced inflammation and insulin resistance without affecting obesity. Cell Metab 6:386–397

Sondergaard L (1993) Homology between the mammalian liver and the Drosophila fat body. Trends Genet 9:193

Song MJ, Kim KH, Yoon JM, Kim JB (2006) Activation of Toll-like receptor 4 is associated with insulin resistance in adipocytes. Biochem Biophys Res Commun 346:739–745

Spiegelman BM (1998) PPAR-gamma: adipogenic regulator and thiazolidinedione receptor. Diabetes 47:507–514

Stoehr JP, Nadler ST, Schueler KL, Rabaglia ME, Yandell BS, Metz SA, Attie AD (2000) Genetic obesity unmasks nonlinear interactions between murine type 2 diabetes susceptibility loci. Diabetes 49:1946–1954

Stratford S, DeWald DB, Summers SA (2001) Ceramide dissociates 3'-phosphoinositide production from pleckstrin homology domain translocation. Biochem J 354:359–368

Strissel KJ, Stancheva Z, Miyoshi H, Perfield JW 2nd, DeFuria J, Jick Z, Greenberg AS, Obin MS (2007) Adipocyte death, adipose tissue remodeling, and obesity complications. Diabetes 56:2910–2918

Stulc T, Sedo A (2010) Inhibition of multifunctional dipeptidyl peptidase-IV: is there a risk of oncological and immunological adverse effects? Diabetes Res Clin Pract 88:125–131

Sun C, Zhang F, Ge X, Yan T, Chen X, Shi X, Zhai Q (2007) SIRT1 improves insulin sensitivity under insulin-resistant conditions by repressing PTP1B. Cell Metab 6:307–319

Summers SA (2006) Ceramides in insulin resistance and lipotoxicity. Prog Lipid Res 45:42–72

Taniguchi CM, Emanuelli B, Kahn CR (2006) Critical nodes in signalling pathways: insights into insulin action. Nat Rev Mol Cell Biol 7:85–96

Thirone AC, Huang C, Klip A (2006) Tissue-specific roles of IRS proteins in insulin signaling and glucose transport. Trends Endocrinol Metab 17:72–78

Tripathy D, Mohanty P, Dhindsa S, Syed T, Ghanim H, Aljada A, Dandona P (2003) Elevation of free fatty acids induces inflammation and impairs vascular reactivity in healthy subjects. Diabetes 52:2882–2887

Tsukumo DM, Carvalho-Filho MA, Carvalheira JB, Prada PO, Hirabara SM, Schenka AA, Araujo EP, Vassallo J, Curi R, Velloso LA, Saad MJ (2007) Loss-of-function mutation in Toll-like receptor 4 prevents diet-induced obesity and insulin resistance. Diabetes 56: 1986–1998

Tzatsos A, Kandror KV (2006) Nutrients suppress phosphatidylinositol 3-kinase/Akt signaling via raptor-dependent mTOR-mediated insulin receptor substrate 1 phosphorylation. Mol Cell Biol 26:63–76

UK Prospective Diabetes Study (UKPDS) Group (1998) Effect of intensive blood-glucose control with metformin on complications in overweight patients with type 2 diabetes (UKPDS 34). Lancet 352:854–865

Um SH, Frigerio F, Watanabe M, Picard F, Joaquin M, Sticker M, Fumagalli S, Allegrini PR, Kozma SC, Auwerx J, Thomas G (2004) Absence of S6K1 protects against age- and diet-induced obesity while enhancing insulin sensitivity. Nature 431:200–205

Unger RH (1995) Lipotoxicity in the pathogenesis of obesity-dependent NIDDM. Genetic and clinical implications. Diabetes 44:863–870

Vestergaard P (2009) Bone metabolism in type 2 diabetes and role of thiazolidinediones. Curr Opin Endocrinol Diabetes Obes 16:125–131

Wahren J, Ekberg K (2007) Splanchnic regulation of glucose production. Annu Rev Nutr 27:329–345

Wang X, Devaiah SP, Zhang W, Welti R (2006) Signaling functions of phosphatidic acid. Prog Lipid Res 45:250–278

Wang B, Wood IS, Trayhurn P (2007) Dysregulation of the expression and secretion of inflammation-related adipokines by hypoxia in human adipocytes. Pflugers Arch 455:479–492

Weisberg SP, McCann D, Desai M, Rosenbaum M, Leibel RL, Ferrante AW Jr (2003) Obesity is associated with macrophage accumulation in adipose tissue. J Clin Invest 112:1796–1808

Weisberg SP, Hunter D, Huber R, Lemieux J, Slaymaker S, Vaddi K, Charo I, Leibel RL, Ferrante AW Jr (2006) CCR2 modulates inflammatory and metabolic effects of high-fat feeding. J Clin Invest 116:115–124

Westermark P, Wernstedt C, O'Brien TD, Hayden DW, Johnson KH (1987) Islet amyloid in type 2 human diabetes mellitus and adult diabetic cats contains a novel putative polypeptide hormone. Am J Pathol 127:414–417

Willson TM, Brown PJ, Sternbach DD, Henke BR (2000) The PPARs: from orphan receptors to drug discovery. J Med Chem 43:527–550

Winter WE, Nakamura M, House DV (1999) Monogenic diabetes mellitus in youth. The MODY syndromes. Endocrinol Metab Clin North Am 28:765–785

Wolfgang MJ, Lane MD (2006) The role of hypothalamic malonyl-CoA in energy homeostasis. J Biol Chem 281:37265–37269

Woltman T, Reidelberger R (1995) Effects of duodenal and distal ileal infusions of glucose and oleic acid on meal patterns in rats. Am J Physiol 269:R7–R14

Xu L, Man CD, Charbonnel B, Meninger G, Davies MJ, Williams-Herman D, Cobelli C, Stein PP (2008a) Effect of sitagliptin, a dipeptidyl peptidase-4 inhibitor, on beta-cell function in patients with type 2 diabetes: a model-based approach. Diabetes Obes Metab 10:1212–1220

Xu Y, Jones JE, Kohno D, Williams KW, Lee CE, Choi MJ, Anderson JG, Heisler LK, Zigman JM, Lowell BB, Elmquist JK (2008b) 5-HT2CRs expressed by pro-opiomelanocortin neurons regulate energy homeostasis. Neuron 60:582–589

Yamauchi T, Kamon J, Waki H, Terauchi Y, Kubota N, Hara K, Mori Y, Ide T, Murakami K, Tsuboyama-Kasaoka N, Ezaki O, Akanuma Y, Gavrilova O, Vinson C, Reitman ML, Kagechika H, Shudo K, Yoda M, Nakano Y, Tobe K, Nagai R, Kimura S, Tomita M, Froguel P, Kadowaki T (2001) The fat-derived hormone adiponectin reverses insulin resistance associated with both lipoatrophy and obesity. Nat Med 7:941–946

Ye J, Gao Z, Yin J, He Q (2007) Hypoxia is a potential risk factor for chronic inflammation and adiponectin reduction in adipose tissue of ob/ob and dietary obese mice. Am J Physiol Endocrinol Metab 293:E1118–E1128

Yu C, Chen Y, Cline GW, Zhang D, Zong H, Wang Y, Bergeron R, Kim JK, Cushman SW, Cooney GJ, Atcheson B, White MF, Kraegen EW, Shulman GI (2002) Mechanism by which fatty acids inhibit insulin activation of insulin receptor substrate-1 (IRS-1)-associated phosphatidylinositol 3-kinase activity in muscle. J Biol Chem 277:50230–50236

Yuan M, Konstantopoulos N, Lee J, Hansen L, Li ZW, Karin M, Shoelson SE (2001) Reversal of obesity- and diet-induced insulin resistance with salicylates or targeted disruption of Ikkbeta. Science 293:1673–1677

Zander M, Madsbad S, Madsen JL, Holst JJ (2002) Effect of 6-week course of glucagon-like peptide 1 on glycaemic control, insulin sensitivity, and beta-cell function in type 2 diabetes: a parallel-group study. Lancet 359:824–830

Zhang X, Zhang G, Zhang H, Karin M, Bai H, Cai D (2008) Hypothalamic IKKbeta/NF-kappaB and ER stress link overnutrition to energy imbalance and obesity. Cell 135:61–73

Zhou L, Sutton GM, Rochford JJ, Semple RK, Lam DD, Oksanen LJ, Thornton-Jones ZD, Clifton PG, Yueh CY, Evans ML, McCrimmon RJ, Elmquist JK, Butler AA, Heisler LK (2007) Serotonin 2C receptor agonists improve type 2 diabetes via melanocortin-4 receptor signaling pathways. Cell Metab 6:398–405

Zhou SY, Lu YX, Owyang C (2008) Gastric relaxation induced by hyperglycemia is mediated by vagal afferent pathways in the rat. Am J Physiol Gastrointest Liver Physiol 294:G1158–G1164

Dual Acting and Pan-PPAR Activators as Potential Anti-diabetic Therapies

Monique Heald and Michael A. Cawthorne

Contents

Abstract The thiazolidinedione PPAR-γ activator drugs rosiglitazone and pioglitazone suppress insulin resistance in type 2 diabetic patients. They lock lipids into adipose tissue triglyceride stores, thereby preventing lipid metabolites from causing insulin resistance in liver and skeletal muscle and β-cell failure. They also reduce the secretion of inflammatory cytokines such as TNFα and increase the plasma level of adiponectin, which increases insulin sensitivity in liver and skeletal muscle. However, they have only a modest effect on dyslipidaemia, and they increase fat mass and plasma volume. Fibrate PPAR-α activator drugs decrease plasma triglycerides and increase HDL-cholesterol levels. PPAR-δ activators increase the capacity for fat oxidation in skeletal muscle.

Clinical experience with bezafibrate, which activates PPAR-δ and -α, and studies on the PPAR-α/δ activator tetradecylthioacetic acid, the PPAR-δ activator GW501516, and combinations of the PPAR-α activator fenofibrate with rosiglitazone or pioglitazone

Monique Heald died suddenly on 29.10.2008

M. Heald and M.A. Cawthorne (✉)
Clore Laboratory, University of Buckingham, Buckingham, UK

M. Schwanstecher (ed.), *Diabetes - Perspectives in Drug Therapy*,
Handbook of Experimental Pharmacology 203,
DOI 10.1007/978-3-642-17214-4_2, © Springer-Verlag Berlin Heidelberg 2011

have encouraged attempts to develop single molecules that activate two or all three PPARs. Most effort has focussed on dual PPAR-α/γ activators. These reduce both hyperglycaemia and dyslipidaemia, but their development has been terminated by issues such as increased weight gain, oedema, plasma creatinine and myocardial infarction or stroke. In addition, the FDA has stated that many PPAR ligands submitted to it have caused increased numbers of tumours in carcinogenicity studies.

Rather than aiming for full potent agonists, it may be best to identify subtype-selective partial agonists or compounds that selectively activate PPAR signalling pathways and use these in combination. Nutrients or modified lipids that are low-affinity agonists may also have potential.

Keywords Fibrate · Insulin sensitiser · Peroxisome proliferator-activated receptor · PPAR-α/γ activator · Thiazolidinedione

1 Introduction

The discovery of the three peroxisome proliferator-activated receptors (PPARs) as nuclear receptors functioning as lipid sensors hinged on the discovery of certain thiazolidinediones as insulin sensitiser agents. The original discovery stemmed from the finding by Takeda toxicologists that a potential triglyceride lowering agent of the fibrate type in which the carboxylate moiety was replaced by the acidic mimetic thiazolidinedione maintained normoglycaemic levels during long-term toxicology studies, whereas ageing control animals developed hyperglycaemia. Subsequent structure activity studies resulted in the compound ciglitazone. Further structure activity work at Sankyo, Beecham, and Takeda resulted in three compounds being progressed to market. These were troglitazone, which was subsequently withdrawn as a result of a liability for liver damage in some patients, rosiglitazone and pioglitazone.

It was the availability of these compounds, particularly the more potent agent rosiglitazone (then called BRL49653), that allowed the identification of the nuclear receptor PPAR-γ as the target for the thiazolidinedione insulin sensitiser drugs (Lehmann et al. 1995).

The three PPAR receptors [PPAR-α, PPAR-β (also called PPAR-δ, fatty acid-activated receptor) and PPAR-γ] form a subfamily of nuclear receptors. They function as lipid sensors and coordinate the regulation of expression of a large number of genes associated with metabolism. Each of the PPARs forms an obligate heterodimer with another nuclear receptor called the retionid X receptor (RXR), which binds to peroxisome proliferator response elements (PPREs) that are located within the regulatory domains of target genes. Activation of the PPAR by an appropriate ligand results in recruitment of co-activators and loss of co-repressors that remodel the chromatin and activate transcription (Desvergne and Wahli 1999).

2 PPAR-γ

Although PPAR-γ is widely expressed in tissues, it is present in high concentrations in adipose tissue (Fajas et al. 1997). It is essential for adipocyte differentiation and promotes lipid accumulation in adipocytes (Tontonoz et al. 1994). Moreover, adipose-specific knock-out of PPAR-γ in mice results in adipocyte hypocellularity and the development of insulin resistance in liver but not in muscle (He et al. 2003).

The anti-diabetic thiazolidinediones suppress endogenous insulin resistance in adipose tissue but also have effects in liver and muscle despite low concentrations of PPAR-γ in these tissues. As noted above, the effect in liver is probably indirect and it is noteworthy that insulin-resistant, muscle-specific PPAR-γ null mice respond to the insulin sensitising effects of PPPRγ activators such as the thiazolidinediones (Hevener et al. 2003; Norris et al. 2003).

Gene expression studies have shown that the thiazolidinedione insulin sensitisers alter the expression of genes involved in lipid uptake, lipid metabolism and insulin action in adipocytes resulting in increased lipid accumulation in adipose tissue and decreased release of free fatty acids. This partitions lipid away from liver and muscle and reverses lipotoxicity-induced insulin resistance in these tissues (Mayerson et al. 2002; Spiegelman 1998).

A consequence of the adipocentric mechanism of action is a gain in fat mass. This is seen in animal models as well as in clinical studies. However, the PPAR-γ activators function as adipose site remodelling agents with a redistribution of fat from large insulin-resistant, lipolytic visceral fat adipocytes to small, newly differentiated insulin-responsive subcutaneous adipocytes (Kawai et al. 1999). This is consistent with human probands with inhibitory PPAR-γ mutations having decreased subcutaneous fat but increased visceral fat together with hyperglycaemia and insulin resistance (Hegele et al. 2002).

In addition to their effects on lipid metabolism, thiazolidinediones have a major effect on the secretion of adipokines. Thus, they reduce the secretion of inflammatory cytokines and chemokines that promote insulin resistance, such as TNFα. These actions occur in both the adipocyte and associated macrophages. Other adipokines are up-regulated, particularly adiponectin, which is known to potentiate insulin sensitivity in liver and skeletal muscle (Berg et al. 2001; Yamauchi et al. 2001). The effects of the thiazolidinedione insulin sensitisers in improving insulin sensitisation in liver and muscle are likely to be mediated in part through alterations in adipokine gene expression through PPAR-γ receptor activation.

Diabetes in animals and humans does not occur unless there is an islet cell malfunction. Thus, in the presence of a fully operational pancreatic islet, obesity-induced insulin resistance will result in impaired glucose tolerance but not frank diabetes. There is growing evidence that lipotoxicity plays an important role in pancreatic islet β-cell failure. By reversing the lipotoxicity, there is an inhibition of apoptosis in the islet cell and an increase in β-cell mass (Han et al. 2008; Ishida et al. 2004; Zeender et al. 2004). Indeed, analyses of diabetes prevention trials have demonstrated that pioglitazone and rosiglitazone are able to reverse β-cell decline in pre-diabetic populations (Defronzo 2009).

In addition to macrophages, PPAR-γ is expressed in endothelial cells, vascular smooth muscle cells and macrophage-derived foam cells that form the cells of atherosclerotic lesions. Consequently, it has been hoped that activating PPAR-γ might have important anti-atherosclerotic effects. Indeed, PPAR-γ ligands have been shown to decrease the size of atherosclerotic lesions in low-density lipoprotein receptor null mice (Li et al. 2000) and in apolipoprotein E null mice (Chen et al. 2001). The mechanism of this effect appears to relate to the anti-inflammatory properties of PPAR-γ activators together with reduced levels of chemotaxis and promotion of apoptosis. In humans, there has been a clear demonstration that treatment of type 2 diabetes mellitus (T2DM) patients with PPAR-γ activators reduces levels of inflammatory biomarkers of cardiovascular disease. However, a reduction in cardiovascular disease has not been categorically shown. Indeed, there have been claims that rosiglitazone increases macrovascular disease, based on a meta-analysis (Nissen and Wolski 2007) study. This analysis has been criticised on the statistical grounds and that it included a high proportion of trials, which had a very low number of cardiovascular incidents and excluded trials where there was no incidence of macrovascular disease or death. The Food and Drug Administration (FDA) analysis reported by Dr Mahoney at the American Diabetes Meeting in 2008 found no evidence of increased cardiovascular events in patients taking either rosiglitazone or pioglitazone.

3 PPAR-α

The first identified PPAR receptor was PPAR-α, activation of which was associated with increased liver weight in rodents but not in humans. PPAR-α is the molecular target for the fibrate hypolipidaemic agents such as fenofibrate and gemfibrozil.

PPAR-α is highly expressed in liver and activation of the receptor results in increased hepatic lipid uptake and oxidation. Thus, the phenotype of the PPAR-α knock-out mouse in the fasted state is hypoglycaemia, hypoketonaemia, hypertri-glyceridaemia and hepatic steatosis (Kersten et al. 1999).

Activators of PPAR-α are used to treat dyslipidaemia. They decrease plasma triglyceride levels and increase high-density lipoprotein cholesterol (HDL-C) levels (Plutzky 2000). The latter effect is probably mediated by augmentation of hepatic production of major components of HDL-C, namely apolipoprotein AI and AII (Vu-Dac et al. 1994, 1995).

It is also possible that, like PPAR-γ activators, PPAR-α activators might have a direct vascular protective effect through action at the PPAR-α receptor in endothe-lial cells resulting in blockade of cytokine-induced cell adhesion. Moreover, by increasing the expression of the HDL receptor CLA-1/SR-BI (Chinetti et al. 2000) and the cholesterol transporter ABCA1 (Chinetti et al. 2001), they promote choles-terol efflux from the macrophages. Through all of these mechanisms, PPAR-α activators have been shown to reduce the progression of atherosclerosis and decrease the incidence of coronary events in several major clinical studies.

4   PPAR-δ

Unlike PPAR-α and PPAR-γ, PPAR-δ is ubiquitously expressed but its pharmacology is less understood than that of the other subtypes. PPAR-δ knock-out mice show an obese phenotype when fed on a high fat diet. Over-expression of PPAR-δ or over-activation by the selective ligand GW501516 resulted in induction of oxidative, mitochondrial rich type 1 muscle fibres that allowed the mice to undertake greater levels of running activity – the so-called marathon mouse (Wang et al. 2004). These transgenic mice were also resistant to diet-induced obesity and insulin resistance. GW501516 also attenuates weight gain and insulin resistance in mice fed on high fat diets. This action appears to result from an increase in the expression of genes in skeletal muscle that promote lipid catabolism and mitochondrial uncoupling resulting in increased β-oxidation of fatty acids in skeletal muscle (Tanaka et al. 2003).

5   Logic for Dual and Triple PPAR Activators
in the Treatment of Diabetes and Insulin Resistance

T2DM patients generally are overweight or obese and may additionally be dysli-pidaemic. The major cause of mortality in diabetic patients is atherosclerotic macrovascular disease culminating in myocardial infarction. These events are linked to the diabetic dyslipidaemia. Unfortunately, the currently available PPAR-γ insulin sensitisers provide only negligible or modest effects on lipid parameters.

In addition to the weak effects on plasma lipids, the thiazolidinediones rosigli-tazone and pioglitazone have been associated with adverse effects including plasma volume expansion, haemodilution, oedema, increased adiposity and weight gain and increased fat deposits in bone marrow (Yki-Jarvinen 2004). These undesirable side effects and the potential to cause congestive heart failure in a subset of diabetic patients with underlying cardiopathies and bone fractures have enhanced the search for PPAR-γ activators with an improved therapeutic window. One approach, based on selective oestrogen receptor modulators, which have equal efficacy, but less toxicity than full agonists at the oestrogen receptor (Miller 2002), has been to seek selective PPAR-γ activators, so-called SPPAR-γ modulator or SPPARMs. An alternative approach is to combine PPAR subtypes to enhance the metabolic effects (see Table 1). Thus, combining PPAR-α and PPAR-γ should lead to additional anti-hyperglycaemic effects by increasing hepatic fatty acid oxidation, alleviation of the dyslipidaemia and enhanced anti-atherosclerotic profile. By combining PPAR-γ

Table 1 Principal location of PPAR subtypes and metabolic effects

	PPAR-α	PPAR-γ	PPAR-δ
Location	Liver endothelial cells	Adipocytes vascular cells	Skeletal muscle
Main actions in target tissues	↑ FA uptake	↑ FA uptake	↑ FA oxidation
	↑ FA oxidation	↓ FA release	↑ Mitochondrial genesis
	↑ Apo AI, Apo AII	↓ Pro-inflammatory cytokines	
		↑ Insulin action	
Consequential effects	↓ Circulating TG	↓ Insulin resistance	↓ Body fat
	↑ HDL-C	↑ Body weight gain	↓ Circulating TG
	↓ Atherosclerosis	↑ Vasoprotection	↑ HDL-C
	↓ Liver fat		↑ Insulin action

FA fatty acid, *TG* triglyceride, *HDL-C* HDL-cholesterol

and PPAR-δ, one might expect a further improvement in insulin sensitivity with less or no weight gain and an improved ability to exercise and gain the beneficial effects of exercise. Clearly, there is also the potential to combine activation of all three PPAR receptors. Work to date has largely been to try to find dual or triple activator activity in a single molecule. This is an enormous challenge in obtaining acceptable therapeutic indices with regard to the potential receptor-mediated adverse effects. However, since there are currently both PPAR-γ activators (rosiglitazone and pioglitazone) and PPAR-α activators (fenofibrate) on the market, it is logical to examine the clinical effects of this combination.

6 The Bezafibrate Experience

Bezafibrate has been available for many years. It has been shown to be a good activator of PPAR-δ and -α but is only a weak activator of PPAR-γ (Krey et al. 1997). Elkeles et al. (1998) examined the effect of bezafibrate in diabetic patients given conventional diabetes treatment (diet and/or oral hypoglycaemic agents – presumably sulphonylureas and/or metformin but not glitazones, as the study was undertaken pre-marketing of these agents). The bezafibrate treatment was associated with significant reductions over 3 years in serum triglycerides, total cholesterol and total to HDL-cholesterol ratio and an increase in HDL-cholesterol. There was a trend to reduce fibrinogen. However, there was no effect on the progression of ultrasonically measured arterial disease. In general, the incidence of coronary heart disease in studies using bezafibrate has tended to be lower, but did not reach statistical significance (Tenenbaum et al. 2005a). However, in patients with metabolic syndrome and a history of recent myocardial infarction and/or stable angina, bezafibrate reduced the incidence of myocardial infarction and cardiac mortality (Tenenbaum et al. 2005b).

Beneficial effects of bezafibrate on glucose and insulin have been demonstrated by showing that there was a decreased incidence and delayed onset of T2DM in patients with impaired fasting glucose concentrations (Tenenbaum et al. 2004) and in obese patients (Tenenbaum et al. 2005c). However, studies on the treatment of patients with T2DM are lacking.

7 Use of Combined Therapy with Fenofibrate and Glitazones

Since fenofibrate is a potent PPAR-α activator, it is logical that the combination of this agent with the marketed glitazones should be examined in clinical studies. These clinical studies followed a mouse study (Carmona et al. 2005) in which C57Bl/6 *ob/ob* mice were given fenofibrate, rosiglitazone or the combination. Co-administration of fenofibrate prevented weight gain and increased fat mass induced by rosiglitazone. Although fenofibrate decreased blood glucose in *ob/ob* mice, it had no effect on plasma insulin, whereas, like rosiglitazone, both glucose and insulin concentrations were reduced by the combined treatment.

The published clinical studies were investigational rather than establishing therapeutic benefit. Thus, Boden et al. (2007) treated eight patients with rosiglitazone (8 mg/day) plus fenofibrate (160 mg/day) for 2 months and compared them with five rosiglitazone patients from an earlier study. The combination produced the benefits of the individual components on glycaemic and lipid parameters and surprisingly showed prevention of the fluid retention associated with rosiglitazone. A better controlled study examined the effect of fenofibrate or pioglitazone for 3 months followed by the addition of the other agent for 3 months in an open-label study (Bajaj et al. 2007). Pioglitazone alone decreased fasting blood glucose and HbA$_{1C}$, increased adiponectin and insulin-stimulated glucose disposal and reduced fasting plasma free fatty acids, triglycerides and hepatic fat content. Fenofibrate had no effect on any glycaemic parameter and the only lipid change was a fall in plasma triglycerides. Addition of pioglitazone to fenofibrate therapy resulted in all the benefits of pioglitazone being shown, whereas addition of fenofibrate to pioglitazone therapy only gave a further lowering of plasma triglycerides.

In a third trial involving 40 T2DM patients with poor metabolic control, the patients received rosiglitazone (4 mg/day) for 12 weeks on top of their existing therapy. Later, 200 mg/day fenofibrate was added for a further 12 weeks. The addition of fenofibrate did not significantly affect the HbA$_{1C}$ level, but the change in LDL-cholesterol level became highly significant. Overall, the concomitant administration of rosiglitazone and fenofibrate did not produce significant improvement in glycaemic control relative to rosiglitazone alone. However, the combination did improve the atherogenic dyslipidaemic profile. Fenofibrate addition did not reverse the effect of rosiglitazone on body mass index (Seber et al. 2006).

Whilst these results are encouraging, large double-blind trials are needed to elucidate any advantage of combining fenofibrate with rosiglitazone or pioglitazone.

8   Dual PPAR-α/γ Activator Drugs

A number of pharmaceutical companies have attempted to develop compounds with dual PPAR-α and -γ activity. It is a difficult task, however, to predict the appropriate balance between these activities without undertaking whole animal, let alone clinical, studies.

Some of the compounds are listed in Table 2. As can be seen, many of the compounds have been discontinued, largely as a result of side effects rather than lack of therapeutic efficacy.

The compounds were all selected for entry into clinical studies following studies in rodents. These studies concentrated on showing efficacy at least similar to rosiglitazone and pioglitazone, although there were few direct comparisons. Some showed additional effects. Thus, Oakes et al. (2005) demonstrated that tesoglitazar gave improved lipid tolerance, reduced hepatic triglyceride secretion and enhanced plasma triglyceride clearance. The same compound was found to increase the clearance of non-esterified fatty acids (NEFA) under both basal and elevated NEFA availability (Hegarty et al. 2004). Their data produced the first direct evidence that a dual activator increased the ability of white fat, liver and skeletal muscle to use fatty acids whilst also improving insulin action in these tissues.

Guo et al. (2004) found that the experimental Merck compound TZD 18 [5-(3-[3-(4-phenoxy-2-propylphenoxy)propoxy]phenyl)-2,4 thiazolidinedione] lowered cholesterol and triglycerides in hamsters and dogs (which are better models for human lipid metabolism) and induced the genes for fatty acid degradation and triglyceride clearance. The authors also demonstrated complete normalisation of glycaemic control in diabetic *db/db* mice. Also, this compound appears generally well balanced between potency at PPAR-α and -γ receptors, but does have some activity at the PPAR-δ receptor, at least against the human receptor. The authors showed 100-fold higher potency for transactivation of both human PPAR-α and -γ versus human PPAR-δ but surprisingly did not test against the rodent receptor.

Table 2 Dual PPAR-α/γ activators that have been in clinical development

Muraglitazar	BMS	Approved then withdrawn from market
Tesoglitazar	AstraZeneca	Discontinued following phase III trials
Ragaglitazar	Dr Reddy	Discontinued 2003
Chiglitazar	Shenzhen Chipscreen, China	Development suspended
MK-767/KRP-297	Merck/Kyorin	Discontinued 2003
TZD 18	Merck	Unknown
PAR-5359	Dong-A, Korea	Pre-clinical
E3030	Eisai, Japan	Phase II?
Cevoglitazar	Novartis	Discontinued 2008
Aleglitazar	Hoffman-La-Roche	Phase III 2010
TAK-559	Takeda	Discontinued 2005
Naveglitazar	Lilly	Phase II?
AVE-0847	Aventis	Phase II?
Sipoglitazar	Takeda	Discontinued 2006

Chira et al. (2007) tested the ability of tesaglitazar to reduce atherosclerosis in a mouse model on the basis that activation of vascular cell PPAR-α and -γ would provide anti-inflammatory and anti-proliferative effects. LDL-receptor null mice fed on a "Western-type" diet for 12 weeks results in marked and predictable atheosclerotic lesions. Co-administration of tesoglitazar with the diet reduced atherosclerosis in female but not male mice without affecting cholesterol or triglyceride levels. Extension of these studies showed that tesaglitazar could reduce the effect of cholesterol on atherosclerosis and block the progression of pre-existing atherosclerosis in APOE*3 Leiden CETP transgenic mice (van der Hoorn et al. 2009; Zadelaar et al. 2006). The authors found that tesaglitazar reduced plasma cholesterol and triglycerides and the mass and activity of cholesterol ester transfer protein (CETP) and increased HDL-cholesterol. Moreover, it reduced vessel wall inflammation, modified lesions to a more stabilised phenotype and completely blocked progression of the pre-existing lesions.

Muraglitazar has a similar potency at human PPAR-α and PPAR-γ receptors in transactivation assays (EC_{50} 0.28 and 0.16 μM, respectively). It has a similar potency to rosiglitazone at hPPAR-γ (EC_{50} 0.06 μM). Rosiglitazone has negligible potency at PPAR-α (Mittra et al. 2007). Pre-clinical studies have largely focussed on animal models of diabetes such as *db/db* mice in which potent anti-diabetic effects, preservation of pancreatic islet insulin content, reduced hyperlipidaemia and hepatic steatosis were shown (Harrity et al. 2006). In follow-up studies, muraglitazar was found to prevent both the development of diabetes in *db/db* mice, including loss of normal β-cell morphology and function, and the deterioration of established diabetes (Tozzo et al. 2007). Treatment of mice with PPAR-γ activators increases weight gain in diabetic animals. This is particularly the case in *db/db* mice and arises from both oedema and adipogenesis. The question whether the addition of PPAR-α activity might reduce weight gain was raised . In fact, muraglitazar had a greater potential than rosiglitazone on weight gain and this involved both oedema and adipogenesis (Mittra et al. 2007). The oedema was coincident with increased expression of mRNA for Enacγ and Na^+, K^+-ATPase in kidneys, mediated by PPAR-γ.

Ragaglitazar also showed similar potency to rosiglitazone in the human PPAR-γ transactivation assay (Chakrabarti et al. 2003). Despite this, it appears more active than rosiglitazone and fenofibrate in head-to-head studies in Zucker *fa/fa* rats, high fat-fed hyperlipidaemic rats and high fat-fed hamsters. Moreover, in a late-stage intervention study in ZDF diabetic rats, ragaglitazar reduced HbA_{1C} by 2.3% compared with 1.1% by rosiglitazone (Brand et al. 2003).

A series of dual activators have been examined in other pre-clinical studies giving similar results. These include chiglitazar (Li et al. 2006), PAR 5359 (Kim et al. 2008), E3030 (Kasai et al. 2008), cevoglitazar (Laurent et al. 2009), and aleglitazar (Benardeau et al. 2009). Takeda attempted to take predictive pre-clinical work a step forward by undertaking rhesus monkey studies in a well-defined colony that is representative of humans and found improvements in glycaemic and lipid parameters without weight gain. Although suggestive that further human trials were warranted, the work was published 2 years after Takeda announced that development was discontinued due to the lack of a sufficiently positive benefit/risk relationship in clinical studies (Ding et al. 2007).

Clinical studies on muraglitazar showed that the 5 mg dose reduced HbA_{1C} levels significantly more than pioglitazone (30 mg) in metformin-treated patients with T2DM in a phase III study (Kendall et al. 2006). Significant improvements over pioglitazone therapy were also seen in plasma triglycerides, apolipoprotein B, non-HDL-cholesterol and in increasing HDL-cholesterol. However, weight gain was greater with muraglitazar as was the oedema incidence.

Analysis of the phase II and phase III trial data in yet another meta-analysis by Nissen et al. (2005) indicated that death, myocardial infarction or stroke occurred in 35 out of 2,374 patients on muraglitazar as opposed to 9 out of 1,351 in the combined placebo- and pioglitazone-treated patients. The incidence of chronic heart failure was 13 out of 2,374 (0.55%) in muraglitazar-treated patients and 1 out of 1,351 in the controls. Both BMS and its marketing partner Merck abandoned the drug.

Measured by number of publications, clinical studies on tesaglitazar have been more extensive than on muraglitazar, although the total number of patients has been of a similar order. Tesaglitazar (0.5 or 1.0 mg/day) gave consistent improvements in glycaemic control and in lipid parameters, but studies reported consistent increases in serum creatinine levels, peripheral oedema and weight gain (Bays et al. 2007; Goke et al. 2007; Goldstein et al. 2006; Ratner et al. 2007; Schuster et al. 2008; Wilding et al. 2007).

As a result of the elevated creatinine levels found in its first four of eight phase III studies (Gallant 6-9) and the associated decrease in glomerula filtration rate, AstraZeneca decided to terminate its development programme on tesaglitazar on the basis that the overall benefit/risk profile was unlikely to offer patients significant advantage over marketed therapies.

9   Outlook for Dual PPAR-α/γ Activators

The data to date show that adding PPAR-α activation to the PPAR-γ profile results in improved lipid profile. However, it is clearly a very difficult task to obtain a balance of two separate properties in a single molecule. The logical approach would be to develop the safest and most appropriate PPAR-α activator and co-administer it with the safest and most efficacious PPAR-γ activator.

It is clear that the therapeutic window for PPAR-γ activation is quite narrow. It seems likely that muraglitazar and tesaglitazar failed largely because of their potency in PPAR-γ activation. The same probably applies to ragaglitazar. With hindsight the liabilities were probably apparent in pre-clinical studies.

Improving insulin sensitivity has added a powerful armamentarium to the treatment of diabetes and as yet the thiazolidinediones such as rosiglitazone and pioglitazone are the only drugs that are clinically proven to suppress pancreatic β-cell failure (Defronzo 2009). Now that the claim of adverse cardiovascular mortality has been discredited, the side effects of these drugs of weight gain and water retention can be managed and the drugs should not be given to potential congestive heart failure patients. However, there still remains the issue of potential fractures in women through a reduction in bone mineral density (Glintborg et al. 2008).

The therapeutic challenge for the pharmaceutical industry is to develop novel PPAR-γ activators with the therapeutic efficacy in improving insulin sensitivity, but with a lower risk of weight gain through adipogenesis and water retention. It seems unlikely that this will be achieved through a conventional full agonist and therefore researchers have focussed on partial agonists or so-called SPPARMs (selective PPAR-γ modulators) such as metaglidasen (Chandalia et al. 2009). This agent is claimed to retain PPAR-γ-related anti-diabetic properties in the absence of weight gain and oedema and selectively modulates a subset of PPAR-γ target genes.

10 PPAR-Pan Activators and PPAR-δ Dual Activators

Earlier in this chapter, it was noted that bezafibrate was a pan-PPAR activator, although its PPAR-γ activation relative to PPAR-α and -δ was weak. This has prompted companies to seek single compounds with all three activities (Evans et al. 2005). Typically, these companies have used high throughput screening systems and adopted their usual approach of seeking compounds with high potency at each receptor. Seeking a compound with high potency at one receptor is a challenge, but seeking one compound with high potency and efficacy at three receptors is an "Everest of a task" and potentially likely to produce toxic liabilities. It may be better to seek low-affinity compounds. One such low-affinity ligand is tetrade-cylthioacetic acid (Bocos et al. 1995). In clinical studies in T2DM patients, it improved the lipid profile but had no effect on glucose metabolism possibly because it is predominantly PPAR-α/-δ with little PPAR-γ activity (Lovas et al. 2009; Rost et al. 2009).

The development of PPAR-α/-δ and PPAR-γ/-δ dual activators has not taken off in the same way as PPAR-α/-γ. This is possibly because the structure–activity around the PPAR-δ receptor has not been fully addressed. However, the studies to date on PPAR-δ suggest that it could be a good target to go alongside PPAR-γ in the treatment of T2DM, which is almost exclusively an obese population (Barish et al. 2006; Lee et al. 2006). Thus, Oliver et al. (2001) found that GW501516 increased the expression of the reverse cholesterol transporter ATP-binding cassette A1 and induced apolipoprotein A1-specific cholesterol efflux. In insulin-resistant, obese, middle-aged rhesus monkeys, GW501516 caused a dramatic and dose-dependent rise in serum HDL-C while lowering the levels of small-dense low-density lipoprotein, fasting triglycerides and insulin. In a recent clinical study (Riserus et al. 2008), GW501516 (10 mg/kg o.d.) given to overweight, but otherwise healthy, men for 2 weeks resulted in significant reductions in fasting plasma triglycerides (−30%), apolipoprotein B (−26%), LDL-cholesterol (−23%), and insulin (−11%).

There was a 20% reduction in liver fat and a 30% reduction in liver isoprostanes; HDL-cholesterol was unchanged. Biopsy samples of skeletal muscle and a 6 h meal tolerance test with stable fatty acid isotopes revealed more exhaled carbon dioxide coming from the meal and increased expression of carnitine palmitoyl transferase 1b. Together, these data support PPAR-δ activators increasing fat oxidation in skeletal muscle.

It is suggested that the identification of a safe and effective PPAR-δ activator would be a good partner for PPAR-γ activators in the treatment of T2DM and the metabolic syndrome.

11 Cancer Liability of PPAR Activators

A large number of PPAR ligands have been submitted to the US FDA over the past 15 years. Many of these, but not all, have been subsequently shown to cause an increased number of tumours in carcinogenicity studies. This involves multiple tumour types in mice and rats of both sexes and multiple strains. The site of tumour development is consistent with the distribution of the PPAR receptors, e.g. adipose, vascular endothelium, bladder, skin, and renal tubules. Consequently, the FDA has been requesting performance of 2-year carcinogenicity studies prior to the initiation of clinical studies longer than 6 months (Aoki 2007).

12 Concluding Remarks

The development of dual and triple activators of PPAR receptors has proved to be difficult and to date no compound that is able to favourably influence the benefit/ risk ratio relative to current treatments for T2DM including the thiazolidinediones, rosiglitazone, and pioglitazone has been identified . The widespread involvement of PPAR receptors as lipid sensors that regulate fatty acid and carbohydrate metabolism, together with knowledge that the natural ligands are almost certainly low-affinity activators, perhaps suggests that the standard pharmaceutical approach of seeking high-affinity ligands might be doomed to failure. This is likely to apply even more to a search for high-affinity dual or triple activators.

Despite these reservations, there appears to be potentially significant clinical benefits in adding either PPAR-α or PPAR-δ activation to the existing profile of the PPAR-γ-mediated insulin sensitisers.

Perhaps the best approach would be to identify subtype-selective partial agonists or SPPARMs for each receptor and use these clinically in appropriate combinations. Meanwhile, there may be scope for identifying nutrients or modified lipids that are low-affinity compounds that could be either used as pharmaceuticals or incorporated into foods such as spreads, ice cream, etc.

References

Aoki T (2007) Current status of carcinogenicity assessment of peroxisome proliferator-activated receptor agonists by the US FDA and a mode-of-action approach to the carcinogenic potential. J Toxicol Pathol 20:197–202

Bajaj M, Suraamornkul S, Hardies LJ, Glass L, Musi N, DeFronzo RA (2007) Effects of peroxisome proliferator-activated receptor (PPAR)-alpha and PPAR-gamma agonists on glucose and lipid metabolism in patients with type 2 diabetes mellitus. Diabetologia 50:1723–1731

Barish GD, Narkar VA, Evans RM (2006) PPAR delta: a dagger in the heart of the metabolic syndrome. J Clin Invest 116:590–597

Bays H, McElhattan J, Bryzinski BS (2007) A double-blind, randomised trial of tesaglitazar versus pioglitazone in patients with type 2 diabetes mellitus. Diab Vasc Dis Res 4:181–193

Benardeau A, Benz J, Binggeli A, Blum D, Boehringer M, Grether U, Hilpert H, Kuhn B, Marki HP, Meyer M, Puntener K, Raab S, Ruf A, Schlatter D, Mohr P (2009) Aleglitazar, a new, potent, and balanced dual PPARalpha/gamma agonist for the treatment of type II diabetes. Bioorg Med Chem Lett 19:2468–2473

Berg AH, Combs TP, Du X, Brownlee M, Scherer PE (2001) The adipocyte-secreted protein Acrp30 enhances hepatic insulin action. Nat Med 7:947–953

Bocos C, Gottlicher M, Gearing K, Banner C, Enmark E, Teboul M, Crickmore A, Gustafsson JA (1995) Fatty acid activation of peroxisome proliferator-activated receptor (PPAR). J Steroid Biochem Mol Biol 53:467–473

Boden G, Homko C, Mozzoli M, Zhang M, Kresge K, Cheung P (2007) Combined use of rosiglitazone and fenofibrate in patients with type 2 diabetes: prevention of fluid retention. Diabetes 56:248–255

Brand CL, Sturis J, Gotfredsen CF, Fleckner J, Fledelius C, Hansen BF, Andersen B, Ye JM, Sauerberg P, Wassermann K (2003) Dual PPARalpha/gamma activation provides enhanced improvement of insulin sensitivity and glycemic control in ZDF rats. Am J Physiol Endocrinol Metab 284:E841–E854

Carmona MC, Louche K, Nibbelink M, Prunet B, Bross A, Desbazeille M, Dacquet C, Renard P, Casteilla L, Penicaud L (2005) Fenofibrate prevents Rosiglitazone-induced body weight gain in ob/ob mice. Int J Obes (Lond) 29:864–871

Chakrabarti R, Vikramadithyan RK, Misra P, Hiriyan J, Raichur S, Damarla RK, Gershome C, Suresh J, Rajagopalan R (2003) Ragaglitazar: a novel PPAR alpha PPAR gamma agonist with potent lipid-lowering and insulin-sensitizing efficacy in animal models. Br J Pharmacol 140:527–537

Chandalia A, Clarke HJ, Clemens LE, Pandey B, Vicena V, Lee P, Lavan BE, Gregoire FM (2009) MBX-102/JNJ39659100, a Novel Non-TZD Selective Partial PPAR-gamma Agonist Lowers Triglyceride Independently of PPAR-alpha Activation. PPAR Res 2009:706852

Chen Z, Ishibashi S, Perrey S, Osuga J, Gotoda T, Kitamine T, Tamura Y, Okazaki H, Yahagi N, Iizuka Y, Shionoiri F, Ohashi K, Harada K, Shimano H, Nagai R, Yamada N (2001) Troglitazone inhibits atherosclerosis in apolipoprotein E-knockout mice: pleiotropic effects on CD36 expression and HDL. Arterioscler Thromb Vasc Biol 21:372–377

Chinetti G, Gbaguidi FG, Griglio S, Mallat Z, Antonucci M, Poulain P, Chapman J, Fruchart JC, Tedgui A, Najib-Fruchart J, Staels B (2000) CLA-1/SR-BI is expressed in atherosclerotic lesion macrophages and regulated by activators of peroxisome proliferator-activated receptors. Circulation 101:2411–2417

Chinetti G, Lestavel S, Bocher V, Remaley AT, Neve B, Torra IP, Teissier E, Minnich A, Jaye M, Duverger N, Brewer HB, Fruchart JC, Clavey V, Staels B (2001) PPAR-alpha and PPAR-gamma activators induce cholesterol removal from human macrophage foam cells through stimulation of the ABCA1 pathway. Nat Med 7:53–58

Chira EC, McMillen TS, Wang S, Haw A 3rd, O'Brien KD, Wight TN, Chait A (2007) Tesaglitazar, a dual peroxisome proliferator-activated receptor alpha/gamma agonist, reduces atherosclerosis in female low density lipoprotein receptor deficient mice. Atherosclerosis 195:100–109

Defronzo RA (2009) Banting Lecture. From the triumvirate to the ominous octet: a new paradigm for the treatment of type 2 diabetes mellitus. Diabetes 58:773–795

Desvergne B, Wahli W (1999) Peroxisome proliferator-activated receptors: nuclear control of metabolism. Endocr Rev 20:649–688

Ding SY, Tigno XT, Braileanu GT, Ito K, Hansen BC (2007) A novel peroxisome proliferator–activated receptor alpha/gamma dual agonist ameliorates dyslipidemia and insulin resistance in prediabetic rhesus monkeys. Metabolism 56:1334–1339

Elkeles RS, Diamond JR, Poulter C, Dhanjil S, Nicolaides AN, Mahmood S, Richmond W, Mather H, Sharp P, Feher MD (1998) Cardiovascular outcomes in type 2 diabetes. A double-blind placebo-controlled study of bezafibrate: the St. Mary's, Ealing, Northwick Park Diabetes Cardiovascular Disease Prevention (SENDCAP) Study. Diab Care 21:641–648

Evans JL, Lin JJ, Goldfine ID (2005) Novel approach to treat insulin resistance, type 2 diabetes, and the metabolic syndrome: simultaneous activation of PPARalpha, PPARgamma, and PPARdelta. Curr Diabetes Rev 1:299–307

Fajas L, Auboeuf D, Raspe E, Schoonjans K, Lefebvre AM, Saladin R, Najib J, Laville M, Fruchart JC, Deeb S, Vidal-Puig A, Flier J, Briggs MR, Staels B, Vidal H, Auwerx J (1997) The organization, promoter analysis, and expression of the human PPARgamma gene. J Biol Chem 272:18779–18789

Glintborg D, Andersen M, Hagen C, Heickendorff L, Hermann AP (2008) Association of pioglitazone treatment with decreased bone mineral density in obese premenopausal patients with polycystic ovary syndrome: a randomized, placebo-controlled trial. J Clin Endocrinol Metab 93:1696–1701

Goke B, Gause-Nilsson I, Persson A (2007) The effects of tesaglitazar as add-on treatment to metformin in patients with poorly controlled type 2 diabetes. Diab Vasc Dis Res 4:204–213

Goldstein BJ, Rosenstock J, Anzalone D, Tou C, Ohman KP (2006) Effect of tesaglitazar, a dual PPAR alpha/gamma agonist, on glucose and lipid abnormalities in patients with type 2 diabetes: a 12-week dose-ranging trial. Curr Med Res Opin 22:2575–2590

Guo Q, Sahoo SP, Wang PR, Milot DP, Ippolito MC, Wu MS, Baffic J, Biswas C, Hernandez M, Lam MH, Sharma N, Han W, Kelly LJ, MacNaul KL, Zhou G, Desai R, Heck JV, Doebber TW, Berger JP, Moller DE, Sparrow CP, Chao YS, Wright SD (2004) A novel peroxisome proliferator-activated receptor alpha/gamma dual agonist demonstrates favorable effects on lipid homeostasis. Endocrinology 145:1640–1648

Han SJ, Kang ES, Hur KY, Kim HJ, Kim SH, Yun CO, Choi SE, Ahn CW, Cha BS, Kang Y, Lee HC (2008) Rosiglitazone inhibits early stage of glucolipotoxicity-induced beta-cell apoptosis. Horm Res 70:165–173

Harrity T, Farrelly D, Tieman A, Chu C, Kunselman L, Gu L, Ponticiello R, Cap M, Qu F, Shao C, Wang W, Zhang H, Fenderson W, Chen S, Devasthale P, Jeon Y, Seethala R, Yang WP, Ren J, Zhou M, Ryono D, Biller S, Mookhtiar KA, Wetterau J, Gregg R, Cheng PT, Hariharan N (2006) Muraglitazar, a novel dual (alpha/gamma) peroxisome proliferator-activated receptor activator, improves diabetes and other metabolic abnormalities and preserves beta-cell function in db/db mice. Diabetes 55:240–248

He W, Barak Y, Hevener A, Olson P, Liao D, Le J, Nelson M, Ong E, Olefsky JM, Evans RM (2003) Adipose-specific peroxisome proliferator-activated receptor gamma knockout causes insulin resistance in fat and liver but not in muscle. Proc Natl Acad Sci USA 100:15712–15717

Hegarty BD, Furler SM, Oakes ND, Kraegen EW, Cooney GJ (2004) Peroxisome proliferator-activated receptor (PPAR) activation induces tissue-specific effects on fatty acid uptake and metabolism in vivo – a study using the novel PPARalpha/gamma agonist tesaglitazar. Endocrinology 145:3158–3164

Hegele RA, Cao H, Frankowski C, Mathews ST, Leff T (2002) PPARG F388L, a transactivation-deficient mutant, in familial partial lipodystrophy. Diabetes 51:3586–3590

Hevener AL, He W, Barak Y, Le J, Bandyopadhyay G, Olson P, Wilkes J, Evans RM, Olefsky J (2003) Muscle-specific Pparg deletion causes insulin resistance. Nat Med 9:1491–1497

Ishida H, Takizawa M, Ozawa S, Nakamichi Y, Yamaguchi S, Katsuta H, Tanaka T, Maruyama M, Katahira H, Yoshimoto K, Itagaki E, Nagamatsu S (2004) Pioglitazone improves insulin secretory capacity and prevents the loss of beta-cell mass in obese diabetic db/db mice: possible protection of beta cells from oxidative stress. Metabolism 53:488–494

Kasai S, Inoue T, Yoshitomi H, Hihara T, Matsuura F, Harada H, Shinoda M, Tanaka I (2008) Antidiabetic and hypolipidemic effects of a novel dual peroxisome proliferator-activated receptor (PPAR) alpha/gamma agonist, E3030, in db/db mice and beagle dogs. J Pharmacol Sci 108:40–48

Kawai T, Takei I, Oguma Y, Ohashi N, Tokui M, Oguchi S, Katsukawa F, Hirose H, Shimada A, Watanabe K, Saruta T (1999) Effects of troglitazone on fat distribution in the treatment of male type 2 diabetes. Metabolism 48:1102–1107

Kendall DM, Rubin CJ, Mohideen P, Ledeine JM, Belder R, Gross J, Norwood P, O'Mahony M, Sall K, Sloan G, Roberts A, Fiedorek FT, DeFronzo RA (2006) Improvement of glycemic control, triglycerides, and HDL cholesterol levels with muraglitazar, a dual (alpha/gamma) peroxisome proliferator-activated receptor activator, in patients with type 2 diabetes inadequately controlled with metformin monotherapy: A double-blind, randomized, pioglitazone-comparative study. Diab Care 29:1016–1023

Kersten S, Seydoux J, Peters JM, Gonzalez FJ, Desvergne B, Wahli W (1999) Peroxisome proliferator-activated receptor alpha mediates the adaptive response to fasting. J Clin Invest 103:1489–1498

Kim MK, Chae YN, Son MH, Kim SH, Kim JK, Moon HS, Park CS, Bae MH, Kim E, Han T, Choi HH, Shin YA, Ahn BN, Lee CH, Lim JI, Shin CY (2008) PAR-5359, a well-balanced PPARalpha/gamma dual agonist, exhibits equivalent antidiabetic and hypolipidemic activities in vitro and in vivo. Eur J Pharmacol 595:119–125

Krey G, Braissant O, L'Horset F, Kalkhoven E, Perroud M, Parker MG, Wahli W (1997) Fatty acids, eicosanoids, and hypolipidemic agents identified as ligands of peroxisome proliferator-activated receptors by coactivator-dependent receptor ligand assay. Mol Endocrinol 11:779–791

Laurent D, Gounarides JS, Gao J, Boettcher BR (2009) Effects of cevoglitazar, a dual PPARalpha/gamma agonist, on ectopic fat deposition in fatty Zucker rats. Diabetes Obes Metab 11:632–636

Lee CH, Olson P, Hevener A, Mehl I, Chong LW, Olefsky JM, Gonzalez FJ, Ham J, Kang H, Peters JM, Evans RM (2006) PPARdelta regulates glucose metabolism and insulin sensitivity. Proc Natl Acad Sci USA 103:3444–3449

Lehmann JM, Moore LB, Smith-Oliver TA, Wilkison WO, Willson TM, Kliewer SA (1995) An antidiabetic thiazolidinedione is a high affinity ligand for peroxisome proliferator-activated receptor gamma (PPAR gamma). J Biol Chem 270:12953–12956

Li AC, Brown KK, Silvestre MJ, Willson TM, Palinski W, Glass CK (2000) Peroxisome proliferator-activated receptor gamma ligands inhibit development of atherosclerosis in LDL receptor-deficient mice. J Clin Invest 106:523–531

Li PP, Shan S, Chen YT, Ning ZQ, Sun SJ, Liu Q, Lu XP, Xie MZ, Shen ZF (2006) The PPARalpha/gamma dual agonist chiglitazar improves insulin resistance and dyslipidemia in MSG obese rats. Br J Pharmacol 148:610–618

Lovas K, Rost TH, Skorve J, Ulvik RJ, Gudbrandsen OA, Bohov P, Wensaas AJ, Rustan AC, Berge RK, Husebye ES (2009) Tetradecylthioacetic acid attenuates dyslipidaemia in male patients with type 2 diabetes mellitus, possibly by dual PPAR-alpha/delta activation and increased mitochondrial fatty acid oxidation. Diabetes Obes Metab 11:304–314

Mayerson AB, Hundal RS, Dufour S, Lebon V, Befroy D, Cline GW, Enocksson S, Inzucchi SE, Shulman GI, Petersen KF (2002) The effects of rosiglitazone on insulin sensitivity, lipolysis, and hepatic and skeletal muscle triglyceride content in patients with type 2 diabetes. Diabetes 51:797–802

Miller CP (2002) SERMs: evolutionary chemistry, revolutionary biology. Curr Pharm Des 8:2089–2111

Mittra S, Sangle G, Tandon R, Sharma S, Roy S, Khanna V, Gupta A, Sattigeri J, Sharma L, Priyadarsiny P, Khattar SK, Bora RS, Saini KS, Bansal VS (2007) Increase in weight induced by muraglitazar, a dual PPARalpha/gamma agonist, in db/db mice: adipogenesis/or oedema? Br J Pharmacol 150:480–487

Nissen SE, Wolski K (2007) Effect of rosiglitazone on the risk of myocardial infarction and death from cardiovascular causes. N Engl J Med 356:2457–2471

Nissen SE, Wolski K, Topol EJ (2005) Effect of muraglitazar on death and major adverse cardiovascular events in patients with type 2 diabetes mellitus. JAMA 294:2581–2586

Norris AW, Chen L, Fisher SJ, Szanto I, Ristow M, Jozsi AC, Hirshman MF, Rosen ED, Goodyear LJ, Gonzalez FJ, Spiegelman BM, Kahn CR (2003) Muscle-specific PPARgamma-deficient mice develop increased adiposity and insulin resistance but respond to thiazolidinediones. J Clin Invest 112:608–618

Oakes ND, Thalen P, Hultstrand T, Jacinto S, Camejo G, Wallin B, Ljung B (2005) Tesaglitazar, a dual PPAR{alpha}/{gamma} agonist, ameliorates glucose and lipid intolerance in obese Zucker rats. Am J Physiol Regul Integr Comp Physiol 289:R938–R946

Oliver WR Jr, Shenk JL, Snaith MR, Russell CS, Plunket KD, Bodkin NL, Lewis MC, Winegar DA, Sznaidman ML, Lambert MH, Xu HE, Sternbach DD, Kliewer SA, Hansen BC, Willson TM (2001) A selective peroxisome proliferator-activated receptor delta agonist promotes reverse cholesterol transport. Proc Natl Acad Sci USA 98:5306–5311

Plutzky J (2000) Emerging concepts in metabolic abnormalities associated with coronary artery disease. Curr Opin Cardiol 15:416–421

Ratner RE, Parikh S, Tou C (2007) Efficacy, safety and tolerability of tesaglitazar when added to the therapeutic regimen of poorly controlled insulin-treated patients with type 2 diabetes. Diab Vasc Dis Res 4:214–221

Riserus U, Sprecher D, Johnson T, Olson E, Hirschberg S, Liu A, Fang Z, Hegde P, Richards D, Sarov-Blat L, Strum JC, Basu S, Cheeseman J, Fielding BA, Humphreys SM, Danoff T, Moore NR, Murgatroyd P, O'Rahilly S, Sutton P, Willson T, Hassall D, Frayn KN, Karpe F (2008) Activation of peroxisome proliferator-activated receptor (PPAR)delta promotes reversal of multiple metabolic abnormalities, reduces oxidative stress, and increases fatty acid oxidation in moderately obese men. Diabetes 57:332–339

Rost TH, Haugan Moi LL, Berge K, Staels B, Mellgren G, Berge RK (2009) A pan-PPAR ligand induces hepatic fatty acid oxidation in PPARalpha-/- mice possibly through PGC-1 mediated PPARdelta coactivation. Biochim Biophys Acta 1791:1076–1083

Schuster H, Fagerberg B, Edwards S, Halmos T, Lopatynski J, Stender S, Birketvedt GS, Tonstad S, Gause-Nilsson I, Halldorsdottir S, Ohman KP (2008) Tesaglitazar, a dual peroxisome proliferator-activated receptor alpha/gamma agonist, improves apolipoprotein levels in non-diabetic subjects with insulin resistance. Atherosclerosis 197:355–362

Seber S, Ucak S, Basat O, Altuntas Y (2006) The effect of dual PPAR alpha/gamma stimulation with combination of rosiglitazone and fenofibrate on metabolic parameters in type 2 diabetic patients. Diabetes Res Clin Pract 71:52–58

Spiegelman BM (1998) PPAR-gamma: adipogenic regulator and thiazolidinedione receptor. Diabetes 47:507–514

Tanaka T, Yamamoto J, Iwasaki S, Asaba H, Hamura H, Ikeda Y, Watanabe M, Magoori K, Ioka RX, Tachibana K, Watanabe Y, Uchiyama Y, Sumi K, Iguchi H, Ito S, Doi T, Hamakubo T, Naito M, Auwerx J, Yanagisawa M, Kodama T, Sakai J (2003) Activation of peroxisome proliferator-activated receptor delta induces fatty acid beta-oxidation in skeletal muscle and attenuates metabolic syndrome. Proc Natl Acad Sci USA 100:15924–15929

Tenenbaum A, Motro M, Fisman EZ, Schwammenthal E, Adler Y, Goldenberg I, Leor J, Boyko V, Mandelzweig L, Behar S (2004) Peroxisome proliferator-activated receptor ligand bezafibrate for prevention of type 2 diabetes mellitus in patients with coronary artery disease. Circulation 109:2197–2202

Tenenbaum A, Motro M, Fisman EZ (2005a) Dual and pan-peroxisome proliferator-activated receptors (PPAR) co-agonism: the bezafibrate lessons. Cardiovasc Diabetol 4:14

Tenenbaum A, Motro M, Fisman EZ, Tanne D, Boyko V, Behar S (2005b) Bezafibrate for the secondary prevention of myocardial infarction in patients with metabolic syndrome. Arch Intern Med 165:1154–1160

Tenenbaum A, Motro M, Fisman EZ, Adler Y, Shemesh J, Tanne D, Leor J, Boyko V, Schwammenthal E, Behar S (2005c) Effect of bezafibrate on incidence of type 2 diabetes mellitus in obese patients. Eur Heart J 26:2032–2038

Tontonoz P, Hu E, Spiegelman BM (1994) Stimulation of adipogenesis in fibroblasts by PPAR gamma 2, a lipid-activated transcription factor. Cell 79:1147–1156

Tozzo E, Ponticiello R, Swartz J, Farrelly D, Zebo R, Welzel G, Egan D, Kunselman L, Peters A, Gu L, French M, Chen S, Devasthale P, Janovitz E, Staal A, Harrity T, Belder R, Cheng PT, Whaley J, Taylor S, Hariharan N (2007) The dual peroxisome proliferator-activated receptor alpha/gamma activator muraglitazar prevents the natural progression of diabetes in db/db mice. J Pharmacol Exp Ther 321:107–115

van der Hoorn JW, Jukema JW, Havekes LM, Lundholm E, Camejo G, Rensen PC, Princen HM (2009) The dual PPARalpha/gamma agonist tesaglitazar blocks progression of pre-existing atherosclerosis in APOE*3Leiden.CETP transgenic mice. Br J Pharmacol 156:1067–1075

Vu-Dac N, Schoonjans K, Laine B, Fruchart JC, Auwerx J, Staels B (1994) Negative regulation of the human apolipoprotein A-I promoter by fibrates can be attenuated by the interaction of the peroxisome proliferator-activated receptor with its response element. J Biol Chem 269:31012–31018

Vu-Dac N, Schoonjans K, Kosykh V, Dallongeville J, Fruchart JC, Staels B, Auwerx J (1995) Fibrates increase human apolipoprotein A-II expression through activation of the peroxisome proliferator-activated receptor. J Clin Invest 96:741–750

Wang YX, Zhang CL, Yu RT, Cho HK, Nelson MC, Bayuga-Ocampo CR, Ham J, Kang H, Evans RM (2004) Regulation of muscle fiber type and running endurance by PPARdelta. PLoS Biol 2:e294

Wilding JP, Gause-Nilsson I, Persson A (2007) Tesaglitazar, as add-on therapy to sulphonylurea, dose-dependently improves glucose and lipid abnormalities in patients with type 2 diabetes. Diab Vasc Dis Res 4:194–203

Yamauchi T, Kamon J, Waki H, Terauchi Y, Kubota N, Hara K, Mori Y, Ide T, Murakami K, Tsuboyama-Kasaoka N, Ezaki O, Akanuma Y, Gavrilova O, Vinson C, Reitman ML, Kagechika H, Shudo K, Yoda M, Nakano Y, Tobe K, Nagai R, Kimura S, Tomita M, Froguel P, Kadowaki T (2001) The fat-derived hormone adiponectin reverses insulin resistance associated with both lipoatrophy and obesity. Nat Med 7:941–946

Yki-Jarvinen H (2004) Thiazolidinediones. N Engl J Med 351:1106–1118

Zadelaar AS, Boesten LS, Jukema JW, van Vlijmen BJ, Kooistra T, Emeis JJ, Lundholm E, Camejo G, Havekes LM (2006) Dual PPARalpha/gamma agonist tesaglitazar reduces atherosclerosis in insulin-resistant and hypercholesterolemic ApoE*3Leiden mice. Arterioscler Thromb Vasc Biol 26:2560–2566

Zeender E, Maedler K, Bosco D, Berney T, Donath MY, Halban PA (2004) Pioglitazone and sodium salicylate protect human beta-cells against apoptosis and impaired function induced by glucose and interleukin-1beta. J Clin Endocrinol Metab 89:5059–5066

GLP-1 Agonists and Dipeptidyl-Peptidase IV Inhibitors

Baptist Gallwitz

Contents

Abstract Novel therapeutic options for type 2 diabetes based on the action of the incretin hormone glucagon-like peptide-1 (GLP-1) were introduced in 2005. Incretin-based therapies consist of two classes: (1) the injectable GLP-1 receptor agonists solely acting on the GLP-1 receptor and (2) dipeptidyl-peptidase inhibitors (DPP-4 inhibitors) as oral medications raising endogenous GLP-1 and other hormone levels by inhibiting the degrading enzyme DPP-4. In type 2 diabetes therapy, incretin-based therapies are attractive and more commonly used due to their action and safety profile. Stimulation of insulin secretion and inhibition of glucagon secretion by the above-mentioned agents occur in a glucose-dependent manner. Therefore, incretin-based therapies have no intrinsic risk for hypoglycemias. GLP-1 receptor agonists

B. Gallwitz
Medizinische Klinik IV, Otfried-Müller-Str. 10, 72076 Tübingen, Germany
e-mail: baptist.gallwitz@med.uni-tuebingen.de

M. Schwanstecher (ed.), *Diabetes - Perspectives in Drug Therapy*,
Handbook of Experimental Pharmacology 203,
DOI 10.1007/978-3-642-17214-4_3, © Springer-Verlag Berlin Heidelberg 2011

allow weight loss; DPP-4 inhibitors are weight neutral. This review gives an overview on the mechanism of action and the substances and clinical data available.

Keywords Antidiabetic therapy · DPP-4 inhibitors · GLP-1 · GLP-1 agonists · Incretins · Type 2 diabetes

1 Type 2 Diabetes, Its Epidemiology, and the Need for Further Treatment Options

Type 2 diabetes incidence and prevalence are increasing tremendously around the world, especially in the countries with lifestyles that go along with less physical activity and high caloric nutrition available at low costs. The prevalence rates are expected to more than double within the next 20 years. Estimates expect 440 million type 2 diabetic people by 2030 (International Diabetes Federation 2009). Additionally, type 2 diabetes changes its prevalence by affecting increasingly younger parts of the population, with higher and growing incidence rates in children and adolescents (Zeitler 2009).

Large parts of the type 2 diabetic population also suffer from ineffective treatment and do not reach the therapeutic goals. This leaves effective and patient-orientated treatment forms to achieve a near-normal HbA1c value as one criterion for an acceptable glycemic control as a major task for diabetes therapy. Besides, further important treatment goals such as body weight reduction or the prevention of hypoglycemia are seldomly accomplished. Insufficient metabolic control in type 2 diabetes is associated with microvascular and macrovascular complications. The cardiovascular mortality risk is increased and 75% of type 2 diabetic patients die from cardiovascular events. The microvascular and macrovascular complication risk can be lowered by an improved metabolic control (Gaede et al. 2003; Holman et al. 2008).

The older insulinotropic treatment regimes with sulfonylureas or metiglinides are associated with an elevated incidence of hypoglycemic events or with an unwanted weight gain. Glitazones are also associated with weight gain, heart failure, and fractures. Insulin therapy leads to weight gain and increases the risk for hypoglycemic episodes (Nathan et al. 2006). These therapies are also not able to alter or stop the disease progression of type 2 diabetes that is caused by the decline of beta-cell function. This functional loss is characterized by an increasing defect in the insulin response to glucose as well as a loss of beta-cell mass over time.

2 Incretin Hormones

The physiological and pharmacological actions of the incretin hormone glucagon-like peptide-1 (GLP-1) were used to develop two novel drug classes for type 2 diabetes treatment: the GLP-1 receptor agonists and the dipeptidyl-peptidase IV inhibitors (DPP-4 inhibitors) (Drucker and Nauck 2006).

The gastrointestinal hormones GLP-1 and GIP (gastric inhibitory polypeptide or glucose-dependent insulinotropic polypeptide) are secreted following a meal from the endocrine L- and K-cells in the intestinal mucosa, respectively (Wellendorph et al. 2009). GLP-1 and GIP contribute to approximately 60% of the insulin secretion postprandially and are responsible for the incretin effect. This effect describes the phenomenon that orally ingested glucose leads to a much larger insulin response than an isoglycemic intravenous glucose load (Creutzfeldt 1979; Nauck et al. 1986). In patients with type 2 diabetes, the incretin effect is diminished (see Fig. 1 for the incretin effect in healthy subjects and type 2 diabetes). One important reason for the diminished incretin effect in type 2 diabetes is that GIP does not act as an insulinotropic hormone under chronic hyperglycemia for reasons that are not completely understood yet. GLP-1, on the other hand, is still able to stimulate insulin secretion under hyperglycemic conditions in type 2 diabetes (Nauck et al. 1993a). It should be noted, however, that hyperglycemia acutely reduces the postprandial levels of GIP and GLP-1, possibly through a deceleration of gastric emptying. Therefore, the reduced incretin hormone concentrations in some patients with type 2 diabetes may be a consequence rather than a cause of type 2 diabetes (Vollmer et al. 2009).

Exogenous GLP-1 application either by subcutaneous or intravenous injection resulting in supraphysiological GLP-1 plasma concentrations restores the incretin effect with an adequate insulin response under hyperglycemic conditions (Nauck et al. 1993b).

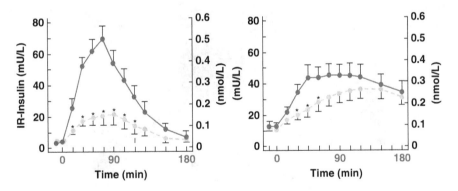

IR = immunoreactive

Fig. 1 The incretin effect in healthy subjects (*left panel*) and in type 2 diabetes (*right panel*). Oral glucose (*closed line*) elicits a much greater insulin response than an intravenous glucose infusion (*dashed line*) mimicking the glucose rise after oral glucose. The differences in the insulin responses after oral and intravenous glucose are due to the physiological effects of the incretin hormones GLP-1 and GIP and are described as incretin effect. Note that the incretin effect is diminished in type 2 diabetes (adapted from Nauck et al. 1986)

2.1 GLP-1 Actions

The glucagon gene encodes for a large peptide that contains the message not only
for glucagon, but also for other peptides that are formed posttranslationally such as
GLP-1. GLP-1 is cleaved posttranslationally from preproglucagon in the neuroendocrine L-cells of the intestinal mucosa and in the central nervous system, but not in
the pancreatic alpha cells of the islets. It binds to highly specific GLP-1 receptors
that belong to the G-protein-coupled receptors (Drucker and Nauck 2006). GLP-1
shows numerous physiological actions in various tissues and a broad therapeutic
potential (see Fig. 2 for details).

   GLP-1 stimulates insulin secretion of the beta cells and additionally inhibits
glucagon secretion from the alpha cells. These two actions occur in a strictly
glucose-dependent manner and lead to a normalization of glycemia either in

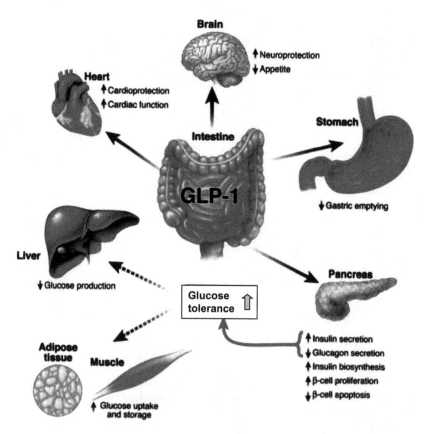

Fig. 2 Multiple physiological effects of GLP-1 (adapted from Baggio and Drucker 2007; Drucker
and Nauck 2006)

the fasting or the postprandial state. Under hypoglycemic conditions, the counter-regulation by glucagon is not affected and insulin secretion is not stimulated. GLP-1 is therefore not able to elicit hypoglycemia by itself (Drucker and Nauck 2006).

In the gastrointestinal tract, GLP-1 slows gastric emptying after a meal. This effect also contributes to a normalization of postprandial hyperglycemia. GLP-1 additionally binds to its receptor on hypothalamic neurons and stimulates satiety by direct action. These two effects explain that long-term treatment with GLP-1 receptor agonists lead to weight loss in the long run (Drucker and Nauck 2006).

Studies in different rodent species and studies in isolated human islets showed beneficial long-term actions of GLP-1: insulin synthesis is stimulated and beta-cell mass is increased (Brubaker and Drucker 2004; Drucker and Nauck 2006; Fehmann and Habener 1992). So far, it is not known whether these findings translate into a benefit in type 2 diabetes therapy with a positive effect on stopping or slowing disease progression. Long-term study data from clinical studies in type 2 diabetes with a sufficient observation time are still not available. Furthermore, there are presently no good validated methods to quantify beta-cell mass in humans in a clinical setting.

Recent studies additionally revealed that pharmacological application of GLP-1 or GLP-1 receptor agonists improved cardiovascular parameters (reduction of systolic blood pressure, beneficial effects on myocardial ischemia in animal models, and positive effects on left ventricular function in heart failure). These promising effects may also have important clinical implications for type 2 diabetes therapy with GLP-1 receptor agonists (Courreges et al. 2008; Klonoff et al. 2008; Sokos et al. 2006).

2.2 Dipeptidyl-Peptidase IV

The ubiquitous enzyme dipeptidyl-peptidase IV (DPP-4) is responsible for GLP-1 degradation resulting in a biological half-life of GLP-1 that amounts to 1–2 min after intravenous injection of GLP-1 (Drucker and Nauck 2006). Subcutaneous injections of GLP-1 also do not result in a sufficiently high and long-lasting elevation of GLP-1 concentrations to use native GLP-1 as a practical therapeutic agent in type 2 diabetes. An animal study in rodents demonstrated that DPP-4 expression in the intestine and the kidneys is also dependent on metabolic factors and is increased with high-fat feeding and type 2 diabetes (Yang et al. 2007). In order to utilize GLP-1 action for type 2 diabetes therapy, two options are presently available (Drucker and Nauck 2006):

1. GLP-1 receptor agonists as injectable compounds
2. Dipeptidyl-peptidase IV (DPP-4) inhibitors as orally active substances

This chapter deals with the approved substances for type 2 diabetes therapy within these classes (state 2010).

3 GLP-1 Receptor Agonists

Exenatide (Byetta®, Amylin and Eli Lilly Pharmaceuticals) was the first GLP-1 receptor agonist for the treatment of type 2 diabetes. It is presently approved in combination with metformin and/or a sulfonylurea in patients failing to reach the therapeutic goals with this oral medication (Gallwitz 2006). Exenatide is the synthetic form of exendin-4, a peptide discovered in the saliva of the gila monster (heloderma suspectum) in 1992. Exenatide has a 53% amino acid sequence similarity to human GLP-1 and is a strong GLP-1 receptor agonist (Eng et al. 1992). In type 2 diabetes treatment, it is injected subcutaneously twice daily. A slow release formulation for once-weekly administration is presently in the approval process after completing important clinical phase III studies (Drucker et al. 2008; Gedulin et al. 2005; Kim et al. 2007). The long-acting human GLP-1 analogue liraglutide (Victoza®, Novo Nordisk Pharmaceuticals) for once-daily injection has been approved and is also available in Europe and the USA (Agerso et al. 2002; Chang et al. 2003; McGill 2009). Figure 3 shows the structural similarities and differences between GLP-1, exenatide, and liraglutide. Further compounds for once-daily or once-weekly application are being developed and in the clinical phase III study

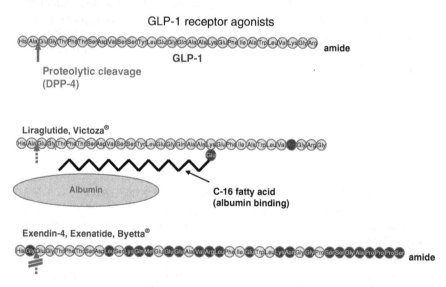

Fig. 3 Native GLP-1, exenatide and liraglutide, and their amino acid sequences. Amino acid sequence similarities are highlighted. Note the *arrows* in the N-terminal position of the peptides between the amino acids 2 and 3: the *top arrow* (*solid line*) on the GLP-1 molecule shows the cleavage site of DPP-4. The *middle arrow* (*dashed line*) on the liraglutide molecule symbolizes the potential DPP-4 cleavage site of the peptide chain. DPP-4 action is inhibited in this area due to albumin binding and heptamer formation of the liraglutide molecule protecting the potential cleavage site. The *arrow* on the bottom of the exendin-4 molecule with the two *crossed lines* symbolizes the lack of a DPP-4 cleavage site due to an amino acid exchange in the N-terminal position 2 of the molecule compared to native GLP-1

programs such as Taspoglutide, Roche Pharmaceuticals (Nauck et al. 2009b; Retterstol 2009); Lixisenatide (AVE0010), Sanofi-Aventis Pharmaceuticals (Werner 2008); Albiglutide, GlaxoSmithKline Pharmaceuticals (Rosenstock et al. 2009b) and more.

3.1 Exenatide

Exenatide has a half-life of approximately 3.5 h, and after subcutaneous injection, sufficient plasma concentrations are reached over a time period of at least 4–6 h (Barnett 2005; Gallwitz 2006; Kolterman et al. 2005). Exenatide reduced the HbA1c by 0.8–1.1% in various clinical studies (Buse et al. 2004; DeFronzo et al. 2005; Kendall et al. 2005). The HbA1c reduction was sustained and remained constant over a period of 3 years in one study (Klonoff et al. 2008). Comparative clinical studies show that the efficacy of exenatide on glycemic parameters is comparable to that of a newly implemented insulin therapy (Barnett et al. 2007; Gallwitz 2006; Heine et al. 2005; Klonoff et al. 2008; Nauck et al. 2007a; Zinman et al. 2007).

Besides the favorable effects on glycemic parameters, exenatide therapy also induced weight loss in patients with type 2 diabetes. In clinical studies, a significant loss in body weight by 1.5–3.0 kg was documented after 30 weeks. This effect continued and led to a further weight loss of 5.3 kg after 3 years (see Fig. 4) (Barnett 2007; Klonoff et al. 2008). Comparing the weight effects of exenatide and insulin therapies, a difference in weight development of 4–5 kg in 30 weeks between the

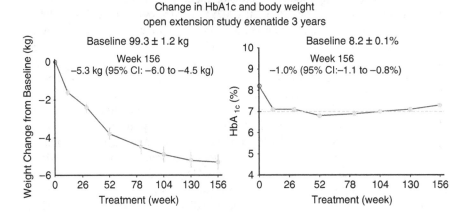

N=217 Mean ± SE

Fig. 4 HbA1c reduction and body weight development over 3 years in patients with type 2 diabetes and an add-on therapy with exenatide (2 × 10 μg/d) to metformin and/or sulfonylurea (adapted from Klonoff et al. 2008)

insulin- and exenatide-treated groups was observed (Barnett et al. 2007; Heine et al. 2005; Nauck et al. 2007a).

Beta-cell function also improved with exenatide and the clinical surrogate parameters insulin secretion rate, HOMA-B [homeostatic modeling assessment of beta-cell function], and the proinsulin/insulin ratio changed in a favorable way. Additionally, the first phase of insulin secretion, which is lost already in the early stages of type 2 diabetes, is restored under treatment with exenatide when examined with an intravenous glucose tolerance test (Barnett 2007; Gallwitz 2006). In a 1-year study with exenatide, however, the improvements of the above-mentioned beta-cell function parameters were no longer present following a washout period of 12 weeks (Bunck et al. 2009).

Exenatide itself has no intrinsic risk for causing hypoglycemias. Severe hypoglycemic events were only observed in exenatide-treated patients who had a combination therapy with a sulfonylurea. The hypoglycemic episodes were caused by the sulfonylurea treatment, and it is generally suggested to reduce the sulfonylurea dose when starting exenatide treatment as add-on therapy. In clinical studies comparing the sulfonylurea plus exenatide combination versus insulin alone, similar rates of hypoglycemic episodes were found. However, the incidence of nocturnal hypoglycemic events was less in the exenatide-treated patients (Barnett 2007; Gallwitz 2006).

The most frequent adverse events associated with exenatide therapy were fullness and nausea. These adverse events were less pronounced when the exenatide dose was titrated from a small dose to the full dose at the beginning of treatment. Dose titration is therefore recommended with exenatide starting with a dose of 5 µg twice daily and an increase to 10 µg twice daily after 4 weeks. Generally, nausea was only mild to moderate and occurred in the first weeks of treatment ceasing with time. Nausea was the most common reason to stop therapy with 2–6.4% dropouts in the clinical studies with exenatide (Barnett 2007; Gallwitz 2006; Klonoff et al. 2008).

Since exenatide is a nonhuman peptide, in approximately 40% of exenatide-treated patients, anti-exenatide antibodies can be detected. These antibodies do not seem to be neutralizing antibodies and do not cross-react with human GLP-1. At least over a time period of 3 years, these antibody titers did not have any obvious effect on glycemic control (Drucker et al. 2008).

After the approval of exenatide, cases of acute pancreatitis were reported (Ahmad and Swann 2008; Cure et al. 2008) and led to the publication of a warning by the Food and Drugs Administration of the United States (FDA). In total, the incidence of pancreatitis is very low and rather corresponds to the elevated risk of pancreatitis in obese type 2 diabetic patients. Type 2 diabetic patients have an elevated pancreatitis risk due to a higher prevalence of gallstones, hypertriglyceridemia, and other factors, which was confirmed in a recent meta-analysis (Dore et al. 2009).

Exenatide is mainly eliminated by glomerular filtration followed by proteolytic degradation (Yoo et al. 2006). Exenatide is not recommended in severe renal impairment (creatinine clearance <30 ml/min). In a study with patients with end-stage renal disease on dialysis, exenatide 5 µg has been poorly tolerated because of gastrointestinal side effects (Barnett 2007; Gallwitz 2006). The FDA published a warning after having observed altered renal function associated with exenatide treatment. A total of 62 patients with acute renal failure and 16 cases of renal insufficiency were found in a time period of 3 years. Mostly, patients with preexisting kidney disease or with one or more risk factors for developing kidney problems showed a deterioration of renal function with exenatide, which might have been triggered by drug-induced nausea, vomiting, and consecutive dehydration (Food and Drug Administration 2009). According to the warning by the FDA, exenatide should not be used in patients with severe renal impairment (creatinine clearance <30 ml/min) or end-stage renal disease.

In a pediatric study, single doses of 2.5 and 5.0 µg exenatide were well tolerated and normalized postprandial glucose and glucagon concentrations compared with placebo. No hypoglycemic events were recorded during the study (Malloy et al. 2009).

3.2 Liraglutide

Liraglutide is the first human GLP-1 analogue. In contrast to native GLP-1 there are two modifications in the amino acid sequence of native GLP-1 and an attachment of a fatty acid side chain to the peptide. It is injected subcutaneously once daily (Agerso et al. 2002).

In rodent models for type 2 diabetes, liraglutide increased beta-cell mass and lowered body weight and food intake in a broad selection of animal models (Sturis et al. 2003). In approval relevant clinical studies in approximately 4,200 type 2 diabetic patients receiving the drug, it is efficacious and safe across all stages of the natural course of type 2 diabetes, in monotherapy, as well as in combination with either one or more oral antidiabetic agents (see Fig. 5) (Garber et al. 2006; Garber and Spann 2008; Nauck et al. 2009a, b; Marre et al. 2009; McGill 2009; Zinman et al. 2009).

In a 2-year study with newly diagnosed type 2 diabetic patients, liraglutide in monotherapy led to a sustained and stable HbA1c reduction of 0.9 or 1.1% in a dose of 1.2 or 1.8 mg once daily, respectively (Garber et al. 2008, 2009).

In further studies, 1.2 or 1.8 mg liraglutide once daily effectively lowered HbA1c in various combinations with oral antidiabetic drugs by approximately 1.0–1.5%. Liraglutide therapy also caused a significant weight loss comparable to that previously observed in studies with exenatide (Deacon 2009; Vilsboll 2009). The weight loss was accompanied by a more pronounced loss in visceral fat than subcutaneous fat (Deacon 2009; Nauck and Marre 2009; Vilsboll 2009).

Furthermore, systolic blood pressure was lowered by 2–6 mmHg in the liraglutide-treated patients. This effect was independent from the weight loss, as the reduction of blood pressure was already observed early on in therapy, when weight loss had not occurred yet (Garber et al. 2008, 2009; Nauck and Marre 2009; Zinman et al. 2009).

Reductions in HbA$_{1c}$ with liraglutide

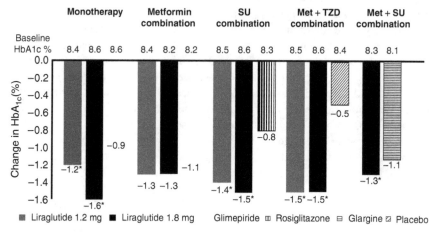

Fig. 5 HbA1c reduction in clinical studies with liraglutide. Liraglutide was given at doses of either 1.2 or 1.8 mg once daily in the different approval studies as either monotherapy or add-on to an oral mono- or combination therapy. The comparator arm of each study is also shown

Hypoglycemia incidence rates were comparable to placebo in all studies, provided no sulfonylurea was used in the combination with liraglutide (Deacon 2009; Vilsboll 2009). Gastrointestinal symptoms were also common in the clinical studies with liraglutide, but in a direct head-to-head study compared to exenatide nausea and vomiting were less frequent and only reported for a shorter period at the beginning of therapy (Buse et al. 2009). In clinical studies, antibodies against liraglutide were detected in no more than 8.6% (Deacon 2009; Garber et al. 2008, 2009; Vilsboll 2009). In a study, directly comparing the clinical efficacy and safety of exenatide and liraglutide, liraglutide proved advantageous with regard to lowering the glycemic parameters HbA1c, fasting glucose, and improving HOMA-B (Buse et al. 2009). Liraglutide improves the first phase of insulin secretion after intravenous glucose as well as the insulin response to a maximal stimulation with arginine (Vilsboll et al. 2008).

Mild-to-moderate renal impairment did not alter the pharmacokinetic profile of liraglutide (Deacon 2009; Vilsboll 2009).

4   DPP-4 Inhibitors

DPP-4 inhibitors are small molecules and orally active. They are tolerated well. After once- or twice-daily dosing they inhibit DPP-4 effectively and lead to a postprandial elevation of endogenous GLP-1 concentrations to the 2- to 3-fold of

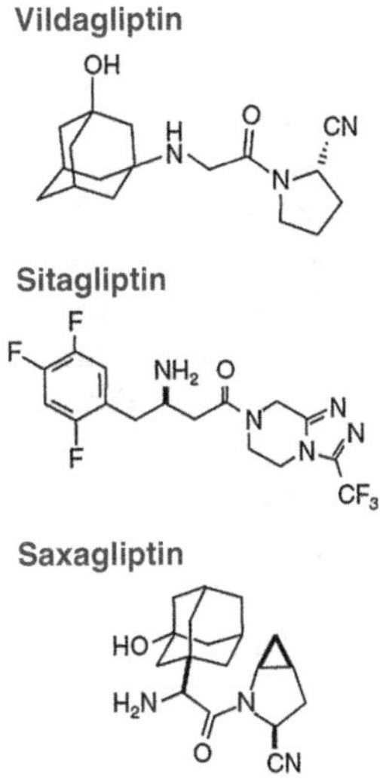

Fig. 6 Structural formulas of the DPP-4 inhibitors saxaglitpin, sitagliptin, and vildagliptin

the normal physiological levels after a meal (Ahren 2008a; Mest 2006). The presently available compounds are Sitagliptin (Januvia[®], Merck Pharmaceuticals), Vildagliptin (Galvus[®], Novartis Pharmaceuticals), and Saxagliptin (Onglyza[®], AstraZeneca and Bristol-Myers Squibb Pharmaceuticals) (see Fig. 6) (Gallwitz 2008). They are approved in combination with metformin, a sulfonylurea or a glitazone or a combination of metformin and a sulfonylurea. Sitagliptin is the first DPP-4 inhibitor with a wider indication that also includes insulin therapy as well as monotherapy (general monotherapy indication USA only, monotherapy indication in Europe for patients with metformin contraindications or intolerance) (Ahren 2008a). There are fixed dose combinations for both sitagliptin and vildagliptin with metformin (sitagliptin plus metformin: Janumet[®], Merck Pharmaceuticals, vilda-glitpin plus metformin: Eucreas[®], Novartis Pharmaceuticals). Further DPP-4 inhibitors are in clinical studies (alogliptin, Takeda Pharmaceuticals (Pratley et al. 2009); linagliptin, Boehringer Ingelheim Pharmaceuticals (Heise et al. 2009) and others) (Pratley 2008). Long-term studies investigating cardiovascular outcomes and a possible positive influence on disease progression of type 2 diabetes are being carried on with the DPP-4 inhibitors.

4.1 Sitagliptin

The first DPP-4 inhibitor that was approved was sitagliptin. In mono- as well as in combination therapy, sitagliptin lowers HbA1c by 0.6–1.1% compared to placebo in a standard dose of 100 mg once daily (Ahren 2008a; Barnett 2009; Karasik et al. 2008; Pratley 2008). It also reduces fasting plasma glucose and postprandial glucose significantly. Sitagliptin was weight neutral in all clinical studies

(Ahren 2008a; Barnett 2009; Karasik et al. 2008; Pratley 2008). As add-on treatment to an existing metformin therapy, sitagliptin lowered the HbA1c by 0.7%. In a primary combination therapy with metformin, a constant and sustained reduction of HbA1c and fasting plasma glucose was observed over a period of 2 years (Green and Feinglos 2008). Hypoglycemia incidence observed under sitagliptin was comparable to that under placebo (Karasik et al. 2008). As a surrogate parameter of beta-cell function, an improvement of the proinsulin/insulin ratio was observed in clinical studies with sitagliptin (Ahren 2008a; Barnett 2009; Green and Feinglos 2008; Pratley 2008). The most common side effects of sitagliptin were unspecific, like headache, arthritis, nasopharyngitis, respiratory or urinary tract infections, and rarely skin reactions (Karasik et al. 2008; Williams-Herman et al. 2008).

The elimination of sitagliptin is mainly renal (75% in the urine as unchanged drug), with a half-time of 12–14 h (Herman et al. 2006a, b, c). Sitagliptin was also generally well tolerated and effective in patients with impaired renal function. In a study with patients with impaired renal function, a dose of 25 mg/d was chosen for patients with a creatinine clearance of <30 ml/min or end-stage renal disease and a dose of 50 mg/dl was given to patients with a creatinine clearance between 30 and 50 ml/min (Chan et al. 2008; Scott et al. 2007).

4.2   Vildagliptin

Vildagliptin is the second available compound of the DPP-4 inhibitors. Its dosage is 50 mg twice daily. In clinical studies testing vildaglitpin in monotherapy or in combination therapies with metformin, glimepiride, pioglitazone, or insulin, vildagliptin decreased the HbA1c by approximately 0.5–1.0% (Ahren 2008b; Barnett 2009; Pratley 2008). As an add-on therapy to metformin, it decreased the HbA1c by 0.65–1.1% (Ahren 2008b). Vildagliptin has a good safety and tolerability profile, and the most common adverse events are unspecific (flue-like symptoms, headache, dizziness, and rarely liver enzyme elevations during the initiation of therapy). The incidence of hypoglycemic episodes is also comparable to placebo. Vildagliptin, like the other DPP-4 inhibitors, is also weight neutral. Acute and medium-term parameters for insulin secretion were improved under vildagliptin treatment (Ahren 2008a, b; Barnett 2009). Similar improvements were observed for HOMA-B, the proinsulin/insulin ratio, and the first phase of insulin secretion after intravenous glucose (Ahren and Foley 2008). Vildagliptin has been tested in an elderly population, where it was shown to be efficacious and safe (Pratley et al. 2007).

4.3   Saxagliptin

Saxagliptin was approved in 2009. It was shown to dose dependently reduce fasting plasma glucose and HbA1c (0.7–0.9%, baseline 7.9%) (Rosenstock et al. 2008).

In a study with drug-naïve patients, saxagliptin lowered the glycemic parameters HBA1c, fasting plasma glucose, and postprandial glucose significantly (Rosenstock et al. 2009a). As add-on medication to a therapy with either metformin or a glitazone, saxagliptin also led to significant metabolic improvements comparable to other DPP-4 inhibitors (Chacra et al. 2009; Deacon and Holst 2009; DeFronzo et al. 2009; Gallwitz 2008; Tahrani et al. 2009). Saxaglitpin also did not cause hypoglycemias, was well tolerated, and was weight neutral just as the other available DPP-4 inhibitors. A meta-analysis of the clinical phase III studies showed favorable data on the development of cardiovascular events (Wolf et al. 2009).

5   Incretin-Based Therapies: Common Characteristics and Differences

GLP-1 receptor agonists and DPP-4 inhibitors offer a good alternative to the established antidiabetic compounds due to their satisfying and glucose-dependent antihyperglycemic efficacy, their lack of risk for causing hypoglycemia, as well as their positive effects on body weight development demonstrating weight loss with GLP-1 receptor agonists and weigh neutrality with DPP-4 inhibitors. A further advantage is their positive effect on surrogate parameters for beta-cell function. At this time, however, it is not clear yet whether incretin-based therapies will lead to a sustained and durable positive effect on beta-cell function and mass under clinical conditions in patients with type 2 diabetes (see above). Animal data suggest that the novel compounds may lead to a retardation or halt of the progression of type 2 diabetes.

The most patient-relevant and striking difference of both incretin-based therapies is that GLP-1 receptor agonists are injectable agents, while DPP-4 inhibitors are effective orally (Table 1). Glycemic control seems to be improved more effectively by GLP-1 receptor agonists in comparison to DPP-4 inhibitors, but the data of a study directly comparing the efficacy and safety of liraglutide with sitagliptin are not available yet. Also, only GLP-1 receptor agonists lead to a reduction in body weight, whereas DPP-4 inhibitors are weight neutral. Furthermore, positive cardiovascular effects have been shown for GLP-1 receptor agonists. Nausea, the most common adverse event observed with GLP-1 receptor agonist therapy, is not observed in the treatment with DPP-4 inhibitors. So far, no characteristic pattern of adverse events has been observed with the DPP-4 inhibitors. DPP-4 is also expressed on the plasma membrane of T-lymphocytes, where it was first described as CD-26 receptor. However, no immunological alterations have been observed with DPP-4 inhibitor therapy. Furthermore, DPP-4 has multiple substrates (all peptides with a penultimate alanine or proline in the N-terminal position); the physiological effect of DPP-4 inhibition on all substrates has not been characterized in detail yet. Further long-term studies should clarify the long-term effects and safety of DPP-4 inhibitors.

Table 1 Differences between GLP-1 receptor agonists and DPP-4 inhibitors

Properties/action	GLP-1 receptor agonists	DPP-4 inhibitors
Application	Subcutaneous	Oral
"GLP-1" levels	Pharmacological ($>5\times$)	Physiological (2–$3\times$)
GLP-1 effects	Interaction with receptors on target organs (hormonal signal pathway)	Interaction with receptors on afferent nerves (mixed neural/ hormonal signal pathway)
Duration of "GLP-1"-elevation	Long, continuously	On–off, postprandially
Other mediators	No	GIP, PACAP, others
Effect on gastric emptying	Yes	No/hardly
Appetite	Reduced	Hardly influenced
Effect on body weight	Weight loss	Weight neutral
Adverse events	Nausea/fullness exenatide: antibodies (?)	No significant effects observed

PACAP pituitary adenylate cyclase activating polypeptide

6 Indications for Incretin-Based Therapies and Their Placement in Treatment Guidelines for Type 2 Diabetes

The DPP-4 inhibitors sitagliptin, vildagliptin, and saxagliptin are approved in many countries for an oral combination therapy, when therapeutic goals are not reached with a lifestyle intervention and metformin monotherapy. The DPP-4 inhibitors have a place in this indication in the German guidelines and a recommendation by the British National Institute for Health and Clinical Excellence (NICE) for patients who should not be treated with sulfonylureas in order to prevent hypoglycemia or further weight gain (Matthaei et al. 2009; National Institute for Health and Clinical Excellence 2009). A recent retrospective study has shown that a higher incidence of hypoglycemia might promote the development of dementia (Whitmer et al. 2009). In this respect, hypoglycemia avoidance as stated by NICE is an important therapeutic goal. It should be noted that DPP-4 inhibitors lower the HbA1c by approximately 1.0% and that other treatment options (namely insulin) should be considered, if the HbA1c is elevated by more than 1.0% or if metabolic control has decompensated. The combination of metformin with DPP-4 inhibitors combined two synergistic treatment principles: metformin acting on insulin resistance and the DPP-4 inhibitor acting on the glucose-dependent stimulation of insulin secretion and inhibition of glucagon secretion (the same synergistic principle of action applies to the combination of a glitazone and a DPP-4 inhibitor). DPP-4 inhibitors are not inferior to sulfonylureas in the combination with metformin regarding glycemic parameters (see Fig. 7) (Nauck et al. 2007b). Theoretically, DPP-4 inhibitors may succeed sulfonylureas as insulinotropic agents, if the above-mentioned advantages are underlined by positive outcomes in long-term studies concerning glycemic and other relevant endpoints as well as safety outcomes.

Therapy with a GLP-1 receptor agonist is a favorable treatment option when an oral therapy with metformin or a combination therapy with metformin and a

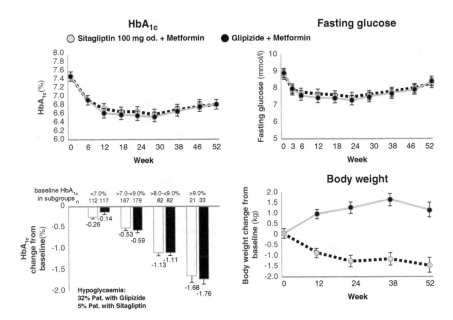

Fig. 7 Effect of sitagliptin (*open circles* and *hatched lines, white bars*) or the sulfonylurea glipizide (*closed circles* and *black lines, black bars*) as add-on therapy to metformin in type 2 diabetic patients not well controlled with metformin monotherapy (adapted from Nauck et al. 2007b)

sulfonylurea are insufficient and a simultaneous loss of body weight is another therapeutic goal (e.g. obesity-associated complications and concomitant morbidity) or hypoglycemia strictly has to be avoided (s.a.). The therapy with a GLP-1 receptor agonist at this stage may be a favorable alternative to initiating insulin treatment. In case sulfonylureas were used before initiation of GLP-1 receptor agonist therapy, the sulfonylurea dose should be at least reduced when adding the GLP-1 receptor agonist. In a large proportion of patients, the sulfonylurea treatment can even be stopped.

Incretin-based therapies may help to bring a larger percentage of patients to their glycemic goals. Fixed dose combinations of a DPP-4 inhibitor with metformin may be a favorable alternative as the patient does not have to take more tablets when intensifying oral antidiabetic therapy with a DPP-4 inhibitor. Obese patients with weight loss as another important therapeutic goal may profit from a therapy with a GLP-1 receptor agonist. The higher price of the novel incretin-based therapies is outweighed in some respect by the possibility to reduce the cost for blood glucose monitoring that is not necessary for safety reasons as long as the patient is not simultaneously treated with sulfonylurea and/or insulin.

Prevention of hypoglycemic events and prevention of further weight gain are important therapeutic goals considering the results of the ACCORD trial that showed an increased mortality in patients with type 2 diabetes who were allocated to the intensified treatment arm with an HbA1c goal <6.0% and were treated with multiple combinations of the classical antidiabetic agents (Gerstein et al. 2008). The increased mortality rate in this group may be explained by the higher gain in body weight and by the increased incidence of hypoglycemic episodes. On the other hand, the 10-year follow-up data of the UKPDS show that an early and effective diabetes treatment lowers not only microvascular complications, but also macrovascular endpoints significantly (Holman et al. 2008). With respect to these study results, patients with a newly diagnosed type 2 diabetes should have a treatment that enables them to reach normoglycemia in a safe way without the risk of hypoglycemia or weight gain.

A consensus statement published in 2008 by the American Diabetes Association (ADA) and the European Association for the Study of Diabetes (EASD) separates the existing antidiabetic compounds and treatment algorithms into well-validated therapies ("tier 1", comprising metformin, sulfonylureas, and insulin) and less-validated therapies ("tier 2", comprising pioglitazone and GLP-1 receptor agonists). In this statement, the established substances are preferred according to their published endpoint and safety data as well as pharmaco-economic data. In the less-validated therapies, GLP-1 receptor agonists, however, have their place as second therapeutic escalation after metformin failure in the same line with the widely used therapy with pioglitazone (Nathan et al. 2009). In the German guidelines, DPP-4 inhibitors and GLP-1 receptor agonists are placed in second line after metformin failure, if the HbA1c does not exceed >7.5% (Matthaei et al. 2009).

Both incretin-based therapies may also have a place in earlier or later stages of type 2 diabetes when effectiveness is shown in these stages. Preliminary data show that the addition of a DPP-4 inhibitor to an existing insulin therapy further reduces HbA1c and may have a positive effect on hypoglycemic events (Vilsboll et al. 2009; Fonseca et al. 2008). Combination studies with insulin and GLP-1 receptor agonists are also carried out and should bring results soon. Furthermore, long-term studies are under way, investigating the effect of incretin-based therapies on disease progression with results being awaited around 2015. If these studies show an effect on disease progression, another argument for using incretin-based therapies early in the disease will be supported by study data. Recently, animal and human studies showed a positive influence of GLP-1 or GLP-1 receptor agonists on the cardiovascular system and on the nervous system describing neuroprotective effects (Müssig et al. 2008; Nikolaidis et al. 2004, 2005a, b; Perry et al. 2007; Sokos et al. 2006). These fields may also open novel indications for incretin-based therapies (Sokos et al. 2006). But long-term studies on hard cardiovascular endpoints and safety finally have to reveal the important data to clarify the efficacy, safety, and placement of incretin-based therapies in type 2 diabetes therapy.

7 Incretin-Based Therapies and Type 1 Diabetes

Since GLP-1 has a positive effect on beta-cell mass in rodents as well as a beneficial effect on survival of isolated human islets in cell culture (Baggio and Drucker 2007; Drucker and Nauck 2006; Farilla et al. 2003), incretin-based therapies might also prove advantageous in type 1 diabetes alone or in combination with immune therapy. Evidence in support of this concept is provided by first preclinical studies (Waldron-Lynch et al. 2008). Besides the positive effect on the beta cells, GLP-1 may also influence glycemic parameters in a favorable way by slowing gastric emptying and affecting glucagon secretion in type 1 diabetes (Raman and Heptulla 2009). A small mechanistic study in type 1 diabetic individuals showed that an intravenous GLP-1 infusion reduced fasting hyperglycemia in the morning that was provoked by omitting the basal insulin injection at night (Creutzfeldt et al. 1996). Preclinical and animal studies should be undertaken to clarify the potential influence of GLP-1 on beta-cell mass in type 1 diabetes and on possible alterations of the autoimmune process. Clinical studies could then be implemented to investigate the metabolic effects of the autoimmune process and in case of positive outcomes should be followed by clinical studies in type 1 diabetes.

References

Agerso H, Jensen LB, Elbrond B, Rolan P, Zdravkovic M (2002) The pharmacokinetics, pharmacodynamics, safety and tolerability of NN2211, a new long-acting GLP-1 derivative, in healthy men. Diabetologia 45:195–202

Ahmad SR, Swann J (2008) Exenatide and rare adverse events. N Engl J Med 358:1970–1971

Ahren B (2008a) Emerging dipeptidyl peptidase-4 inhibitors for the treatment of diabetes. Expert Opin Emerg Drugs 13:593–607

Ahren B (2008b) Novel combination treatment of type 2 diabetes DPP-4 inhibition + metformin. Vasc Health Risk Manag 4:383–394

Ahren B, Foley JE (2008) The islet enhancer vildagliptin: mechanisms of improved glucose metabolism. Int J Clin Pract Suppl 175:8–14

Baggio LL, Drucker DJ (2007) Biology of incretins: GLP-1 and GIP. Gastroenterology 132:2131–2457

Barnett AH (2005) Exenatide. Drugs Today (Barc) 41:563–578

Barnett A (2007) Exenatide. Expert Opin Pharmacother 8:2593–2608

Barnett AH (2009) New treatments in type 2 diabetes: a focus on the incretin-based therapies. Clin Endocrinol (Oxf) 70:343–353

Barnett AH, Burger J, Johns D, Brodows R, Kendall DM, Roberts A, Trautmann ME (2007) Tolerability and efficacy of exenatide and titrated insulin glargine in adult patients with type 2 diabetes previously uncontrolled with metformin or a sulfonylurea: a multinational, randomized, open-label, two-period, crossover noninferiority trial. Clin Ther 29:2333–2348

Brubaker PL, Drucker DJ (2004) Minireview: glucagon-like peptides regulate cell proliferation and apoptosis in the pancreas, gut, and central nervous system. Endocrinology 145:2653–2659

Bunck MC, Diamant M, Corner A, Eliasson B, Malloy JL, Shaginian RM, Deng W, Kendall DM, Taskinen MR, Smith U, Yki-Jarvinen H, Heine RJ (2009) One-year treatment with exenatide improves beta-cell function, compared with insulin glargine, in metformin-treated type 2 diabetic patients: a randomized, controlled trial. Diabetes Care 32:762–768

Buse JB, Henry RR, Han J, Kim DD, Fineman MS, Baron AD (2004) Effects of exenatide (exendin-4) on glycemic control over 30 weeks in sulfonylurea-treated patients with type 2 diabetes. Diabetes Care 27:2628–2635

Buse JB, Rosenstock J, Sesti G, Schmidt WE, Montanya E, Brett JH, Zychma M, Blonde L (2009) Liraglutide once a day versus exenatide twice a day for type 2 diabetes: a 26-week randomised, parallel-group, multinational, open-label trial (LEAD-6). Lancet 374:39–47

Chacra AR, Tan GH, Apanovitch A, Ravichandran S, List J, Chen R (2009) Saxagliptin added to a submaximal dose of sulphonylurea improves glycaemic control compared with uptitration of sulphonylurea in patients with type 2 diabetes: a randomised controlled trial. Int J Clin Pract 63:1395–1406

Chan JC, Scott R, Arjona Ferreira JC, Sheng D, Gonzalez E, Davies MJ, Stein PP, Kaufman KD, Amatruda JM, Williams-Herman D (2008) Safety and efficacy of sitagliptin in patients with type 2 diabetes and chronic renal insufficiency. Diabetes Obes Metab 10:545–555

Chang AM, Jakobsen G, Sturis J, Smith MJ, Bloem CJ, An B, Galecki A, Halter JB (2003) The GLP-1 derivative NN2211 restores beta-cell sensitivity to glucose in type 2 diabetic patients after a single dose. Diabetes 52:1786–1791

Courreges JP, Vilsboll T, Zdravkovic M, Le-Thi T, Krarup T, Schmitz O, Verhoeven R, Buganova I, Madsbad S (2008) Beneficial effects of once-daily liraglutide, a human glucagon-like peptide-1 analogue, on cardiovascular risk biomarkers in patients with Type 2 diabetes. Diabet Med 25:1129–1131

Creutzfeldt W (1979) The incretin concept today. Diabetologia 16:75–85

Creutzfeldt WO, Kleine N, Willms B, Orskov C, Holst JJ, Nauck MA (1996) Glucagonostatic actions and reduction of fasting hyperglycemia by exogenous glucagon-like peptide I(7-36) amide in type I diabetic patients. Diabetes Care 19:580–586

Cure P, Pileggi A, Alejandro R (2008) Exenatide and rare adverse events. N Engl J Med 358:1969–1970

Deacon CF (2009) Potential of liraglutide in the treatment of patients with type 2 diabetes. Vasc Health Risk Manag 5:199–211

Deacon CF, Holst JJ (2009) Saxagliptin: a new dipeptidyl peptidase-4 inhibitor for the treatment of type 2 diabetes. Adv Ther 26:488–499

DeFronzo RA, Ratner RE, Han J, Kim DD, Fineman MS, Baron AD (2005) Effects of exenatide (exendin-4) on glycemic control and weight over 30 weeks in metformin-treated patients with type 2 diabetes. Diabetes Care 28:1092–1100

DeFronzo RA, Hissa MN, Garber AJ, Luiz Gross J, Yuyan Duan R, Ravichandran S, Chen RS (2009) The efficacy and safety of saxagliptin when added to metformin therapy in patients with inadequately controlled type 2 diabetes with metformin alone. Diabetes Care 32:1649–1655

Dore DD, Seeger JD, Arnold Chan K (2009) Use of a claims-based active drug safety surveillance system to assess the risk of acute pancreatitis with exenatide or sitagliptin compared to metformin or glyburide. Curr Med Res Opin 25:1019–1027

Drucker DJ, Nauck MA (2006) The incretin system: glucagon-like peptide-1 receptor agonists and dipeptidyl peptidase-4 inhibitors in type 2 diabetes. Lancet 368:1696–1705

Drucker DJ, Buse JB, Taylor K, Kendall DM, Trautmann M, Zhuang D, Porter L (2008) Exenatide once weekly versus twice daily for the treatment of type 2 diabetes: a randomised, open-label, non-inferiority study. Lancet 372:1240–1250

Eng J, Kleinman WA, Singh L, Singh G, Raufman JP (1992) Isolation and characterization of exendin-4, an exendin-3 analogue, from Heloderma suspectum venom. Further evidence for an exendin receptor on dispersed acini from guinea pig pancreas. J Biol Chem 267:7402–7405

Farilla L, Bulotta A, Hirshberg B, Li Calzi S, Khoury N, Noushmehr H, Bertolotto C, Di Mario U, Harlan DM, Perfetti R (2003) Glucagon-like peptide 1 inhibits cell apoptosis and improves glucose responsiveness of freshly isolated human islets. Endocrinology 144:5149–5158

Fehmann HC, Habener JF (1992) Insulinotropic hormone glucagon-like peptide-I(7-37) stimulation of proinsulin gene expression and proinsulin biosynthesis in insulinoma beta TC-1 cells. Endocrinology 130:159–166

Fonseca V, Baron M, Shao Q, Dejager S (2008) Sustained efficacy and reduced hypoglycemia during one year of treatment with vildagliptin added to insulin in patients with type 2 diabetes mellitus. Horm Metab Res 40:427–430

Food and Drug Administration (FDA) (2009) Byetta (exenatide) – Renal Failure. http://www.fda. gov/Safety/MedWatch/SafetyInformation/SafetyAlertsforHumanMedicalProducts/ucm188703. htm. 2 Nov 2009

Gaede P, Vedel P, Larsen N, Jensen GV, Parving HH, Pedersen O (2003) Multifactorial intervention and cardiovascular disease in patients with type 2 diabetes. N Engl J Med 348:383–393

Gallwitz B (2006) Exenatide in type 2 diabetes: treatment effects in clinical studies and animal study data. Int J Clin Pract 60:1654–1661

Gallwitz B (2008) Saxagliptin, a dipeptidyl peptidase IV inhibitor for the treatment of type 2 diabetes. IDrugs 12:906–917

Garber AJ, Spann SJ (2008) An overview of incretin clinical trials. J Fam Pract 57(Suppl): S10–S18

Garber A, Klein E, Bruce S, Sankoh S, Mohideen P (2006) Metformin-glibenclamide versus metformin plus rosiglitazone in patients with type 2 diabetes inadequately controlled on metformin monotherapy. Diabetes Obes Metab 8:156–163

Garber A, Henry R, Ratner R, Garcia-Hernandez PA, Rodriguez-Pattzi H, Olvera-Alvarez I, Hale PM, Zdravkovic M, Bode B (2008) Liraglutide versus glimepiride monotherapy for type 2 diabetes (LEAD-3 Mono): a randomised, 52-week, phase III, double-blind, parallel-treatment trial. Lancet 373:473–481

Garber A, Henry RR, Ratner RE, Hale P, Chang CT, Bode B (2009) Monotherapy with Liraglutide, a Once-Daily Human GLP-1 analog, provides sustained reductions in A1C, FPG, and weight compared with Glimepiride in type 2 diabetes: LEAD-3 Mono 2-Year Results. Diabetes 58, Suppl 1: 162-OR

Gedulin BR, Smith P, Prickett KS, Tryon M, Barnhill S, Reynolds J, Nielsen LL, Parkes DG, Young AA (2005) Dose-response for glycaemic and metabolic changes 28 days after single injection of long-acting release exenatide in diabetic fatty Zucker rats. Diabetologia 48:1380–1385

Gerstein HC, Miller ME, Byington RP, Goff DC Jr, Bigger JT, Buse JB, Cushman WC, Genuth S, Ismail-Beigi F, Grimm RH Jr, Probstfield JL, Simons-Morton DG, Friedewald WT (2008) Effects of intensive glucose lowering in type 2 diabetes. N Engl J Med 358:2545–2559

Green J, Feinglos M (2008) New combination treatments in the management of diabetes: focus on sitagliptin-metformin. Vasc Health Risk Manag 4:743–751

Heine RJ, Van Gaal LF, Johns D, Mihm MJ, Widel MH, Brodows RG (2005) Exenatide versus insulin glargine in patients with suboptimally controlled type 2 diabetes: a randomized trial. Ann Intern Med 143:559–569

Heise T, Graefe-Mody EU, Huttner S, Ring A, Trommeshauser D, Dugi KA (2009) Pharmacokinetics, pharmacodynamics and tolerability of multiple oral doses of linagliptin, a dipeptidyl peptidase-4 inhibitor in male type 2 diabetes patients. Diabetes Obes Metab 11:786–794

Herman GA, Bergman A, Liu F, Stevens C, Wang AQ, Zeng W, Chen L, Snyder K, Hilliard D, Tanen M, Tanaka W, Meehan AG, Lasseter K, Dilzer S, Blum R, Wagner JA (2006a) Pharmacokinetics and pharmacodynamic effects of the oral DPP-4 inhibitor sitagliptin in middle-aged obese subjects. J Clin Pharmacol 45:876–886

Herman GA, Bergman A, Stevens C, Kotey P, Yi B, Zhao P, Dietrich B, Golor G, Schrodter A, Keymeulen B, Lasseter KC, Kipnes MS, Snyder K, Hilliard D, Tanen M, Cilissen C, De Smet M, de Lepeleire I, Van Dyck K, Wang AQ, Zeng W, Davies MJ, Tanaka W, Holst JJ, Deacon CF, Gottesdiener KM, Wagner JA (2006b) Effect of single oral doses of sitagliptin, a dipeptidyl peptidase-4 inhibitor, on incretin and plasma glucose levels after an oral glucose tolerance test in patients with type 2 diabetes. J Clin Endocrinol Metab 91:4612–4619

Herman GA, Bergman A, Yi B, Kipnes M (2006c) Tolerability and pharmacokinetics of metformin and the dipeptidyl peptidase-4 inhibitor sitagliptin when co-administered in patients with type 2 diabetes. Curr Med Res Opin 22:1939–1947

Holman RR, Paul SK, Bethel MA, Matthews DR, Neil HA (2008) 10-year follow-up of intensive glucose control in type 2 diabetes. N Engl J Med 359:1577–1589

International Diabetes Federation (IDF) (2009) Diabetes atlas. http://www.diabetesatlas.org

Karasik A, Aschner P, Katzeff H, Davies MJ, Stein PP (2008) Sitagliptin, a DPP-4 inhibitor for the treatment of patients with type 2 diabetes: a review of recent clinical trials. Curr Med Res Opin 24:489–496

Kendall DM, Riddle MC, Rosenstock J, Zhuang D, Kim DD, Fineman MS, Baron AD (2005) Effects of exenatide (exendin-4) on glycemic control over 30 weeks in patients with type 2 diabetes treated with metformin and a sulfonylurea. Diabetes Care 28:1083–1091

Kim D, MacConell L, Zhuang D, Kothare PA, Trautmann M, Fineman M, Taylor K (2007) Effects of once-weekly dosing of a long-acting release formulation of exenatide on glucose control and body weight in subjects with type 2 diabetes. Diabetes Care 30:1487–1493

Klonoff DC, Buse JB, Nielsen LL, Guan X, Bowlus CL, Holcombe JH, Wintle ME, Maggs DG (2008) Exenatide effects on diabetes, obesity, cardiovascular risk factors and hepatic biomarkers in patients with type 2 diabetes treated for at least 3 years. Curr Med Res Opin 24:275–286

Kolterman OG, Kim DD, Shen L, Ruggles JA, Nielsen LL, Fineman MS, Baron AD (2005) Pharmacokinetics, pharmacodynamics, and safety of exenatide in patients with type 2 diabetes mellitus. Am J Health Syst Pharm 62:173–181

Malloy J, Capparelli E, Gottschalk M, Guan X, Kothare P, Fineman M (2009) Pharmacology and tolerability of a single dose of exenatide in adolescent patients with type 2 diabetes mellitus being treated with metformin: a randomized, placebo-controlled, single-blind, dose-escalation, crossover study. Clin Ther 31:806–815

Marre M, Shaw J, Brandle M, Bebakar WM, Kamaruddin NA, Strand J, Zdravkovic M, Le Thi TD, Colagiuri S (2009) Liraglutide, a once-daily human GLP-1 analogue, added to a sulphonylurea over 26 weeks produces greater improvements in glycaemic and weight control compared with adding rosiglitazone or placebo in subjects with Type 2 diabetes (LEAD-1 SU). Diabet Med 26:268–278

Matthaei S, Bierwirth R, Fritsche A, Gallwitz B, Häring HU, Joost HG, Kellerer M, Kloos C, Kunt T, Nauck MA, Schernthaner G, Siegel E, Thienel F (2009) Medicinal antihyperglycemic therapy of type 2 diabetes. Guidelines of the German Diabetes Association. Exp Clin Endocrinol Diabetes 117:522–557

McGill JB (2009) Insights from the Liraglutide Clinical Development Program – the Liraglutide Effect and Action in Diabetes (LEAD) studies. Postgrad Med 121:16–25

Mest JH (2006) Dipeptidyl peptidase-IV inhibitors can restore glucose homeostasis in type 2 diabetics via incretin enhancement. Curr Opin Investig Drugs 7:338–343

Müssig K, Oncu A, Lindauer P, Heininger A, Aebert H, Unertl K, Ziemer G, Häring HU, Holst JJ, Gallwitz B (2008) Effects of intravenous glucagon-like peptide-1 on glucose control and hemodynamics after coronary artery bypass surgery in patients with type 2 diabetes. Am J Cardiol 102:646–647

Nathan DM, Buse JB, Davidson MB, Heine RJ, Holman RR, Sherwin R, Zinman B (2006) Management of hyperglycaemia in type 2 diabetes: a consensus algorithm for the initiation and adjustment of therapy. A consensus statement from the American Diabetes Association and the European Association for the Study of Diabetes. Diabetologia 49:1711–1721

Nathan DM, Buse JB, Davidson MB, Ferrannini E, Holman RR, Sherwin R, Zinman B (2009) Medical management of hyperglycemia in type 2 diabetes: a consensus algorithm for the initiation and adjustment of therapy. Diabetes Care 32:193–203

National Institute for Health and Clinical Excellence (2009) http://www.guidance.nice.org.uk/CG87

Nauck M, Marre M (2009) Adding liraglutide to oral antidiabetic drug monotherapy: efficacy and weight benefits. Postgrad Med 121:5–15

Nauck M, Stockmann F, Ebert R, Creutzfeldt W (1986) Reduced incretin effect in type 2 (non-insulin-dependent) diabetes. Diabetologia 29:46–52

Nauck MA, Heimesaat MM, Orskov C, Holst JJ, Ebert R, Creutzfeldt W (1993a) Preserved incretin activity of glucagon-like peptide 1 [7-36 amide] but not of synthetic human gastric inhibitory polypeptide in patients with type-2 diabetes mellitus. J Clin Invest 91:301–307

Nauck MA, Kleine N, Orskov C, Holst JJ, Willms B, Creutzfeldt W (1993b) Normalization of fasting hyperglycaemia by exogenous glucagon-like peptide 1 (7-36 amide) in type 2 (non-insulin-dependent) diabetic patients. Diabetologia 36:741–744

Nauck MA, Duran S, Kim D, Johns D, Northrup J, Festa A, Brodows R, Trautmann M (2007a) A comparison of twice-daily exenatide and biphasic insulin aspart in patients with type 2 diabetes who were suboptimally controlled with sulfonylurea and metformin: a non-inferiority study. Diabetologia 50:259–267

Nauck MA, Meininger G, Sheng D, Terranella L, Stein PP (2007b) Efficacy and safety of the dipeptidyl peptidase-4 inhibitor, sitagliptin, compared with the sulfonylurea, glipizide, in patients with type 2 diabetes inadequately controlled on metformin alone: a randomized, double-blind, non-inferiority trial. Diabetes Obes Metab 9:194–205

Nauck M, Frid A, Hermansen K, Shah NS, Tankova T, Mitha IH, Zdravkovic M, During M, Matthews DR (2009a) Efficacy and safety comparison of liraglutide, glimepiride, and placebo, all in combination with metformin, in type 2 diabetes: the LEAD (liraglutide effect and action in diabetes)-2 study. Diabetes Care 32:84–90

Nauck MA, Ratner RE, Kapitza C, Berria R, Boldrin M, Balena R (2009b) Treatment with the human once-weekly glucagon-like peptide-1 analog taspoglutide in combination with metformin improves glycemic control and lowers body weight in patients with type 2 diabetes inadequately controlled with metformin alone: a double-blind placebo-controlled study. Diabetes Care 32:1237–1243

Nikolaidis LA, Elahi D, Hentosz T, Doverspike A, Huerbin R, Zourelias L, Stolarski C, Shen YT, Shannon RP (2004) Recombinant glucagon-like peptide-1 increases myocardial glucose uptake and improves left ventricular performance in conscious dogs with pacing-induced dilated cardiomyopathy. Circulation 110:955–961

Nikolaidis LA, Doverspike A, Hentosz T, Zourelias L, Shen YT, Elahi D, Shannon RP (2005a) Glucagon-like peptide-1 limits myocardial stunning following brief coronary occlusion and reperfusion in conscious canines. J Pharmacol Exp Ther 312:303–308

Nikolaidis LA, Elahi D, Shen YT, Shannon RP (2005b) Active metabolite of GLP-1 mediates myocardial glucose uptake and improves left ventricular performance in conscious dogs with dilated cardiomyopathy. Am J Physiol Heart Circ Physiol 289:H2401–H2408

Perry T, Holloway HW, Weerasuriya A, Mouton PR, Duffy K, Mattison JA, Greig NH (2007) Evidence of GLP-1-mediated neuroprotection in an animal model of pyridoxine-induced peripheral sensory neuropathy. Exp Neurol 203:293–301

Pratley RE (2008) Overview of glucagon-like peptide-1 analogs and dipeptidyl peptidase-4 inhibitors for type 2 diabetes. Medscape J Med 10:171

Pratley RE, Rosenstock J, Pi-Sunyer FX, Banerji MA, Schweizer A, Couturier A, Dejager S (2007) Management of type 2 diabetes in treatment-naive elderly patients: benefits and risks of vildagliptin monotherapy. Diabetes Care 30:3017–3022

Pratley RE, Reusch JE, Fleck PR, Wilson CA, Mekki Q (2009) Efficacy and safety of the dipeptidyl peptidase-4 inhibitor alogliptin added to pioglitazone in patients with type 2 diabetes: a randomized, double-blind, placebo-controlled study. Curr Med Res Opin 25:2361–2371

Raman VS, Heptulla RA (2009) New potential adjuncts to treatment of children with type 1 diabetes mellitus. Pediatr Res 65:370–374

Retterstol K (2009) Taspoglutide: a long acting human glucagon-like polypeptide-1 analogue. Expert Opin Investig Drugs 18:1405–1411

Rosenstock J, Sankoh S, List JF (2008) Glucose-lowering activity of the dipeptidyl peptidase-4 inhibitor saxagliptin in drug-naive patients with type 2 diabetes. Diabetes Obes Metab 10:376–386

Rosenstock J, Aguilar-Salinas C, Klein E, Nepal S, List J, Chen R (2009a) Effect of saxagliptin monotherapy in treatment-naive patients with type 2 diabetes. Curr Med Res Opin 25:2401–2411

Rosenstock J, Reusch J, Bush M, Yang F, Stewart M (2009b) The potential of Albiglutide, a long-acting GLP-1 receptor agonist, in type 2 diabetes: a randomized controlled trial exploring weekly, biweekly, and monthly dosing. Diabetes Care 32:1880–1886

Scott R, Wu M, Sanchez M, Stein P (2007) Efficacy and tolerability of the dipeptidyl peptidase-4 inhibitor sitagliptin as monotherapy over 12 weeks in patients with type 2 diabetes. Int J Clin Pract 61:171–180

Sokos GG, Nikolaidis LA, Mankad S, Elahi D, Shannon RP (2006) Glucagon-like peptide-1 infusion improves left ventricular ejection fraction and functional status in patients with chronic heart failure. J Card Fail 12:694–699

Sturis J, Gotfredsen CF, Romer J, Rolin B, Ribel U, Brand CL, Wilken M, Wassermann K, Deacon CF, Carr RD, Knudsen LB (2003) GLP-1 derivative liraglutide in rats with beta-cell deficiencies: influence of metabolic state on beta-cell mass dynamics. Br J Pharmacol 140:123–132

Tahrani AA, Piya MK, Barnett AH (2009) Saxagliptin: a new DPP-4 inhibitor for the treatment of type 2 diabetes mellitus. Adv Ther 26:249–262

Vilsboll T (2009) Liraglutide: a new treatment for type 2 diabetes. Drugs Today (Barc) 45:101–113

Vilsboll T, Brock B, Perrild H, Levin K, Lervang HH, Kolendorf K, Krarup T, Schmitz O, Zdravkovic M, Le-Thi T, Madsbad S (2008) Liraglutide, a once-daily human GLP-1 analogue, improves pancreatic B-cell function and arginine-stimulated insulin secretion during hyperglycaemia in patients with Type 2 diabetes mellitus. Diabet Med 25:152–156

Vilsboll T, Rosenstock J, Yki-Jarvinen H, Cefalu WT, Chen Y, Ling Y, Meehan AG, Katz L, Engel SS, Kaufman KD, Amatruda JM (2009) Sitagliptin, a selective DPP-4 inhibitor, improves glycemic control when added to insulin, with or without Metformin, in patients with type 2. Diabetes 58(Suppl 1):588

Vollmer K, Gardiwal H, Menge BA, Goetze O, Deacon CF, Schmidt WE, Holst JJ, Meier JJ (2009) Hyperglycemia acutely lowers the postprandial excursions of glucagon-like Peptide-1 and gastric inhibitory polypeptide in humans. J Clin Endocrinol Metab 94:1379–1385

Waldron-Lynch F, von Herrath M, Herold KC (2008) Towards a curative therapy in type 1 diabetes: remission of autoimmunity, maintenance and augmentation of beta cell mass. Novartis Found Symp 292:146–155

Wellendorph P, Johansen LD, Brauner-Osborne H (2009) Molecular pharmacology of promiscuous seven transmembrane receptors sensing organic nutrients. Mol Pharmacol 76:453–465

Werner U (2008) Preclinical pharmacology of the new GLP-1 receptor agonist AVE0010. Ann Endocrinol (Paris) 69:164–165

Whitmer RA, Karter AJ, Yaffe K, Quesenberry CP Jr, Selby JV (2009) Hypoglycemic episodes and risk of dementia in older patients with type 2 diabetes mellitus. JAMA 301:1565–1572

Williams-Herman D, Round E, Swern AS, Musser B, Davies MJ, Stein PP, Kaufman KD, Amatruda JM (2008) Safety and tolerability of sitagliptin in patients with type 2 diabetes: a pooled analysis. BMC Endocr Disord 8:14

Wolf R, Frederich R, Fiedorek FT, Donovan M, Xu Z, Harris S, Chen R (2009) Evaluation of CV risk in the Saxagliptin clinical trials. Diabetes 59(Suppl 1):8-LB

Yang J, Campitelli J, Hu G, Lin Y, Luo J, Xue C (2007) Increase in DPP-IV in the intestine, liver and kidney of the rat treated with high fat diet and streptozotocin. Life Sci 81:272–279

Yoo BK, Triller DM, Yoo DJ (2006) Exenatide: a new option for the treatment of type 2 diabetes. Ann Pharmacother 40:1777–1784

Zeitler P (2009) Update on nonautoimmune diabetes in children. J Clin Endocrinol Metab 94:2215–2220

Zinman B, Hoogwerf BJ, Duran Garcia S, Milton DR, Giaconia JM, Kim DD, Trautmann ME, Brodows RG (2007) The effect of adding exenatide to a thiazolidinedione in suboptimally controlled type 2 diabetes: a randomized trial. Ann Intern Med 146:477–485

Zinman B, Gerich J, Buse JB, Lewin A, Schwartz S, Raskin P, Hale PM, Zdravkovic M, Blonde L (2009) Efficacy and safety of the human glucagon-like peptide-1 analog liraglutide in combination with metformin and thiazolidinedione in patients with type 2 diabetes (LEAD-4 Met+ TZD). Diabetes Care 32:1224–1230

Cannabinoids and Endocannabinoids in Metabolic Disorders with Focus on Diabetes

Vincenzo Di Marzo, Fabiana Piscitelli, and Raphael Mechoulam

Contents

Abstract The cannabinoid receptors for Δ^9-THC, and particularly, the CB_1 receptor, as well as its endogenous ligands, the endocannabinoids anandamide and 2-arachidonoylglycerol, are deeply involved in all aspects of the control of energy balance in mammals. While initially it was believed that this endocannabinoid signaling system would only facilitate energy intake, we now know that perhaps even more important functions of endocannabinoids and CB_1 receptors in this context are to enhance energy storage into the adipose tissue and reduce energy expenditure by influencing both lipid and glucose metabolism. Although normally well controlled by hormones and neuropeptides, both central and peripheral aspects

This article is dedicated to the loving memory of Ester Fride (1953–2010).

V. Di Marzo (✉) and F. Piscitelli
Endocannabinoid Research Group, Institute of Biomolecular Chemistry, National Research Council, Via Campi Flegrei 34, Comprensorio Olivetti, 80078 Pozzuoli (NA), Italy

R. Mechoulam (✉)
Medicinal Chemistry and Natural Products Department, Medical Faculty, Hebrew University of Jerusalem, Ein Kerem, Jerusalem 91120, Israel

M. Schwanstecher (ed.), *Diabetes - Perspectives in Drug Therapy*,   75
Handbook of Experimental Pharmacology 203,
DOI 10.1007/978-3-642-17214-4_4, © Springer-Verlag Berlin Heidelberg 2011

of endocannabinoid regulation of energy balance can become dysregulated and contribute to obesity, dyslipidemia, and type 2 diabetes, thus raising the possibility that CB_1 antagonists might be used for the treatment of these metabolic disorders. On the other hand, evidence is emerging that some nonpsychotropic plant cannabinoids, such as cannabidiol, can be employed to retard β-cell damage in type 1 diabetes. These novel aspects of endocannabinoid research are reviewed in this chapter, with emphasis on the biological effects of plant cannabinoids and endocannabinoid receptor antagonists in diabetes.

Keywords CB_1 receptor · Endocannabinoid · Lipids · Phytocannabinoid · Rimonabant

1   A Brief Introduction to the Endocannabinoid System

The discovery of the endogenous signaling system that is now referred to as the endocannabinoid system started with the chemical identification in the mid-1960s of the major psychoactive component of *Cannabis sativa* and marijuana, $Δ^9$-tetrahydrocannabinol ($Δ^9$-THC) (Gaoni and Mechoulam 1964). Almost three decades later, another milestone was the finding that $Δ^9$-THC owes its psychotropic and immunomodulatory effects to its capability to bind to and activate specific plasma membrane proteins: (1) the cannabinoid CB_1 receptor, one of the most abundant G-protein-coupled receptors in the central nervous system (Devane et al. 1988); and (2) the cannabinoid CB_2 receptor, expressed abundantly in several immune cells and tissues (Munro et al. 1993). In fact, brain CB_1 receptors are coupled, among other things, to inhibition of neurotransmitter release (Schlicker and Kathmann 2001), whereas CB_2 receptors participate in the regulation of cytokine release and function (Klein 2005).

The tissue and organ distributions of CB_1 and CB_2 receptors are not as segregated as they were originally thought after their identification, and it is now becoming increasingly accepted that while CB_1 receptors play important functions in peripheral tissues, CB_2 receptors are also present in the brain and in nonimmune cells (Ashton et al. 2006; Gong et al. 2006; Van Sickle et al. 2005). In addition, the existence of other receptors for cannabimimetic compounds has been suggested based on pharmacologic data, but these putative proteins have not been fully characterized yet (Di Marzo and De Petrocellis 2005).

The discovery of CB_1 and CB_2 receptors suggested the existence of endogenous compounds capable of binding to and activating them, the *endocannabinoids*, the two best studied of which are anandamide (*N*-arachidonoylethanolamine) (Devane et al. 1992) and *2-arachidonoylglycerol* (2-AG) (Mechoulam et al. 1995; Sugiura et al. 1995) (Fig. 1). These, as well as other proposed endocannabinoids, derive from the nonoxidative metabolism of the essential ω6-polyunsaturated fatty acid, arachidonic acid. The cannabinoid receptors, the endocannabinoids, and the proteins catalyzing endocannabinoid biosynthesis and inactivation constitute the endocannabinoid system (Di Marzo et al. 2004).

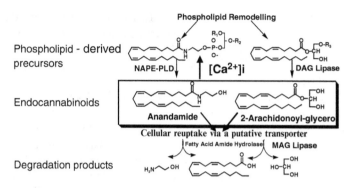

Fig. 1 Schematic representation of the biosynthetic pathways of the two major endocannabinoids, anandamide and 2-arachidonoylglycerol, through which (1) direct biosynthetic precursors are produced "on demand" following Ca^{2+} elevation in the cell and subsequent phospholipid remodeling and to be then converted into endocannabinoid; (2) endocannabinoids, after activation of cannabinoid receptors, are immediately taken up by cells to be degraded via hydrolytic reactions to cannabinoid-receptor-inactive metabolites. Abbreviations: *DAG* diacylglycerol, *FAAH* fatty acid amide hydrolase, *MAG* monoacylglycerol, *NAPE-PLD* N-acyl-phosphatidylethanolamine-sepcific phospholiapse D

The endocannabinoids produce different cellular effects through their ability to differentially activate the different G-proteins that are normally coupled to CB_1 and CB_2 receptors. For example, 2-AG induces and exhibits maximal stimulation of both G_o and G_i, whereas anandamide seems to be able to activate only G_i. Both cannabinoid receptor types are coupled to inhibition of adenylate cyclase via G_i, but whereas 2-AG always acts as a full agonist for the inhibition of cyclic adenosine monophosphate (cAMP) formation, anandamide is significantly more efficacious at CB_1 than CB_2 receptors (Glass and Northup 1999). CB_1 activation is also coupled, again via $G_{i/o}$ proteins, to the inhibition of voltage-dependent N-, P-, and Q-type calcium channels and to the activation of inwardly rectifying potassium channels (McAllister and Glass 2002), and, in certain cells, to $G_{q/11}$ proteins with subsequent stimulation of intracellular Ca^{2+} transients (De Petrocellis et al. 2007). Some of these intracellular effects may underlie the neuromodulatory effects of endocannabinoids in the central and peripheral nervous systems as well as in their intervention in retrograde signaling (Wilson and Nicoll 2002).

It is now well established that the two most studied endocannabinoids, anandamide and 2-AG, are not prestored into secretory vesicles, but are biosynthesized de novo following an increase of the intracellular concentration of Ca^{2+}, within a framework of phospholipid metabolic reactions (Fig. 1). In fact, both the formation of the two direct and distinct biosynthetic precursors of anandamide and 2-AG and their conversion into the two endocannabinoids are catalyzed by Ca^{2+}-sensitive enzymes (Di Marzo et al. 2004). This means that the whole cascade of endocannabinoid production is triggered "on demand," thus leading to endocannabinoid levels that will also ultimately depend on the availability of arachidonic acid on the *sn*-1 or -2 position of phosphoglycerides for anandamide and 2-AG, respectively, and, as the levels of arachidonic acid depend on essential fatty acids, also on the type of the diet. Endocannabinoids are then released from the cell immediately after their biosynthesis in order to activate their targets and then rapidly removed from the extracellular space by rapid and selective uptake into the cell and intracellular enzymatic hydrolysis (Di Marzo et al. 2004).

The enzymes most likely responsible for formation of anandamide and 2-AG biosynthesis from their direct biosynthetic precursors (Fig. 1), the N-acyl-phosphatidyl-ethanolamine-selective phospholipase D (NAPE-PLD) and the sn-1-selective diacylglycerol lipases (DAGLα and β), respectively, have been recently cloned (Bisogno et al. 2003; Okamoto et al. 2004). N-Arachidonoyl-phosphatidyl-ethanolamine (NArPE) is converted into anandamide by NAPE-PLD, although other pathways exist for the processing of NArPE (Leung et al. 2006; Sun et al. 2004). NArPE in turn is produced from phospholipid remodeling and N-arachidonoylation of phosphatidylethanolamine. The formation of 2-AG in the brain occurs instead mostly through the hydrolysis of the sn-2-arachidonate-containing diacylglycerols (DAGs), which are catalyzed by the sn-1-selective diacylglycerol lipases (DAGLα and β) (Bisogno et al. 2003). Different types of 2-arachidonate-containing phosphoglyceride precursors, including phosphoinositide, phosphatidylcholine, and phosphatidic acid, act as precursors of 2-arachidonate-containing DAGs via Ca^{2+}-sensitive phospholipases C or phosphatidic acid hydrolase, respectively (Bisogno et al. 1999; Di Marzo et al. 1996).

Also, the enzymes mostly involved in the degradation of anandamide and 2-AG have been identified and cloned (Fig. 1). An intracellular integral membrane protein of 597 amino acids belonging to the amidase family of enzymes, known as "fatty acid amide hydrolase" (FAAH), catalyzes the hydrolysis of anandamide and 2-AG (Cravatt et al. 1996). It is found in the brain (Thomas et al. 1997; Tsou et al. 1998), but also, among various peripheral tissues, in the vasculature (Bilfinger et al. 1998), pancreas, and adipose tissue (Engeli et al. 2005; Matias et al. 2006). The monoacylglycerol lipase (MAGL), another hydrolase belonging to a different family of Ser proteases, plays a key role, in most cases a more important one than FAAH, in the enzymatic hydrolysis of 2-AG (Dinh et al. 2002).

2   Central Endocannabinoid Control of Energy Balance

Activation of CB_1 receptors is responsible for the well-known appetite-inducing actions of Δ^9-THC and, perhaps more importantly, also of anandamide and 2-AG (Kirkham et al. 2002; Williams and Kirkham 1999). While the consumption of palatable foods and drinks can be stimulated by CB_1 agonists and blocked by CB_1 antagonists even in satiated animals, the intake of "normal" food is reduced by CB_1 receptor blockade more efficiently when animals have been deprived of food for a few hours (Colombo et al. 1998; Di Marzo et al. 2001; Gallate et al. 1999; Koch 2001; Williams et al. 1998). These observations suggested a potential role of the endocannabinoid system in the control of both hedonic and appetitive aspects of food intake and that an endocannabinoid "tone" can be triggered by brief food deprivation as well as by exposure to palatable foods. In fact, CB_1 receptors have been detected in the brain nuclei involved in all aspects of food intake, including the hypothalamus (Gonzalez et al. 1999), which "senses" a negative energy balance, and the nucleus accumbens (Robbe et al. 2002), which contains circuitries involved in the "liking" and "wanting" of food and hence mediating its incentive and motivational value (Cota et al. 2006). Furthermore, the endocannabinoid system is also active in the vagus nerve and its termination at the level of the nodose ganglion (Burdyga et al. 2004), as well as in other brainstem areas "sensing" gastrointestinal content and controlling satiety and emesis, such as the nucleus tractus solitarius and area postrema (Partosoedarso et al. 2003). On the other hand, as previously observed for several other proposed orexigenic signals, it has been found that the levels of anandamide and/or 2-AG decrease during food consumption and increase during food deprivation in both the limbic forebrain (which contains the nucleus accumbens) and hypothalamus, whereas CB_1 receptor expression increases in the nodose ganglion during food deprivation and decreases, under the negative control of cholecystokinin, following food intake (Burdyga et al. 2004; Gomez et al. 2002; Kirkham et al. 2002). It is important to emphasize that the central effects of endocannabinoids on energy balance reflect the high degree of interactions among the neural circuitries belonging to the hypothalamus, mesolimbic system, and brainstem. For example, direct or "indirect" (i.e., via inhibition of endocannabinoid degradation) activation of CB_1 receptors in the nucleus accumbens, while inducing food intake, also activates several neurons in the hypothalamus (Soria-Gómez et al. 2007). Likewise, direct or "indirect" activation of CB_1 receptors in the pontine parabrachial nucleus, which is located in the brainstem and gates neurotransmission associated with, but not limited to, the gustatory properties of food, selectively induces intake of palatable food (DiPatrizio and Simansky 2008a, b).

Brain CB_1 receptors control energy intake mainly in two ways, depending on their subcellular distribution (Fig. 2). When they activate presynaptic CB_1 receptors, postsynaptically derived endocannabinoids act as retrograde signals (Wilson and Nicoll 2002), which normally leads to rapid modulation of glutamate or GABA release, with subsequent tuning of the activity of orexigenic or anorectic postsynaptic neurons. Conversely, when postsynaptic CB_1 receptors are stimulated, this might lead to long-term regulation of the expression of genes encoding for orexigenic or anorectic neuropeptides or to modulation of the postsynaptic receptors of such mediators. In the former case, electrophysiological data support the involvement of endocannabinoids, under the negative control of leptin, as modulators of short-term synaptic plasticity in the hypothalamus (Jo et al. 2005), with subsequent decrease of GABA release onto orexigenic melanin concentrating hormone (MCH)-releasing neurons of the lateral hypothalamus (LHA). Since MCH is an orexigenic neuropeptide controlling both hypothalamic and neucleus accumbens activity, this effect is consistent with increased feeding behavior. Likewise, in the paraventricular nucleus (PVN), endocannabinoids acting retrogradely on CB_1, expressing parvocellular neurons and produced following activation of ghrelin or

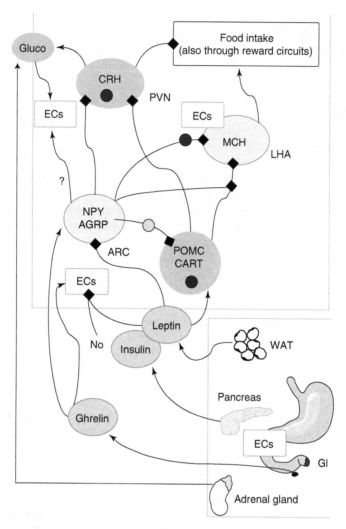

Fig. 2 Regulation and function of hypothalamic endocannabinoids (ECs). In the hypothalamic nuclei of the midbrain, EC levels under the positive control of glucocorticoids (Gluco) and ghrelin (possibly also via NPY) and under the negative control of leptin, but not insulin. CB_1 receptors control negatively the expression of corticotropin-releasing hormone (CRH) and of cocaine- and amphetamine-regulated transcript (CART) in periventricular nuclei (PVN) and arcuate nucleus (ARC), respectively, as well as MCH release in the lateral hypothalamus (LHA) and, possibly, CRH release in the PVN via retrograde signaling. Finally, they both enhance and reduce the activity of POMC-expressing neurons in the ARC via retrograde inhibition of GABA-ergic and glutamatergic inputs, respectively. Adapted from Matias and Di Marzo (2007). Symbols: *dark circles* denote endocannabinoid inhibitory actions; *light circles* denote stimulatory action; *black diamonds* denote inhibition by non-endocannabinoid mediators; *black arrows* denote stimulation/ production

"fast" glucocorticoid receptors in postsynaptic PVN neurons, inhibit glutamatergic signaling, and subsequently, the activity of these neurons, which often produce the anorectic corticotropin-releasing hormone (CRH) (Kola et al. 2008; Malcher-Lopes et al. 2006). These effects are often under the negative control of postsynaptic ob receptors for leptin, which, by decreasing intracellular calcium, reduce endocannabinoid biosynthesis (Di Marzo et al. 2001; Jo et al. 2005). On the other hand, CB_1 receptors potentially localized postsynaptically in anorectic CRH-expressing neurons of the PVN and preopiomelanocortin (POMC) and cocaine- and amphetamine-regulated transcript (CART)-expressing neurons of the arcuate nucleus (ARC) appear to control negatively the expression of CRH and CART. In fact, CRH mRNA levels in the PVN are higher in CB_1-deficient mice (Cota et al. 2003, 2007), whereas elevation of endogenous anandamide levels, obtained by knocking out the FAAH enzyme, is accompanied by reduced CART release in several hypothalamic regions, an effect antagonized by a CB_1 antagonist (Hilairet et al. 2003; Osei-Hyiaman et al. 2005a).

In agreement with the former finding, blood corticosterone levels are increased following CB_1 antagonism in obese Zucker rats (Doyon et al. 2006), which contain high hypothalamic levels of endocannabinoids (Di Marzo et al. 2001). Finally, another type of effect by postsynaptic CB_1 receptors might be to enhance orexin A signaling. In fact, it has been shown that orexin A receptors can be sensitized by CB_1 receptors coexpressed in the same cells (Ellis et al. 2006; Hilairet et al. 2003), and unpublished immunohistochemical work carried out in Di Marzo's laboratory showed that, in the mouse hypothalamus, CB_1 and orexin A receptors are indeed strongly colocalized in the ARC and PVN, at both the presynaptic and postsynaptic level.

The effect of endocannabinoids in milk on pup growth has also been examined. While the levels of anandamide are negligible, milk – the only food of young mammalians – contains considerable amounts of 2-AG (Di Marzo et al. 1998; Fride et al. 2001). This finding led to an examination of the role of the CB_1 receptor in suckling. The cannabinoid CB_1 receptor antagonist, rimonabant, completely inhibited the physical growth of mouse pups and caused death within 1 week, by depriving them of the essential benefits of suckling. This effect of rimonabant was seen either after a single injection of the compound, administered within the first 24 h after birth, or after daily injections between days 2 and 8. Thus, the first 24 h of life seem to be most critical for the endocannabinoid-induced suckling-promoting effect, which is compatible with the levels of 2-AG in milk. Accordingly, 2-AG partly reversed the effect of rimonabant on suckling, and its activity was significantly enhanced by 2-palmitoyl glycerol and 2-linoleoyl glycerol, which have no activity per se – an effect named "entourage effect" (Ben-Shabat et al. 1998), which is a unique route to enhance biological activity. These data strongly suggest that the anti-suckling and growth-inhibiting effects of rimonabant are mediated at least in part by blockage of CB_1 receptors. Surprisingly, however, milk intake and survival were also impaired upon administration of the CB_1 receptor antagonist in CB_1 receptor-deficient pups, although not as dramatically as in wild-type pups. These results support evidence for the existence of additional cannabinoid receptors.

In fact, Zimmer et al. (1999) found that cannabinoid CB_1 receptor knockout mice survive the initial stages of life, which obviously involve suckling. Presumably, other mechanisms compensate for the lack of CB_1 receptor-based suckling. These mechanisms may involve endogenous opioids or lysophosphatidic acid as both types of compounds have been shown to be involved in suckling. Indeed, the chemical structures of 2-AG and lysophosphatidic acid (with 2-arachidonoyl as the acyl moiety) only differ by the absence of a phosphate group in 2-AG (Mechoulam et al. 2006).

3 Peripheral Endocannabinoid Control of Energy Balance

Endocannabinoids and cannabinoid CB_1 receptors are present in peripheral cells and tissues involved in the control of energy homeostasis, including the mesenteric neurons and epithelial cells of the small and large intestine (Coutts and Izzo 2004), the liver and hepatocytes (Osei-Hyiaman et al. 2005b), the white adipose tissue (Engeli et al. 2005) and adipocytes (Bensaid et al. 2003; Cota et al. 2003; Roche et al. 2006), the skeletal muscle (Pagotto et al. 2006), and the pancreas (Juan-Pico et al. 2006; Matias et al. 2006). In fact, the endocannabinoid system is emerging as one of the key players in the peripheral control of metabolism, including nutrient assimilation, processing, and storage.

Cota and colleagues (Cota et al. 2003) showed for the first time that wild-type mice exhibit significantly higher amounts of fat mass than CB_1 receptor-null mice, as well as a trend toward lower energy expenditure, even when fed with the same amount of food. Furthermore, it was shown that functional CB_1 receptors coupled to stimulation of lipoprotein lipase activity are present in adipocytes. These findings suggested for the first time that the endocannabinoid system contributes to fat accumulation, independently from the amount of food ingested and by acting directly on the adipose tissue. Indeed, studies carried out in mouse 3T3 adipocyte cell lines have shown that (1) stimulation and blockade of CB_1 receptors arrest and stimulate adipocyte proliferation, respectively (Gary-Bobo et al. 2006; Bellocchio et al. 2008); (2) formation of endocannabinoids precedes preadipocyte differentiation into mature adipocytes; and (3) chronic stimulation of CB_1 receptors during adipocyte differentiation enhances the expression of an early marker of differentiation, the peroxisome proliferator-activated receptor (PPAR)-γ, while inducing accumulation of lipid droplets (Matias et al. 2006). These findings indicate that the endocannabinoid system directly participates in adipogenesis and adipocyte lipogenesis. Importantly, in this model of adipocytes, CB_1 activation and blockade also enhance and reduce, respectively, fatty acid synthase (FAS), acetyl-CoA carboxylase (ACC), stearoyl-CoA desaturase, and diacylglycerol transferase-2 (Bellocchio et al. 2008). Since the endocannabinoid anandamide binds to PPAR-γ and stimulates its transcriptional activity also independently from CB_1 (Bouaboula et al. 2005), it is possible that both CB_1 and non-CB_1 receptors might be involved in these phenomena. Interestingly, the stimulatory effect of CB_1 stimulation on

PPAR-γ expression has also been recently confirmed in human adipocytes (Pagano et al. 2007). Furthermore, data exist in support of other mechanisms through which CB_1 stimulation can ensure de novo lipogenesis in adipocytes. In fact, studies carried out in both mouse 3T3 and human adipocytes recently showed that CB_1 agonists can stimulate glucose uptake, very probably via translocation of the glucose 4 transporter to the plasma membrane, and that these effects are reversed by the CB_1 receptor antagonist rimonabant (Gasperi et al. 2007; Pagano et al. 2007). Enhanced intracellular glucose, following glucose oxidation, might provide the adipocytes with the biosynthetic precursors for de novo fatty acid biosynthesis, especially in the presence of upregulated FAS and of inhibited cAMP formation and AMP kinase activity, two lipogenetic responses that (endo)cannabinoids have been reported to induce in the adipose tissue in vivo (Kola et al. 2005; Matias et al. 2006; Osei-Hyiaman et al. 2005b).

The endocannabinoid system also stimulates *lipogenesis* in the liver. CB_1 receptors are expressed in hepatocytes, where they stimulate the expression of the important transcription factor sterol response element binding protein (SREBP)-1c and of its targets ACC and FAS. These effects are likely to explain why CB_1 receptor stimulation causes fatty acid synthesis and lipogenesis in these cells (Osei-Hyiaman et al. 2005b), although this mechanism is unlikely to occur in the healthy liver, where CB_1 expression is relatively low.

Preliminary findings suggest that endocannabinoids are also involved in the control of *metabolism* by regulating insulin release from β-cells as well as glucose uptake and utilization by tissues, with subsequent impact on glucose tolerance (Bermudez-Silva et al. 2006; Juan-Pico et al. 2006). Juan-Pico et al. (2006), using mouse islets of Langerhans, showed that stimulation of CB_2 – and, to a lesser extent, CB_1 – receptors reduces insulin release via inhibition of calcium transients. Accordingly, little, if any, CB_1 receptors are expressed in mouse β-cells (Starowicz et al. 2008). However, Matias et al. (2006) and Bermúdez-Silva et al. (2008) later found that in rat insulinoma and human β-cells CB_1 receptors are indeed expressed, a finding also confirmed by Starowicz and colleagues (Starowicz et al. 2008). It was suggested that in both rat insulinoma and human β-cells, CB_1 stimulation enhances glucose-induced insulin release, whereas in human β-cells CB_2 activation reduces this effect. However, it was also reported that systemic CB_1 or CB_2 receptor stimulation reduces or enhances plasma glucose clearance in rats (Bermudez-Silva et al. 2006), two effects that are opposite to those expected from the above-mentioned actions of the two receptor types on insulin release. This suggests that the endocannabinoid system might affect glucose utilization at the level of insulin sensitivity rather than of its release, with possible impact on glucose uptake by the skeletal muscle and liver and subsequent energy expenditure.

Indirect evidence for a food intake-independent and tonic retarding action of the endocannabinoid system and CB_1 receptors on metabolism came recently from studies using two distinct selective CB_1 receptor antagonists/inverse agonists, i.e., *rimonabant* (Kunz et al. 2008) and AVE1625 (Herling et al. 2007). In the former study, when compared with vehicle-treated rats, rats administered 3 and 10 mg/kg rimonabant together with their food showed an 18 and 49% increase in O_2

consumption, respectively, after 3 h. Respiratory quotients revealed no effect of rimonabant on the relative rate of carbohydrate and fat oxidation. Analysis of the correlation between O_2 consumption and physical activity indicated that factors other than increased physical activity contributed to the increase in O_2 consumption. Similar studies in mice demonstrated that wild-type but not $CB_1^{-/-}$ mice showed a change in O_2 consumption and physical activity following rimonabant administration, suggesting that these effects were mediated by the cannabinoid CB_1 receptor. These studies suggested that rimonabant stimulates significant acute energy expenditure in nonobese rodents, which is not completely accounted for by an increase in physical activity (Kunz et al. 2008). In the second study, the authors reported that AVE1625, when acutely administered postprandially to rats, causes instead slight as well as rapid increase in basal lipolysis and, 6 h after administration, also strong glycogenolysis. These two metabolic effects were accompanied by immediate increase in energy expenditure, a long-lasting increase of fat oxidation and a transient increase of glucose oxidation. These latter findings agree with the aforementioned potential tonic inhibition by CB_1 receptors of lipolysis and insulin sensitivity and are also supported by the recent finding that CB1 receptor inactivation with rimonabant or siRNA selectively increases glucose uptake by human differentiated L6 myotubes in a time- and dose-dependent manner (Esposito et al. 2008). This latter effect was due to activation of the phosphoinsotide-3-kinase pathway, which might have sensitized skeletal muscle cells to the action of insulin.

In summary, from the results described in this section it is clear that the endocannabinoid system affects energy metabolism not only via a central control of food intake, but also by enhancing lipogenesis and reducing lipid and glucose oxidation. This system might be one of the proposed "thrifty" mechanisms aimed at optimizing energy intake and storage, and minimizing energy expenditure, following food consumption and after periods of food deprivation.

4 Regulation and Dysregulation of the Endocannabinoid System in the Control of Metabolism

Several hormones coordinate the local effects of endocannabinoids on energy balance in organs and tissues as the hypothalamus, nucleus accumbens (Fig. 2), duodenum, adipose tissue, and liver. Higher hypothalamic levels of endocannabinoids are observed in rats following brief food deprivation (Kirkham et al. 2002), and this might be due to the fact that leptin, an adipocyte-derived hormone the levels of which are increased following food consumption and decreased following food deprivation, significantly reduces the levels of anandamide and 2-AG in the rat hypothalamus (Di Marzo et al. 2001) much in the same way it reduces the levels of hypothalamic orexigenic mediators and increases those of anorexic ones. Another group (Hanus et al. 2003) confirmed the significant enhancement of hypothalamic

2-AG levels after 24 h of fasting in mice. However, diet restriction over 12 days, instead, lowered the levels of 2-AG in both the hippocampus and the hypothalamus. Thus, while Kirkham et al. apparently measured the effects of hunger, Hanus et al. presumably recorded the effect of semistarvation, which might be partly due to the shortage of crucial ultimate biosynthetic precursors for 2-AG biosynthesis, which originate from the diet. If these observations in mice parallel the human condition, we can expect that diet restriction self-imposed by humans, as in anorexia, may cause lowering of hypothalamic 2-AG levels leading to further reduction of food consumption, thus perpetuating the clinical condition.

Ghrelin, another important peripheral hormone involved in food intake and released into the bloodstream from the stomach during food deprivation to stimulate energy intake, instead affects hypothalamic endocannabinoid levels positively. In fact, blockade of CB_1 receptors with rimonabant strongly reduces the orexigenic action of intrahypothalamic injections of the hormone, which has also been shown to act through elevated 2-AG levels (Tucci et al. 2004). Direct elevation of 2-AG levels in the hypothalamus by ghrelin was also demonstrated (Kola et al. 2008).

Glucocorticoid is another likely candidate for the regulation of pre- and postprandial hypothalamic endocannabinoids. In fact, activation of fast plasma membrane glucocorticoid receptors was found to stimulate the biosynthesis of endocannabinoids in this brain area (Malcher-Lopes et al. 2006), and the circulating levels of corticosterone are known to increase after food deprivation and decrease after food consumption. The tonic inhibition by CB_1 receptors of the release of corticosterone into the bloodstream (Doyon et al. 2006) might thus represent a negative feedback loop on glucocorticoid-stimulated endocannabinoid formation.

Leptin also decreases endocannabinoid levels in other cells or tissues that express the leptin receptors, such as T lymphocytes and the uterus (Maccarrone et al. 2003, 2005). This widespread tonic inhibition of endocannabinoid levels by the hormone also likely occurs in humans, since a negative correlation between blood leptin and anandamide levels was found in normoweight and anorexic women (Monteleone et al. 2005). In the latter case, the low levels of leptin typical of this eating disorder were consequently accompanied by significantly increased blood anandamide levels, possibly in the attempt of overcoming the effect of the shortage of dietary endocannabinoid precursors, mentioned above.

Indeed, endocannabinoid levels are also regulated in several peripheral tissues. In the duodenum, both anandamide and 2-AG levels are increased after food deprivation possibly in order to act on sensory and vagal nerves terminating in the brainstem and regulating satiety (Gomez et al. 2002; Izzo et al. 2009). In the adipose tissue, instead, no changes in endocannabinoid levels are observed following food deprivation (Izzo et al. 2009), as expected from the fact that this condition should be accompanied by activation of lipolytic, rather than lipogenic, mechanisms in this organ. Indeed, several mechanisms exist that ensure a strict regulation of endocannabinoid levels in adipocytes. Firstly, PPAR-γ stimulation with ciglitazone or rosiglitazone can (1) inhibit the levels of 2-AG in mature, but not hypertrophic, 3T3 F442A mouse adipocytes (Matias et al. 2006); and (2) downregulate CB_1 expression and upregulate FAAH expression in human adipocytes (Pagano et al. 2007). Thus,

CB_1 receptors act as an early inducer of adipocyte differentiation, perhaps even upstream of PPAR-γ and are subsequently "turned off" by this nuclear receptor once that differentiation is complete and PPAR-γ expression is maximal. Secondly, PPAR-δ, which can be activated following physical exercise, also inhibits CB_1 receptor expression (Yan et al. 2007). Finally, both insulin and leptin, the blood levels of which are increased after a meal or after fat accumulation in adipocytes, respectively, both inhibit endocannabinoid levels in these cells (Matias et al. 2006) (D'Eon et al. 2008), possibly by enhancing FAAH expression (Murdolo et al. 2007).

Also in RIN-m5F, rat insulinoma β-pancreatic cells *insulin* inhibits glucose-stimulated elevation of anandamide and 2-AG levels (Matias et al. 2006), thus suggesting that the negative control by this hormone over endocannabinoid tone might occur in all insulin-sensitive cells and be impaired during conditions of insulin resistance. Indeed, evidence for this has been recently provided for humans (Di Marzo et al. 2009b). In the rat liver, elevated endocannabinoid levels are found following food deprivation (Izzo et al. 2009) and this phenomenon, given the coupling of CB_1 receptors to reduced fatty acid oxidation and increased fatty acid synthesis (Osei-Hyiaman et al. 2005b), might prevent the liver from oxidizing fatty acids that are needed by other tissues as an alternative to glucose. On the other hand, following food intake, when easily utilizable "fuel" becomes again available, hepatic 2-AG (but not anandamide) levels remain elevated to make sure that at least some of this fuel is directed into fatty acid synthesis.

The same mechanisms that are used to keep the endocannabinoid system under control in various central and peripheral tissue, if disrupted, might determine dysregulation of anandamide and 2-AG levels or of CB_1 receptor expression. For example, as a consequence of leptin or leptin receptor deficiency, in the hypothalamus of obese and hyperglycemic *ob/ob* mice, or of rodents that are characterized by impaired leptin receptor (*db/db* mice and *fa/fa* Zucker rats), permanently elevated endocannabinoid levels are found (Di Marzo et al. 2001). This situation might also occur in obese human beings, in whom central leptin insensitivity develops, and might represent one of the reasons why, although blockade of CB_1 receptors inhibits food intake and decreases body weight in lean animals, it undoubtedly does so more efficaciously in genetically obese rodents or in rodents that become obese because of a prolonged high-fat diet. In fact, in lean animals, CB_1 receptor antagonists always appear to be more efficacious in the presence of a demonstrated higher tone of the endocannabinoid system in those brain areas controlling food intake, that is following brief periods of food deprivation or when the animals are exposed to palatable foods (McLaughlin et al. 2003). Therefore, the higher efficacy observed with CB_1 antagonists like rimonabant (SR141716A) (Rinaldi-Carmona et al. 1994) or AM251 (Chambers et al. 2004) in obese *vs.* lean animals, or the fact that obese rodents do not need to be deprived of food to be shown to be sensitive to treatments with these antagonists, is suggestive of a similarly higher endocannabinoid tone in obesity. Furthermore, since daily CB_1 receptor antagonism in obese rodents causes a transient (~1 week) inhibition of food intake, as opposed to effects on body weight that persist for several weeks (Ravinet Trillou et al. 2003), it is possible that it is the part of the endocannabinoid system that controls the peripheral aspects of energy balance (e.g., fat accumulation and energy expenditure, rather than food intake) that becomes more active during obesity and for a longer time and which, therefore, better responds to CB_1 antagonists. This hypothesis is supported by the observation that congenital blockade of CB_1 receptor expression, as in CB_1 knockout mice (Ravinet Trillou et al. 2004), or chronic treatment with CB_1 antagonists, prevents in mice not only the development of high-fat diet-induced obesity (DIO) (Ravinet Trillou et al. 2003), but also the metabolic consequences of these conditions such as high triglycerides, low HDL cholesterol, hyperglycemia, and hyperinsulinemia (Poirier et al. 2005; Ravinet Trillou et al. 2003, 2004), whereas pair-feeding (i.e., giving control animals the same amount of food consumed by animals in which CB_1 receptors is pharmacologically blocked or genetically impaired) produces significantly smaller effects. Accordingly, the dramatic alterations of the expression of white and brown adipose tissue enzymes and proteins involved in metabolism and energy expenditure in DIO mice are not observed if CB_1 receptors are genetically or pharmacologically impaired (Jbilo et al. 2005). Furthermore, in mice, selective genetical impairment of CB_1 receptors in hepatocytes prevents the development of high-fat diet-induced fatty liver, low HDL cholesterol, hyperglycemia, insulin resistance, and, to some extent, high triglycerides (Osei-Hyiaman et al. 2008), whereas, in rats, peripheral administration of rimonabant enhances *lipolysis* in adipocytes in a food intake-independent way (Nogueiras et al. 2008). These observations are in agreement with the idea that, during obesity and hyperglycemia, CB_1 signaling might become overactive in peripheral organs controlling energy accumulation, transformation, and expenditure, thereby contributing to high-fat diet-induced metabolic alterations.

That the levels of either endocannabinoids or CB_1 receptors, or both, are permanently upregulated in several peripheral organs and tissues of obese/hyperglycemic animals and humans is now a well-accepted concept. Evidence of endocannabinoid upregulation has been recently reported in adipocytes and β-cells (Matias et al. 2006). When treated with a high concentration of insulin, under conditions mimicking hyperglycemia and leading to insulin resistance and adipocyte hypertrophy, the levels of endocannabinoids and of CB_1 receptors are still significantly higher than in preadipocytes and mature adipocytes (D'Eon et al. 2008; Matias et al. 2006). The effect on 2-AG levels appears to be the result of changes in the regulation of 2-AG biosynthesis by PPAR-γ when passing from mature to hypertrophic adipocytes (Matias et al. 2006). In agreement with these findings in isolated adipocytes, enhanced levels of 2-AG, but not anandamide, have been reported in the epididymal fat of DIO mice compared with mice fed a normal diet (Matias et al. 2006). However, in both DIO mice and Zucker rats, when compared to lean animals of the same age, endocannabinoid levels are strongly *decreased* in the subcutaneous fat (Izzo et al. 2009; Starowicz et al. 2008), thus indicating that in obese/hyperglycemic rodents there is clearly a hypoactive endocannabinoid system in this adipose depot, whereas in the mesenteric fat of DIO mice, endocannabinoid levels are unaltered (Starowicz et al. 2008). In RIN-m5F rat insulinoma β-pancreatic cells, where a high glucose "pulse" elevates both anandamide and 2-AG levels, insulin keeps this latter effect under negative control when the cells are maintained in a relatively low concentration of glucose. However, under conditions mimicking hyperglycemia, insulin does not inhibit glucose-induced endocannabinoid levels and it even stimulates these levels per se (Matias et al. 2006). Accordingly, enhanced levels of both anandamide and 2-AG were observed in the pancreas of DIO mice as compared with mice fed a normal diet (Matias et al. 2006) and in that of Zucker rats as compared with lean rats (Izzo et al. 2009). Elevation of endocannabinoid levels is also observed in the liver of DIO mice and Zucker rats (Izzo et al. 2009; Osei-Hyiaman et al. 2005b), in the heart and kidneys of DIO mice (Matias et al. 2008), and in the duodenum of Zucker rats (Izzo et al. 2009). An early elevation of endocannabinoid levels was also found in the brown adipose tissue and in the skeletal muscle of DIO mice (Matias et al. 2008), where this phenomenon, like in the liver (Osei-Hyiaman et al. 2005b), is also accompanied by upregulation of CB_1 receptors (Pagotto et al. 2006).

4.1 Role of Dysregulated Endocannabinoid Signaling in Type 2 Diabetes and Obesity-Related Metabolic and Cardiovascular Disorders

The dysregulation of endocannabinoid tone in so many key organs for the control of metabolism cannot but have a strong impact on metabolism itself (Fig. 3). For example, overstimulation of CB_1 receptors in hypertrophic adipocytes leads to inhibition of adiponectin expression (Matias et al. 2006; Bellocchio et al. 2008) and this explains why (1) CB_1 receptor antagonism causes elevation of adiponectin

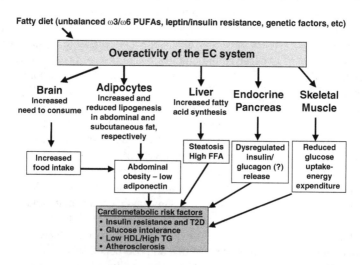

Fig. 3 Potential causes of the dysregulation of the endocannabinoid (EC) system in various peripheral tissues, and their consequences on metabolic risk factors. Abbreviations: *FFA* free fatty acids, *HDL* high-density lipoprotein, *T2D* type 2 diabetes, *TG* triglycerides

expression in adipocytes much more effectively in obese than in lean mice (Bensaid et al. 2003); and (2) the expression of several adiponectin-depending genes, which is strongly dysregulated in the adipose tissue of DIO mice, is restored to that of a lean phenotype following blockade of CB_1 receptors (Jbilo et al. 2005). Given the important protective role played by adiponectin against insulin resistance and atherogenic inflammation, it is tempting to speculate that endocannabinoid overactivity in some adipose depots might contribute to reduced levels of this hormone in obesity and, hence, to insulin resistance and atherosclerosis. By contrast, a lower endocannabinoid tone in subcutaneous *vs.* visceral (e.g., epidydimal or mesenteric) fat is likely to eventually contribute to excessive accumulation of this latter adipose depot at the expense of the more "beneficial" subcutaneous depots.

In view of the data described in the previous section, the overactivity of the endocannabinoid system in the pancreas is likely to have a strong impact on insulin levels and perhaps underlies the hyperinsulinemia that characterizes obesity. This is likely to lead to β-cell hypertrophy and damage, thus eventually contributing to the development of type 2 diabetes. In agreement with this hypothesis, a recent study showed how, in isolated pancreatic islets from Zucker diabetic rats or in islets from lean rats incubated with high glucose and palmitic acid, rimonabant decreases basal insulin hypersecretion without affecting glucose-stimulated insulin secretion (Getty-Kaushik et al. 2009). In the mouse liver, overactivity of CB_1 receptors alone might determine large part of the insulin resistance and hepatosteatosis, which are consequences of obesity (Osei-Hyiaman et al. 2008). In the skeletal muscle, given the observation that CB_1 receptor blockade improves glucose uptake and AMP kinase expression (Cavuoto et al. 2007; Esposito et al. 2008), the overactivity of CB_1 receptors might contribute to reduced insulin sensitivity, glucose uptake, and fatty acid oxidation.

That endocannabinoid dysregulation contributes not only to hyperphagia and increased body weight but also to the metabolic consequences of obesity, starting with reduced energy expenditure, insulin resistance, dyslipidemia, and dyslipoproteinemia and ending with type 2 diabetes and atherosclerosis, is being more and more confirmed by in vivo data obtained using CB_1 receptor antagonists in obese rodents. For example, CB_1 receptor blockade causes enhanced glucose uptake by the soleus muscle and increases oxygen consumption in *ob/ob* mice (Liu et al. 2005). In female candy-fed Wistar rats treated with rimonabant (10 mg/kg) and matched with pair-fed rats to distinguish between hypophagic action and hypothesized effects on energy expenditure, rimonabant reduced body weight nearly to levels of standard rat chow-fed rats within the first week of treatment. Evaluation of energy balance (energy expenditure measured by indirect calorimetry in relation to metabolizable energy intake calculated by bomb calorimetry) revealed that increased fat oxidation contributed more to sustained body weight reduction than reduced food intake. The acute effect of rimonabant on lipolysis was further investigated in postprandial male rats, demonstrating an inherent pharmacological activity of rimonabant to induce lipolysis in a way not secondary to postabsorptive reduced food intake. The authors concluded that the weight-reducing effect of rimonabant is due to continuously elevated energy expenditure based on increased fat oxidation driven by lipolysis from fat tissue as long as fat stores are elevated, whereas when the amount of endogenous fat stores decline, rimonabant-induced increased energy expenditure is maintained by a re-increase in food intake (Herling et al. 2008a). Importantly, in the same model, it was also found that treatment with rimonabant for the last 6 weeks of a 12-week candy-feeding causes a preferential loss of visceral vs. subcutaneous fat, and a corresponding loss of hepatic and skeletal muscle fat (Herling et al. 2008b), in agreement with the aforementioned higher tone of the endocannabinoid system in nonsubcutaneous fat depots and in the liver and skeletal muscle and with its likely role in determining reduced energy expenditure in obesity.

Three important studies have been carried out with rimonabant in Zucker rats at an age in which these obese rodents exhibit several of the features of the "metabolic syndrome." First, it was shown that oral treatment of these rats with rimonabant (30 mg/kg) daily for 8 weeks abolished obesity-associated hepatic steatosis and related features of metabolic syndrome: inflammation (elevated plasma levels of tumor necrosis factor alpha [TNF-α]), dyslipidemia, and reduced plasma levels of adiponectin (Gary-Bobo et al. 2007). The treatment also reduced hepatomegaly, elevation of plasma levels of enzyme markers of hepatic damage (alanine amino-transferase, gamma glutamyltransferase, and alkaline phosphatase), and the high levels of circulating TNF-α associated with steatohepatitis. Finally, rimonabant treatment also improved dyslipidemia by both decreasing plasma levels of trigly-cerides, free fatty acids, and total cholesterol and increasing the HDL/LDL choles-terol ratio. Importantly, all these effects of rimonabant were not or only slightly observed in pair-fed obese animals, highlighting the additional beneficial effects of treatment with rimonabant compared to diet. In the second study, the effect of CB_1 receptor antagonism on mortality and chronic renal failure associated with obesity in Zucker rats was investigated (Janiak et al. 2007). The rats received either rimonabant or vehicle for 12 months and were compared to a pair-fed but untreated group of obese rats. Mortality in the obese rats was significantly reduced by rimonabant along with a sustained decrease in body weight, transient reduction in food intake, and an increase in plasma adiponectin. As expected from previous studies, this was associated with significant reduction not only in plasma total cholesterol, LDL/HDL cholesterol ratio, triglycerides, and glucose, but also in norepinephrine, plasminogen activator inhibitor 1, and preservation of pancreatic weight and β-cell mass index. The CB_1 antagonist attenuated the increase in proteinuria, urinary N-acetylglucosaminidase excretion, plasma creatinine, and urea nitrogen levels while improving creatinine clearance and reduced renal hyper-trophy and glomerular and tubulointerstitial lesions. Although the drug did not modify hemodynamics, it normalized the pressor response to angiotensin II. Again, most of these effects were induced only to a smaller extent by diet restriction, pointing to the existence of direct effects of CB_1 receptor antagonism in peripheral organs. In the third study, Schäfer et al. (2008) investigated the effect of rimonabant on inflammation and enhanced platelet reactivity. The CB_1 antagonist (10 mg/kg by gavage) was fed for 2 weeks to 3-month-old male obese Zucker rats as an impaired glucose tolerance model and for 10 weeks to 6-month-old male obese Zucker rats as a model of the metabolic syndrome. In these rats, RANTES (Regulated upon Activation, Normal T-cell Expressed, and Secreted) and MCP-1 (monocyte chemotactic protein-1) serum levels were increased in obese *vs.* lean rats and significantly reduced by long-term treatment with rimonabant, which slowed weight gain in rats with the metabolic syndrome. Neutrophils and monocytes were significantly increased in young and old obese *vs.* lean Zucker rats and again lowered by rimonabant. Platelet-bound fibrinogen was significantly enhanced in obese *vs.* lean Zucker rats of both age, and this effect was also reduced by rimonabant, which also attenuated thrombin-induced aggregation and adhesion to fibrinogen of platelets from obese rats. The authors suggested that these effects of CB_1 antagonism, if translated to human beings with a metabolic syndrome, may potentially contribute to a reduction of cardiovascular risk. This conclusion is also supported by a very recent investigation carried out using an animal model of atherosclerosis, the low-density lipoprotein receptor-null [LDLR(−/−)] mouse. Rimonabant (50 mg/kg/d in the diet) significantly reduced food intake, weight gain, serum total cholesterol, and atherosclerotic lesion development in the aorta and aortic sinus of LDLR(−/−) mice fed a western-type diet for 3 months. Rimonabant also strongly reduced plasma levels of the proinflammatory cytokines MCP-1 and interleukin (IL)-12 and decreased lipopolysaccharide- and IL-1β-induced proinflammatory gene expression in mouse peritoneal macrophages in vitro as well as thioglycollate-induced recruitment of macrophages in vivo. Pair-fed animals had reduced weight gain, but developed atherosclerotic lesions that were as large as those of untreated animals, showing that the antiatherosclerotic effect of rimonabant is not related to reduced food intake. Interestingly, rimonabant at a lower dose (30 mg/kg/d in the diet) reduced atherosclerosis development in the aortic sinus, without affecting serum total cholesterol. The findings of these studies, taken together, support the role of a peripherally, rather than centrally, overactive endocannabinoid system in determining some deleterious consequences of obesity and substantiate the possibility that CB_1 receptor antagonists can be used for the treatment of obesity-associated metabolic and cardiovascular disorders.

In agreement with the above observations, a recent study used a different approach to investigate the consequences of the overactive endocannabinoid system on obesity-related metabolic dysfunction. Ruby et al. (2008) used a chemical approach to evaluate the direct effects of increased endocannabinoid signaling in mice by inducing acute elevations of endogenously produced anandamide and 2-AG through pharmacological inhibition of their enzymatic hydrolysis by isopropyl dodecylfluorophosphonate (IDFP). Acute IDFP treatment increased plasma levels of triglyceride (2.0- to 3.1-fold) and cholesterol (1.3- to 1.4-fold) in conjunction with an accumulation in plasma of apolipoprotein (apo) E-depleted triglyceride-rich lipoproteins. These changes did not occur in either $CB_1^{-/-}$ or $apoE^{-/-}$ mice, were prevented by pretreatment with CB_1 antagonists, and were not associated with reduced hepatic apoE gene expression. Although IDFP treatment increased hepatic mRNA levels of some lipogenic genes, there was no effect on triglyceride secretion into plasma. Instead, IDFP treatment impaired clearance of an intravenously administered triglyceride emulsion, despite increased lipoprotein lipase activity.

Importantly, these effects were not observed following inhibition of FAAH alone, indicating either that elevation of anandamide levels alone is not sufficient to affect triglyceride levels or that the endocannabinoid uniquely responsible for this effect is 2-AG (see below). The authors suggested that overactive endocannabinoid signaling elicits an increase in plasma triglyceride levels, which is associated with reduced plasma triglyceride clearance and an accumulation in plasma of apoE-depleted triglyceride-rich lipoproteins rather than enhanced lipogenesis, thus underscoring the potential efficacy of CB_1 antagonists in treating hypertriglyceridemia.

4.2 Endocannabinoid Dysregulation in Human Abdominal Obesity and Hyperglycemia: Relationship with Cardiometabolic Risk Factors and Type 2 Diabetes

The association between high intra-abdominal adiposity (IAA) and type 2 diabetes and between atherogenic inflammation and increased risk of experiencing serious cardiovascular disorders has been well documented by the medical literature (Després and Lemieux 2006). Data are now available that suggest that an upregulation of endocannabinoid signaling also occurs in humans and in association not only with obesity, but also with high visceral adiposity and hyperglycemia. In agreement with the aforementioned observations in DIO mice, significantly higher levels of 2-AG, but not anandamide, were detected in the visceral, but not subcutaneous, fat of obese patients (Matias et al. 2006). A recent study by Pagano and coworkers (2007) strengthened the link between endocannabinoid dysregulation and abdominal adiposity by showing that in both the abdominal subcutaneous and visceral fat of obese patients, both CB_1 receptor expression and endocannabinoid turnover (in terms of the expression of endocannabinoid biosynthetic and degrading enzymes) were upregulated in comparison to analogous tissues from lean individuals, whereas a reduction was observed in the gluteal subcutaneous fat, again similar to what described in the subcutaneous fat of DIO mice (Starowicz et al. 2008). Also obese patients with type 2 diabetes were recently described to exhibit an overall lower tone of endocannabinoids in the subcutaneous adipose tissue as compared to normoweight volunteers (Annuzzi et al. 2010). This phenomenon, in view of the aforementioned prolipogenetic role of the endocannabinoid system in the adipose tissue, might eventually result in more and more fat being stored in the abdominal depots and less and less in the subcutaneous depots, with potential deleterious consequences on cardiometabolic risk factors (Després and Lemieux 2006).

A direct relationship between IAA and high circulating 2-AG levels was established by two independent studies (Bluher et al. 2006; Cote et al. 2007). When examining two cohorts of obese patients with quite different anthropometric features, both groups of investigators found a strong direct correlation between the concentration of 2-AG, but not of the other endocannabinoid anandamide, in the blood and the amount of IAA determined by computer tomography. More importantly, high 2-AG levels also directly correlated with several cardiometabolic risk factors, including low HDL cholesterol, high triglycerides, low insulin sensitivity, and glucose tolerance and, in the case of the study by Cote et al. (2007), also low plasma adiponectin levels. Accordingly, in nonobese ($28 < BMI < 32$) patients with partially corrected hyperglycemia caused by type 2 diabetes, the levels of both anandamide and 2-AG were again significantly higher than in age-, BMI-, and gender-matched nondiabetic volunteers (Matias et al. 2006). In a very recent study, the changes in plasma endocannabinoid levels in a cohort of viscerally obese men following a 1-year lifestyle modification program were measured and correlated with changes in visceral adipose tissue and metabolic risk factors (Di Marzo et al. 2009a). Forty-nine viscerally obese men underwent a 1-year lifestyle modification program including healthy eating and physical activity. As a result, most risk factors were improved by the intervention, including body weight, waist circumference, and visceral adipose tissue, and these changes were accompanied by a strong reduction of plasma anandamide (-7.1%, $p = 0.005$) and, particularly, 2-AG (-62.3%, $p < 0.0001$) levels. Importantly, only the decrease of 2-AG levels correlated with decreases of visceral adipose tissue and triglyceride levels, with the increase of HDL3 cholesterol levels and with the decrease of a parameter of insulin resistance. Multivariate analyses suggested that decreases in 2-AG and visceral adipose tissue were both independently associated with decreases in triglycerides. These findings represent strong evidence of a link of visceral obesity-related metabolic risk factors, including hyperglycemia and type 2 diabetes, with peripheral endocannabinoid dysregulation. However, it must be emphasized that endocannabinoids are not normally released from tissues into the bloodstream to act as hormone-like molecules, and, therefore, it remains to be clarified whether and to what extent circulating 2-AG and anandamide levels reflect an overproduction by peripheral tissues ("spillover" effect).

5 Clinical Use of CB_1 Receptor Antagonists/Inverse Agonists Against Type 2 Diabetes

The study of the role of the endocannabinoids and their receptors in food intake and energy balance led, only little more than a decade after the identification of this signaling system, to the development of several new therapeutic drugs. The CB_1 receptor antagonist rimonabant has successfully completed four phase III and two phase IIIB clinical trials in obese patients with and without comorbidities (dyslipidemia and/or type 2 diabetes) (Després et al. 2005; Pi-Sunyer et al. 2006; Scheen et al. 2006; Van Gaal et al. 2005) and was on the market in Europe until October 23, 2008, under the trademark name of *Acomplia*^{TM}, as a therapeutic aid to dietary restriction and exercise for the treatment of obesity (in patients with BMI > 30) or of abdominal obesity accompanied by metabolic disorders such as dyslipidemia and type 2 diabetes (in patients with BMI > 27). The pooled results of the four Rimonabant in Obesity (RIO) trials have been very recently analyzed (Van Gaal et al. 2008). Other compounds with the same mechanism of action, such as taranabant and otenabant, have just completed or are completing phase III trials (see Di Marzo 2008a, for review). Such compounds not only reduce food intake and body weight in obese patients, but also significantly ameliorate the signs of the metabolic syndrome in overweight/viscerally obese and/or type 2 diabetes patients. In particular, in obese patients with type 2 diabetes, three clinical studies using rimonabant at the daily oral dose of 20 mg have been published. The RIO-Diabetes trial enrolled 1,045 overweight and obese subjects with type 2 diabetes. Subjects in this trial had to have taken metformin or sulfonylurea monotherapy for at least 6 months and to have fasting plasma glucose levels between 100 and 271 mg/dL and hemoglobin A1c levels (HbA1c, a standard blood measure value that is indicative of a patients' glucose for about 2 months) between 6.5 and 10%. All subjects continued treatment with metformin or sulfonylurea throughout the study. Subjects treated with rimonabant (20 mg/d for 1 year, intention-to-treat population) had significant decreases in body weight and improvements in glycemic control compared with placebo (Scheen et al. 2006). This latter effect of rimonabant appeared to be partly independent of weight loss. The SERENADE trial (Study Evaluating Rimonabant Efficacy in Drug-NAïve DiabEtic Patients) was conducted on 278 patients with T2D, HbA1c levels >7% and <10%, and not adequately controlled by diet alone for a period of 6 months in the absence of other medications. Baseline HbA1c (7.9%) was reduced by −0.8% with rimonabant vs. −0.3% with placebo, with a larger rimonabant effect in patients with baseline HbA(1c) ≥ 8.5% (Delta HbA(1c) −1.25%; $P = 0.0009$). Weight loss from baseline was −6.7 kg with rimonabant vs. −2.8 kg with placebo. Rimonabant induced improvements from baseline in waist circumference, fasting plasma glucose, triglycerides, and HDL cholesterol. Again, statistical analyses suggested that approximately 57% of the improvements in HbA1c were independent of weight loss (Rosenstock et al. 2008). The results of the ARPEGGIO trial, the first trial of rimonabant in patients with type 2 diabetes, not adequately controlled with insulin therapy, were recently published (Hollander et al. 2010). The 368 type 2 diabetic patients participating in this 11-month trial had been treated with insulin for an average duration of 6 years prior to entering the 48-week treatment with the CB_1 antagonist. Rimonabant 20 mg significantly improved HbA1c by 0.89% from the baseline value and 0.64% over the control group ($p < 0.0001$). Rimonabant tripled the number of diabetic patients reaching the 7% HbA1c level recommended by the international medical guidelines (18.4% for the rimonabant group below 7% and 6.75% patients for the control group). A statistically significant reduction of fasting plasma glucose over control, resulting in a mean treatment difference of −0.88 mmol/l in favor of rimonabant 20 mg/day, consistent with the HbA1c reduction, was observed. This action on glucose control was three times more pronounced when rimonabant was added than insulin and lifestyle advice alone. Also in this case, rimonabant demonstrated a statistically significant body weight loss over placebo, resulting in a mean treatment difference of −2.56 kg (Hollander et al. 2010). Concern has been raised regarding the side effects of chronic treatment with rimonabant 20 mg/day in obese and/or type 2 diabetes patients. While such treatment increases the odds of experiencing nausea, dizziness, and diarrhea (Di Marzo 2008b; Van Gaal et al. 2008), it is the worsening effects on anxiety and depression that caused the strongest preoccupation (see below), also due to the fact that obesity and type 2 diabetes are associated with increased depression, and weight loss per se is also cause of depressed mood. One way to circumvent this problem is to treat only those patients with no previous history of strong anxiety and depression. In fact, it has been estimated that in this case depressive disorders are observed in 3.2% of the patients, and anxiety in 5.6% of the patients, as compared to 1.6 and 2.4% in placebos, respectively. In the ARPEGGIO trial, fewer patients in the rimonabant group compared with the control group experienced serious treatment emergent adverse events (16.8% vs. 19.3%, respectively). Anxiety was reported in 5% of the patients in the control arm vs. 14% in the rimonabant arm. Depression (including depressed mood) was 7.5% in the control group vs. 14% in the rimonabant group. However, in this trial, a previous history of depression was not an exclusion criterion. On the other hand, similar numbers of severe hypoglycemia were reported with rimonabant 20 mg/day and control. In the SERENADE study, these adverse events occurred less frequently than in the ARPEGGIO study, i.e., anxiety was reported in 5.8% vs. 3.6% and depressed mood in 5.8% vs. 0.7% in rimonabant vs. the placebo groups.

As shown by these clinical trials, the metabolic effects of rimonabant are elicited in a way that seems to be only partially due to weight loss, in agreement with the hypothesis that CB_1 receptor antagonists target a potentially overactive endocannabinoid system acting directly on peripheral cells and organs. Further studies had been planned to explore the possibility that rimonabant, alone or in combination, can be used as a therapy to ensure glycemic control in patients with type 2 diabetes. For example, the safety and effectiveness of combining drugs with different mechanisms of action for the treatment of obesity and diabetes were under investigation in the REASSURE and ALLEGRO trials. In these trials, obese patients with type 2 diabetes would have received concomitant treatment with rimonabant, metformin, and sulfonylureas. In the REASSURE trial, rimonabant was being administered to subjects not adequately controlled on metformin and sulfonylures, whereas in the ALLEGRO trial, glimepiride treatment would have been compared with rimonabant plus metformin combination treatment (ClinicalTrials.gov Identifiers NCT00546325 and NCT00449605). However, these two trials have been interrupted due to the discontinuation of rimonabant clinical development by Sanofi-Aventis.

Indeed, during the preparation of this article the European Medicine Agency (EMEA) decided to suspend the marketing of Acomplia in the EU since it has become the opinion of the EMEA's Committee for Medicinal Products for Human Use (CHMP) that the benefits of this drug in its "real life" use (i.e., outside of the well-controlled boundaries of the clinical trial) "no longer outweigh its risks," possibly also due to the fact that "available data indicate that patients generally take Acomplia only for a short period." This decision prompted Sanofi-Aventis to interrupt, in the fall of 2008, the marketing and further clinical development of rimonabant and other pharmaceutical companies to suspend the development of their CB_1 receptor inverse agonists/antagonists. However, the suspension of Acomplia for obesity should not preclude the development of a second generation of CB_1 antagonists, for example those that, because of lesser penetration in the brain (see Tam et al. 2010, for a recent example), should be devoid of the psychiatric side effects that have led to the discontinuation of rimonabant. As discussed by Di Marzo & Després (2010), rather than global obesity, which is very often associated with depressive disorders, the indication of such novel compounds, as well as of other strategies aiming at reducing endocannabinoid overactivity, should be instead for those metabolic dysfunctions that are associated with abdominal obesity. For example, the administration of second-generation CB_1 receptor antagonists to moderately obese patients ($27 < BMI < 33$) with high IAA, low HDL cholesterol, high triglycerides, and high fasting glycemia, is likely to improve considerably the benefit/risk ratio of these compounds and result in long-term beneficial outcomes on type 2 diabetes.

6 Plant Cannabinoids and Type 1 Diabetes

Cannabis sativa is the unique source of a group of terpeno-phenols known as cannabinoids. (Mechoulam 1970; Hanus and Mechoulam 2008) Only two of them – Δ^9-THC and *cannabidiol* (CBD) – have been studied extensively. The propyl homologue of Δ^9-THC (named Δ^9-tetrahydrocannabivarin, Δ^9-THCV) has recently been found to act as cannabinoid antagonist (Mechoulam 2005; Thomas et al. 2005). While some of the relevant activities of Δ^9-THC are discussed above, the major nonpsychoactive constituent CBD has only recently become the object of detailed studies (Mechoulam and Hanus 2002; Mechoulam et al. 2002, 2007). This compound causes a myriad of pharmacological effects. Its binding to the cannabinoid receptors is extremely weak and its effects are apparently based on numerous mechanisms, such as inhibition of adenosine uptake, anti-oxidant activity, action on 5-HT receptors (Mechoulam et al. 2007), or on TRPV2 and TRPA1 channels (De Petrocellis et al. 2008; Qin et al. 2008) and possibly others. Its relevance to the topics discussed in this review is mainly its activity on diabetes type 1.

6.1 Type 1 Diabetes Mellitus

Type 1 diabetes mellitus (insulin-dependent diabetes) is an autoimmune disease resulting in destruction of insulin-producing pancreatic β-cells, a process which is assumed to be mediated mainly by CD4 Th1 and CD8 T lymphocytes (Atkinson and Leiter 1999; Mandrup-Poulsen 2003). The nonobese diabetes-prone (NOD) mouse is the animal used most often for preclinical evaluation of prophylactic and therapeutic treatments of type 1 diabetes. When NOD mice are not treated, they develop a disease with characteristics similar to autoimmune insulitis in humans (Anderson and Bluestone 2005). Insulitis is an inflammation of the islets of Langerhans and is the initial lesion of insulin-dependent diabetes mellitus, during which leukocytes, and lymphocytes in particular, surround and infiltrate the islets.

As CBD has been shown to suppress cell-mediated autoimmune joint destruction in an animal model of rheumatoid arthritis, and as diabetes type 1 has an autoimmune basis, its activity was studied on an animal model of this disease (Weiss et al. 2006). It was found that CBD treatment of 6- to 12-week-old NOD mice significantly reduces the incidence of diabetes from 86% in nontreated control mice to 30% in CBD-treated mice. CBD treatment also resulted in the significant reduction of plasma levels of the proinflammatory cytokines, IFN-γ and TNF-α. Th1-associated cytokine production of in vitro-activated T-cells and peritoneal macrophages was also significantly reduced in CBD-treated mice, whereas production of the Th2-associated cytokines, IL-4 and IL-10, was increased when compared to untreated control mice. Histological examination of the pancreatic islets of CBD-treated mice revealed significantly reduced insulitis. These results indicated that CBD can inhibit and delay destructive insulitis and inflammatory Th1-associated cytokine production in NOD mice, resulting in a decreased incidence of diabetes possibly through an immunomodulatory mechanism shifting the immune response from Th1 to Th2 dominance.

While numerous approaches have successfully been used to prevent diabetes type 1 in NOD young mice, the amelioration of diabetes after onset, or around onset time, is much more difficult to achieve. Importantly, however, some of the therapies initiated in this model around onset time were found to show efficacy in clinical trials. Hence, Weiss and coworkers repeated the CBD treatment in 11- to 14-week-old female NOD mice, which were either in a latent diabetes stage or with initial symptoms of diabetes. It was noted that CBD ameliorates the manifestations of the disease (Weiss et al. 2008). At the end of the treatment, diabetes was diagnosed in only 32% of the mice in the CBD-treated group, compared to 86 and 100% in the emulsifier-treated and untreated groups, respectively. In addition, the levels of the proinflammatory cytokine IL-12 produced by splenocytes was significantly reduced, whereas the levels of the anti-inflammatory IL-10 was significantly elevated following CBD treatment. Histological examination of the pancreas of CBD-treated mice revealed more intact islets than in the controls. These data strengthen the previous assumption that CBD, known to be safe in humans, can possibly be used as a therapeutic agent for treatment of type 1 diabetes.

7   Concluding Remarks

The endocannabinoid system has been in the center of interest in numerous fields over the last few years. In this review, we have summarized its involvement in feeding and energy metabolism. Originally, it was believed that the endocannabinoid system mostly affected appetite. Additional research, however, led to findings that strongly indicate its involvement in energy storage in adipose tissue and reduction of energy expenditure. It was also found that the endocannabinoid system affects both lipid and glucose metabolism. Its dysregulation is now known to cause effects leading to a number of pathological metabolic conditions, such as obesity, dyslipidemia, and type 2 diabetes. A drug, rimonabant, acting as an antagonist/inverse agonist at the CB_1 receptor was developed and used in the clinic mostly as an anti-obesity drug rather than to correct the metabolic effects of endocannabinoid dysregulation. Unfortunately, its side effects associated with the actions of the endocannabinoid system on mood led to its withdrawal. The possibility of using lower (and safer) doses of rimonabant and other second-generation CB_1 antagonists/inverse agonists was not so much to reduce total body weight, but instead a special type of obesity, known as abdominal ("central") obesity, and its direct consequence, i.e., dyslipidemia, as well as against prediabetes and type 2 diabetes still exists. Indeed, peripherally restricted CB_1 antagonists, which seem to be devoid of central side effects in rodents, are being developed (Tam et al. 2010). Finally, the use of nonpsychotropic plant cannabinoids, such as CBD, against type 1 diabetes should now be tested in specific clinical trials.

References

Anderson MS, Bluestone JA (2005) The NOD mouse: a model of immune dysregulation. Annu Rev Immunol 23:447–485

Annuzzi G, Piscitelli F, Di Marino L et al (2010) Differential alterations of the concentrations of endocannabinoids and related lipids in the subcutaneous adipose tissue of obese diabetic patients. Lipids Health Dis 9:43

Ashton JC, Friberg D, Darlington CL et al (2006) Expression of the cannabinoid CB2 receptor in the rat cerebellum: an immunohistochemical study. Neurosci Lett 396:113–116

Atkinson MA, Leiter EH (1999) The NOD mouse model of type 1 diabetes: as good as it gets? Nat Med 5:601–604

Bellocchio L, Cervino C, Vicennati V et al. (2008) Cannabinoid type 1 receptor: another arrow in the adipocytes' bow. J Neuroendocrinol Suppl 1:130–138 (Review)

Bensaid M, Gary-Bobo M, Esclangon A et al (2003) The cannabinoid CB1 receptor antagonist SR141716 increases Acrp30 mRNA expression in adipose tissue of obese fa/fa rats and in cultured adipocyte cells. Mol Pharmacol 63:908–914

Ben-Shabat S, Fride E, Sheskin T et al (1998) An entourage effect: inactive endogenous fatty acid glycerol esters enhance 2-arachidonoyl-glycerol cannabinoid activity. Eur J Pharmacol 353:23–31

Bermudez-Silva FJ, Serrano A, Diaz-Molina FJ et al (2006) Activation of cannabinoid CB(1) receptors induces glucose intolerance in rats. Eur J Pharmacol 531:282–284

Bermúdez-Silva FJ, Suárez J, Baixeras E et al (2008) Presence of functional cannabinoid receptors in human endocrine pancreas. Diabetologia 51:476–487

Bilfinger TV, Salzet M, Fimiani C, Deutsch DG, Tramu G, Stefano GB (1998) Pharmacological evidence for anandamide amidase in human cardiac and vascular tissues. Int J Cardiol 64 (Suppl 1):S15–S22

Bisogno T, Melck D, De Petrocellis L et al (1999) Phosphatidic acid as the biosynthetic precursor of the endocannabinoid 2-arachidonoylglycerol in intact mouse neuroblastoma cells stimulated with ionomycin. J Neurochem 72:2113–2119

Bisogno T, Howell F, Williams G et al (2003) Cloning of the first sn1-DAG lipases points to the spatial and temporal regulation of endocannabinoid signaling in the brain. J Cell Biol 163:463–468

Bluher M, Engeli S, Kloting N et al (2006) Dysregulation of the peripheral and adipose tissue endocannabinoid system in human abdominal obesity. Diabetes 55:3053–3060

Bouaboula M, Hilairet S, Marchand J et al (2005) Anandamide induced PPARgamma transcriptional activation and 3T3-L1 preadipocyte differentiation. Eur J Pharmacol 517:174–181

Burdyga G, Lal S, Varro A et al (2004) Expression of cannabinoid CB1 receptors by vagal afferent neurons is inhibited by cholecystokinin. J Neurosci 24:2708–2715

Cavuoto P, McAinch AJ, Hatzinikolas G et al (2007) Effects of cannabinoid receptors on skeletal muscle oxidative pathways. Mol Cell Endocrinol 267:63–69

Chambers AP, Sharkey KA, Koopmans HS (2004) Cannabinoid (CB)1 receptor antagonist, AM 251, causes a sustained reduction of daily food intake in the rat. Physiol Behav 82:863–869

Colombo G, Agabio R, Diaz G et al (1998) Appetite suppression and weight loss after the cannabinoid antagonist SR 141716. Life Sci 63:113–117

Cota D, Marsicano G, Tschop M et al (2003) The endogenous cannabinoid system affects energy balance via central orexigenic drive and peripheral lipogenesis. J Clin Invest 112:423–431

Cota D, Tschop MH, Horvath TL et al (2006) Cannabinoids, opioids and eating behavior: the molecular face of hedonism? Brain Res Rev 51:85–107

Cota D, Steiner MA, Marsicano G et al (2007) Requirement of cannabinoid receptor type 1 for the basal modulation of hypothalamic-pituitary-adrenal axis function. Endocrinology 148:1574–1581

Cote M, Matias I, Lemieux I et al (2007) Circulating endocannabinoid levels, abdominal adiposity and related cardiometabolic risk factors in obese men. Int J Obes (Lond) 31:692–699

Coutts AA, Izzo AA (2004) The gastrointestinal pharmacology of cannabinoids: an update. Curr Opin Pharmacol 4:572–579

Cravatt BF, Giang DK, Mayfield SP et al (1996) Molecular characterization of an enzyme that degrades neuromodulatory fatty-acid amides. Nature 384:83–87

De Petrocellis L, Marini P, Matias I et al (2007) Mechanisms for the coupling of cannabinoid receptors to intracellular calcium mobilization in rat insulinoma beta-cells. Exp Cell Res 313:2993–3004

De Petrocellis L, Vellani V, Schiano-Moriello A et al (2008) Plant-derived cannabinoids modulate the activity of transient receptor potential channels of ankyrin type-1 and melastatin type-8. J Pharmacol Exp Ther 325:1007–1015

D'Eon TM, Pierce KA, Roix JJ et al (2008) The role of adipocyte insulin resistance in the pathogenesis of obesity-related elevations in endocannabinoids. Diabetes 57:1262–1268

Després JP, Lemieux I (2006) Abdominal obesity and metabolic syndrome. Nature 444:881–887

Després JP, Golay A, Sjöström L et al (2005) Effects of rimonabant on metabolic risk factors in overweight patients with dyslipidemia. N Engl J Med 353:2121–2134

Devane WA, Dysarz FA, Johnson MR et al (1988) Determination and characterization of a cannabinoid receptor in rat brain. Mol Pharmacol 34:605–613

Devane WA, Hanus L, Breuer A et al (1992) Isolation and structure of a brain constituent that binds to the cannabinoid receptor. Science 258:1946–1949

Di Marzo V (2008a) Targeting the endocannabinoid system: to enhance or reduce? Nat Rev Drug Discov 7:438–455

Di Marzo V (2008b) The endocannabinoid system in obesity and type 2 diabetes. Diabetologia 51:1356–1367

Di Marzo V, Després JP (2010) CB1 antagonists for obesity–what lessons have we learned from rimonabant? Nat Rev Endocrinol 5:633–638

Di Marzo V, De Petrocellis L, Sugiura T et al (1996) Potential biosynthetic connections between the two cannabimimetic eicosanoids, anandamide and 2-arachidonoyl-glycerol, in mouse neuroblastoma cells. Biochem Biophys Res Commun 227:281–288

Di Marzo V, Sepe N, De Petrocellis L et al (1998) Trick or treat from food endocannabinoids? Nature 396:636–637

Di Marzo V, Goparaju SK, Wang L et al (2001) Leptin-regulated endocannabinoids are involved in maintaining food intake. Nature 410:822–825

Di Marzo V, Bifulco M, De Petrocellis L (2004) The endocannabinoid system and its therapeutic exploitation. Nat Rev Drug Discov 3:771–784

Di Marzo V, De Petrocellis L, Di Marzo V, De Petrocellis L (2005) Non-CB1, non-CB2 receptors for endocannabinoids. In: Onaivi ES, Sugiura T, Di Marzo V (eds) Endocannabinoids: The Brain and Body's Marijuana and Beyond. CRC Press, Taylor & Francis Group, Boca Raton, FL, pp 151–174

Di Marzo V, Côté M, Matias I et al (2009a) Changes in plasma endocannabinoid levels in viscerally obese men following a one-year lifestyle modification program and waist circumference reduction: associations with changes in metabolic risk factors. Diabetologia 52:213–217

Di Marzo V, Verrijken A, Hakkarainen A et al (2009b) Role of insulin as a negative regulator of plasma endocannabinoid levels in obese and nonobese subjects. Eur J Endocrinol 161:715–722

Dinh TP, Carpenter D, Leslie FM et al (2002) Brain monoglyceride lipase participating in endocannabinoid inactivation. Proc Natl Acad Sci USA 99:10819–10824

DiPatrizio NV, Simansky KJ (2008a) Activating parabrachial cannabinoid CB1 receptors selectively stimulates feeding of palatable foods in rats. J Neurosci 28:9702–9709

Dipatrizio NV, Simansky KJ (2008) Inhibiting parabrachial fatty acid amide hydrolase activity selectively increases the intake of palatable food via cannabinoid CB1 receptors. Am J Physiol Regul Integr Comp Physiol [Epub ahead of print]

Doyon C, Denis RG, Baraboi ED et al (2006) Effects of rimonabant (SR141716) on fasting-induced hypothalamic-pituitary-adrenal axis and neuronal activation in lean and obese Zucker rats. Diabetes 55:3403–3410

Ellis J, Pediani JD, Canals M et al (2006) Orexin-1 receptor-cannabinoid CB1 receptor heterodimerization results in both ligand-dependent and -independent coordinated alterations of receptor localization and function. J Biol Chem 281:38812–38824

Engeli S, Bohnke J, Feldpausch M et al (2005) Activation of the peripheral endocannabinoid system in human obesity. Diabetes 54:2838–2843

Esposito I, Proto MC, Gazzerro P et al (2008) The cannabinoid CB1 receptor antagonist Rimonabant stimulates 2-deoxyglucose uptake in skeletal muscle cells by regulating phosphatidylinositol-3-kinase activity. Mol Pharmacol 74(6):1678–1686

Fride E, Ginzburg Y, Breuer A et al (2001) Critical role of the endogenous cannabinoid system in mouse pup suckling and growth. Eur J Pharmacol 419:207–214

Gallate JE, Saharov T, Mallet PE et al (1999) Increased motivation for beer in rats following administration of a cannabinoid CB1 receptor agonist. Eur J Pharmacol 370:233–240

Gaoni Y, Mechoulam R (1964) Isolation, structure, and partial synthesis of an active constituent of hashish. J Am Chem So 86:1646–1647

Gary-Bobo M, Elachouri G, Scatton B et al (2006) The cannabinoid CB1 receptor antagonist rimonabant (SR141716) inhibits cell proliferation and increases markers of adipocyte maturation in cultured mouse 3T3 F442A preadipocytes. Mol Pharmacol 69:471–478

Gary-Bobo M, Elachouri G, Gallas JF et al (2007) Rimonabant reduces obesity-associated hepatic steatosis and features of metabolic syndrome in obese Zucker fa/fa rats. Hepatology 46:122–129

Gasperi V, Fezza F, Pasquariello N et al (2007) Endocannabinoids in adipocytes during differentiation and their role in glucose uptake. Cell Mol Life Sci 64:219–229

Getty-Kaushik L, Richard A-MT, Deeney JT, Shirihai O, Corkey B (2009) The CB1 antagonist, rimonabant, decreases insulin hypersecretion in rat pancreatic islets. Obesity 17:1856–1860

Glass M, Northup JK (1999) Agonist selective regulation of G proteins by cannabinoid CB(1) and CB(2) receptors. Mol Pharmacol 56:1362–1369

Gomez R, Navarro M, Ferrer B et al (2002) A peripheral mechanism for CB1 cannabinoid receptor-dependent modulation of feeding. J Neurosci 22:9612–9617

Gong JP, Onaivi ES, Ishiguro H et al (2006) Cannabinoid CB2 receptors: immunohistochemical localization in rat brain. Brain Res 1071:10–23

Gonzalez S, Manzanares J, Berrendero F et al (1999) Identification of endocannabinoids and cannabinoid CB(1) receptor mRNA in the pituitary gland. Neuroendocrinology 70:137–145

Hanus L, Mechoulam R (2008) Plant and brain cannabinoids: The chemistry of major new players in physiology. In: Ikan R (ed) Selected topics in the chemistry of natural products. World Scientific Publishing Company. Imperial College Press, London, pp 49–75

Hanus L, Avraham Y, Ben-Shushan D et al (2003) Short term fasting and prolonged semistarvation have opposite effect on 2-AG levels in mouse brain. Brain Res 983:144–151

Herling AW, Gossel M, Haschke G et al (2007) CB1 receptor antagonist AVE1625 affects primarily metabolic parameters independently of reduced food intake in Wistar rats. Am J Physiol Endocrinol Metab 293:E826–E832

Herling AW, Kilp S, Elvert R et al (2008a) Increased energy expenditure contributes more to the body weight-reducing effect of rimonabant than reduced food intake in candy-fed wistar rats. Endocrinology 149:2557–2566

Herling AW, Kilp S, Juretschke HP et al (2008b) Reversal of visceral adiposity in candy-diet fed female Wistar rats by the CB1 receptor antagonist rimonabant. Int J Obes (Lond) 32:1363–1372

Hilairet S, Bouaboula M, Carrière D et al (2003) Hypersensitization of the Orexin 1 receptor by the CB1 receptor: evidence for cross-talk blocked by the specific CB1 antagonist, SR141716. J Biol Chem 278:23731–23737

Hollander PA, Amod A, Litwak LE, Chaudhari U; ARPEGGIO Study Group (2010) Effect of rimonabant on glycemic control in insulin-treated type 2 diabetes: the ARPEGGIO trial. Diabetes Care 33:605–607

Izzo AA, Piscitelli F, Capasso R et al (2009) Peripheral endocannabinoid dysregulation in two experimental models of obesity: potential relationships with intestinal motility and food deprivation/refeeding-induced energy processing. Br J Pharmacol 158:451–461

Janiak P, Poirier B, Bidouard JP et al (2007) Blockade of cannabinoid CB1 receptors improves renal function, metabolic profile, and increased survival of obese Zucker rats. Kidney Int 72 (11):1345–1357

Jbilo O, Ravinet-Trillou C, Arnone M et al (2005) The CB1 receptor antagonist rimonabant reverses the diet-induced obesity phenotype through the regulation of lipolysis and energy balance. FASEB J 19:1567–1569

Jo YH, Chen YJ, Chua SC Jr et al (2005) Integration of endocannabinoid and leptin signaling in an appetite-related neural circuit. Neuron 48:1055–1066

Juan-Pico P, Fuentes E, Bermudez-Silva FJ et al (2006) Cannabinoid receptors regulate Ca(2+) signals and insulin secretion in pancreatic beta-cell. Cell Calcium 39:155–162

Kirkham TC, Williams CM, Fezza F et al (2002) Endocannabinoid levels in rat limbic forebrain and hypothalamus in relation to fasting, feeding and satiation: stimulation of eating by 2-arachidonoyl glycerol. Br J Pharmacol 136:550–557

Klein TW (2005) Cannabinoid-based drugs as anti-inflammatory therapeutics. Nat Rev Immunol 5:400–411

Koch JE (2001) Delta(9)-THC stimulates food intake in Lewis rats: effects on chow, high-fat and sweet high-fat diets. Pharmacol Biochem Behav 68:539–543

Kola B, Hubina E, Tucci SA et al (2005) Cannabinoids and ghrelin have both central and peripheral metabolic and cardiac effects via AMP-activated protein kinase. J Biol Chem 280:25196–25201

Kola B, Farkas I, Christ-Crain M et al (2008) The orexigenic effect of ghrelin is mediated through central activation of the endogenous cannabinoid system. PLoS ONE 3:1797

Kunz I, Meier MK, Bourson A et al (2008) Effects of rimonabant, a cannabinoid CB1 receptor ligand, on energy expenditure in lean rats. Int J Obes (Lond) 32:863–70

Leung D, Saghatelian A, Simon GM et al (2006) Inactivation of N-acyl phosphatidylethanolamine phospholipase d reveals multiple mechanisms for the biosynthesis of endocannabinoids. Biochemistry 45:4720–4726

Liu YL, Connoley IP, Wilson CA et al (2005) Effects of the cannabinoid CB1 receptor antagonist SR141716 on oxygen consumption and soleus muscle glucose uptake in Lep(ob)/Lep(ob) mice. Int J Obes (Lond) 29:183–187

Maccarrone M, Di Rienzo M, Finazzi-Agro A et al (2003) Leptin activates the anandamide hydrolase promoter in human T lymphocytes through STAT3. J Biol Chem 278:13318–13324

Maccarrone M, Fride E, Bisogno T et al (2005) Up-regulation of the endocannabinoid system in the uterus of leptin knockout (ob/ob) mice and implications for fertility. Mol Hum Reprod 11:21–28

Malcher-Lopes R, Di S, Marcheselli VS et al (2006) Opposing crosstalk between leptin and glucocorticoids rapidly modulates synaptic excitation via endocannabinoid release. J Neurosci 26:6643–6650

Mandrup-Poulsen T (2003) Beta cell death and protection. Ann NY Acad Sci 1005:32–42

Matias I, Di Marzo V (2007) Endocannabinoids and the control of energy balance. Trends Endocrinol Metab 18:27–37

Matias I, Gonthier MP, Orlando P et al (2006) Regulation, function and dysregulation of endo-cannabinoids in models of adipose and β-pancreatic cells and in obesity and hyperglycemia. J Clin Endocrinol Metab 91:3171–80

Matias I, Petrosino S, Racioppi A et al (2008) Dysregulation of peripheral endocannabinoid levels in hyperglycemia and obesity: Effect of high fat diets. Mol Cell Endocrinol 286:S66–78

McAllister SD, Glass M (2002) CB(1) and CB(2) receptor-mediated signaling: a focus on endocannabinoids. Prostaglandins Leukot Essent Fatty Acids 66:161–171

McLaughlin PJ, Winston K, Swezey L et al (2003) The cannabinoid CB1 antagonists SR 141716A and AM 251 suppress food intake and food-reinforced behavior in a variety of tasks in rats. Behav Pharmacol 14:583–588

Mechoulam R (1970) Marihuana chemistry. Science 168:1159–66

Mechoulam R (2005) Plant cannabinoids: a neglected pharmacological treasure trove. Br J Pharmacol 146:913–915

Mechoulam R, Hanus L (2002) Cannabidiol: An overview of some chemical and pharmacological aspects. Part I: Chemical Aspects Chem Phys Lipids 121:35–43

Mechoulam R, Ben-Shabat S, Hanus L et al (1995) Identification of an endogenous 2-monoglyc-eride, present in canine gut, that binds to cannabinoid receptors. Biochem Pharmacol 50:83–90

Mechoulam R, Parker LA, Gallily R (2002) Cannabidiol: An overview of some pharmacological aspects. J Clin Pharmacol 42:11S–19S

Mechoulam R, Berry EM, Avraham Y et al (2006) Endocannabinoids, feeding and suckling – from our perspective. Int J Obes (Lond) 30:S24–S28

Mechoulam R, Peters M, Murillo-Rodriguez E et al (2007) Cannabidiol – recent advances. Chem Biodivers 4:1678–1692

Monteleone P, Matias I, Martiadis V et al (2005) Blood levels of the endocannabinoid anandamide are increased in anorexia nervosa and in binge-eating disorder, but not in bulimia nervosa. Neuropsychopharmacology 30:1216–1221

Munro S, Thomas KL, Abu-Shaar M (1993) Molecular characterization of a peripheral receptor for cannabinoids. Nature 365:61–65

Murdolo G, Kempf K, Hammarstedt A, Murdolo G, Kempf K, Hammarstedt A et al (2007) Insulin differentially modulates the peripheral endocannabinoid system in human subcutaneous abdominal adipose tissue from lean and obese individuals. J Endocrinol Invest 30:RC17–RC21

Nogueiras R, Veyrat-Durebex C, Suchanek PM et al (2008) Peripheral, but not central, CB1 antagonism provides food intake independent metabolic benefits in diet-induced obese rats. Diabetes 57(11):2977–2991

Okamoto Y, Morishita J, Tsuboi K et al (2004) Molecular characterization of a phospholipase D generating anandamide and its congeners. J Biol Chem 279:5298–5305

Osei-Hyiaman D, Depetrillo M, Harvey-White J et al (2005a) Cocaine- and amphetamine-related transcript is involved in the orexigenic effect of endogenous anandamide. Neuroendocrinology 81:273–282

Osei-Hyiaman D, DePetrillo M, Pacher P et al (2005b) Endocannabinoid activation at hepatic CB1 receptors stimulates fatty acid synthesis and contributes to diet-induced obesity. J Clin Invest 115:1298–1305

Osei-Hyiaman D, Liu J, Zhou L et al (2008) Hepatic CB1 receptor is required for development of diet-induced steatosis, dyslipidemia, and insulin and leptin resistance in mice. J Clin Invest 118:3160–3169

Pagano C, Pilon C, Calcagno A et al (2007) The endogenous cannabinoid system stimulates glucose uptake in human fat cells via PI3-kinase and calcium-dependent mechanisms. J Clin Endocrinol Metab 92(12):4810–4819

Pagotto U, Marsicano G, Cota D et al (2006) The emerging role of the endocannabinoid system in endocrine regulation and energy balance. Endocr Rev 27:73–100

Partosoedarso ER, Abrahams TP, Scullion RT et al (2003) Cannabinoid1 receptor in the dorsal vagal complex modulates lower oesophageal sphincter relaxation in ferrets. J Physiol 550:149–158

Pi-Sunyer FX, Aronne LJ, Heshmati HM et al (2006) Effect of rimonabant, a cannabinoid-1 receptor blocker, on weight and cardiometabolic risk factors in overweight or obese patients: RIO-North America: a randomized controlled trial. JAMA 295:761–775

Poirier B, Bidouard JP, Cadrouvele C et al (2005) The anti-obesity effect of rimonabant is associated with an improved serum lipid profile. Diabetes Obes Metab 7:65–72

Qin N, Neeper MP, Liu Y et al (2008) TRPV2 is activated by cannabidiol and mediates CGRP release in cultured rat dorsal root ganglion neurons. J Neurosci 28:6231–6238

Ravinet Trillou C, Arnone M, Delgorge C et al (2003) Anti-obesity effect of SR141716, a CB1 receptor antagonist, in diet-induced obese mice. Am J Physiol Regul Integr Comp Physiol 284:345–353

Ravinet Trillou C, Delgorge C, Menet C et al (2004) CB1 cannabinoid receptor knockout in mice leads to leanness, resistance to diet-induced obesity and enhanced leptin sensitivity. Int J Obes Relat Metab Disord 28:640–648

Rinaldi-Carmona M, Barth F, Heaulme M et al (1994) SR141716A, a potent and selective antagonist of the brain cannabinoid receptor. FEBS Lett 350:240–244

Robbe D, Kopf M, Remaury A et al (2002) Endogenous cannabinoids mediate long-term synaptic depression in the nucleus accumbens. Proc Natl Acad Sci USA 99:8384–8388

Roche R, Hoareau L, Bes-Houtmann S et al (2006) Presence of the cannabinoid receptors, CB1 and CB2, in human omental and subcutaneous adipocytes. Histochem Cell Biol 4:1–11

Rosenstock J, Hollander P, Chevalier S et al (2008) SERENADE trial: effects of monotherapy with rimonabant, the first selective cb1 receptor antagonist, on glycemic control, body weight and lipid profile in drug-naive type 2 diabetes. Diabetes Care 31(11):2169–2176

Ruby MA, Nomura DK, Hudak CS et al (2008) Overactive endocannabinoid signaling impairs apolipoprotein E-mediated clearance of triglyceride-rich lipoproteins. Proc Natl Acad Sci USA 105:14561–14566

Schäfer A, Pfrang J, Neumüller J et al (2008) The cannabinoid receptor-1 antagonist rimonabant inhibits platelet activation and reduces pro-inflammatory chemokines and leukocytes in Zucker rats. Br J Pharmacol 154:1047–1054

Scheen AJ, Finer N, Hollander P et al (2006) Efficacy and tolerability of rimonabant in overweight or obese patients with type 2 diabetes: a randomised controlled study. Lancet 368:1660–1672

Schlicker E, Kathmann M (2001) Modulation of transmitter release via presynaptic cannabinoid receptors. Trends Pharmacol Sci 22:565–572

Soria-Gómez E, Matias I, Rueda-Orozco PE et al (2007) Pharmacological enhancement of the endocannabinoid system in the nucleus accumbens shell stimulates food intake and increases c-Fos expression in the hypothalamus. Br J Pharmacol 151:1109–1116

Starowicz K, Cristino L, Matias I et al (2008) Endocannabinoid dysregulation in the pancreas and adipose tissue of mice fed a high fat diet. Obesity 16(3):553–565

Sugiura T, Kondo S, Sukagawa A et al (1995) 2-Arachidonoylglycerol: a possible endogenous cannabinoid receptor ligand in brain. Biochem Biophys Res Commun 215:89–97

Sun YX, Tsuboi K, Okamoto Y et al (2004) Biosynthesis of anandamide and N-palmitoylethanolamine by sequential actions of phospholipase A2 and lysophospholipase D. Biochem J 380:749–756

Tam J, Vemuri VK, Liu J et al (2010) Peripheral CB1 cannabinoid receptor blockade improves cardiometabolic risk in mouse models of obesity. J Clin Invest 120:2953–2966

Thomas EA, Cravatt BF, Danielson PE et al (1997) Fatty acid amide hydrolase, the degradative enzyme for anandamide and oleamide, has selective distribution in neurons within the rat central nervous system. J Neurosci Res 50:1047–1052

Thomas A, Stevenson LA, Wease KN et al (2005) Evidence that the plant cannabinoid Δ9-tetrahydrocannabivarin is a cannabinoid CB1 and CB2 receptor antagonist. Br J Pharmacol 146:917–926

Tsou K, Nogueron MI, Muthian S et al (1998) Fatty acid amide hydrolase is located preferentially in large neurons in the rat central nervous system as revealed by immunohistochemistry. Neurosci Lett 254:137–140

Tucci SA, Rogers EK, Korbonits M et al (2004) The cannabinoid CB1 receptor antagonist SR141716 blocks the orexigenic effects of intrahypothalamic ghrelin. Br J Pharmacol 143:520–533

Van Gaal LF, Rissanen AM, Scheen AJ et al (2005) Effects of the cannabinoid-1 receptor blocker Rimonabant on weight reduction and cardiovascular risk factors in overweight patients: 1-year experience from the RIO-Europe study. Lancet 365:1389–1397

Van Gaal L, Pi-Sunyer X, Després JP et al (2008) Efficacy and safety of rimonabant for improvement of multiple cardiometabolic risk factors in overweight/obese patients: pooled 1-year data from the Rimonabant in Obesity (RIO) program. Diabetes Care 31:S229–240

Van Sickle MD, Duncan M, Kingsley PJ et al (2005) Identification and functional characterization of brainstem cannabinoid CB2 receptors. Science 310:329–332

Weiss L, Zeira M, Reich S et al (2006) Cannabidiol lowers incidence of diabetes in non-obese diabetic mice. Autoimmunity 39:143–151

Weiss L, Zeira M, Reich S et al (2008) Cannabidiol arrests onset of autoimmune diabetes in NOD mice. Neuropharmacol 54:244–249

Williams CM, Kirkham TC (1999) Anandamide induces overeating: mediation by central cannabinoid (CB1) receptors. Psychopharmacol (Berl) 143:315–317

Williams CM, Rogers PJ, Kirkham TC (1998) Hyperphagia in pre-fed rats following oral delta⁹-THC. Physiol Behav 65:343–346

Wilson RI, Nicoll RA (2002) Endocannabinoid signaling in the brain. Science 296:678–682

Yan ZC, Liu DY, Zhang LL et al (2007) Exercise reduces adipose tissue via cannabinoid receptor type 1 which is regulated by peroxisome proliferator-activated receptor-delta. Biochem Biophys Res Commun 354:427–33

Zimmer A, Zimmer AM, Hochmann AG (1999) Increased mortality, hypoactivity and hypoalgesia in cannabinoid CB1 receptor knockout mice. PNAS 96:5780–5785

SGLT Inhibitors as New Therapeutic Tools in the Treatment of Diabetes

Rolf K.H. Kinne and Francisco Castaneda

Contents

Abstract Recently, the idea has been developed to lower blood glucose levels in diabetes by inhibiting sugar reabsorption in the kidney. The main target is thereby the early proximal tubule where secondary active transport of the sugar is mediated by the sodium-D-glucose cotransporter SGLT2. A model substance for the inhibitors is the O-glucoside phlorizin which inhibits transport competitively. Its binding to the transporter involves at least two different domains: an aglucone binding site at the transporter surface, involving extramembranous loops, and the sugar binding/ translocation site buried in a hydrophilic pocket of the transporter. The properties of

R.K.H. Kinne (✉) and F. Castaneda
Max-Planck-Institute of Molecular Physiology, Otto-Hahn-Str. 11, 44227 Dortmund, Germany
e-mail: rolf.kinne@mpi-dortmund.mpg.de

M. Schwanstecher (ed.), *Diabetes - Perspectives in Drug Therapy*,
Handbook of Experimental Pharmacology 203,
DOI 10.1007/978-3-642-17214-4_5, © Springer-Verlag Berlin Heidelberg 2011

these binding sites differ between SGLT2 and SGLT1, which mediates sugar absorption in the intestine. Various O-, C-, N- and S-glucosides have been synthesized with high affinity and high specificity for SGLT2. Some of these glucosides are in clinical trials and have been proven to successfully increase urinary glucose excretion and to decrease blood sugar levels without the danger of hypoglycaemia during fasting in type 2 diabetes.

Keywords Blood glucose control · Diabetes · Glucosides · Renal sugar transport · Sodium-ᴅ-glucose cotransporter

1 Preface

One of the main features of diabetes is the elevation of blood sugar with its deleterious consequences in a variety of tissues (Ceriello 2005). Thus, control of the plasma glucose level is of utmost importance in the treatment of this disease. In recent years, the idea has evolved that affecting glucose absorption in the intestine and/or the glucose reabsorption in the kidney might be a possible way to control the sugar level. Therefore, initially inhibitors of sugar absorption have been developed which inhibit the hydrolysis of sucrose and lactose by disaccharidases in the intestinal lumen. Examples that have been successfully introduced in the market are acarbose (Precose R or Glucobay R), voglibose (Basen R) and miglitol (Glyset R) (Asano 2003; de Melo et al. 2006). As the molecular understanding of sugar transport progressed, inhibitors of the transport molecule itself have been synthesized, some of which are currently undergoing preclinical and clinical testing. The main emphasis was thereby placed on specific inhibitors of the sugar reabsorption in the kidney. The following chapter briefly reviews the cellular and molecular basis of transepithelial sugar transport. It then summarizes the properties of the lead substance phlorizin and defines the pharmacologically important regions of the molecule. A description of the "phlorizin receptor" and its binding sites for phlorizin follows. Then screening procedures for sodium-ᴅ-glucose cotransporter (SGLT) inhibitors are introduced and the results of preclinical and clinical tests are compiled. In a synopsis, the current state of art in the development of SGLT inhibitors and their potential therapeutic use are discussed.

2 The Sodium-ᴅ-Glucose Cotransporter as Target

2.1 Role of Sodium-ᴅ-Glucose Cotransport in Transepithelial Sugar Transport

The SGLT plays a pivotal role in the translocation of sugars across epithelial membranes. In the small intestine and the renal tubule, transport of the sugar is active and requires the coupling of cellular energy metabolism to the transepithelial translocation. The SGLT is the site where such coupling occurs. The coupling is not direct, i.e. there is no hydrolysis of ATP involved as in other so-called primary active transport events such as those mediated by ion translocating ATPases. Instead, the transporter uses the energy "stored" in an ion gradient to transport sugars against their concentration difference. Such processes are called secondary active and are used widely in unicellular and multicellular organisms. In mammalian species, the most prominent ion whose gradient across the cell membrane is used as driving force is sodium. The secondary active transport of organic substances – sugars, amino acids, carboxylic acids and inorganic ions such as chloride and phosphate involves the simultaneous movement of sodium ions – one, two or three – mostly in the same direction as the substrate in a symport mode. For vectorial transcellular transport also an asymmetry of the cell must be established, so that the plasma membrane facing one compartment has to contain different transporters than the membrane facing the other compartment into which translocation occurs (Kinne 1991).

Both of these elements are incorporated in current models on active transepithelial sugar transport. Only the apical membrane of the epithelial cell (termed brush border in the small intestine and in the renal proximal tubule) contains the SGLT. The sodium gradient across the brush border is generated by the sodium–potassium stimulated ATPase, a primary active ion transport ATPase which removes sodium from the cell interior in exchange to potassium ions. Thus, D-glucose can be accumulated in the epithelial cells uphill from the intestinal or renal tubular lumen above the sugar concentration in the blood. The sugar leaves the cell along its concentration difference in a carrier-mediated, sodium-independent, passive movement (Wright et al. 2007).

Sugar absorption in the human gut occurs in the first segments and there it has a high affinity and high velocity. As of the colon, sugar absorption ceases and sugar required for the intracellular metabolism enters the intestinal cells from the cell side exposed to the blood. Studies on the presence of SGLT1 mRNA and protein expression in the various intestinal segments confirmed and extended the knowledge on the intra-intestinal distribution. It should be noted that sugar transport in the intestine is also a mode to absorb sodium across the epithelium; therefore, the enteral application of a sodium–sugar solution is one of the most effective ways to compensate for the fluid and electrolyte loss in diarrhoea (Wright et al. 2007).

The other main organ where active sugar transport occurs is the kidney. In the early part of the proximal tubule, bulk reabsorption of filtered glucose against a small gradient occurs, in the late part residual glucose is removed against a steep concentration difference. Transport studies in the early and late segments of the proximal tubule as well as vesicle studies showed that the kidney contains two D-glucose cotransporters which differ in their stoichiometry for sodium in the early proximal tubule, one sodium and one sugar molecule are translocated together across the luminal membrane whereas in the late part two sodium ions are translocated with one sugar molecule. The two transporters also differ in their affinity for D-glucose; in the early part the apparent transport affinity (K_m) is about 2 mM, in the late part the affinity is higher – K_m less than 0.5 mM. The two SGLTs also exhibit a different substrate specificity – distinction between D-galactose in the early part but not in the late part (Burckhardt and Kinne 1992).

The transporter in the late part is similar to the one found in the intestine termed SGLT1 whereas the one in the early proximal tubule is now referred to as SGLT2. The inherited disease of familial renal glucosuria can indeed be traced to a lack in SGLT2 in the early proximal tubule (Feld 2001; Calado et al. 2008). Interestingly, in these patients the reabsorptive capacity of the late proximal tubule seems to suffice to maintain a normal plasma glucose level. Thus, the only symptom is the increased urinary glucose excretion (Kleta et al. 2004). A lack of SGLT1 causes glucose–galactose malabsorption in the intestine with severe diarrhoea and salt and fluid loss (Wright et al. 2007).

2.2 Molecular Basis of Sodium-D-Glucose Cotransport

In 1987, Ernie Wright and his associates were able to identify the genetic message coding for the human intestinal SGLT by expression cloning using functional expression in oocytes and fractionation of the mRNA as a tool (Hediger et al. 1987). The gene codes for a protein of ~75 kDa which is heavily glycosylated. The glycosylation apparently does not affect its function but is probably related to its presence in the luminal brush border membrane, which is covered by an extensive glycocalyx.

Availability of the SGLT1 gene led thereafter to the identification of an identical gene in the kidney (Morrison et al. 1991) and to the similarity cloning of SGLT2 (Kanai et al. 1994). It also allowed performing mutagenesis studies to further elucidate the substrate binding sites, phosphorylation sites and inhibitor binding sites (Wright et al. 2007). In addition, overexpression in cells could be achieved which in the end led to the isolation of the transporter after heterologous expression in yeast (Tyagi et al. 2005).

The molecule responsible for the sodium-independent translocation of sugars has also been identified – there is a whole GLUT family whose properties have been reviewed recently. There is only a very limited sequence homology between SGLT and GLUT (Stuart and Trayhurn 2003).

Knowledge of the sequence also allowed for prediction and experimental analysis of the membrane topology of the transporter. The results of these studies consistently show that the N-terminus is pointing to the outside of the cell and that the molecule has 14 transmembrane segments. A major discrepancy exists concerning the topology at the C-terminus. Wright and associates position the loop between the 13th and 14th transmembrane segment (loop 13–14) into the cytoplasm (Hediger et al. 1987) whereas Puntheeranurak et al. have evidence for a location of at least the late part of the loop on the extracellular surface of the transporter (Puntheeranurak et al. 2006). The location of loop 13–14 is essential because, as will be shown later, this loop appears to be part of the phlorizin binding site of the transporter.

The C-terminus also constitutes the transmembrane segments X–XIII of the transporter which are involved in sugar recognition and sugar translocation (Wright et al. 2007); the N-terminal segments IV–V are involved in sodium recognition as well as glucose binding and coupling of the sodium gradient to the sugar movement during cotransport.

2.3   Sugar Binding Sites of the SGLT

2.3.1   Substrate Specificity of the Sodium-D-Glucose Cotransporters

Extensive investigations on the essential features of the D-glucose molecule were undertaken to define the substrate specificity of transport. Only sugars in a hexose and the D-configuration are translocated whereas fructose as well as L-glucose and other sugars with the L-conformation are not transported. The positioning of the hydroxyl group at C1 is important as beta sugars are transported more avidly than alpha sugars. Short aliphatic and aromatic residues as in alpha-methyl D-glucose and arbutin (beta-phenyl D-glucoside) are also tolerated (Wright et al. 2007). Beta-glucosides with larger aromatic aglucones and the aglucones themselves are bound to the transporter and not translocated; they act as inhibitors of the transport.

The hydroxyl group at C2 has to be present in an equatorial position, thus 2-deoxy D-glucose, mannose and N-glucosamine are not transported. The presence of the hydroxyl group at C3 and its positioning has also some effect on transport, but 3-deoxy D-glucose, allose and 3-O-methyl D-glucose are transported although with a lower affinity and velocity. Galactose with a modification at the hydroxyl group at C4 is transported by SGLT1 but only to a limited extent by SGLT2. The hydroxyl group at C6 is of minor importance, 6-deoxy-glucose is transported by both SGLTs. At position C6, the SGLT2 seems to have an interesting additional hydrophobic binding site with a rather high binding capacity. Thus, C6 alkyl residues have a higher affinity for SGLT2 than for SGLT1 (Kipp et al. 1997).

It has to be pointed out that the substrate specificity at the inner or cytoplasmic side of the carrier is different. At the cytoplasmic side, the selectivity with regard to L-glucose is reduced, the affinity for sugars is very much lower and the sodium sensitivity is also considerably decreased (Firnges et al. 2001).

There are differences in the substrate specificity of the overall transport process and the initial binding events at the outer surface of the transporter that precede the translocation. At the initial binding site, the primary sorting of the sugars according to the D- or L-configuration and the presence of the hydroxyl group at C3 occurs. At the entry to the translocation step, a further selection of the sugars with regard to the position and presence of the OH-groups at the other C-atoms takes place (Puntheeranurak et al. 2007b; Tyagi et al. 2011).

O-glucosides

1

2

3

4

5

6

7

8

C-arylglucosides

9

10

S-glycosides

11

12

Fig. 1 Chemical structure of phlorizin and some SGLT inhibitors including O-glucosides, C-arylglucosides and S-glycosides. For details see text

2.3.2 Sugar Binding Site(s) of the Sodium-D-Glucose Cotransporter

According to mutagenesis studies, amino acids involved in sugar transport are clustered in transmembrane segments X, XI, XII and XIII. These amino acids are accessible from the outside of the membrane and are located in a hydrophilic pocket (Tyagi et al. 2007). The binding pocket has a depth of about 7 Å. A recent publication on the structure of an archetype of a sugar sodium cotransporter confirms the assumption of the presence of a hydrophilic pocket in the outside-facing conformation and in the inside-facing conformation. The gates between these two pockets are formed by clusters of hydrophobic amino acids (Faham et al. 2008).

Atomic force microscopy (AFM) studies combined with accessibility and muta-genesis studies have recently shown that disulfide bridges between extramembra-nous loops are also important in forming the initial sugar binding site at the carrier. Thus, extramembranous loops 6–7 and 13–14 seem to be interconnected by a disulfide bridge to form a binding pocket which also brings loop 8–9 and TMs VI to XIII closely together, thereby facilitating the formation of the sugar translocation pathway (Puntheeranurak et al. 2007a; Tyagi et al. 2011).

3 The Prototype of SGLT Inhibitors: Phlorizin

3.1 General Remarks

The structure of the most frequently studied O-glucoside inhibitor of the SGLT, phlorizin, is shown in Fig. 1 (compound 1). Phlorizin, a beta glucoside derived from the bark of apple roots inhibits glucose transport in the intestine and in the kidney (Ehrenkranz et al. 2005). The affinity of phlorizin in transport studies, i.e. in the presence of D-glucose is about 1 μM for hSGLT1 and 20 nM for hSGLT2 (Pajor et al. 2008). Phlorizin was also shown to competitively inhibit the transport without evidence that it is translocated across the membrane. In a series of transport studies in microperfused rat renal tubules (Vick et al. 1973), it was found that the affinity of phlorizin congeners mirrored the stereospecific requirements for sugar transloca-tion, supporting the view that interaction of phlorizin with the transporter occurs at the sugar binding/translocation site of the transporter. On the other hand, the nature of the aglucone influenced the inhibitory potency. It had also been observed previously that phloretin, the aglucone of phlorizin inhibits sugar transport but in a non-competitive way and with a low affinity (Vick et al. 1973). Thus, it was concluded that phlorizin binds to the outside of the SGLT at two domains, one represents the sugar binding site and the other acts as an aglucone binding site. These multiple interactions can explain the high affinity of phlorizin compared to the substrate D-glucose.

3.2   SGLT as Phlorizin Receptor

3.2.1   Binding Studies on Brush Border Membranes

The availability of radioactively labelled phlorizin made it possible to study the interaction between phlorizin and the SGLT directly. On isolated brush border membranes of rat kidney high affinity binding of phlorizin could be demonstrated, which was sodium dependent and competitively inhibited by D-glucose but not by L-glucose. The maximum amount of phlorizin bound specifically to the brush border membrane was found to be 10 nmol/mg membrane protein and the number of receptors (transporters) per apical cell surface was estimated to be about 6,000 per cell (Bode et al. 1970). Sodium-dependent, D-glucose-inhibitable binding of phlorizin was also used in our laboratory as a signal in initial attempts to purify the transporter, as was the interaction with a phlorizin polymer.

It should be noted here that for the binding studies renal but not intestinal brush border membranes had to be used, because the renal membranes do not contain the disaccharidases that hydrolyze O-glycosides in the intestine.

3.2.2   Interactions of Phlorizin with the Isolated Transporter and Its Subdomains

Heterologous expression of hSGLT1 in yeast and isolation of the protein by affinity chromatography made it possible to study the interactions of the phlorizin receptor with phlorizin in more detail. As determined by tryptophan fluorescence experiments, phlorizin is bound to the isolated receptor in a sodium-dependent manner with an affinity of about 5 µM. It also could be shown that phlorizin but not phloretin induces the same conformational changes in the substrate binding site as D-glucose, adding proof to the assumption of an interaction of the glucose moiety of phlorizin with the sugar binding site of the phlorizin receptor (Tyagi et al. 2007).

The location of the aglucone binding site was investigated initially in rabbit SGLT1 by mutagenesis and transport studies in transfected cells which showed that the region between aa 602 and 610 is critically involved in phlorizin binding but not in sugar binding (Novakova et al. 2001). The same conclusion was reached in Trp fluorescence studies on the isolated transporter (hSGLT1) which showed that positions 602 and 609 undergo major changes in conformation when phlorizin or phloretin is present in the solution but undergo only minor changes in the presence of D-glucose (Tyagi et al. 2007). Evidence for a location on the surface of the transporter was derived from AFM studies using cantilever tips primed with an antibody against aa 606 to 630. The antibody interacted specifically with rbSGLT1 in intact cells under physiological conditions suggesting that this region of the receptor is accessible from the extracellular site. In the presence of phlorizin, the probability of binding of the antibody was drastically reduced indicating a strong conformational change in this region of the transporter. The aa 606 to 630 represent the late part of loop 13–14 – presumably one of the extramembranous loops of the transporter – as discussed earlier (Puntheeranurak et al. 2007a).

The molecular mechanism of the binding of phlorizin was further investigated on isolated subdomains of rbSGLT1. Binding studies with loop 13 in solution confirmed a specific interaction of phlorizin with this domain of the transporter. Phlorizin unfolds part of the loop and then brings the two small helical segments closely together, leading to a condensed conformation of the loop (Raja et al. 2003). Such condensation – increased hydrophobicity of the peptides – is also observed in the complete isolated carrier (hSGLT1) in the presence of phlorizin. Alkylglucosides which have also been found to inhibit SGLT bind with a high affinity to the isolated loop, they induce a different but also condense conformation (Kipp et al. 1997).

3.2.3 Differences Between hSGLT1 and hSGLT2

For the discussion in this chapter, it is of interest to compare the sequence of SGLT1 and SGLT2. As expected there are many sequence similarities or conservative replacements and identities in particular in the transmembrane segments. The extramembranous loops are more variable, in particular in the C-terminus where the aglucone binding site of the phlorizin receptor is located (Althoff et al. 2006). Thus, loop 13–14 of the hSGLT2 has ten more amino acids than hSGLT1 and two more cysteines close to the assumed binding site. In addition, several hydrophilic amino acids have been replaced by hydrophobic residues. Only recently, the two human transporters were compared with regard to the properties and the importance of the amino acids in the aglucone binding region. Pajor et al. (2008) described that hSGLT2 has a higher affinity for the aglucone phloretin than hSGLT1, confirming differences in the properties of the binding site and explaining the higher affinity of hSGLT2 for phlorizin. Furthermore, mutation of the conserved cysteine in position 610 in hSGLT1 led, as expected, to a decrease in the affinity of the transporter to phlorizin whereas mutation of the equivalent Cys 615 in hSGLT2 caused an increase in affinity. Thus, the two transporters differ significantly not only in their sugar binding sites but also in the conformation and the physical chemical properties of their aglucone binding sites.

3.3 Pharmacophore Analysis and Dimensions of the Phlorizin Binding Pocket

In a study combining 2D-NMR, molecular dynamics and pharmacophore analysis, the essential elements for the interaction of phlorizin with its binding pocket in nonhuman SGLTs were determined. The most probable phlorizin conformation shows a nearly perpendicular arrangement of the two aromatic rings (A and B) with the ring B situated above the sugar ring. As shown in Fig. 2, hydrogen bonding via

Fig. 2 Pharmacophore
analysis of phlorizin (*white
bars* carbon atoms, *grey bars*
hydrogen atoms, *black bars*
oxygen atoms and *circles*
pharmacophoric elements)
(reprinted from Wielert-Badt
et al. 2000)

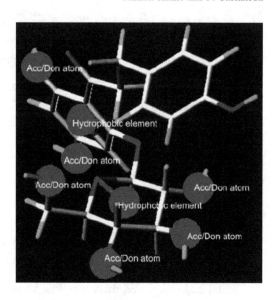

hydroxyl groups of the glucose moiety at C(2), C(3), C(4) and C(6) and at C(4) and
C(6) of aromatic ring A and hydrophobic interactions via the pyranoside ring and
aromatic ring A are the essential features. From these conformational features of the
pharmacophore, the dimension of the phlorizin binding site on the SGLT was
estimated to be $13 \times 10 \times 7$ Å (Wielert-Badt et al. 2000). Combined with the
fluorescence studies mentioned earlier, the 7 Å probably correspond to the hydrophilic pocket into which the sugar molecules and the sugar moiety of phlorizin
orient themselves. The area above the pocket probably represents the (mostly
hydrophobic) aglucone binding site on the surface of the transporter (late part of
loop 13–14). A tentative model of the phlorizin binding site of hSGLT1 is shown in
Fig. 3. The sugar moiety thereby interacts with amino acids in the sugar binding
pocket of the transporter whereas the aglucone moiety interacts with amino acids
located on the surface of the transporter in the late part of loop 13–14 (Tyagi et al.
2007, 2011). In rat kidney also the hydroxyl group at the 6 position of the B ring
proved to be important for the action of various compounds on the glucose
reabsorption (Hongu et al. 1998b).

4 Synthesis and Screening of Derivates of Phlorizin

4.1 O-glucosides, C-arylglucosides, N-glucosides and S-glycosides

Phlorizin   (glucose,   1-[2-(β-ᴅ-glucopyranosyloxy)-4,6-dihydroxyphenyl]-3-(4-
hydroxyphe-nyl)-1-propanone, compound **1** in Fig. 1) can be considered as the

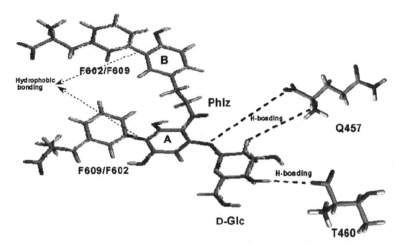

Fig. 3 Hypothetical scheme of major interaction sites between phlorizin (Phlz) and hSGLT1. The sugar moiety of phlorizin interacts with residues Gln457 and Thr460 present in transmembrane helix XI probably by the same hydrogen bond interactions as D-glucose does; the aromatic ring A of the aglucone interacts with Phe609/Phe602, and ring B makes contact with Phe602/Phe609; both are present in the extracellular loop 13 of hSGLT1 (reprinted from Tyagi et al. 2005)

lead substance for inhibitors of SGLT. Despite the observed anti-diabetic effect of phlorizin, it is not suited as a potential drug for the treatment of diabetes because of its low absorption caused by the hydrolysis of the O-glycosidic bond by intestinal disaccharidases. As a consequence, other glycosides have been developed some of which are shown in Fig. 1. One of these substances is T-1095 (compound **2**: 3-(benzo [*b*]furan-5-yl)-2,6-dihydroxy-4-methyl propiophenone 2-*O*-(6-*O* methoxy-carbonyl-β-D-glucopyranoside), a phlorizin-based SGLT inhibitor (Adachi et al. 2000; Arakawa et al. 2001; Oku et al. 1999, 2000a; Tsujihara et al. 1999). T-1095 has been shown to correct elevated blood glucose levels in rodents (Oku et al. 1999, 2000c; Ueta et al. 2005, 2006).

In glucosylated dihydrochalcones modification of the benzofuran moiety to benzodioxane (compound **3**) or 4-ethoxyphenyl (compound **4**), the 4-substituent of the phenyl ring improve SGLT2 selectivity. Another example is the modification of the ketone/phenol moiety of compound 2 leading to the indole-O-glucosides (compound **5** and compound **6**). Recently, canagliflozin, a novel C-glucoside with a thiophene ring has been added as potential drug. Also N-glucosides have been considered as interesting compounds (Washburn 2009).

Finally, thioglycosides, such as phenyl-1-thio-β-D-glucopyranoside (compound **11**) and 2-hydroxymethyl-phenyl-1-thio-β-D-galacto-pyranoside (compound **12**), have been shown to exert a pronounced inhibitory effect on SGLT2 and SGLT1, respectively (Castaneda et al. 2007)

Several SGLT inhibitors are currently in clinical trials. This is the case for sergliflozin (compound **7**) and dapagliflozin. Sergliflozin is a benzylphenol glucoside (Katsuno et al. 2007) and dapagliflozin is a C-aryl glucoside (Meng et al. 2008).

Dapagliflozin is the SGLT2 inhibitor with the most clinical data available to date, with other SGLT2 inhibitors currently in the developmental pipeline. It has demonstrated sustained, dose-dependent glucosuria over 24 h with once-daily dosing in clinical trials (Hanefeld and Forst 2010; Neumiller et al. 2010).

4.2 Screening Methods

4.2.1 Cellular Assays

The activity of SGLT has been analyzed by different methods, including transport assays with radioactively labelled sugars in brush border membrane vesicles (Kinne et al. 1975; Murer et al. 1974), *Xenopus* oocytes (Parent et al. 1992a) and transiently or permanently transfected cells (Brot-Laroche et al. 1987). The *Xenopus* oocyte expression system in combination with voltage-clamp techniques has yielded important details on the mechanisms underlying the binding and translocation reactions of the transporter (Kimmich 1990; Parent et al. 1992b; Sakhrani et al. 1984; Wright et al. 2007). One of the most widely used cell line is the Chinese Hamster Ovary (CHO) cell line (Lin et al. 1998; Sakhrani et al. 1984). Stably transfected CHO cells expressing hSGLT1 or hSGLT2 (Castaneda and Kinne 2005) or transiently transfected CHO (Katsuno et al. 2007) and COS-7 cells (Pajor et al. 2008) have also been used for functional characterization and screening of mutants and analysis of inhibitors.

Current methods that use radioactive labelled substances, however, are expensive due to the amount of radioactive compounds required and the large amount of substance necessary for inhibitor-type studies. For these reasons, we developed an alternative 96-well automated method to study the activity of hSGLT1 and hSGLT2 using stably transfected CHO cells (Castaneda and Kinne 2005). The advantage of using the 96-well method is the low amount of radioactive compounds and inhibitory substances required, as well as the ability to establish reproducibility because of the repetition built into the assay. Furthermore, this method can easily be performed semi-automatically to yield quantitative data regarding key aspects of glucose membrane transport and kinetic studies of potential inhibitors of human SGLT1 and SGLT2.

In addition to electrophysiology and radioactivity, fluorescence methods have also been reported. Fluorescence resonance energy transfer has been used to monitor changes of membrane potential as an indicator of uptake caused by SGLT (Castaneda et al. 2007; Weinglass et al. 2008). Also, fluorophore-conjugated SGLT2 inhibitors have been synthesized (compound **10**, Fig. 1) (Landsdell et al. 2008). It has been suggested that these compounds could be used in binding studies. Fluorescence presents some advantages over radioactivity, specifically with regard to waste disposal and safety. Furthermore, fluorescence assays can be used for high-throughput screening (Weinglass et al. 2008).

4.2.2   Animal Models

Animals have been used widely as experimental models in diabetes research. The importance of animal models in this type of research was demonstrated in studies using pancreatectomised dogs. These studies revealed the essential role of the pancreas and insulin in glucose homeostasis (Finkelstein et al. 1975). They also provide important information about pharmacokinetics, pharmacodynamics, ADME and toxicity. All of this information is mandatory to continue with clinical trials, which is the next and last step in drug development.

Each of these models is characterized by the presence of specific metabolic alterations found in diabetes such as hyperglycaemia, obesity, early hyperinsulinaemia and insulin resistance (Berglund et al. 1978; Herberg and Coleman 1977). One of the most used animal models is the Zucker diabetic fatty (ZDF) rat model (Zhang et al. 2006). However, other specific models for the study of glucose intolerance and diabetic complications have also been developed. For example, Goto-Kakizaki (GK) rats have been developed by selectively breeding them from non-diabetic Wistar rat with glucose intolerance (Goto et al. 1988). In addition, the C57BL/KsJ-db/db mouse represents a diabetic nephropathy animal model that is often used (Arakawa et al. 2001).

Several studies using animal models reported the effect of SGLT inhibitors for the treatment of diabetes. For example, chronic subcutaneous administration of phlorizin has been shown to reduce plasma glucose levels in diabetic rodents (Jonas et al. 1999). T-1095 reduces blood glucose levels in both type 1 (Adachi et al. 2000; Oku et al. 1999) and type 2 (Arakawa et al. 2001; Oku et al. 1999; Ueta et al. 2005) diabetic animal models. Canagliflozin also showed pronounced anti-hyperglycaemic effects in high-fat fed KK-mice (Nomura et al. 2010).

5   Therapeutic Efficacy of SGLT Inhibitors

5.1   In Vitro Studies on Sugar Transport by Cultured Transfected Cells

The first evidence for the potential of glucosides in inhibiting glucose transport is obtained in in vitro cell-based studies (Ohsumi et al. 2003; Oku et al. 1999). These studies represent the first stage in drug discovery and development. The concentration required for 50% inhibition of the $[^{14}C]$AMG-uptake rate (IC_{50}) represents an important indicator of the potential therapeutic effect of SGLT inhibitors for the control of hyperglycaemia. The SGLT inhibitors exert their action by competitive inhibition. For this reason, the inhibitory constant (K_i) can also be used as an indicator of the effect of SGLT inhibitors. The IC_{50} values and K_i values of SGLT inhibitors described in the literature are given in Table 1. In addition to their affinity to the transporter – which determines the dose of the compound – the

Table 1 Effect of O-glucosides, C-arylglucosides and S-glucosides based on IC50 values obtained in in vitro studies. For structure of compounds see Fig. 1

Compound	hSGLT1 (IC_{50}), µM	hSGLT2 (IC_{50}), µM	SGLT1/ SGLT2	Therapeutic effect	References
O-glucosides					
1	0.16	0.16	1	–	Oku et al. 1999
2	0.20	0.05	4	+	Oku et al. 1999[b]
3	6.28	0.015	418.7	n.d.	Dudash et al. 2004[a]
4	8.39	0.034	246.8	n.d.	Dudash et al. 2004[a]
5	0.145	0.024	6	n.d.	Zhang et al. 2006
6	2.14	0.028	76.4	n.d.	Zhang et al. 2006
7	2.1	0.010	210	+	Pajor et al. 2008[b]
8	4.5	0.012	364.5	+	Fujimori et al. 2008
C-arylglucosides					
9	1.4	0.001	1,263.6	+	Meng et al. 2008[b]
10	n.d.	0.055	n.d.	n.d.	Landsdell et al. (2008)
S-glucosides					
11	30	10	3	n.d.	Castaneda et al. 2007
12	14	88	0.2	n.d.	Castaneda et al. 2007

[a]K_i
[b]Used in clinical trials
Recently, also the results of transport studies with canagliflozin have been published (Nomura et al. 2010). The IC_{50} for hSGLT2 is 2.2 nmol and the specificity ratio is 413

selectivity for SGLT2 is important, since inhibition of glucose reabsorption in the kidney but not of the sugar absorption in the intestine has to be achieved. Therefore, the ratio of the affinities for SGLT1 and SGLT2 is also given in the table. According to both criteria compound **9** seems to be optimal, it shows a high affinity to SGLT2 combined with high selectivity for SGLT2. For these reasons, compound **9** (Fig. 1) seems to be a promising therapeutic substance for the treatment of hyperglycaemia.

5.2 Effect of SGLT Inhibitors in Preclinical and Clinical Studies

5.2.1 Urinary Glucose Excretion

The SGLT inhibitors induce a dose-dependent increase of urinary glucose excretion in different diabetic animal models including rats, mice and dogs (Adachi et al. 2000; Arakawa et al. 2001; Ellsworth et al. 2008; Oku et al. 1999, 2000b; Ueta et al. 2005). Inhibition of SGLT2 activity reduces tubular glucose reabsorption and due to the reduction of TmG, excess plasma glucose is excreted into the urine (see Fig. 4), thus reducing hyperglycaemia after glucose loading (Katsuno et al. 2007; Tsujihara et al. 1996, 1999). Due to the selective inhibition of SGLT2 a residual reabsorption of about 20–25% of the filtered load for glucose remains, attributable to the operation of SGLT1 in the late proximal tubule. The effect of SGLT inhibitors on urinary glucose excretion has also been associated with reduction of

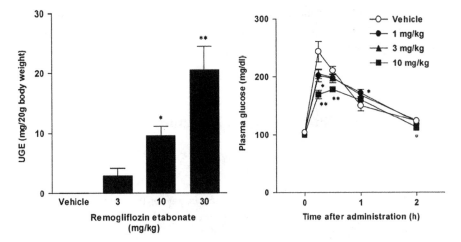

Fig. 4 Effect of SGLT2 inhibitor remogliflozin on urinary glucose excretion (UGE) and on plasma glucose during an oral glucose tolerance test (reprinted from Fujimori et al. 2008). The obtained UGE for vehicle was zero

body weight gain in diabetic rats resulting in a sustained improvement of hyperglycaemia (Ueta et al. 2005).

5.2.2 Effect on Plasma Glucose Levels

The SGLT inhibitors regulate blood glucose levels in a dose-dependent manner (Adachi et al. 2000; Katsuno et al. 2007). This effect is similar to that observed with other anti-diabetic drugs, such as insulin secretagogues or glucosidases, and results in improvements of postprandial hyperglycaemia (Ichikawa et al. 2002; Ikenoue et al. 1997). In addition to reducing blood glucose, SGLT inhibitors also decrease HbA_{1c} levels. This effect improves glucose intolerance and insulin resistance, and prevents the development of diabetic neuropathy in Goto-Kakizaki rats (Ueta et al. 2005).

5.2.3 Effect on Fasting Hypoglycaemia

One of the most severe potential side effects of anti-diabetic drugs is the induction of fasting hypoglycaemia. Animal studies demonstrated that SGLT inhibitors, such as T-1095 (Adachi et al. 2000; Oku et al. 2000c) and sergliflozin (Katsuno et al. 2007), do not cause hypoglycaemia (equivalent to plasma glucose level below 70 mg/dl) even in a 6-h fasting condition. The reason for this behaviour is that the inhibitors used mainly inhibit SGLT2 and not SGLT1. As evident from the familial renal glucosuria, even a complete absence of SGLT2 does not lead to hypoglycaemia (Kleta et al. 2004) because the residual reabsorption of glucose in the late proximal tubule mediated by SGL1 suffices to maintain close to normal plasma glucose level. Therefore, the risk of hypoglycaemia with these drugs is predicted to be low, making them potentially safe drugs even in the instance of an overdose.

6 Benefits and Pitfalls of SGLT Inhibitors

6.1 Benefits

The SGLT2 represents an important molecular target for the treatment of diabetes because it plays a major role in renal glucose reabsorption and its tissue distribution is limited to the kidney, thus reducing side effects.

It has been suggested that the effect of SGLT inhibitors on blood glucose control via an increase in urinary glucose excretion results in negative energy balance with body weight control and preservation of insulin secretion (Katsuno et al. 2007). Weight reduction and improved insulin sensitivity caused by SGLT inhibitors would offer some advantages over other anti-diabetic drugs, such as sulfonylureas, α-glucosidase inhibitors, thiazolidinediones and biguanides. These effects would be particularly important for the treatment of type 2 diabetes.

The SGLT inhibitors correct both hyperglycaemia and energy imbalance (Hongu et al. 1998a, b; Tsujihara et al. 1996). A negative energy balance contributes to a reduction of diabetes complications without necessarily decreasing body weight or the risk of hypoglycaemia (DeFronzo 2004). In addition to the effect on hyperglycaemia with reduction of diabetic complications observed with the use of SGLT inhibitors, these inhibitors have also been found to be associated with restoration of impaired insulin secretion from pancreatic ß-cells (DeFronzo 2004).

Another important characteristic of SGLT inhibitors is that they do not alter the glucose transport via GLUT. As a consequence, glucose delivery to the brain, liver and muscle is not affected. Moreover, the high specificity of some SGLT inhibitors, such as sergliflozin (Katsuno et al. 2007), dapagliflozin (Meng et al. 2008) and canagliflozin (Nomura et al. 2010), precludes any negative gastrointestinal side effects associated with SGLT1 inhibition.

6.2 Pitfalls

The expression of SGLT1 in other tissues in addition to the kidney represents a restriction for the use of SGLT inhibitors that are not selective to SGLT2. The SGLT1 is expressed in the capillary endothelial cells of rat heart and muscle (Wright et al. 2007), in capillary endothelial cells of the blood brain barrier in rat (Elfeber et al. 2004) as well as in other tissues (trachea, testis, prostate, mammary and salivary glands) (Kinne unpublished; Wright et al. 2007). Therefore, the effect of SGLT inhibitors on other tissues in which SGLT is expressed must be further investigated.

Another important factor that remains to be evaluated is the effect of glycosuria and osmotic diuresis induced by SGLT inhibitors. Glucosuria has been reported as a risk factor associated with bacterial urinary tract infection. Interestingly, the concentration of glucose excreted into the urine is higher if bacterial growth has been observed (Tsujihara et al. 1999). However, patients with familial renal glucosuria due to a defect in the SGLT2 gene remain asymptomatic (van den Heuvel et al. 2002; Scholl-Burgi et al. 2004). Furthermore, altered bladder function secondary to diabetes seems to play an additional role in the development of urinary tract infections (Patterson and Andriole 1995) and not only the amount of glucose excreted into the urine. For that reason, clinical trials are necessary to evaluate the role of SGLT inhibitors in the development of urinary tract infections.

The SGLT inhibitors, such as sergliflozin, produce a slight and temporary diuretic effect when they are used at high concentrations (equivalent to 30 mg/kg) (Katsuno et al. 2007). At low concentration (of about 3 or 10 mg/kg), however, no significant change in urinary volume has been reported. Osmotic diuresis in normoglycaemic conditions remains asymptomatic but under hyperglycaemic conditions might be associated with adverse reactions (Venkatraman and Singhi 2006). Osmotic diuresis leads to depletion of intravascular volume and decreased renal perfusion. As a result, the amount of glucose that can be excreted is reduced (Adrogue 1992). The resultant hyperosmolarity and dehydration stimulates the production of cortisol, catecholamines and glucagon, and further increases the degree of hyperglycaemia (Kitabchi et al. 2001). These side effects need to be investigated further to better understand the safety of SGLT inhibitors.

7   Current State and Future Developments

Based on the observed effects of different SGLT inhibitors in diabetes and hyperglycaemia, they have been proposed as potential therapeutic substances and new compounds are currently under investigation. One of these compounds was a benzylpyrazole glucoside known as remogliflozin (compound **8**, Fig. 1) (Fujimori et al. 2008). Studies performed in vitro and using diabetic animal models have demonstrated that remogliflozin is a potent and highly selective SGLT2 inhibitor, as shown by increased urinary glucose excretion, reduced fasting plasma glucose and glycated haemoglobin levels, improved hyperglycaemia, hyperinsulinaemia, hypertriglyceridaemia and insulin resistance in rodents (Fujimori et al. 2008). However, its further development has been discontinued by GlaxoSmithKline (Kissei Pharmaceutical Co. Ltd 2009). The most promising candidate is currently dapagliflozin (Hanefeld and Forst 2010).

Another substance that currently is under investigation is desoxyrhaponticin, an O-glucoside, found in extracts of the rhubarb plant. Studies in diabetic rats demonstrated that desoxyrhaponticin has an inhibitory effect on SGLT (Li et al. 2007). The exact mechanism of action of this substance remains to be determined.

The SGLT2 represents a promising molecular target for the development of new alternatives in the treatment of diabetes based on the following aspects: (a) SGLT2 is expressed exclusively in the renal proximal tubules, and thus a selective SGLT2 inhibitor should not affect other tissues and (b) enhancement of urinary glucose excretion via SGLT2 inhibition leads to a negative energy balance, which currently is not really achieved by any existing clinical pharmacological intervention. Inhibition of glucose transport in the kidney through SGLT inhibitors represents a different mechanism of action from other hypoglycaemic agents. For these reasons, SGLT inhibitors should be considered as a potential new therapeutic alternative for the treatment of diabetes either alone or in combination with other anti-diabetic drugs. Moreover, therapeutic alternatives to achieve a reduction of glucose reabsorption in the proximal tubules, which represents the last step in the fate of glucose in the organism, represent an additional strategy for the treatment of diabetes that needs to be investigated further.

Acknowledgments The authors would like to thank Christine Riemer, for her outstanding organizational skill and computer proficiency, as well as her endurance and patience. Without her, the writing of this chapter while travelling between two continents would not have been possible. Also, the enthusiastic and untiring support of the staff of the library of the Max Planck Institute, particularly Christiane Berse and Jürgen Block is gratefully acknowledged. In the analysis of the differences of the aglucone binding site of hSGLT1 and hSGLT2, Dr. Thorsten Althoff, now at the Max Planck Institute of Biophysics in Frankfurt, Germany, provided essential help.

References

Adachi T, Yasuda K, Okamoto Y, Shihara N, Oku A, Ueta K, Kitamura K, Saito A, Iwakura I, Yamada Y, Yano H, Seino Y, Tsuda K (2000) T-1095, a renal Na+-glucose transporter inhibitor, improves hyperglycemia in streptozotocin-induced diabetic rats. Metabolism 49: 990–995

Adrogue HJ (1992) Glucose homeostasis and the kidney. Kidney Int 42:1266–1282

Althoff T, Hentschel H, Luig J, Schutz H, Kasch M, Kinne RK (2006) Na+-D-glucose cotransporter in the kidney of Squalus acanthias: molecular identification and intrarenal distribution. Am J Physiol Regul Integr Comp Physiol 290:R1094–R1104

Arakawa K, Ishihara T, Oku A, Nawano M, Ueta K, Kitamura K, Matsumoto M, Saito A (2001) Improved diabetic syndrome in C57BL/KsJ-db/db mice by oral administration of the Na+-glucose cotransporter inhibitor T-1095. Br J Pharmacol 132:578–586

Asano N (2003) Glycosidase inhibitors: update and perspectives on practical use. Glycobiology 13:93R–104R

Berglund O, Frankel BJ, Hellman B (1978) Development of the insulin secretory defect in genetically diabetic (db/db) mouse. Acta Endocrinol (Copenh) 87:543–551

Bode F, Baumann K, Frasch W, Kinne R (1970) Binding of phlorhizin to the brushborder fraction of rat kidney. Pflugers Arch 315:53–65

Brot-Laroche E, Supplisson S, Delhomme B, Alcalde AI, Alvarado F (1987) Characterization of the D-glucose/Na+ cotransport system in the intestinal brush-border membrane by using the specific substrate, methyl alpha-D-glucopyranoside. Biochim Biophys Acta 904:71–80

Burckhardt G, Kinne RKH (1992) Transport proteins. Cotransporters and countertransporters. In: Seldin DW, Giebisch G (eds) The kidney: physiology and pathophysiology. New York, Raven, pp 537–586

Calado J, Sznajer Y, Metzger D, Rita A, Hogan MC, Kattamis A, Scharf M, Tasic V, Greil J, Brinkert F, Kemper MJ, Santer R (2008) Twenty-one additional cases of familial renal glucosuria: absence of genetic heterogeneity, high prevalence of private mutations and further evidence of volume depletion. Nephrol Dial Transplant 12:1–6

Castaneda F, Kinne RKH (2005) A 96-well automated method to study inhibitors of human sodium-dependent D-glucose transport. Mol Cell Biochem 280:91–98

Castaneda F, Burse A, Boland W, Kinne RK (2007) Thioglycosides as inhibitors of hSGLT1 and hSGLT2: potential therapeutic agents for the control of hyperglycemia in diabetes. Int J Med Sci 4:131–139

Ceriello A (2005) PROactive Study: (r)evolution in the therapy of diabetes? Diabet Med 22: 1463–1464

de Melo EB, Gomes AD, Carvalho I (2006) alpha- and beta-Glucosidase inhibitors: chemical structure and biological activity. Tetrahedron 62:10277–10302

DeFronzo RA (2004) Pathogenesis of type 2 diabetes mellitus. Med Clin North Am 88:787

Dudash J Jr, Zhang X, Zeck RE, Johnson SG, Cox GG, Conway BR, Rybczynski PJ, Demarest KT (2004) Glycosylated dihydrochalcones as potent and selective sodium glucose co-transporter 2 (SGLT2) inhibitors. Bioorg Med Chem Lett 14(20):5121–5125

Ehrenkranz JRL, Lewis NG, Kahn CR, Roth J (2005) Phlorizin: a review. Diab Metab Res Rev 21:31–38

Elfeber K, Stumpel F, Gorboulev V, Mattig S, Deussen A, Kaissling B, Koepsell H (2004) Na+-D-glucose cotransporter in muscle capillaries increases glucose permeability. Biochem Biophys Res Commun 314:301–305

Ellsworth BA, Meng W, Patel M, Girotra RN, Wu G, Sher PM, Hagan DL, Obermeier MT, Humphreys WG, Robertson JG, Wang A, Han S, Waldron TL, Morgan NN, Whaley JM, Washburn WN (2008) Aglycone exploration of C-arylglucoside inhibitors of renal sodium-dependent glucose transporter SGLT2. Bioorg Med Chem Lett 18:4770–4773

Faham S, Watanabe A, Besserer GM, Cascio D, Specht A, Hirayama BA, Wright EM, Abramson J (2008) The crystal structure of a sodium galactose transporter reveals mechanistic insights into Na$^+$/sugar symport. Science 3:1–5

Feld LG (2001) Renal glycosuria. e Med J 2:1–6

Finkelstein SM, Bleicher MA, Batthany S, Tiefenbrun J (1975) In vivo modeling for glucose homeostasis. IEEE Trans Biomed Eng 22:47–52

Firnges MA, Lin JT, Kinne RKH (2001) Functional asymmetry of the sodium-D-glucose cotransporter expressed in yeast secretory vesicles. J Membr Biol 179:143–153

Fujimori Y, Katsuno K, Nakashima I, Ishikawa-Takemura Y, Fujikura H, Isaji M (2008) Remogliflozin etabonate, in a novel category of selective low-affinity / high-capacity sodium glucose cotransporter (SGLT2) inhibitors, exhibits antidiabetic efficacy in rodent models. J Pharmacol Exp Ther 327:268–276

Goto Y, Suzuki K, Ono T, Sasaki M, Toyota T (1988) Development of diabetes in the non-obese NIDDM rat (GK rat). Adv Exp Med Biol 246:29–31

Hanefeld M, Forst T (2010) Dapagliflozin, an SGLT2 inhibitor, for diabetes. Lancet 375: 2196–2198

Hediger MA, Coady MJ, Ikeda TS, Wright EM (1987) Expression cloning and cDNA sequencing of the Na$^+$/glucose co-transporter. Nature 330:379–381

Herberg L, Coleman DL (1977) Laboratory animals exhibiting obesity and diabetes syndromes. Metabolism 26:59–99

Hongu M, Funami N, Takahashi Y, Saito K, Arakawa K, Matsumoto M, Yamakita H, Tsujihara K (1998a) Na+-glucose cotransporter inhibitors as antidiabetic agents. III. Synthesis and pharmacological properties of 4'-dehydroxyphlorizin derivatives modified at the OH groups of the glucose moiety. Chem Pharm Bull 46:1545–1555

Hongu M, Tanaka T, Funami N, Saito K, Arakawa K, Matsumoto M, Tsujihara K (1998b) Na+-glucose cotransporter inhibitors as antidiabetic agents. II. Synthesis and structure-activity relationships of 4'-dehydroxyphlorizin derivatives. Chem Pharm Bull 46:22–33

Ichikawa K, Yamato T, Ojima K, Tsuji A, Ishikawa K, Kusama H, Kojima M (2002) Effect of KAD-1229, a novel hypoglycaemic agent, on plasma glucose levels after meal load in type 2 diabetic rats. Clin Exp Pharmacol Physiol 29:423–427

Ikenoue T, Okazaki K, Fujitani S, Tsuchiya Y, Akiyoshi M, Maki T, Kondo N (1997) Effect of a new hypoglycemic agent, A-4166 [(-)-N-(trans-4-isopropyl-cyclohexanecarbonyl)-D-phenyl-alanine], on postprandial blood glucose excursion: Comparison with voglibose and glibenclamide. 7. Biol Pharm Bull 20:354–359

Jonas JC, Sharma A, Hasenkamp W, Ilkova H, Patane G, Laybutt R, Bonner-Weir S, Weir GC (1999) Chronic hyperglycemia triggers loss of pancreatic beta cell differentiation in an animal model of diabetes. J Biol Chem 274:14112–14121

Kanai Y, Lee WS, You G, Brown D, Hediger MA (1994) The human kidney low affinity Na+/glucose cotransporter SGLT2. Delineation of the major renal reabsorptive mechanism for D-glucose. J Clin Invest 93:397–404

Katsuno K, Fujimori Y, Takemura Y, Hiratochi M, Itoh F, Komatsu Y, Fujikura H, Isaji M (2007) Sergliflozin, a novel selective inhibitor of low-affinity sodium glucose cotransporter (SGLT2), validates the critical role of SGLT2 in renal glucose reabsorption and modulates plasma glucose level. J Pharmacol Exp Ther 320:323–330

Kimmich GA (1990) Membrane potentials and the mechanism of intestinal Na(+)-dependent sugar transport. J Membr Biol 114(1):1–27. Review

Kinne RKH (1991) Selectivity and direction – plasma-membranes in renal transport. Am J Physiol 260:F153–F162

Kinne R, Murer H, Kinne-Saffran E, Thees M, Sachs G (1975) Sugar-transport by renal plasma-membrane vesicles – Characterization of systems in brush-border microvilli and basal-lateral plasma-membranes. J Membr Biol 21:375–395

Kipp H, Kinne-Saffran E, Bevan C, Kinne RK (1997) Characteristics of renal Na+-D-glucose cotransport in the skate (Raja erinacea) and shark (Squalus acanthias). Am J Physiol 273: R134–R142

Kissei Pharmaceutical Co. Ltd. (2009) Discontinuation of the development of "Remogliflozin" by GlaxoSmithKline, News Release July 3

Kitabchi AE, Umpierrez GE, Murphy MB, Barrett EJ, Kreisberg RA, Malone JI, Wall BM (2001) Management of hyperglycemic crises in patients with diabetes. Diab Care 24:131–153

Kleta R, Stuart C, Gill FA, Gahl WA (2004) Renal glucosuria due to SGLT2 mutations. Mol Genet Metab 82:56–58

Landsdell MI, Burring DJ, Hepworth D, Strawbridge M, Graham E, Guyot T, Betson MS, Hart JD (2008) Design and synthesis of fluorescent SGLT2 inhibitors. Bioorg Med Chem Lett 18:4944–4947

Li JM, Che CT, Lau CBS, Leung PS, Cheng CHK (2007) Desoxyrhaponticin (3, 5-dihydroxy-4'-methoxystilbene 3-O-beta-D-glucoside) inhibits glucose uptake in the intestine and kidney: In vitro and in vivo studies. J Pharmacol Exp Ther 320:38–46

Lin JT, Kormanec J, Wehner F, Wielert-Badt S, Kinne RKH (1998) High-level expression of Na+/D-glucose cotransporter (SGLT1) in a stably transfected Chinese hamster ovary cell line. Biochim Biophys Acta: Biomembr 1373:309–320

Meng W, Ellsworth BA, Nirschl AA, McCann PJ, Patel M, Girotra RN, Wu G, Sher PM, Morrison EP, Biller SA, Zahler R, Deshpande PP, Pullockaran A, Hagan DL, Morgan N, Taylor JR, Obermeier MT, Humphreys WG, Khanna A, Discenza L, Robertson JG, Wang A, Han S, Wetterau JR, Janovitz EB, Flint OP, Whaley JM, Washburn WN (2008) Discovery of dapagliflozin: a potent, selective renal sodium-dependent glucose cotransporter 2 (SGLT2) inhibitor for the treatment of type 2 diabetes. J Med Chem 51:1145–1149

Morrison AI, Panayotova-Heiermann M, Feigl G, Scholermann B, Kinne RKH (1991) Sequence comparison of the sodium-D-glucose cotransport systems in rabbit renal and intestinal epithelia. Biochim Biophys Acta 1089:121–123

Murer H, Hopfer U, Kinne-Saffran E, Kinne R (1974) Glucose transport in isolated brush-border and lateral-basal plasma-membrane vesicles from intestinal epithelial cells. Biochim Biophys Acta 345:170–179

Neumiller JJ, White JR Jr, Campbell RK (2010) Sodium-glucose co-transport inhibitors: progress and therapeutic potential in type 2 diabetes mellitus. Drugs 70:377–385

Nomura S, Sakamaki S, Hongu M, Kawanishi E, Koga Y, Sakamoto T, Yamamoto Y, Ueta K, Kimata H, Nakayama K, Tsuda-Tsukimoto M (2010) Discovery of canagliflozin, a novel C-glucoside with thiophene ring, as sodium-dependent glucose cotransporter 2 inhibitor for the treatment of type 2 diabetes mellitus. J Med Chem 53:6355–6360

Novakova R, Homerova D, Kinne RK, Kinne-Saffran E, Lin JT (2001) Identification of a region critically involved in the interaction of phlorizin with the rabbit sodium-D-glucose cotransporter SGLT1. J Membr Biol 184:55–60

Ohsumi K, Matsueda H, Hatanaka T, Hirama R, Umemura T, Oonuki A, Ishida N, Kageyama Y, Maezono K, Kondo N (2003) Pyrazole-O-glucosides as novel Na^+-glucose cotransporter (SGLT) inhibitors. Bioorg Med Chem Lett 13:2269–2272

Oku A, Ueta K, Arakawa K, Ishihara T, Nawano M, Kuronuma Y, Matsumoto M, Saito A, Tsujihara K, Anai M, Asano T, Kanai Y, Endou H (1999) T-1095, an inhibitor of renal Na+-glucose cotransporters, may provide a novel approach to treating diabetes. Diabetes 48:1794–1800

Oku A, Ueta K, Arakawa K, Kano-Ishihara T, Matsumoto M, Adachi T, Yasuda K, Tsuda K, Saito A (2000a) Antihyperglycemic effect of T-1095 via inhibition of renal Na+-glucose cotransporters in streptozotocin-induced diabetic rats. Biol Pharm Bull 23:1434–1437

Oku A, Ueta K, Arakawa K, Kano-Ishihara T, Matsumoto T, Adachi T, Yasuda K, Tsuda K, Ikezawa K, Saito A (2000b) Correction of hyperglycemia and insulin sensitivity by T-1095, an inhibitor of renal Na+-glucose cotransporters, in streptozotocin-induced diabetic rats. Jpn J Pharmacol 84:351–354

Oku A, Ueta K, Nawano M, Arakawa K, Kano-Ishihara T, Matsumoto M, Saito A, Tsujihara K, Anai M, Asano T (2000c) Antidiabetic effect of T-1095, an inhibitor of Na+-glucose cotransporter, in neonatally streptozotocin-treated rats. Eur J Pharmacol 391:183–192

Pajor AM, Randolph KM, Kerner SA, Smith CD (2008) Inhibitor binding in the human renal low- and high-affinity Na+/glucose cotransporters. J Pharmacol Exp Ther 324:985–991

Parent L, Supplisson S, Loo DDF, Wright EM (1992a) Electrogenic properties of the cloned Na^+ glucose cotransporter. 1. Voltage-clamp studies. J Membr Biol 125:49–62

Parent L, Supplisson S, Loo DDF, Wright EM (1992b) Electrogenic properties of the cloned Na+ glucose cotransporter. 2. A transport model under nonrapid equilibrium conditions. J Membr Biol 125:63–79

Patterson JE, Andriole VT (1995) Bacterial urinary-tract infections in diabetes. Infect Dis Clin North Am 9:25–51

Puntheeranurak T, Wildling L, Gruber HJ, Kinne RK, Hinterdorfer P (2006) Ligands on the string: single-molecule AFM studies on the interaction of antibodies and substrates with the Na+-glucose co-transporter SGLT1 in living cells. J Cell Sci 119:2960–2967

Puntheeranurak T, Kasch M, Xia X, Hinterdorfer P, Kinne RK (2007a) Three surface subdomains form the vestibule of the Na^+/glucose cotransporter SGLT1. J Biol Chem 282:25222–25230

Puntheeranurak T, Wimmer B, Castaneda F, Gruber HJ, Hinterdorfer P, Kinne RKH (2007b) Substrate specificity of sugar transport by rabbit SGLT1: single molecule AFM versus transport. Biochemistry 46:2797–2804

Raja MM, Tyagi NK, Kinne RKH (2003) Phlorizin recognition in a C-terminal fragment of SGLT1 studied by tryptophan scanning and affinity labeling. J Biol Chem 278:49154–49163

Sakhrani LM, Badiedezfooly B, Trizna W, Mikhail N, Lowe AG, Taub M, Fine LG (1984) Transport and metabolism of glucose by renal proximal tubular cells in primary culture. Am J Physiol 246:F757–F764

Scholl-Burgi S, Santer R, Ehrich JHH (2004) Long-term outcome of renal glucosuria type 0: the original patient and his natural history. Nephrol Dial Transplant 19:2394–2396

Stuart IS, Trayhurn P (2003) Glucose transporters (GLUT and SGLT): expanded families of sugar transport proteins. Br J Nutr 89:3–9

Tsujihara K, Hongu M, Saito K, Inamasu M, Arakawa K, Oku A, Matsumoto M (1996) Na+-glucose cotransporter inhibitors as antidiabetics. 1. Synthesis and pharmacological properties of 4'-Dehydroxyphlorizin derivatives based on a new concept. Chem Pharm Bull 44:1174–1180

Tsujihara K, Hongu M, Saito K, Kawanishi H, Kuriyama K, Matsumoto H, Oku A, Ueta K, Tsuda M, Saito A (1999) Na+-glucose cotransporter (SGLT) inhibitors as antidiabetic agents. 4. Synthesis and pharmacological properties of 4'-dehydroxyphlorizin derivatives substituted on the B ring. J Med Chem 42:5311–5324

Tyagi NK, Goyal P, Kumar A, Pandey D, Siess W, Kinne RK (2005) High-yield functional expression of human sodium-D-glucose cotransporter1 in Pichia pastoris and characterization of ligand-induced conformational changes as studied by tryptophan fluorescence. Biochemistry 44:15514–15524

Tyagi NK, Kumar A, Goyal P, Pandey D, Siess W, Kinne RK (2007) D-glucose-recognition and phlorizin-binding sites in human sodium-D-glucose cotransporter 1 (hSGLT1): a tryptophan scanning study. Biochemistry 46:13616–13628

Tyagi NK, Puntheeranurak T, Raja M, Kumar A, Wimmer B, Neundlinger I, Gruber H, Hinterdorfer P, Kinne RK (2011) A biophysical glance at the outer surface of the membrane transporter SGLT1. Biochim Biophys Acta 1808(1):1–18, Epub 2010 Aug 6

Ueta K, Ishihara T, Matsumoto Y, Oku A, Nawano M, Fujita T, Saito A, Arakawa K (2005) Long-term treatment with the Na+-glucose cotransporter inhibitor T-1095 causes sustained improvement in hyperglycemia and prevents diabetic neuropathy in Goto-Kakizaki Rats. Life Sci 76:2655–2668

Ueta K, Yoneda H, Oku A, Nishiyama S, Saito A, Arakawa K (2006) Reduction of renal transport maximum for glucose by inhibition of Na+-glucose cotransporter suppresses blood glucose elevation in dogs. Biol Pharm Bull 29:114–118

van den Heuvel LP, Assink K, Willemsen M, Monnens L (2002) Autosomal recessive renal glucosuria attributable to a mutation in the sodium glucose cotransporter (SGLT2). Hum Genet 111:544–547

Venkatraman R, Singhi SC (2006) Hyperglycemic hyperosmolar nonketotic syndrome. Indian J Pediatr 73:55–60

Vick H, Diedrich DF, Baumann K (1973) Reevaluation of renal tubular glucose transport inhibition by phlorizin analogs. Am J Physiol 224:552–557

Washburn WN (2009) Evolution of sodium glucose co-transporter 2 inhibitors as anti-diabetic agents. Expert Opin Ther Pat 19:1485–1499

Weinglass AB, Swensen AM, Liu J, Schmalhofer W, Thomas A, Williams B, Ross L, Hashizume K, Kohler M, Kaczorowski GJ, Garcia ML (2008) A high-capacity membrane potential FRET-based assay for the sodium-coupled glucose co-transporter SGLT1. Assay Drug Dev Technol 6:255–262

Wielert-Badt S, Lin JT, Lorenz M, Fritz S, Kinne RK (2000) Probing the conformation of the sugar transport inhibitor phlorizin by 2D-NMR, molecular dynamics studies, and pharmacophore analysis. J Med Chem 43:1692–1698

Wright EM, Hirayama BA, Loo DF (2007) Active sugar transport in health and disease. J Intern Med 261:32–43

Zhang XY, Urbanski M, Patel M, Cox GG, Zeck RE, Bian HY, Conway BR, Beavers MP, Rybczynski PJ, Demarest KT (2006) Indole-glucosides as novel sodium glucose co-transporter 2 (SGLT2) inhibitors. Part 2. Bioorg Med Chem Lett 16:1696–1701

Inhibitors of 11β-Hydroxysteroid Dehydrogenase Type 1 in Antidiabetic Therapy

Minghan Wang

Contents

Abstract Glucocorticoid action is mediated by glucocorticoid receptor (GR), which upon cortisol binding is activated and regulates the transcriptional expression of target genes and downstream physiological functions. 11β-Hydroxysteroid dehydrogenase type 1 (11β-HSD1) catalyzes the conversion of inactive cortisone to active cortisol. Since cortisol is also produced through biosynthesis in the adrenal glands, the total cortisol level in a given tissue is determined by both the circulating cortisol concentration and the local 11β-HSD1 activity. 11β-HSD1 is expressed in liver, adipose, brain, and placenta. Since it contributes to the local cortisol levels in these tissues, 11β-HSD1 plays a critical role in glucocorticoid action. The metabolic symptoms caused by glucocorticoid excess in Cushing's syndrome overlap with the characteristics of the metabolic syndrome, suggesting that increased glucocorticoid activity may play a role in the etiology of the metabolic syndrome. Consistent with this notion, elevated adipose expression of 11β-HSD1 induced metabolic syndrome-like phenotypes in mice. Thus, 11β-HSD1 is a proposed

M. Wang
Department of Metabolic Disorders, Amgen Inc., One Amgen Center Drive, Mail Stop 29-1-A, Thousand Oaks, CA 91320, USA
e-mail: mwang@amgen.com

M. Schwanstecher (ed.), *Diabetes - Perspectives in Drug Therapy*,
Handbook of Experimental Pharmacology 203,
DOI 10.1007/978-3-642-17214-4_6, © Springer-Verlag Berlin Heidelberg 2011

therapeutic target to normalize glucocorticoid excess in a tissue-specific manner and mitigate obesity and insulin resistance. Selective inhibitors of 11β-HSD1 are under development for the treatment of type 2 diabetes and other components of the metabolic syndrome.

Keywords 11β-Hydroxysteroid dehydrogenase · Cortisol · Cushing's syndrome · Glucocorticoid excess · HPA axis · Metabolic syndrome

1 Glucocorticoid Metabolism and Action

The active glucocorticoid cortisol in humans (or corticosterone in rodents) is produced through both adrenal biosynthesis and tissue-specific regeneration. In contrast, the inert glucocorticoid cortisone [or 11-dehydrocorticosterone (11-DHC) in rodents] is only generated in kidney through enzymatic conversion of cortisol. The interconversion of cortisol and cortisone is catalyzed by two isozymes of 11β-hydroxysteroid dehydrogenase (11β-HSD), type 1 (11β-HSD1) and type 2 (11β-HSD2). 11β-HSD1 is expressed in liver, adipose, brain, and placenta; 11β-HSD2 is mainly expressed in kidney. Both enzymes have reductase and dehydrogenase activities in vitro. But in vivo, 11β-HSD1 is primarily a reductase and 11β-HSD2 is a dehydrogenase (Bujalska et al. 1997; Jamieson et al. 1995, 2000). 11β-HSD1 is responsible for tissue-specific regeneration of cortisol using cortisone as substrate, whereas 11β-HSD2 catalyzes the opposite reaction as the main source of endogenous cortisone production. Cortisol mediates diverse physiological actions as endogenous ligand of glucocorticoid receptor (GR), which upon ligand-induced activation triggers transcriptional activation or suppression of downstream target genes. GR is a nuclear receptor with ubiquitous tissue expression. Both the local cortisol concentration and the GR expression level in a given tissue dictate the degree of glucocorticoid action. There are two sources of cortisol in liver and adipose where 11β-HSD1 is expressed, the fraction diffused from blood circulation and that produced by 11β-HSD1 within the tissue. Unlike cortisone, which is completely in the free unbound form, the circulating cortisol is protein bound with 6% and 90% to albumin and corticosteroid binding globulin (CBG), respectively (Dunn et al. 1981). Since only free cortisol can penetrate tissues and drive glucocorticoid action, CBG and albumin binding restrict the access of cortisol to target tissues. Taken together, there are several levels of regulation that are important determinants of glucocorticoid action: the adrenal cortisol production rate, the protein binding by cortisol, the density of GR in tissues, and the local cortisol production by 11β-HSD1 within a tissue.

The adrenal cortisol biosynthesis is regulated by a neuroendocrine feedback circuit involving the hypothalamic, pituitary, and adrenal functions. This circuit is activated by physiological stimuli such as stress (Fig. 1), resulting in the release of the hypothalamic peptide corticotrophin-releasing hormone (CRH), which subsequently acts on the anterior pituitary to stimulate the secretion of adrenocorticotropic hormone (ACTH), a peptide hormone that augments adrenal cortisol production and release. The circulating cortisol level itself is a regulatory feedback

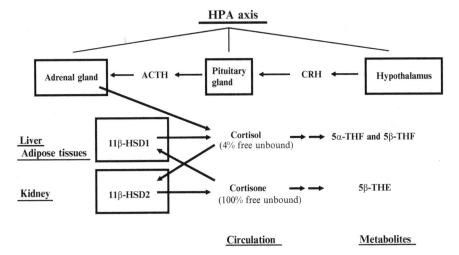

Fig. 1 Schematic representation of the HPA axis and glucocorticoid metabolism. For simplicity, only liver and adipose tissues are included as main tissues expressing 11β-HSD1. *ACTH* adrenocorticotropic hormone, *CRH*, corticotrophin-releasing hormone; *5α- and 5β-THF*, 5α- and 5β-tetrahydrocortisol; *5β-THE*, 5β-tetrahydrocortisone

signal for CRH and ACTH release. This endocrine regulatory network constitutes the hypothalamic–pituitary–adrenal (HPA) axis (Delbende et al. 1992). The circulating cortisol follows a diurnal pattern throughout the day, reaching its maximum concentration in the morning and nadir in late afternoon in humans (Walker et al. 1992). In contrast, the blood cortisone level remains almost constant throughout the day (Walker et al. 1992). The HPA axis maintains the circulating cortisol concentration by responding to stimuli such as diurnal changes. Since 11β-HSDs also contribute to cortisol and cortisone production at the tissue level, the systemic glucocorticoid homeostasis is achieved by the activities of the HPA axis and 11β-HSDs (Fig. 1). The metabolism of both cortisol and cortisone occurs in liver involving A-ring reductases (Walker and Stewart 2003). The main final metabolites of cortisol and cortisone, 5α- and 5β-tetrahydrocortisol (5α- and 5β-THF) and 5β-tetrahydrocortisone (THE) (Fig. 1), are eliminated through urinary excretion along with much lower levels of free cortisol and cortisone (Stewart et al. 1999; Tomlinson et al. 2002). Since only free cortisol is available for catabolism, CBG and albumin binding plays a role in its metabolic clearance.

2 Glucocorticoid Excess and the Development of the Metabolic Syndrome

Most patients that develop cardiovascular disease share several common risk factors. The clustering of these factors was first named Syndrome X by Reaven (1988) and in recent years more often referred to as the metabolic syndrome

(Grundy et al. 2004). The metabolic syndrome is defined as a collection of metabolic abnormalities including central obesity, insulin resistance, atherogenic dyslipidemia, hyperglycemia, and hypertension (Grundy et al. 2004). Not all of the listed characteristics may be found in the same patient. The diagnosis of the metabolic syndrome can be made when an individual presents with three of the five characteristics (Grundy et al. 2004). Although insulin resistance was postulated as a major underlying factor (Reaven 1988), the etiology of the metabolic syndrome is still not well defined. The prevalence of the metabolic syndrome among the US adult population is more than 20% (Ford et al. 2002). Further, about 80% of type 2 diabetics meet the criteria for the diagnosis of this disorder (Isomaa et al. 2001). The metabolic syndrome is a major factor that increases the risk of cardiovascular disease and type 2 diabetes. However, it has not been recognized as a therapeutic indication by regulatory authorities. Separate treatments of the individual features of the metabolic syndrome have been common practice, but complete mitigation of all the metabolic abnormalities can hardly be achieved even with the combination of existing individual treatments.

The role of glucocorticoid action in the development of the metabolic syndrome is implicated by findings in Cushing's syndrome. Patients with Cushing's syndrome have elevated circulating cortisol levels mainly due to increased ACTH secretion caused by pituitary adenoma (Arnaldi et al. 2003). The elevated cortisol exposure, namely glucocorticoid excess in a systemic manner, leads to symptoms that overlap with those of the metabolic syndrome (Arnaldi et al. 2003). Further, clinical use of glucocorticoids leads to metabolic complications that resemble the characteristics of the metabolic syndrome (Davis 1986; Gallant and Kenny 1986). Upon surgical removal of the pituitary adenoma in Cushing's syndrome, plasma cortisol levels were normalized and as a result, blood pressure, wait-to-hip ratio (WHR) and LDL cholesterol decreased (Faggiano et al. 2003). These observations suggest that there may be a role of glucocorticoid excess in the development of the metabolic syndrome and suppression of glucocorticoid action can correct some of the metabolic disturbances. However, it is noteworthy that the patients were still at high cardiovascular risk after surgery, perhaps due to established vascular damage and atherosclerotic plaques that cannot be reversed by normalizing cortisol levels (Faggiano et al. 2003). Treatment of Cushing's syndrome with antiglucocorticoids also improved metabolic functions. Administration of GR antagonist RU 486 reduced central obesity and glycosylated hemoglobin A_{1c} and reversed heart failure in patients with Cushing's syndrome (Chu et al. 2001; Nieman et al. 1985). In addition, ketoconazole, a potent inhibitor of 11β-hydroxylase and cholesterol side-chain cleavage and therefore a blocker of adrenal cortisol production, was also effective in treating Cushing's syndrome (Sonino 1987). These clinical experience indicate that blockade of glucocorticoid action is a viable approach to improve metabolic functions in humans, especially under conditions of glucocorticoid excess. In light of these findings, one relevant question is whether type 2 diabetics have glucocorticoid excess so that glucocorticoid blockade can be used as a treatment strategy. Several reports indicate that there is hypercortisolism in some subjects with type 2 diabetes (Cameron et al. 1987; Lee et al. 1999; Oltmanns et al. 2006).

However, these studies were conducted with small numbers of subjects. A general conclusion needs to be substantiated by a larger study. It seems that only 2–5% of type 2 diabetics have hypercortisolism (Findling and Raff 2005), suggesting that glucocorticoid excess is not a common factor in the disease. Despite this caveat, the question still remains whether suppression of glucocorticoid action is metabolically beneficial to patients with type 2 diabetes. Such a treatment strategy can be implemented with GR antagonists and ketoconazole, a potent inhibitor of cortisol biosynthesis. However, the use of these agents is not clinically desirable because they cause HPA axis activation (Lamberts et al. 1991; Laue et al. 1990; Pont et al. 1984). Further, ketoconazole is not a selective inhibitor of cortisol biosynthesis; it also inhibits testosterone biosynthesis leading to unacceptable clinical side effects (Pont et al. 1984; Sonino 1987). Since the HPA axis activation caused by RU 456 treatment is thought to be triggered by GR antagonism in brain, an alternative approach is liver-specific GR blockade. GR antagonists with liver-selective distribution were designed by conjugating bile acids to RU 486 (von Geldern et al. 2004). One liver-selective GR antagonist from this effort suppressed hepatic glucose production and improved insulin sensitivity in animal models of type 2 diabetes (Jacobson et al. 2005; Zinker et al. 2007). These data suggest that blockade of hepatic glucocorticoid action can improve metabolic characteristics in patients with type 2 diabetes. However, to this date, there has been no reported clinical success with this approach, perhaps due to limited bioavailability of the conjugated molecules. A new strategy is to target tissue-specific glucocorticoid action by inhibiting 11β-HSD1 in subjects with type 2 diabetes. This begs the question of whether 11β-HSD1 expression is dysregulated in type 2 diabetes and further, if inhibition of glucocorticoid action at the tissue level is sufficient to improve insulin sensitivity.

3   Association of 11β-HSD1 with Obesity and Insulin Resistance

The expression and activity of 11β-HSD1 have been closely examined in animal models of type 2 diabetes and obese human subjects. The hepatic expression of 11β-HSD1 is reduced in *ob/ob* mice (Liu et al. 2003). A similar impairment was seen in obese Zucker rats, but the adipose expression was elevated in these animals (Livingstone et al. 2000a, b). Tissue-specific cortisol metabolism was examined in three groups of age-matched men with low, middle and high body mass index (BMI) values. The study found that the hepatic 11β-HSD1 activity decreased with the degree of obesity while the adipose 11β-HSD1 activity was positively correlated with BMI (Rask et al. 2001). These data are consistent with the results from several additional studies (Rask et al. 2002; Stewart et al. 1999; Wake et al. 2003). In addition, the omental 11β-HSD1 activity in obese women is positively correlated with larger omental adipocytes, increased omental lipolysis, increased lipoprotein lipase activity, decreased high-density lipoprotein cholesterol, decreased adiponectin levels, and increased insulin resistance (Veilleux et al. 2009). The adipose 11β-HSD1 expression was also examined in 17 young adult monozygotic twin pairs with an intrapair difference in BMI of 3.8 kg/m^2 (Kannisto et al. 2004). This study was intended to investigate the relationship of adipose 11β-HSD1 expression and the degree of obesity independent of genetic influence. The adipose 11β-HSD1 expression is higher in the obese twin than the lean counterpart and the degree of elevation in expression is positively correlated with the difference in BMI in this study (Kannisto et al. 2004), suggesting that 11β-HSD1 expression in adipose tissue is increased in acquired obesity.

The physiological role of elevated adipose 11β-HSD1 expression in obese human subjects was examined in animal models. Transgenic mice were created with selective overexpression of 11β-HSD1 in adipose tissue at a level similar to that observed in human obesity (Masuzaki et al. 2001). These animals had increased adipose corticosterone regeneration and developed central obesity, insulin resistance, hyperlipidemia and hypertension (Masuzaki et al. 2001, 2003). This collection of phenotypes resembles those in human metabolic syndrome. Like obese humans, these mice developed hyperleptinemia (Masuzaki et al. 2001). The insulin resistance in these animals correlated with impaired glucose, reduced insulin sensitivity and increased mesenteric and subcutaneous adipocyte size (Masuzaki et al. 2001). Selective overexpression of 11β-HSD1 in liver only caused mild insulin resistance without altering body fat mass (Paterson et al. 2004). These animals had elevated liver triglyceride and serum non-esterified fatty acid (NEFA) levels (Paterson et al. 2004). They also developed hypertension paralleled by increased hepatic angiotensinogen expression (Paterson et al. 2004). These data demonstrate the important metabolic role of 11β-HSD1 in both liver and adipose tissue. Further, the results suggest that the adipose glucocorticoid regeneration by 11β-HSD1 appears to be more critical in mediating metabolic effects in animals. Mice deficient in 11β-HSD1 were unable to regenerate corticosterone from 11-DHC [i.e. the rodent cortisone analogue, see above] and had attenuated activation of the key hepatic gluconeogenic enzymes phosphoenolpyruvate carboxykinase (PEPCK) and glucose-6-phosphatase (G6Pase) (Kotelevtsev et al. 1997). These animals exhibited HPA axis activation and developed adrenal hyperplasia (Kotelevtsev et al. 1997). Despite the potential neutralizing effects of increased glucocorticoid exposure caused by HPA axis activation, these animals had improved glucose tolerance, elevated HDL, and reduced triglyceride levels (Morton et al. 2001). Moreover, 11β-HSD1 knockout mice exhibited increased adipose expression of adiponectin, PPARγ, and UCP-2 but reduced expression of leptin, resistin, and TNFα in the same tissue (Morton et al. 2004a). Consistent with this finding, these animals had reduced visceral fat accumulation upon high fat feeding, improved insulin sensitivity, and increased energy expenditure (Morton et al. 2004a). These data suggest that inhibition of 11β-HSD1 is an interesting strategy to improve insulin sensitivity and mitigate other characteristics of the metabolic syndrome. Since the adipose 11β-HSD1 seems to play a more important role in metabolic homeostasis and its expression in obese humans is elevated, inhibition of the adipose 11β-HSD1 is thought to be more therapeutically relevant than the liver enzyme. Mice with selective overexpression of 11β-HSD2 were generated to suppress glucocorticoid action specifically in adipose tissue (Kershaw et al. 2005). Since the two 11β-HSD isozymes catalyze the opposite reactions of glucocorticoid interconversion, the biological effect of 11β-HSD2 overexpression is equivalent to inhibition of the adipose 11β-HSD1. As expected, the systemic glucocorticoid exposure did not change in these animals (Kershaw et al. 2005). They were resistant to high fat diet-induced weight gain and had increased energy expenditure as well as improved insulin sensitivity (Kershaw et al. 2005). Like the 11β-HSD1 knockout mice (Morton et al. 2004a), the 11β-HSD2 adipose transgenic mice on high fat diet had increased adipose expression of adiponectin, PPARγ, and UCP-2 (Kotelevtsev et al. 1997). These data support the premise that suppression of glucocorticoid action by inhibiting the adipose 11β-HSD1 is a viable strategy to improve insulin sensitivity and treat type 2 diabetes.

Pharmacological studies with 11β-HSD1 inhibitors in animal models of obesity and insulin resistance produced results similar to those observed in the adipose-specific 11β-HSD2 transgenic mice. BVT.2733 is a selective mouse 11β-HSD1 inhibitor identified through medicinal chemistry efforts (Barf et al. 2002). After administration to *ob/ob* and KKAy mice, this compound lowered blood glucose, improved glucose tolerance and utilization, and suppressed endogenous glucose production (Alberts et al. 2002, 2003). Compound 544 is a selective 11β-HSD1 inhibitor from a different chemical class. Treatment of diet-induced obesity (DIO) mice with this compound led to the reduction of body weight gain, retroperitoneal fat mass, fasting glucose, serum insulin, triglycerides, and cholesterol (Hermanowski-Vosatka et al. 2005). Further, the same compound decreased fasting glucose, serum insulin, triglycerides and NEFAs, and improved glucose tolerance in high fat-fed and streptozotocin-treated (HF/STZ) mice (Hermanowski-Vosatka et al. 2005). Interestingly, compound 544 also decreased aortic lesion areas as well as the cholesterol content in atherosclerotic lesions in apoE knockout mice (Hermanowski-Vosatka et al. 2005), suggesting that 11β-HSD1 inhibitors may be effective in treating atherosclerosis.

4 11β-HSD1 Biochemistry and Regulation

11β-HSD1 is anchored to the endoplasmic reticulum (ER) via a short N-terminal transmembrane region with the bulk of the protein in the ER lumen (Ozols 1995). Several putative N-linked glycosylation sites in the luminal bulk region were identified (Oppermann et al. 1995; Ozols 1995), and mutations of these sites were reportedly to affect the enzymatic activity (Agarwal et al. 1995). However, a later study by direct deglycosylation demonstrated that the N-linked glycosylation is not required for enzymatic activity (Blum et al. 2000). 11β-HSD1 exists as a homodimer (Maser et al. 2002), although a monomeric form was also observed in human liver extracts (Walker et al. 2001). 11β-HSD1 utilizes cortisone in human and 11-DHC in rodents as substrates. It also catalyzes the conversion of the synthetic steroidal substrate prednisone to prednisolone. 11β-HSD1 is a typical member of the small chain dehydrogenase/reductase (SDR) superfamily with 3,000 known members (Oppermann et al. 2003). The catalytic domain in a SDR contains the YXXXK signature sequence, which is represented as YSASK in 11β-HSD1. The two serines are not critical in substrate binding but involved in determining catalytic rate (Obeyesekere et al. 1998). The reductase activity of 11β-HSD1 requires the presence of NADPH as a co-factor. The regeneration of NADPH by hexose-6-phosphate dehydrogenase (H6PDH), which is also within the ER lumen, drives the reductase activity of 11β-HSD1 (Walker et al. 2007). Glucose-6-phosphate (G6P) is the main substrate of H6PDH and therefore a key component for NADPH regeneration. G6P must be translocated from the cytosol to the ER lumen via a G6P transporter (G6PT). After reaching the ER lumen, G6P can be hydrolyzed by G6Pase-α (van Schaftingen and Gerin 2002). Therefore, the trafficking and accumulation of G6P in the ER is important for NADPH accumulation and regulates 11β-HSD1 activity (Walker et al. 2007). Since G6P is an intermediate product in gluconeogenesis and glycogenolysis and cortisol augments glucose production, the regulation of 11β-HSD1 activity by H6PDH represents a link between glucose metabolism and glucocorticoid action (Fig. 2).

The expression of 11β-HSD1 is regulated at multiple levels. HIV patients on highly active antiretroviral therapy (HAART) develop poorly understood lipodystrophy, hypertriglyceridemia, and insulin resistance (Carr and Cooper 2000). This disorder is named pseudo-Cushing's syndrome due to its symptomatic resemblance to Cushing's syndrome. For instance, the distribution of fat accumulation in these patients is similar to that induced by glucocorticoid excess in Cushing's syndrome (Miller et al. 1998). However, the circulating cortisol levels are not elevated in these patients (Miller et al. 1998). Interestingly, the adipose expression of

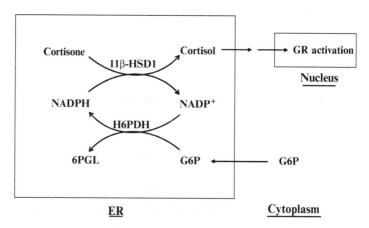

Fig. 2 The reaction catalyzed by 11β-HSD1 and its dependence on the NADPH regeneration by H6PDH. *G6P*, glucose-6-phosphate; *6PGL*, 6-phosphogluconate; *ER*, endoplasmic reticulum; *GR*, glucocorticoid receptor

11β-HSD1 in these patients is elevated implicating tissue glucocorticoid excess in the development of the disease. In addition, the hepatic 11β-HSD1 expression is elevated by fivefold in alcoholic liver disease (ALD) (Ahmed et al. 2008), suggesting that hepatic glucocorticoid excess may play a role in the development of insulin resistance in this disease. Further, the adipose 11β-HSD1 expression is suppressed by rosiglitazone in obese human subjects with impaired glucose tolerance (Mai et al. 2007), which may in part account for the beneficial effects of thiazolidinediones (TZDs). One interesting aspect of the adipose 11β-HSD1 expression is that it responds to obesity differently in humans and rodents. The expression of 11β-HSD1 in subcutaneous adipose tissue is elevated with human obesity (Rask et al. 2001; Wake et al. 2003). However, its adipose expression is reduced in DIO animals compared with lean controls (Drake et al. 2005; Morton et al. 2004b). This could represent an adaptation mechanism in rodents that protects against metabolic complications of obesity.

5 Non-selective 11β-HSD1 Inhibitors as Tools

Glycyrrhetinic acid (GA) is a naturally occurring 11β-HSD1 inhibitor that is typically used for allergic or infectious inflammation of the skin. Its synthetic hemisuccinate ester, carbenoxolone (CBX), is a licensed drug in the UK for oesophageal ulceration and inflammation and the treatment of oral and perioral lesions. Neither CBX nor GA is a selective 11β-HSD1 inhibitor because they also inhibit 11β-HSD2 (Diederich et al. 2000; Hult et al. 1998). Despite this limitation, CBX has been used as an 11β-HSD1 inhibitor in animal and human studies. CBX attenuated fasting plasma lipid and insulin levels, decreased liver triglyceride and free cholesterol, and reduced atherosclerotic lesion areas in severely obese mice derived from heterozygous agouti ($A^{y/a}$) and homozygous $LDLR^{-/-}$ breeding pairs ($A^{y/a};LDLR^{-/-}$ mice) (Nuotio-Antar et al. 2007). In type 2 diabetic human subjects, CBX reduced hepatic glucose production through suppression of glycogenolysis (Andrews et al. 2003). CBX also decreased lipolysis in human subjects (Tomlinson et al. 2007). These data suggest that 11β-HSD1 inhibition may lead to beneficial metabolic effects. Interestingly, CBX improved cognitive function in healthy elderly men and type 2 diabetics (Sandeep et al. 2004). This is consistent with a similar observation made in 11β-HSD1 knockout mice (Yau et al. 2001), suggesting that 11β-HSD1 inhibitors could be used to treat neurological disorders. However, it is important to stress that CBX is not a selective 11β-HSD1 inhibitor. Thus, these findings should be validated with studies using selective 11β-HSD1 inhibitors. In addition to licorice derivatives, several bile acids are also 11β-HSD1 inhibitors but with much lower potency values (Diederich et al. 2000). The 11β-hydroxylase inhibitor metyrapone is also a weak inhibitor of 11β-HSD1 with a potency value in the sub to single mM range (Diederich et al. 2000; Sampath-Kumar et al. 1997). The in vivo effect of metyrapone on 11β-HSD1 may be therefore limited.

CBX has been a useful tool to better understand the active site structure of 11β-HSD1 and even design selective inhibitors. The cocrystal structure of CBX with human 11β-HSD1 demonstrated that the enzyme is a homodimer (PDB code: 2BEL.pdb). In the cocrystal structure, the C-terminus appears to have some flexibility and its role in CBX binding by the dimer partner is not clear (Kim et al. 2007). We have conducted mutagenesis studies and found the C-terminus is not important in CBX binding (Kim et al. 2007). The binding of GA by 11β-HSD1 is substantially different from that of CBX. We carried out modeling studies and discovered that residue Y177 is involved in the binding of the A ring of GA through van der Waals interactions (Kim et al. 2006).

6   Studies with Selective 11β-HSD1 Inhibitors

Different chemical classes of selective 11β-HSD1 inhibitors have been made with medicinal chemistry efforts. The first published selective inhibitors were from Biovitrum (Barf et al. 2002). One interesting finding is the species selectivity of compounds with respect to potency. BVT.2733 is more potent against the mouse enzyme than the human enzyme (Barf et al. 2002). In contrast, BVT.14225, a close analog of BVT.2733, is more potent against human 11β-HSD1 than the mouse enzyme (Barf et al. 2002). This is not surprising because based on the crystal structures of the human and mouse 11β-HSD1 (Hosfield et al. 2005; Zhang et al. 2005), there are significant differences at the steroidal binding sites where the binding of a compound could be favored by one enzyme relative to the other. This phenomenon is further exemplified by the cross-species potency shifts of several other compounds summarized in Table 1. The first three compounds in Table 1 exhibit higher potency for rat 11β-HSD1 than the human enzyme (Richards et al. 2006). The other three compounds, however, have higher potency for human 11β-HSD1 than the mouse enzyme (Rohde et al. 2007; Sorensen et al. 2006; St Jean et al. 2007). Not all 11β-HSD1 inhibitors have species selectivity with respect to human and rodent activities; there are many compounds with comparable human and rodent potency values.

Several well studied 11β-HSD1 inhibitors include compound 544 from Merck (Hermanowski-Vosatka et al. 2005), compound 2922 from Amgen (Hale et al. 2008; St Jean et al. 2007), and compound PF-915275 from Pfizer (Bhat et al. 2008). The selectivity of both compounds 544 and 2922 over other human targets were demonstrated by counter screens or assays (Hale et al. 2008; Hermanowski-Vosatka et al. 2005). Compound 544 was tested in ex vivo assays using mouse adipose, liver, and brain. The assay method is based on the notion that the compound is distributed to tissues via circulation after oral dosing. When a harvested tissue piece is incubated with a substrate (i.e., cortisone) in an ex vivo assay, the conversion of the substrate to product is mediated by the remaining 11β-HSD1 activity in the tissue. The reduction in total activity compared with control tissues should represent the inhibitory effect of the compound. This ex vivo

Table 1 Representative 11β-HSD1 inhibitors with differential cross-species potency values

Structure	Company	h-HSD1 K_i (nM)	Rodent 11β-HSD1 K_i (nM)	References
	Abbott	700	20 (r-HSD1)	Richards et al. (2006)
	Abbott	560	9 (r-HSD1)	Richards et al. (2006)
	Abbott	1,100	17 (r-HSD1)	Richards et al. (2006)
	Abbott	7	500 (m-HSD1)	Sorensen et al. (2006)
	Abbott	8	34 (m-HSD1)	Rohde et al. (2007)

(continued)

Table 1 (continued)

Structure	Company	h-HSD1 K_i (nM)	Rodent 11β-HSD1 K_i (nM)	References
	Amgen	22	130 (m-HSD1)	St Jean et al. (2007)

h-HSD1, human 11β-HSD1; *m-HSD1*, mouse 11β-HSD1; *r-HSD1*, rat 11β-HSD1.

method generates results consistent with those from in vivo enzymatic assays, where the synthetic glucocorticoid prednisone was given to compound-treated animals and the conversion of prednisone to prednisolone in vivo is monitored to assess the whole body 11β-HSD1 activity (Bhat et al. 2008; Hale et al. 2008). Under such a condition, the measured enzymatic activity is primarily the hepatic 11β-HSD1 activity. Amgen tested compound 2922 in an adipose ex vivo assay and an in vivo prednisone to prednisolone conversion assay (Hale et al. 2008). Pfizer tested their compound PF-915275 in monkeys with an in vivo prednisone to prednisolone conversion assay (Bhat et al. 2008). All these data demonstrated that these compounds inhibited 11β-HSD1 activity. In addition to the pharmacological studies conducted with compound 544 and BVT.2733 as described above (Alberts et al. 2002, 2003; Hermanowski-Vosatka et al. 2005), compound 2922 from Amgen was also demonstrated to improve insulin sensitivity in DIO mice (Véniant et al. 2009). Interestingly, we noticed that the time of the day for compound administration plays a role in efficacy due to the circadian variation of the circulating glucocorticoids (Véniant et al. 2009). Moreover, Merck's compound A was tested in DIO rats and it decreased the expression of genes involved in lipid synthesis and fatty acid cycling in mesenteric fat (Berthiaume et al. 2007a). It also reduced hepatic triglyceride secretion and increased lipid oxidation in muscle and heart (Berthiaume et al. 2007b). Further, this compound increased fatty acid oxidation and the expression levels of several enzymes involved in mitochondrial and peroxisomal β-oxidation in the liver of rats fed an obesogenic diet (Berthiaume et al. 2010). In Ldlr 3KO (Ldlr(−/−)Apob(100/100)Lep(ob/ob)) mice, a genetic model of obesity, insulin resistance, dyslipidemia, and atherosclerosis, we demonstrated that compound 2922 improved glucose metabolism (Hermanowski-Vosatka et al. 2005). Unlike compound 544 which decreased atherosclerotic lesions in apoE knockout mice (Hermanowski-Vosatka et al. 2005), compound 2922 did not have any effect on the atherosclerotic lesions in the 3KO model (Lloyd et al. 2009). These data indicate that inhibition of 11β-HSD1 is metabolically beneficial but the improved metabolic profiles are likely to depend on the models used.

7   Clinical Experience with 11β-HSD1 Inhibitors

The first reported clinical study with an 11β-HSD1 inhibitor is the evaluation of CBX in a small number of type 2 diabetic patients. Although CBX marginally improved insulin sensitivity in type 2 diabetics (Andrews et al. 2003), the study is not ideal because CBX is not a selective 11β-HSD1 inhibitor. The Pfizer compound PF-915275 was tested in a phase 1 clinical trial with 60 healthy adult volunteers. After multiple oral doses, this compound was generally safe and well tolerated (Courtney et al. 2008). Using an in vivo prednisone to prednisolone conversion assay, it was demonstrated that 11β-HSD1 activity was inhibited by this compound in the individuals that received the drug (Courtney et al. 2008). In the meantime, the ratio of key cortisol and cortisone metabolites (5αTHF + 5βTHF:THE), a bio-marker for 11β-HSD1 activity, was also reduced (Courtney et al. 2008). These data demonstrate that PF-915275 exhibited pharmacodynamic effect on 11β-HSD1 in vivo and may be suitable for a study in type 2 diabetic patients. To date, there is no report on whether this compound improves insulin sensitivity in human subjects with type 2 diabetes.

Incyte Corporation released data from a phase 2a study in type 2 diabetics using their 11β-HSD1 inhibitor INCB13739. In this study, obese type 2 diabetic patients were treated with the 11β-HSD1 inhibitor for 4 weeks and at the end of the treatment, there was a trend of reduced fasting glucose and glucose production and increased glucose disposal (Incyte 2007, 2008). In a phase 2b trial involving 302 type 2 diabetic patients on metformin but with inadequate glycemic control, daily oral administration of 200 mg of INCB13739 for 12 weeks resulted in significant reductions in HbA1c, fasting plasma glucose and HOMA-IR with additional beneficial effects on plasma lipids and body weight (Rosenstock et al. 2010). Although the compound caused elevation of ACTH, these results demonstrate that inhibition of 11β-HSD1 is a valid approach to treat elements of the metabolic syndrome.

8   Potential Therapeutic Challenges with 11β-HSD1 Inhibitors

Accumulating evidence in preclinical animal models of obesity and insulin resistance suggest that 11β-HSD1 is an interesting target for the treatment of type 2 diabetes. This notion is further supported by elevated expression of 11β-HSD1 in the adipose tissue of obese human subjects. Although systematic glucocorticoid excess has been observed in some type 2 diabetics (Cameron et al. 1987; Lee et al. 1999; Oltmanns et al. 2006), the percentage of patients with this abnormality is fairly small (Findling and Raff 2005), suggesting that glucocorticoid excess is not the major underlying mechanism for the metabolic abnormalities in these patients. However, suppression of glucocorticoid action in these patients may still provide beneficial metabolic effects and therefore represents a viable treatment strategy. The phase 2a and 2b studies with patients suffering from type 2 diabetes by Incyte Corporation (Incyte 2007, 2008; Rosenstock et al. 2010) demonstrated concrete evidence of human efficacy with 11β-HSD1 inhibitors as antidiabetic therapy.

One major concern with 11β-HSD1 inhibitors is potential HPA activation. 11β-HSD1 is an enzyme that generates the active glucocorticoid in tissues. Since it is expressed in brain, inhibition of the enzyme could result in reduced local glucocorticoid in certain brain regions and consequently leads to compensatory feedback response by the HPA axis. HPA axis activation has been observed in 11β-HSD1 knockout mice, where the 24-h circadian concentration of corticosterone remains elevated throughout most of the day (Harris et al. 2001). Mild HPA axis activation was observed in humans deficient in 11β-HSD1 (Jamieson et al. 1999; Phillipou and Higgins 1985). Further, a 3-week repeated treatment of rats with an 11β-HSD1 inhibitor increased adrenal weight by 38%, an indication of HPA axis activation (Berthiaume et al. 2007a). Although no signs of HPA axis activation were observed in either the Incyte phase 2a or the Pfizer 11β-HSD1 clinical study (Incyte 2007, 2008; Courtney et al. 2008), there is clear evidence of ACTH elevation in the Incyte phase 2b study (Rosenstock et al. 2010), confirming that activation of the HPA axis is a potential effect associated with 11β-HSD1 inhibition.

9 Concluding Remarks

The link of glucocorticoid action to the development of multiple metabolic disorders in the metabolic syndrome has been implicated in Cushing's syndrome. This led to attempts to explain the etiology of metabolic syndrome with potential glucocorticoid excess. However, glucocorticoid excess is not largely prevalent in type 2 diabetic patients and therefore, it is unlikely to be the major underlying mechanism for the development of obesity and insulin resistance. However, tissue-specific glucocorticoid excess has been observed in obese humans in the form of elevated adipose 11β-HSD1 expression. It is possible that tissue-specific glucocorticoid excess caused multiple metabolic defects in the metabolic syndrome. This premise is supported by genetic studies with 11β-HSD1 transgenic and knockout mice as well as pharmacologic studies with 11β-HSD1 inhibitors. These data strongly suggest that 11β-HSD1 inhibitors are potential treatments for type 2 diabetes and other metabolic disorders. A large variety of chemical series of 11β-HSD1 inhibitors have been identified through independent medicinal chemistry efforts. With the encouraging clinical results from the Incyte phase 2b study (Rosenstock et al. 2010), additional clinical testing of some of these compounds may continue to generate promising results in the near future in support of this class of antidiabetic drugs.

References

Agarwal AK, Mune T, Monder C, White PC (1995) Mutations in putative glycosylation sites of rat 11β-hydroxysteroid dehydrogenase affect enzymatic activity. Biochim Biophys Acta 1248:70–74

Ahmed A, Saksena S, Sherlock M, Olliff SP, Elias E, Stewart PM (2008) Induction of hepatic 11β-hydroxysteroid dehydrogenase type 1 in patients with alcoholic liver disease. Clin Endocrinol (Oxf) 68:898–903

Alberts P, Engblom L, Edling N, Forsgren M, Klingstrom G, Larsson C, Ronquist-Nii Y, Ohman B, Abrahmsen L (2002) Selective inhibition of 11β-hydroxysteroid dehydrogenase type 1 decreases blood glucose concentrations in hyperglycaemic mice. Diabetologia 45:1528–1532

Alberts P, Nilsson C, Selen G, Engblom LO, Edling NH, Norling S, Klingstrom G, Larsson C, Forsgren M, Ashkzari M, Nilsson CE, Fiedler M, Bergqvist E, Ohman B, Bjorkstrand E, Abrahmsen LB (2003) Selective inhibition of 11β-hydroxysteroid dehydrogenase type 1 improves hepatic insulin sensitivity in hyperglycemic mice strains. Endocrinology 144:4755–4762

Andrews RC, Rooyackers O, Walker BR (2003) Effects of the 11β-hydroxysteroid dehydrogenase inhibitor carbenoxolone on insulin sensitivity in men with type 2 diabetes. J Clin Endocrinol Metab 88:285–291

Arnaldi G, Angeli A, Atkinson AB, Bertagna X, Cavagnini F, Chrousos GP, Fava GA, Findling JW, Gaillard RC, Grossman AB, Kola B, Lacroix A, Mancini T, Mantero F, Newell-Price J, Nieman LK, Sonino N, Vance ML, Giustina A, Boscaro M (2003) Diagnosis and complications of Cushing's syndrome: a consensus statement. J Clin Endocrinol Metab 88:5593–5602

Barf T, Vallgarda J, Emond R, Haggstrom C, Kurz G, Nygren A, Larwood V, Mosialou E, Axelsson K, Olsson R, Engblom L, Edling N, Ronquist-Nii Y, Ohman B, Alberts P, Abrahmsen L (2002) Arylsulfonamidothiazoles as a new class of potential antidiabetic drugs. Discovery of potent and selective inhibitors of the 11β-hydroxysteroid dehydrogenase type 1. J Med Chem 45:3813–3815

Berthiaume M, Laplante M, Festuccia W, Gelinas Y, Poulin S, Lalonde J, Joanisse DR, Thieringer R, Deshaies Y (2007a) Depot-specific modulation of rat intraabdominal adipose tissue lipid metabolism by pharmacological inhibition of 11β-hydroxysteroid dehydrogenase type 1. Endocrinology 148:2391–2397

Berthiaume M, Laplante M, Festuccia WT, Cianflone K, Turcotte LP, Joanisse DR, Olivecrona G, Thieringer R, Deshaies Y (2007b) 11β-HSD1 inhibition improves triglyceridemia through reduced liver VLDL secretion and partitions lipids toward oxidative tissues. Am J Physiol Endocrinol Metab 293:E1045–E1052

Berthiaume M, Laplante M, Festuccia WT, Berger JP, Thieringer R, Deshaies Y (2010) Preliminary report: pharmacologic 11beta-hydroxysteroid dehydrogenase type 1 inhibition increases hepatic fat oxidation in vivo and expression of related genes in rats fed an obesogenic diet. Metabolism 59:114–117

Bhat BG, Hosea N, Fanjul A, Herrera J, Chapman J, Thalacker F, Stewart PM, Rejto PA (2008) Demonstration of proof of mechanism and pharmacokinetics and pharmacodynamic relationship with 4'-cyano-biphenyl-4-sulfonic acid (6-amino-pyridin-2-yl)-amide (PF-915275), an inhibitor of 11β-hydroxysteroid dehydrogenase type 1, in cynomolgus monkeys. J Pharmacol Exp Ther 324:299–305

Blum A, Martin HJ, Maser E (2000) Human 11β-hydroxysteroid dehydrogenase type 1 is enzymatically active in its nonglycosylated form. Biochem Biophys Res Commun 276:428–434

Bujalska I, Shimojo M, Howie A, Stewart PM (1997) Human 11β-hydroxysteroid dehydrogenase: studies on the stably transfected isoforms and localization of the type 2 isozyme within renal tissue. Steroids 62:77–82

Cameron OG, Thomas B, Tiongco D, Hariharan M, Greden JF (1987) Hypercortisolism in diabetes mellitus. Diab Care 10:662–664

Carr A, Cooper DA (2000) Adverse effects of antiretroviral therapy. Lancet 356:1423–1430

Chu JW, Matthias DF, Belanoff J, Schatzberg A, Hoffman AR, Feldman D (2001) Successful long-term treatment of refractory Cushing's disease with high-dose mifepristone (RU 486). J Clin Endocrinol Metab 86:3568–3573

Courtney R, Stewart PM, Toh M, Ndongo MN, Calle RA, Hirshberg B (2008) Modulation of 11β-hydroxysteroid dehydrogenase (11βHSD) activity biomarkers and pharmacokinetics of PF-00915275, a selective 11βHSD1 inhibitor. J Clin Endocrinol Metab 93:550–556

Davis GF (1986) Adverse effects of corticosteroids: II. Systemic. Clin Dermatol 4:161–169

Delbende C, Delarue C, Lefebvre H, Bunel DT, Szafarczyk A, Mocaer E, Kamoun A, Jegou S, Vaudry H (1992) Glucocorticoids, transmitters and stress. Br J Psychiatry Suppl 15:24–35

Diederich S, Grossmann C, Hanke B, Quinkler M, Herrmann M, Bahr V, Oelkers W (2000) In the search for specific inhibitors of human 11β-hydroxysteroid-dehydrogenases (11β-HSDs): chenodeoxycholic acid selectively inhibits 11β-HSD-I. Eur J Endocrinol 142:200–207

Drake AJ, Livingstone DE, Andrew R, Seckl JR, Morton NM, Walker BR (2005) Reduced adipose glucocorticoid reactivation and increased hepatic glucocorticoid clearance as an early adaptation to high-fat feeding in Wistar rats. Endocrinology 146:913–919

Dunn JF, Nisula BC, Rodbard D (1981) Transport of steroid hormones: binding of 21 endogenous steroids to both testosterone-binding globulin and corticosteroid-binding globulin in human plasma. J Clin Endocrinol Metab 53:58–68

Faggiano A, Pivonello R, Spiezia S, De Martino MC, Filippella M, Di Somma C, Lombardi G, Colao A (2003) Cardiovascular risk factors and common carotid artery caliber and stiffness in patients with Cushing's disease during active disease and 1 year after disease remission. J Clin Endocrinol Metab 88:2527–2533

Findling JW, Raff H (2005) Screening and diagnosis of Cushing's syndrome. Endocrinol Metab Clin North Am 34:385–402, ix–x

Ford ES, Giles WH, Dietz WH (2002) Prevalence of the metabolic syndrome among US adults: findings from the third National Health and Nutrition Examination Survey. JAMA 287:356–359

Gallant C, Kenny P (1986) Oral glucocorticoids and their complications. A review. J Am Acad Dermatol 14:161–177

Grundy SM, Brewer HB Jr, Cleeman JI, Smith SC Jr, Lenfant C (2004) Definition of metabolic syndrome: Report of the National Heart, Lung, and Blood Institute/American Heart Association conference on scientific issues related to definition. Circulation 109:433–438

Hale C, Veniant M, Wang Z, Chen M, McCormick J, Cupples R, Hickman D, Min X, Sudom A, Xu H, Matsumoto G, Fotsch C, St. Jean DJ Jr, Wang M (2008) Structural characterization and pharmacodynamic effects of an orally active 11β-hydroxysteroid dehydrogenase type 1 inhibitor. Chem Biol Drug Des 71:36–44

Harris HJ, Kotelevtsev Y, Mullins JJ, Seckl JR, Holmes MC (2001) Intracellular regeneration of glucocorticoids by 11β-hydroxysteroid dehydrogenase (11β-HSD)-1 plays a key role in regulation of the hypothalamic-pituitary-adrenal axis: analysis of 11β-HSD-1-deficient mice. Endocrinology 142:114–120

Hermanowski-Vosatka A, Balkovec JM, Cheng K, Chen HY, Hernandez M, Koo GC, Le Grand CB, Li Z, Metzger JM, Mundt SS, Noonan H, Nunes CN, Olson SH, Pikounis B, Ren N, Robertson N, Schaeffer JM, Shah K, Springer MS, Strack AM, Strowski M, Wu K, Wu T, Xiao J, Zhang BB, Wright SD, Thieringer R (2005) 11β-HSD1 inhibition ameliorates metabolic syndrome and prevents progression of atherosclerosis in mice. J Exp Med 202:517–527

Hosfield DJ, Wu Y, Skene RJ, Hilgers M, Jennings A, Snell GP, Aertgeerts K (2005) Conformational flexibility in crystal structures of human 11β-hydroxysteroid dehydrogenase type I provide insights into glucocorticoid interconversion and enzyme regulation. J Biol Chem 280:4639–4648

Hult M, Jornvall H, Oppermann UC (1998) Selective inhibition of human type 1 11β-hydroxysteroid dehydrogenase by synthetic steroids and xenobiotics. FEBS Lett 441:25–28

Incyte (2007) Incyte corporation presentation. 2007 UBS global life sciences conference

Incyte (2008) Incyte presentation. 26th annual JPMorgan healthcare conference

Isomaa B, Almgren P, Tuomi T, Forsen B, Lahti K, Nissen M, Taskinen MR, Groop L (2001) Cardiovascular morbidity and mortality associated with the metabolic syndrome. Diab Care 24:683–689

Jacobson PB, von Geldern TW, Ohman L, Osterland M, Wang J, Zinker B, Wilcox D, Nguyen PT, Mika A, Fung S, Fey T, Goos-Nilsson A, Grynfarb M, Barkhem T, Marsh K, Beno DW, Nga-Nguyen B, Kym PR, Link JT, Tu N, Edgerton DS, Cherrington A, Efendic S, Lane BC, Opgenorth TJ (2005) Hepatic glucocorticoid receptor antagonism is sufficient to reduce elevated hepatic glucose output and improve glucose control in animal models of type 2 diabetes. J Pharmacol Exp Ther 314:191–200

Jamieson PM, Chapman KE, Edwards CR, Seckl JR (1995) 11β-hydroxysteroid dehydrogenase is an exclusive 11β- reductase in primary cultures of rat hepatocytes: effect of physicochemical and hormonal manipulations. Endocrinology 136:4754–4761

Jamieson A, Wallace AM, Andrew R, Nunez BS, Walker BR, Fraser R, White PC, Connell JM (1999) Apparent cortisone reductase deficiency: a functional defect in 11β-hydroxysteroid dehydrogenase type 1. J Clin Endocrinol Metab 84:3570–3574

Jamieson PM, Walker BR, Chapman KE, Andrew R, Rossiter S, Seckl JR (2000) 11β-hydro-xysteroid dehydrogenase type 1 is a predominant 11β-reductase in the intact perfused rat liver. J Endocrinol 165:685–692

Kannisto K, Pietilainen KH, Ehrenborg E, Rissanen A, Kaprio J, Hamsten A, Yki-Jarvinen H (2004) Overexpression of 11β-hydroxysteroid dehydrogenase-1 in adipose tissue is associated with acquired obesity and features of insulin resistance: studies in young adult monozygotic twins. J Clin Endocrinol Metab 89:4414–4421

Kershaw EE, Morton NM, Dhillon H, Ramage L, Seckl JR, Flier JS (2005) Adipocyte-specific glucocorticoid inactivation protects against diet-induced obesity. Diabetes 54:1023–1031

Kim KW, Wang Z, Busby J, Tsuruda T, Chen M, Hale C, Castro VM, Svensson S, Nybo R, Xiong F, Wang M (2006) The role of tyrosine 177 in human 11β-hydroxysteroid dehydrogenase type 1 in substrate and inhibitor binding: an unlikely hydrogen bond donor for the substrate. Biochim Biophys Acta 1764:824–830

Kim KW, Wang Z, Busby J, Tsuruda T, Chen M, Hale C, Castro VM, Svensson S, Nybo R, Xiong F, Wang M (2007) The selectivity of tyrosine 280 of human 11β-hydroxysteroid dehydroge-nase type 1 in inhibitor binding. FEBS Lett 581:995–999

Kotelevtsev Y, Holmes MC, Burchell A, Houston PM, Schmoll D, Jamieson P, Best R, Brown R, Edwards CR, Seckl JR, Mullins JJ (1997) 11β-hydroxysteroid dehydrogenase type 1 knockout mice show attenuated glucocorticoid-inducible responses and resist hyperglycemia on obesity or stress. Proc Natl Acad Sci USA 94:14924–14929

Lamberts SW, Koper JW, de Jong FH (1991) The endocrine effects of long-term treatment with mifepristone (RU 486). J Clin Endocrinol Metab 73:187–191

Laue L, Lotze MT, Chrousos GP, Barnes K, Loriaux DL, Fleisher TA (1990) Effect of chronic treatment with the glucocorticoid antagonist RU 486 in man: toxicity, immunological, and hormonal aspects. J Clin Endocrinol Metab 71:1474–1480

Lee ZS, Chan JC, Yeung VT, Chow CC, Lau MS, Ko GT, Li JK, Cockram CS, Critchley JA (1999) Plasma insulin, growth hormone, cortisol, and central obesity among young Chinese type 2 diabetic patients. Diab Care 22:1450–1457

Liu Y, Nakagawa Y, Wang Y, Li R, Li X, Ohzeki T, Friedman TC (2003) Leptin activation of corticosterone production in hepatocytes may contribute to the reversal of obesity and hyper-glycemia in leptin-deficient ob/ob mice. Diabetes 52:1409–1416

Livingstone DE, Jones GC, Smith K, Jamieson PM, Andrew R, Kenyon CJ, Walker BR (2000a) Understanding the role of glucocorticoids in obesity: tissue-specific alterations of corticoste-rone metabolism in obese Zucker rats. Endocrinology 141:560–563

Livingstone DE, Kenyon CJ, Walker BR (2000b) Mechanisms of dysregulation of 11β-hydro-xysteroid dehydrogenase type 1 in obese Zucker rats. J Endocrinol 167:533–539

Lloyd DJ, Helmering J, Cordover D, Bowsman M, Chen M, Hale C, Fordstrom P, Zhou M, Wang M, Kaufman SA, Véniant MM (2009) Antidiabetic effects of 11beta-HSD1 inhibition in a mouse model of combined diabetes, dyslipidaemia and atherosclerosis. Diabetes Obes Metab 11:688–699

Mai K, Andres J, Bobbert T, Maser-Gluth C, Mohlig M, Bahr V, Pfeiffer AF, Spranger J, Diederich S (2007) Rosiglitazone decreases 11β-hydroxysteroid dehydrogenase type 1 in subcutaneous adipose tissue. Clin Endocrinol (Oxf) 67:419–425

Maser E, Volker B, Friebertshauser J (2002) 11β-hydroxysteroid dehydrogenase type 1 from human liver: dimerization and enzyme cooperativity support its postulated role as glucocorticoid reductase. Biochemistry 41:2459–2465

Masuzaki H, Paterson J, Shinyama H, Morton NM, Mullins JJ, Seckl JR, Flier JS (2001) A transgenic model of visceral obesity and the metabolic syndrome. Science 294:2166–2170

Masuzaki H, Yamamoto H, Kenyon CJ, Elmquist JK, Morton NM, Paterson JM, Shinyama H, Sharp MG, Fleming S, Mullins JJ, Seckl JR, Flier JS (2003) Transgenic amplification of glucocorticoid action in adipose tissue causes high blood pressure in mice. J Clin Invest 112:83–90

Miller KK, Daly PA, Sentochnik D, Doweiko J, Samore M, Basgoz NO, Grinspoon SK (1998) Pseudo-Cushing's syndrome in human immunodeficiency virus-infected patients. Clin Infect Dis 27:68–72

Morton NM, Holmes MC, Fievet C, Staels B, Tailleux A, Mullins JJ, Seckl JR (2001) Improved lipid and lipoprotein profile, hepatic insulin sensitivity, and glucose tolerance in 11β-hydroxysteroid dehydrogenase type 1 null mice. J Biol Chem 276:41293–41300

Morton NM, Paterson JM, Masuzaki H, Holmes MC, Staels B, Fievet C, Walker BR, Flier JS, Mullins JJ, Seckl JR (2004a) Novel adipose tissue-mediated resistance to diet-induced visceral obesity in 11β-hydroxysteroid dehydrogenase type 1-deficient mice. Diabetes 53:931–938

Morton NM, Ramage L, Seckl JR (2004b) Down-regulation of adipose 11β-hydroxysteroid dehydrogenase type 1 by high-fat feeding in mice: a potential adaptive mechanism counteracting metabolic disease. Endocrinology 145:2707–2712

Nieman LK, Chrousos GP, Kellner C, Spitz IM, Nisula BC, Cutler GB, Merriam GR, Bardin CW, Loriaux DL (1985) Successful treatment of Cushing's syndrome with the glucocorticoid antagonist RU 486. J Clin Endocrinol Metab 61:536–540

Nuotio-Antar AM, Hachey DL, Hasty AH (2007) Carbenoxolone treatment attenuates symptoms of metabolic syndrome and atherogenesis in obese, hyperlipidemic mice. Am J Physiol Endocrinol Metab 293:E1517–E1528

Obeyesekere VR, Trzeciak WH, Li KX, Krozowski ZS (1998) Serines at the active site of 11β-hydroxysteroid dehydrogenase type I determine the rate of catalysis. Biochem Biophys Res Commun 250:469–473

Oltmanns KM, Dodt B, Schultes B, Raspe HH, Schweiger U, Born J, Fehm HL, Peters A (2006) Cortisol correlates with metabolic disturbances in a population study of type 2 diabetic patients. Eur J Endocrinol 154:325–331

Oppermann UC, Netter KJ, Maser E (1995) Cloning and primary structure of murine 11β-hydroxysteroid dehydrogenase/microsomal carbonyl reductase. Eur J Biochem 227:202–208

Oppermann U, Filling C, Hult M, Shafqat N, Wu X, Lindh M, Shafqat J, Nordling E, Kallberg Y, Persson B, Jornvall H (2003) Short-chain dehydrogenases/reductases (SDR): the 2002 update. Chem Biol Interact 143–144:247–253

Ozols J (1995) Lumenal orientation and post-translational modifications of the liver microsomal 11β-hydroxysteroid dehydrogenase. J Biol Chem 270:2305–2312

Paterson JM, Morton NM, Fievet C, Kenyon CJ, Holmes MC, Staels B, Seckl JR, Mullins JJ (2004) Metabolic syndrome without obesity: hepatic overexpression of 11β-hydroxysteroid dehydrogenase type 1 in transgenic mice. Proc Natl Acad Sci USA 101:7088–7093

Phillipou G, Higgins BA (1985) A new defect in the peripheral conversion of cortisone to cortisol. J Steroid Biochem 22:435–436

Pont A, Graybill JR, Craven PC, Galgiani JN, Dismukes WE, Reitz RE, Stevens DA (1984) High-dose ketoconazole therapy and adrenal and testicular function in humans. Arch Intern Med 144:2150–2153

Rask E, Olsson T, Soderberg S, Andrew R, Livingstone DE, Johnson O, Walker BR (2001) Tissue-specific dysregulation of cortisol metabolism in human obesity. J Clin Endocrinol Metab 86:1418–1421

Rask E, Walker BR, Soderberg S, Livingstone DE, Eliasson M, Johnson O, Andrew R, Olsson T (2002) Tissue-specific changes in peripheral cortisol metabolism in obese women: increased adipose 11β-hydroxysteroid dehydrogenase type 1 activity. J Clin Endocrinol Metab 87:3330–3336

Reaven GM (1988) Banting lecture 1988. Role of insulin resistance in human disease. Diabetes 37:1595–1607

Richards S, Sorensen B, Jae HS, Winn M, Chen Y, Wang J, Fung S, Monzon K, Frevert EU, Jacobson P, Sham H, Link JT (2006) Discovery of potent and selective inhibitors of 11β-HSD1 for the treatment of metabolic syndrome. Bioorg Med Chem Lett 16:6241–6245

Rohde JJ, Pliushchev MA, Sorensen BK, Wodka D, Shuai Q, Wang J, Fung S, Monzon KM, Chiou WJ, Pan L, Deng X, Chovan LE, Ramaiya A, Mullally M, Henry RF, Stolarik DF, Imade HM, Marsh KC, Beno DW, Fey TA, Droz BA, Brune ME, Camp HS, Sham HL, Frevert EU, Jacobson PB, Link JT (2007) Discovery and metabolic stabilization of potent and selective 2-amino-N-(adamant-2-yl) acetamide 11β-hydroxysteroid dehydrogenase type 1 inhibitors. J Med Chem 50:149–164

Rosenstock J, Banarer S, Fonseca VA, Inzucchi SE, Sun W, Yao W, Hollis G, Flores R, Levy R, Williams WV, Seckl JR, Huber R (2010) The 11-beta-hydroxysteroid dehydrogenase type 1 inhibitor INCB13739 improves hyperglycemia in patients with type 2 diabetes inadequately controlled by metformin monotherapy. Diab Care 33(7):1516–1522

Sampath-Kumar R, Yu M, Khalil MW, Yang K (1997) Metyrapone is a competitive inhibitor of 11β-hydroxysteroid dehydrogenase type 1 reductase. J Steroid Biochem Mol Biol 62:195–199

Sandeep TC, Yau JL, MacLullich AM, Noble J, Deary IJ, Walker BR, Seckl JR (2004) 11β-hydroxysteroid dehydrogenase inhibition improves cognitive function in healthy elderly men and type 2 diabetics. Proc Natl Acad Sci USA 101:6734–6739

Sonino N (1987) The use of ketoconazole as an inhibitor of steroid production. N Engl J Med 317:812–818

Sorensen B, Rohde J, Wang J, Fung S, Monzon K, Chiou W, Pan L, Deng X, Stolarik D, Frevert EU, Jacobson P, Link JT (2006) Adamantane 11-β-HSD-1 inhibitors: application of an isocyanide multicomponent reaction. Bioorg Med Chem Lett 16:5958–5962

St Jean DJ Jr, Yuan C, Bercot EA, Cupples R, Chen M, Fretland J, Hale C, Hungate RW, Komorowski R, Veniant M, Wang M, Zhang X, Fotsch C (2007) 2-(S)-phenethylaminothiazolones as potent, orally efficacious inhibitors of 11β-hydroxysteriod dehydrogenase type 1. J Med Chem 50:429–432

Stewart PM, Boulton A, Kumar S, Clark PM, Shackleton CH (1999) Cortisol metabolism in human obesity: impaired cortisone– > cortisol conversion in subjects with central adiposity. J Clin Endocrinol Metab 84:1022–1027

Tomlinson JW, Draper N, Mackie J, Johnson AP, Holder G, Wood P, Stewart PM (2002) Absence of Cushingoid phenotype in a patient with Cushing's disease due to defective cortisone to cortisol conversion. J Clin Endocrinol Metab 87:57–62

Tomlinson JW, Sherlock M, Hughes B, Hughes SV, Kilvington F, Bartlett W, Courtney R, Rejto P, Carley W, Stewart PM (2007) Inhibition of 11β-hydroxysteroid dehydrogenase type 1 activity in vivo limits glucocorticoid exposure to human adipose tissue and decreases lipolysis. J Clin Endocrinol Metab 92:857–864

van Schaftingen E, Gerin I (2002) The glucose-6-phosphatase system. Biochem J 362:513–532

Veilleux A, Rhéaume C, Daris M, Luu-The V, Tchernof A (2009) Omental adipose tissue type 1 11 beta-hydroxysteroid dehydrogenase oxoreductase activity, body fat distribution, and metabolic alterations in women. J Clin Endocrinol Metab 94:3550–3557

Véniant MM, Hale C, Komorowski R, Chen MM, St. Jean DJ Jr, Fotsch C, Wang M (2009) Time of the day for 11β-HSD1 inhibition plays a role in improving glucose homeostasis in DIO mice. Diabetes Obes Metab 11:109–117

von Geldern TW, Tu N, Kym PR, Link JT, Jae HS, Lai C, Apelqvist T, Rhonnstad P, Hagberg L, Koehler K, Grynfarb M, Goos-Nilsson A, Sandberg J, Osterlund M, Barkhem T, Hoglund M, Wang J, Fung S, Wilcox D, Nguyen P, Jakob C, Hutchins C, Farnegardh M, Kauppi B, Ohman L, Jacobson PB (2004) Liver-selective glucocorticoid antagonists: a novel treatment for type 2 diabetes. J Med Chem 47:4213–4230

Wake DJ, Rask E, Livingstone DE, Soderberg S, Olsson T, Walker BR (2003) Local and systemic impact of transcriptional up-regulation of 11β-hydroxysteroid dehydrogenase type 1 in adipose tissue in human obesity. J Clin Endocrinol Metab 88:3983–3988

Walker EA, Stewart PM (2003) 11β-hydroxysteroid dehydrogenase: unexpected connections. Trends Endocrinol Metab 14:334–339

Walker BR, Campbell JC, Fraser R, Stewart PM, Edwards CR (1992) Mineralocorticoid excess and inhibition of 11β-hydroxysteroid dehydrogenase in patients with ectopic ACTH syndrome. Clin Endocrinol (Oxf) 37:483–492

Walker EA, Clark AM, Hewison M, Ride JP, Stewart PM (2001) Functional expression, characterization, and purification of the catalytic domain of human 11-β-hydroxysteroid dehydrogenase type 1. J Biol Chem 276:21343–21350

Walker EA, Ahmed A, Lavery GG, Tomlinson JW, Kim SY, Cooper MS, Ride JP, Hughes BA, Shackleton CH, McKiernan P, Elias E, Chou JY, Stewart PM (2007) 11β-Hydroxysteroid dehydrogenase type 1 regulation by intracellular glucose 6-phosphate provides evidence for a novel link between glucose metabolism and hypothalamo-pituitary-adrenal axis function. J Biol Chem 282:27030–27036

Yau JL, Noble J, Kenyon CJ, Hibberd C, Kotelevtsev Y, Mullins JJ, Seckl JR (2001) Lack of tissue glucocorticoid reactivation in 11β-hydroxysteroid dehydrogenase type 1 knockout mice ameliorates age-related learning impairments. Proc Natl Acad Sci USA 98:4716–4721

Zhang J, Osslund TD, Plant MH, Clogston CL, Nybo RE, Xiong F, Delaney JM, Jordan SR (2005) Crystal structure of murine 11β-hydroxysteroid dehydrogenase 1: an important therapeutic target for diabetes. Biochemistry 44:6948–6957

Zinker B, Mika A, Nguyen P, Wilcox D, Ohman L, von Geldern TW, Opgenorth T, Jacobson P (2007) Liver-selective glucocorticoid receptor antagonism decreases glucose production and increases glucose disposal, ameliorating insulin resistance. Metabolism 56:380–387

Nampt and Its Potential Role in Inflammation and Type 2 Diabetes

Antje Garten, Stefanie Petzold, Susanne Schuster, Antje Körner, Jürgen Kratzsch, and Wieland Kiess

Contents

Abstract Nicotinamide phosphoribosyltransferase (Nampt) is a key nicotinamide adenine dinucleotide (NAD) biosynthetic enzyme in mammals, converting nicotinamide into nicotinamide mononucleotide (NMN), an NAD intermediate. First identified in humans as a cytokine pre-B-cell colony enhancing factor (PBEF) and subsequently described as an insulin-mimetic hormone visfatin, Nampt has recently excited the scientific interest of researchers from diverse fields, including NAD biology, metabolic regulation, and inflammation. As an NAD biosynthetic enzyme, Nampt regulates the activity of NAD-consuming enzymes such as sirtuins and influences a variety of metabolic and stress responses. Nampt plays an important role in the regulation of insulin secretion in pancreatic β-cells. Nampt also functions as an immunomodulatory cytokine and is involved in the regulation of inflammatory responses. This chapter summarizes the various functional aspects of Nampt and discusses its potential roles in diseases, with special focus on type 2 diabetes mellitus (T2DM).

A. Garten, S. Petzold, S. Schuster, A. Körner, and W. Kiess (✉)
University of Leipzig, Hospital for Children and Adolescents, Liebigstr. 20a, 04103 Leipzig, Germany
e-mail: wieland.kiess@medizin.uni-leipzig.de

J. Kratzsch
University of Leipzig, Institute for Laboratory Medicine, Clinical Chemistry and Molecular Diagnostics, Paul-List-Str.13a, 04103 Leipzig, Germany

M. Schwanstecher (ed.), *Diabetes - Perspectives in Drug Therapy*,   147
Handbook of Experimental Pharmacology 203,
DOI 10.1007/978-3-642-17214-4_7, © Springer-Verlag Berlin Heidelberg 2011

Keywords Nampt · NAD · PBEF · Visfatin · Sirtuin

1 Introduction

In the pathogenesis of type 2 diabetes mellitus (T2DM), the tight regulation of insulin sensitivity and secretion is dysbalanced. This is caused by both environmental and genetic factors.

While the physiological significance of nicotinamide phosphoribosyltransferase (Nampt) in obesity, T2DM, and other metabolic disorders is still unclear, the intracellular nicotinamide adenine dinucleotide (NAD) biosynthetic function of Nampt is well characterized. In the NAD biosynthetic pathway from nicotinamide, Nampt (EC 2.4.2.12) catalyzes the rate-limiting step, namely the transfer of a phosphoribosyl group from 5-phosphoribosyl-1-pyrophosphate (PRPP) to nicotinamide, forming nicotinamide mononucleotide (NMN) and pyrophosphate (PP_i) (Rongvaux et al. 2002) (Fig. 1). NMN is then converted into NAD by the isoenzymes nicotinamide mononucleotide adenylyltransferase (Nmnat, EC 2.7.7.1) 1 – 3 (Fig. 1). While tryptophan and nicotinic acid are also precursors for NAD biosynthesis in mammals, nicotinamide is predominantly used to synthesize NAD (Magni et al. 1999; Rongvaux et al. 2003). NAD is a coenzyme with a well-established role in cellular redox reactions. Recently, several lines of evidence have implicated NAD biochemistry in a broad range of biological functions. For example, NAD is used as a substrate in a number of important signaling pathways in mammalian cells, including poly(ADP-ribosyl)ation in DNA repair (Ménissier de Murcia et al. 2003), mono-ADP-ribosylation in both the immune response and G-protein-coupled signaling (Corda and Di Girolamo 2003), and synthesis of cyclic ADP-ribose and nicotinate adenine dinucleotide phosphate (NAADP) in intracellular calcium signaling (Lee 2001). Furthermore, NAD and its derivatives also play important roles in transcriptional regulation (Lin and Guarente 2003). In particular,

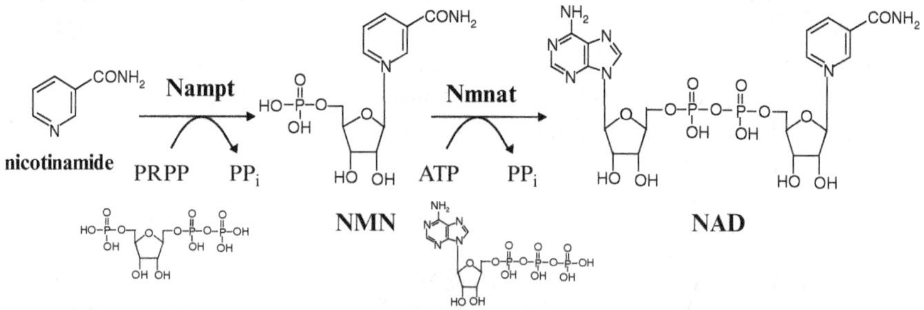

Fig. 1 NAD biosynthetis from nicotinamide. The rate-limiting step in mammalian NAD biosynthesis from nicotinamide is the transfer of a phosphoribosyl residue from 5-phosphoribosyl-1-pyrophosphate (PRPP) to nicotinamide catalyzed by nicotinamide phosphoribosyltransferase (Nampt) to produce nicotinamide mononucleotide (NMN), which is then converted into NAD by nicotinamide mononucleotide adenylyltransferase (Nmnat)

the discovery that yeast and mammalian Sir2 (*silent* *i*nformation *r*egulator 2) proteins require NAD for their deacetylase activity (Imai et al. 2000) has drawn much attention to this novel regulatory role for NAD.

Although Nampt enzymatic activity was reported and characterized as early as 1957 (Preiss and Handler 1957), the gene encoding Nampt was first identified in *Haemophilus ducreyi* in 2001 (Martin et al. 2001). Since then, several groups have characterized the enzymological features of mammalian Nampt (Revollo et al. 2004; van der Veer et al. 2005). The K_m value of Nampt for nicotinamide is ~1 μM, and it does not use nicotinic acid as a substrate (Revollo et al. 2004). The crystal structure of Nampt has been determined, which clearly demonstrates that this protein belongs to the dimeric class of type II phosphoribosyltransferases (Khan et al. 2006; Kim et al. 2006; Wang et al. 2006).

Whereas the biochemical and structural basis of Nampt as an intracellular NAD biosynthetic enzyme has been well established, this protein seems to have several other physiological functions. Human Nampt was originally characterized as a cytokine named pre-B-cell colony enhancing factor (PBEF) (Samal et al. 1994). Nampt was also claimed to function as an insulin-mimetic adipocytokine, predominantly secreted from visceral fat and therefore named visfatin (Fukuhara et al. 2005). We will discuss these different functional aspects of Nampt and its potential roles in a variety of pathophysiological conditions including type 2 diabetes.

2 Nampt and Metabolic Disorders

The function of intracellular Nampt (iNampt) as NAD biosynthetic enzyme has been well characterized. In contrast, the physiological role of extracellular Nampt (eNampt) has been a matter of much debate. At least three different functions have been assigned to eNampt – an insulin-mimetic, a cytokine-like, and a function as an NMN producing enzyme (Pilz et al. 2007; Revollo et al. 2007a; Sethi 2007; Yang et al. 2006). It is not clear which of these functions is important in what physiological context. There is also no study that has demonstrated in what way eNampt enters extracellular space and what distinguishes iNampt and eNampt on the molecular level. This makes it difficult to investigate the role that iNampt and eNampt play in various pathophysiological metabolic states as is discussed in more detail below.

2.1 eNampt and iNampt: Regulation of Pancreatic β-Cell Function

The most controversial function assigned to the Nampt protein was the insulin-mimetic activity as an adipocytokine named "visfatin" described by Fukuhara et al. (2005). This study found that eNampt acted like insulin by binding to the insulin receptor, activating the associated downstream signaling pathway and eliciting similar biological reponses in vitro and in vivo on adipogenesis, cellular glucose uptake, and blood glucose levels. Their results immediately drew attention to a possible connection between eNampt and metabolic complications, such as obesity and T2DM. However, this paper has been retracted (Fukuhara et al. 2007). Three subsequent studies have so far provided indirect evidence for the connection between eNampt and insulin signaling (Dahl et al. 2007; Song et al. 2008; Xie et al. 2007). One group has reported an insulin-like action of eNampt on osteoblasts, which was blocked by the pretreatment with the insulin receptor kinase inhibitor HNMPA-(AM)₃ (Xie et al. 2007). In another study, it has been reported that the same inhibitor blocked the effect of eNampt on matrix metalloproteinase (MMP)-9 activity in THP-1 cells and the production of TNF-α and IL-8 in PBMC (Dahl et al. 2007). A third group has found multiple actions of eNampt on cultured kidney mesangial cells. eNampt induced uptake of glucose, glucose transporter (GLUT)-1 protein expression, and synthesis of profibrotic molecules, including transforming growth factor (TGF)-β1, plasminogen activator inhibitor (PAI)-1, and type I collagen (Song et al. 2008). Whereas a potent Nampt inhibitor FK866 (Hasmann and Schemainda 2003) blocked this eNampt-mediated glucose uptake, knockdown of the insulin receptor also inhibited this effect (Song et al. 2008). Unfortunately, none of the above studies has examined whether eNampt directly binds to the insulin receptor of respective target cells and whether the NAD biosynthetic activity is required for the observed effect.

In another study, it was demonstrated that eNampt does not exert insulin-mimetic effects in vitro or in vivo, but rather exhibits robust NMN biosynthetic activity and that this eNampt-mediated NMN biosynthesis plays a critical role in the regulation of glucose-stimulated insulin secretion (GSIS) in pancreatic β-cells in vitro and in vivo (Revollo et al. 2007b). To address the physiological significance of Nampt function in vivo, Nampt-deficient mice were generated. While Nampt homozygous (Nampt$^{-/-}$) mice are embryonic lethal likely due to failure of adequate NAD biosynthesis, Nampt heterozygous (Nampt$^{+/-}$) mice do not differ visibly from wild-type mice but show significant decreases in total NAD levels in tissues (Revollo et al. 2007b). Nampt$^{+/-}$ female mice revealed moderately impaired glucose tolerance and a significant defect in GSIS. Whereas islet morphology and size in Nampt$^{+/-}$ mice do not differ from control mice, further analyses of isolated primary islets revealed that Nampt$^{+/-}$ islets have functional defects in NAD biosynthesis and GSIS. Remarkably, insulin secretion defects in Nampt$^{+/-}$ mice and islets can be corrected by administration of NMN, confirming that the defects observed in Nampt$^{+/-}$ mice and islets are due to a lack of the NAD biosynthetic activity of Nampt. Furthermore, FK866, a potent chemical inhibitor of Nampt, significantly inhibited NAD biosynthesis and GSIS in isolated wild-type primary islets. Again, administration of NMN ameliorated defects in NAD biosynthesis and GSIS in FK866-treated wild-type islets. Thus, pancreatic β-cells require Nampt-mediated NAD biosynthesis to maintain normal NAD biosynthesis and GSIS and are capable of incorporating NMN from the extracellular space.

These findings are supported by a study on the mouse pancreatic β-cell line βTC6. Incubation with recombinant Nampt caused significant changes in the mRNA expression of several key diabetes-related genes, including upregulation of insulin, hepatocyte nuclear factor (HNF)-1β, HNF-4α, and nuclear factor-κB. Significant downregulation was seen in angiotensin-converting enzyme and UCP2. Insulin secretion was increased compared to that of control at low glucose and this increase was blocked by coincubation with the specific Nampt inhibitor FK866. Both Nampt and NMN induced activation of insulin receptor and extracellular signal-regulated kinase (ERK)1/2. Nampt-induced insulin receptor and ERK1/2 activation were inhibited by FK866 (Brown et al. 2010).

The contradictory results on the insulin-mimetic activity of Nampt could be explained by a possible crosstalk between eNampt-mediated and insulin signaling pathways in a cell type-dependent manner. eNampt could possibly bind to and activate an unidentified receptor that might indirectly affect insulin signaling. Another possibility is that Nampt-mediated NAD biosynthesis might have an impact on the insulin signaling pathway. To address these possibilities in future studies, it is critical to examine (1) whether the existence of the insulin receptor is necessary for the observed insulin-mimetic effects by using mutant cells that lack the insulin receptor, (2) whether nicotinamide, which is usually included at very high concentrations in cell culture media, is necessary to observe those effects, and (3) whether mutant Nampt proteins that lack NAD biosynthetic activity can still mediate the activity of interest in each condition. These analyses will resolve contradictory results around the claimed insulin-mimetic activity of Nampt.

In humans, it has been reported that individuals who carry specific single nucleotide polymorphism variants in the *Nampt* gene promoter region have lower fasting plasma insulin levels (Bailey et al. 2006; Mirzaei et al. 2009), suggesting that Nampt-mediated NAD biosynthesis might also regulate insulin secretion in humans.

Aging is one of the greatest risk factors for developing T2DM and other metabolic complications (Chang and Halter 2003; Moller et al. 2003). Sirtuins are a group of NAD-dependent enzymes that regulate metabolic responses to nutritional availability in different tissues and cellular responses to a variety of stresses and are also involved in the regulation of aging. Sirtuins deacetylate and/or ADP-ribosylate lysine residues of many target regulatory factors (Blander and Guarente 2004; Schwer and Verdin 2008). In their deacetylation reactions, sirtuins produce acetyl-ADP-ribose, nicotinamide, and deacetylated proteins. Silent information regulator 2 (Sir2), as the prototypical enzyme of this group, regulates the replicative life span of yeast mother cells (Kaeberlein et al. 1999). Strikingly, Sir2 homologues also regulate life span in worms and flies (Rogina and Helfand 2004; Tissenbaum and Guarente 2001) and, depending on the genetic background, mediate life span extension caused by caloric restriction, the only dietary regimen that can retard aging and extend life span in a wide variety of organisms (Guarente 2005). It is not yet known whether Sirt1, the mammalian Sir2 orthologue, regulates aging and longevity in mammals. Because sirtuins absolutely require NAD for their function, NAD biosynthesis and Nampt play a critical role in the regulation of mammalian sirtuin activity.

A recent study found that Nampt-mediated systemic NAD biosynthesis and Sirt1 activity are affected in pancreatic β-cells of aged mice. Pancreatic β-cell-specific Sirt1-overexpressing (BESTO) mice exhibited significantly enhanced GSIS and improved glucose tolerance (Moynihan et al. 2005). However, these phenotypes were completely lost in both BESTO males and females when they reached 18–24 months of age (Ramsey et al. 2008). Plasma NMN levels and Sirt1 activity in pancreatic islets were significantly reduced in those aged BESTO mice.

Consistent with this finding, NMN administration could restore enhanced GSIS and improved glucose tolerance in aged BESTO females but not in males (Ramsey et al. 2008), although the reason for the observed sex-dependent difference remained unclear. These findings suggest that an age-dependent decline in Nampt-mediated systemic NAD biosynthesis contributes to reduced Sirt1 activity in aged pancreatic islets and likely in other aged tissues. Because sirtuins have recently emerged as promising pharmaceutical targets to develop therapeutic interventions against age-associated diseases (Milne et al. 2007; Westphal et al. 2008), the systemic enhancement of NAD biosynthesis might provide another pharmacological means to activate Sirt1 and to convey benefits against metabolic diseases like T2DM (Imai and Kiess 2009).

2.2   eNampt in Human Circulation: A Biomarker for Obesity and T2DM?

Nampt possesses neither a signal sequence nor a caspase I cleavage site (Rongvaux et al. 2002). Therefore, several studies have suggested that eNampt might be released simply by cell lysis or cell death (Hug and Lodish 2005; Stephens and Vidal-Puig 2006). On the other hand, it was shown that eNampt release is governed by a highly regulated positive secretory process in a cell type-dependent manner (Revollo et al. 2007b). Fully differentiated mouse and human adipocytes are capable of secreting eNampt through a nonclassical secretory pathway, which is not blocked by inhibitors of the classical ER–Golgi secretory pathway, such as brefeldin A and monensin (Revollo et al. 2007b; Tanaka et al. 2007). There is evidence that other cell types, such as human primary hepatocytes (Garten et al. 2010) and leukocytes (D. Friebe, personal communication) are also a significant source of eNampt in human circulation.

Numerous studies have been published, addressing possible associations between plasma eNampt levels and anthropometric and metabolic parameters in obesity and T2DM. So far, results have been conflicting, showing positive, negative, or no association (Arner 2006; Revollo et al. 2007a; Sethi 2007; Stephens and Vidal-Puig 2006). For example, one study reported a positive correlation of plasma eNampt concentrations with body mass index (BMI) and percent body fat (Berndt et al. 2005), while another study found that plasma eNampt is reduced in human obesity and not related to insulin resistance (Pagano et al. 2006). Two other studies reported that higher plasma eNampt levels are independently and significantly associated with T2DM even after adjusting for known biomarkers (Chen et al. 2006; Retnakaran et al. 2008). These contradictory findings appear to be in part due to significant differences in immunoassays and the treatment and type of samples. Freeze–thaw cycles and different sample additives have considerable influence on the measurement of eNampt concentrations (Nüsken et al. 2007). Also, commercially available immunoassays differ considerably in the specificity and sensitivity of eNampt detection in human serum and plasma (Körner et al. 2007).

Looking at recent studies (as summarized in Table 1), it appears that most groups detected an elevation of plasma Nampt levels in obese subjects. However, the association between plasma eNampt levels and metabolic disorders, such as obesity and T2DM, is still unclear, and careful assessments with highly accurate assays for the measurement of eNampt will be necessary to address this critical issue.

Interestingly, in a recent study, Nampt levels in cerebrospinal fluid (CSF) were found to decrease with increasing plasma Nampt concentrations, BMI, body fat mass, and insulin resistance, indicating that the transport of Nampt across the blood–brain barrier might be impaired in obesity and that central nervous Nampt insufficiency or resistance might be linked to pathogenetic mechanisms of obesity (Hallschmid et al. 2009).

2.3   Role of Hepatic Nampt in T2DM

Hepatic insulin resistance is an underlying cause of the metabolic syndrome and the development of T2DM. In particular, the failure of insulin to suppress hepatic gluconeogenesis leads to hyperglycemia and, consequently, to the persistent stimulation of insulin production (Leclercq et al. 2007).

Sirt1 has been implicated in the regulation of hepatic glucose metabolism through the induction of gluconeogenic genes (Rodgers et al. 2005). Nampt regulates Sirt1 activity and also seems to be involved in the regulation of insulin signaling in the liver, as has been shown in animal studies. One group reported upregulation of Sirt1 and PPAR-α together with increasing NAD levels and Nampt activity in fasting mice (Hayashida et al. 2010). In rats that were injected with a Nampt overexpression plasmid insulin sensitivity was increased, while total cholesterol plasma levels decreased. The animals also displayed an enhanced insulin receptor substrate (IRS)-1 tyrosine phosphorylation in response to insulin as well as increased mRNA expression of peroxisome proliferator-activated receptor gamma (PPAR-γ) and sterol regulatory element-binding protein 2 (SREBP-2) in the liver and adipose tissues (Sun et al. 2009). SIRT1 protein and activity was increased by treatment with metformin through an adenosine monophosphate kinase (AMPK)-mediated increase in gene expression of Nampt in db/db mice (Caton et al. 2010).

Hepatic insulin resistance and a dysregulation of hepatic glucose and lipid metabolism are also major reasons for the development of nonalcoholic fatty liver disease (NAFLD), which has been recognized as a component of the

Table 1 Most recent studies concerning correlation of plasma eNampt concentrations with anthropometric and metabolic parameters

Plasma eNampt concentrations	Correlation	Study
↑ in obese children	+ with visceral adipose tissue area	Araki et al. (2008)
↑ in obese adolescents	± with anthropometric or lipid parameters in nonobese − with age + with high-density lipoprotein (HDL)-cholesterol in obese	Jin et al. (2008)
↑ in type 2 diabetes mellitus (T2DM) without macroangiopathy	± with body mass index (BMI), insulin, glucose, or HOMA-insulin resistance index (HOMA-IR) − with high sensitive C-reactive protein (hsCRP) and IL-6 plasma concentration	Alghasham and Barakat (2008)
± in patiens with coronary heart disease	± with any variables of the metabolic syndrome	Choi et al. (2008)
↑ in T2DM	+ with proteinuria	Yilmaz et al. (2008)
−	+ with HDL-cholesterol − with triglycerides	Wang et al. (2007)
↑ in obese women	+ with epicardial fat thickness	Malavazos et al. (2008)
↑ in preeclampsia	+ with CRP − with HOMA-IR	Fasshauer et al. (2008)
−	+ with HDL in women − with LDL in women and BMI in males ± with height, weight, body mass index, waist and hip circumferences, waist-to-hip ratio, blood pressure, fasting serum insulin and fasting plasma glucose, lipid profiles, and uric acid levels	Chen et al. (2007)
↑ in coronary artery disease (CAD) patients	−	Cheng et al. (2008)
	± with fat mass and bone mineral density in men	Peng et al. (2008)
↑ in obese women	+ with BMI	Choi et al. (2007)
↓ after exercise		
	+ with IL-6 plasma concentration and diastolic blood pressure	Seo et al. (2008)
↑ in obese patients with impaired fasting glucose and diabetes	+ with leptin plasma concentration	García-Fuentes et al. (2007)
↓ after exercise	+ with plasma insulin concentration and glucose AUC	Haus et al. (2009)

(*continued*)

Table 1 (continued)

Plasma eNampt concentrations	Correlation	Study
↑ in lean glucose-tolerant patients with PCOS	–	Yildiz et al. (2010)
↑ in obese women	+ with fat mass and AUC insulin during OGTT	Unluturk et al. (2010)
↔ in morbidly obese patients with and without T2DM ↑ transiently during OGTT	–	Hofsø et al. (2009)
↑ in HD patients	–	Nüsken et al. (2009)
	– with TNF-α plasma concentration in patients with impaired fasting glucose	de Luis et al. (2009)
↓ in patients with T1DM	– with HbA1c	Toruner et al. (2009)
↑ in GDM patients	–	Gok et al. (2010)
↑ in patients with GDM and pre-GDM	–	Coskun et al. (2010)
↑ in pregnant women	± with glucose tolerance	Szamatowicz et al. (2009)
↓ in GDM patients at term	+ with IL-6 and TNF-α mRNA in SAT and placental tissue	Telejko et al. (2009)

↑ Increase, ↓ Decrease, + Positive correlation, – Negative correlation, ± No correlation

metabolic syndrome and can lead to liver cirrhosis. There are few studies on Nampt expression in human livers. In patients with liver cirrhosis, it was found that hepatic *Nampt* mRNA expression and Nampt levels in circulation were decreased compared with healthy controls (de Boer et al. 2009). In severely obese patients with NAFLD, Nampt serum levels were shown to be increased. After bariatric surgery, Nampt expression in livers and Nampt serum levels decreased as the patients lost weight (Moschen et al. 2009). Another report demonstrated a correlation of serum Nampt levels with liver histology in NAFLD. Nampt levels predicted the presence of portal inflammation in NAFLD patients (Aller et al. 2009).

From these results in animal and human studies, one can speculate that Nampt is involved in regulating glucose metabolism in hepatocytes through activation of Sirt1. The liver also seems to be a major source for circulating Nampt, and Nampt levels in circulation are consequently influenced by impaired liver function.

3   eNampt: A Link Between T2DM and Inflammation

Human Nampt was originally identified by a screen of a human peripheral blood lymphocyte cDNA library and named pre-B-cell colony enhancing factor (PBEF, see above). This 52 kDa protein was reported to act as a presumptive cytokine that increased pre-B-cell colony forming activity together with interleukin (IL)-7 and stem cell factor (SCF) (Samal et al. 1994). iNampt is also highly expressed in human fetal membranes, amnion, and placenta. *Nampt* mRNA expression increases in fetal membranes after labor and in severely infected amnion membranes. Interestingly, eNampt treatment upregulates the expression of inflammatory cytokines, such as IL-6 and IL-8, in amnion-like epithelial cells. Thus, it has been speculated that eNampt might have a cytokine-like function involved in the regulation of labor and in infection-induced preterm birth (Ognjanovic and Bryant-Greenwood 2002; Ognjanovic et al. 2005).

In the pathophysiology of obesity and T2DM, it has been revealed that chronic inflammation plays an important role in the development of insulin resistance and other associated complications, such as atherosclerosis (Guest et al. 2008; Poirier et al. 2006; Schenk et al. 2008; Sowers 2003). In this regard, one might speculate that eNampt could show an association with the development of vascular inflammation induced by obesity and T2DM, instead of an association with anthropometric and metabolic parameters. Indeed, it has been shown that eNampt induces the adhesion of leukocytes to endothelial cells and aortic endothelium by activating intercellular adhesion molecule (ICAM)-1 and vascular cell adhesion molecule (VCAM)-1 (Adya et al. 2008). This phenomenon appears to be mediated through the proinflammatory transcription factor nuclear factor-κB (NF-κB) in a reactive oxygen species (ROS)-dependent manner. The same study also revealed that eNampt significantly increases the transcriptional activity of NF-κB in human vascular endothelial cells, resulting in the activation of the MMP-2/9.

A study that was aimed at investigating the expression of Nampt in circulating blood monocytes in obese and/or type 2 diabetic human subjects found that Nampt expression was significantly upregulated in obese type 2 diabetic patients, whereas obese nondiabetics exhibited similar levels compared to lean controls (Laudes et al. 2010).

In contrast, an upregulation of *Nampt* mRNA expression levels in peripheral blood cells (PBCs) of obese compared with lean subjects was reported, along with a correlation of Nampt plasma levels to cholesterol, triglycerides, and hepatic enzymes in circulation (Catalán et al. 2010). Another group determined Sirt1 expression in peripheral PBMC and found that insulin resistance and metabolic syndrome were associated with low PBMC *Sirt1* gene and protein expression. Sirt1 gene expression was negatively correlated with carotid intima-media thickness (CIMT). In the monocytic THP-1 cell line, high glucose and palmitate reduced Sirt1 and Nampt expression and reduced the levels of intracellular NAD through oxidative stress. These effects on Nampt and Sirt1 were prevented by resveratrol in vitro (de Kreutzenberg et al. 2010). Moreover, a direct association of Nampt with advanced carotid atherosclerosis and CIMT in patients with T2DM was determined with increased Nampt serum levels, especially in patients with carotid plaques (Kadoglou et al. 2010). Therefore, eNampt might play an important role in the progression and/or the associated complications, especially inflammatory complications, of obesity and T2DM.

eNampt has also been shown to be involved in the regulation of apoptosis as a cytokine. In human neutrophils, eNampt inhibits their apoptosis in response to various inflammatory stimuli, although this particular effect of eNampt requires the presence of iNampt to some extent (Jia et al. 2004). Recently, Nampt was identified as an essential enzyme mediating granulocyte colony-stimulating factor (G-CSF)-triggered granulopoiesis in healthy individuals and in individuals with severe congenital neutropenia. Both extracellularly and intracellularly administered Nampt induced granulocytic differentiation of CD34(+) hematopoietic progenitor cells by NAD-dependent Sirt1 activation and subsequent upregulation of G-CSF synthesis and G-CSF receptor expression (Skokowa et al. 2009).

Furthermore, serum eNampt levels are found to be upregulated in sepsis patients, and rates of neutrophil apoptosis are found to be profoundly reduced in those patients (Jia et al. 2004). In amniotic epithelial cells, eNampt treatment appears to confer protection from apoptosis as a stretch-responsive cytokine (Kendal-Wright et al. 2008). In patients with inflammatory bowel disease (Crohn's disease and ulcerative colitis), *Nampt* mRNA levels are increased in colon biopsy samples, and plasma eNampt levels are elevated (Moschen et al. 2007). Because eNampt stimulates the production of proinflammatory cytokines in PBMCs and upregulates IL-6 mRNA and serum levels in vivo when given intraperitoneally to mice, it has been suggested that eNampt itself also functions as a proinflammatory cytokine (Moschen et al. 2007). Stimulation of the tumor necrosis factor family member TNF- and APOL-related leukocyte-expressed ligand (TALL)-1, which is involved in lupus-like autoimmune diseases, increased *Nampt* mRNA (Xu et al. 2002). Additionally, Nampt has been found to be upregulated in a variety of other immunological disorders including acute lung injury, rheumatoid arthritis, and myocardial infarction and is considered a novel mediator of innate immunity (Luk et al. 2008).

These studies all suggest that eNampt might function as an inflammatory cytokine. In most studies, it has not been fully addressed which activity of eNampt, NAD biosynthetic activity vs. cytokine-like activity, is responsible for the observed effects of eNampt. However, two studies have reported an enzyme-dependent proinflammatory action of Nampt: in inflammatory cells in vitro and in joints affected with rheumatoid arthritis in vivo (Busso et al. 2008) and in cultured human aortic smooth muscle cells (Romacho et al. 2009). In contrast, one group has demonstrated that eNampt protects macrophages from ER stress-induced apoptosis through its cytokine-like activity that is totally separated from its NAD biosynthetic activity (Li et al. 2008).

There is evidence that Nampt exerts its proinflammatory actions through the upregulation of monocyte chemoattractant protein (MCP)-1. MCP-1 levels were significantly increased after Nampt administration in cultured human adipocyte supernatant and serum of mice. The detectability of Nampt in serum predicted circulating MCP-1 independent of age and gender in humans (Sommer et al. 2010).

Therefore, although how eNampt exerts its cytokine-like activity in different cellular contexts still needs to be investigated, there is clear evidence that it functions as an immunomodulatory cytokine.

4 Concluding Remarks and Future Aspects

Nampt functions as an intra- and extracellular NAD biosynthetic enzyme that is important for the regulation of metabolism and stress resistance through sirtuins and other NAD-consuming regulators. On the other hand, eNampt appears to act as a cytokine, independent of its enzymatic activity, and plays a major role in the regulation of immune responses. The cytokine-like and the enzymatic functions of Nampt still need to be investigated extensively. Thorough assessments of these two roles will presumably greatly enhance our knowledge about its role in different physiological contexts. In this regard, it will also be important to analyze the mechanism and regulation of eNampt secretion and identify its putative receptor and upstream signaling.

To further clarify the effects of iNampt on cellular metabolism, it will also be important to understand the subcellular compartmentalization of Nampt and other enzymes involved in the biosynthesis and the breakdown of NAD and the regulation of their localization. Additionally, very little is known about the flux of NAD substrates, intermediates, and metabolites. Therefore, it will be critical to study not only the regulation of NAD biosynthesis and breakdown in each cellular compartment, but also the spatial and temporal dynamics of NAD metabolism at a systemic level using a metabolomics approach. For example, it will be of great interest to clarify how NMN as a product from the eNampt enzymatic reaction is distributed to target tissues, e.g., pancreatic β-cells, and how it mediates its physiological and pharmacological effects in these tissues.

To date, a substantial part of the studies on the biological functions of Nampt has been conducted in cell culture and mouse models, and the studies on possible correlations between human plasma eNampt levels and metabolic parameters have been contradictory. Therefore, more work needs to be done to elucidate the physiological relevance of the eNampt function in normal individuals and patients with metabolic and other diseases in humans.

Nampt itself or any component in Nampt-mediated systemic NAD biosynthesis could be an effective therapeutic target/reagent for the prevention and the treatment of metabolic disorders including obesity and T2DM, inflammation, and cancer. Because downstream regulators, such as sirtuins and PARPs, have pleiotropic functions, more rigorous investigations will be necessary to clarify possible benefits from the manipulation of Nampt-mediated NAD biosynthesis.

References

Adya R, Tan BK, Chen J, Randeva HS (2008) Nuclear factor B induction by visfatin in human vascular endothelial cells: role in MMP-2/9 production and activation{kappa}. Diab Care 31:758–760

Alghasham AA, Barakat YA (2008) Serum visfatin and its relation to insulin resistance and inflammation in type 2 diabetic patients with and without macroangiopathy. Saudi Med J 29:185–192

Aller R, de Luis DA, Izaola O, Sagrado MG, Conde R, Velasco MC (2009) Influence of visfatin on histopathological changes of non-alcoholic fatty liver disease. Dig Dis Sci 54:1772–1777

Araki S, Dobashi K, Kubo K, Kawagoe R, Yamamoto Y, Kawada Y, Asayama K, Shirahata A (2008) Plasma visfatin concentration as a surrogate marker for visceral fat accumulation in obese children. Obesity (Silver Spring) 16:384–388

Arner P (2006) Visfatin - a true or false trail to type 2 diabetes mellitus. J Clin Endocrinol Metab 91:28–30

Bailey SD, Loredo-Osti JC, Lepage P, Faith J, Fontaine J, Desbiens KM, Hudson TJ, Bouchard C, Gaudet D, Pérusse L, Vohl MC, Engert JC (2006) Common polymorphisms in the promoter of the visfatin gene (PBEF1) influence plasma insulin levels in a French-Canadian population. Diabetes 55:2896–2902

Berndt J, Klöting N, Kralisch S, Kovacs P, Fasshauer M, Schön MR, Stumvoll M, Blüher M (2005) Plasma visfatin concentrations and fat depot-specific mRNA expression in humans. Diabetes 54:2911–2916

Blander G, Guarente L (2004) The Sir2 family of protein deacetylases. Annu Rev Biochem 73:417–435

Brown JE, Onyango DJ, Ramanjaneya M, Conner AC, Patel ST, Dunmore SJ, Randeva HS (2010) Visfatin regulates insulin secretion, insulin receptor signalling and mRNA expression of diabetes-related genes in mouse pancreatic beta-cells. J Mol Endocrinol 44:171–178

Busso N, Karababa M, Nobile M, Rolaz A, Van Gool F, Galli M, Leo O, So A, DeSmedt T (2008) Pharmacological inhibition of nicotinamide phosphoribosyltransferase/visfatin enzymatic activity identifies a new inflammatory pathway linked to NAD. PLoS ONE 3:e2267

Catalán V, Gómez-Ambrosi J, Rodríguez A, Ramírez B, Silva C, Rotellar F, Cienfuegos JA, Salvador J, Frühbeck G (2010) Association of increased Visfatin/PBEF/NAMPT circulating concentrations and gene expression levels in peripheral blood cells with lipid metabolism and fatty liver in human morbid obesity. Nutr Metab Cardiovasc Dis. doi:10.1016/j.numecd.2009.09.008

Caton PW, Nayuni N, Kieswich J, Khan N, Yaqoob M, Corder R (2010) Metformin suppresses hepatic gluconeogenesis through induction of SIRT1 and GCN5. J Endocrinol 205:97–106

Chang AM, Halter JB (2003) Aging and insulin secretion. Am J Physiol Endocrinol Metab 284:E7–E12

Chen CC, Li TC, Li CI, Liu CS, Lin WY, Wu MT, Lai MM, Lin CC (2007) The relationship between visfatin levels and anthropometric and metabolic parameters: association with cholesterol levels in women. Metabolism 56(9):1216–1220

Chen MP, Chung FM, Chang DM, Tsai JC, Huang HF, Shin SJ, Lee YJ (2006) Elevated plasma level of visfatin/pre-B cell colony-enhancing factor in patients with type 2 diabetes mellitus. J Clin Endocrinol Metab 91:295–299

Cheng KH, Chu CS, Lee KT, Lin TH, Hsieh CC, Chiu CC, Voon WC, Sheu SH, Lai WT (2008) Adipocytokines and proinflammatory mediators from abdominal and epicardial adipose tissue in patients with coronary artery disease. Int J Obes (Lond) 32(2):268–274

Choi KM, Kim JH, Cho GJ, Baik SH, Park HS, Kim SM (2007) Effect of exercise training on plasma visfatin and eotaxin levels. Eur J Endocrinol 157:437–442

Choi KM, Lee JS, Kim EJ, Baik SH, Seo HS, Choi DS, Oh DJ, Park CG (2008) Implication of lipocalin-2 and visfatin levels in patients with coronary heart disease. Eur J Endocrinol 158:203–207

Corda D, Di Girolamo M (2003) Functional aspects of protein mono-ADP-ribosylation. EMBO J 22:1953–1958

Coskun A, Ozkaya M, Kiran G, Kilinc M, Arikan DC (2010) Plasma visfatin levels in pregnant women with normal glucose tolerance, gestational diabetes and pre-gestational diabetes mellitus. J Matern Fetal Neonatal Med 23:1014–1018

Dahl TB, Ynestad A, Skjelland M, Øie E, Dahl A, Michelsen A, Damås JK, Tunheim SH, Ueland T, Smith C, Bendz B, Tonstad S, Gullestad L, Frøland SS, Krohg-Sørensen K, Russell D, Aukrust P, Halvorsen B (2007) Increased expression of visfatin in macrophages of human unstable carotid and coronary atherosclerosis: possible role in inflammation and plaque destabilization. Circulation 115:972–980

de Boer JF, Bahr MJ, Böker KH, Manns MP, Tietge UJ (2009) Plasma levels of PBEF/Nampt/visfatin are decreased in patients with liver cirrhosis. Am J Physiol Gastrointest Liver Physiol 296:G196–G210

de Kreutzenberg SV, Ceolotto G, Papparella I, Bortoluzzi A, Semplicini A, Dalla Man C, Cobelli C, Fadini GP, Avogaro A (2010) Downregulation of the longevity-associated protein SIRT1 in insulin resistance and metabolic syndrome. Potential biochemical mechanisms. Diabetes 59:1006–1015

de Luis DA, Sagrado MG, Conde R, Aller R, Izaola O (2009) Relation of visfatin to cardiovascular risk factors and adipocytokines in patients with impaired fasting glucose. Nutrition. doi:10.1016/j.nut.2008.11.005

Fasshauer M, Waldeyer T, Seeger J, Schrey S, Ebert T, Kratzsch J, Lossner U, Bluher M, Stumvoll M, Faber R, Stepan H (2008) Serum levels of the adipokine visfatin are increased in preeclampsia. Clin Endocrinol (Oxf) 69:69–73

Fukuhara A, Matsuda M, Nishizawa M, Segawa K, Tanaka M, Kishimoto K, Matsuki Y, Murakami M, Ichisaka T, Murakami H, Watanabe E, Takagi T, Akiyoshi M, Ohtsubo T, Kihara S, Yamashita S, Makishima M, Funahashi T, Yamanaka S, Hiramatsu R, Matsuzawa Y, Shimomura I (2005) Visfatin: a protein secreted by visceral fat that mimics the effects of insulin. Science 307:426–430

Fukuhara A, Matsuda M, Nishizawa M, Segawa K, Tanaka M, Kishimoto K, Matsuki Y, Murakami M, Ichisaka T, Murakami H, Watanabe E, Takagi T, Akiyoshi M, Ohtsubo T, Kihara S, Yamashita S, Makishima M, Funahashi T, Yamanaka S, Hiramatsu R, Matsuzawa Y, Shimomura I (2007) Retraction. Science 318:565

García-Fuentes E, García-Almeida JM, García-Arnés J, García-Serrano S, Rivas-Marín J, Gallego-Perales JL, Rojo-Martínez G, Garrido-Sánchez L, Bermudez-Silva FJ, Rodríguez de Fonseca F, Soriguer F (2007) Plasma visfatin concentrations in severely obese subjects are increased after intestinal bypass. Obesity (Silver Spring) 15:2391–2395

Garten A, Petzold S, Barnikol-Oettler A, Körner A, Thasler W, Kratzsch J, Kiess W, Gebhardt R (2010) Nicotinamide phosphoribosyltransferase (NAMPT/PBEF/visfatin) is constitutively released from human hepatocytes. Biochem Biophys Res Commun 391:376–381

Gok DE, Yazici M, Uckaya G, Bolu SE, Basaran Y, Ozgurtas T, Kilic S, Kutlu M (2010) The role of visfatin in the pathogenesis of gestational diabetes mellitus. J Endocrinol Invest. doi:10.3275/6902

Guarente L (2005) Calorie restriction and Sir2 genes – towards a mechanism. Mech Ageing Dev 126:923–928

Guest CB, Park MJ, Johnson DR, Freund GG (2008) The implication of proinflammatory cytokines in type 2 diabetes. Front Biosci 13:5187–5194

Hallschmid M, Randeva H, Tan BK, Kern W, Lehnert H (2009) Relationship between cerebrospinal fluid visfatin (PBEF/Nampt) levels and adiposity in humans. Diabetes 58:637–640

Hasmann M, Schemainda I (2003) FK866, a highly specific noncompetitive inhibitor of nicotinamide phosphoribosyltransferase, represents a novel mechanism for induction of tumor cell apoptosis. Cancer Res 63:7436–7442

Haus JM, Solomon TP, Marchetti CM, O'Leary VB, Brooks LM, Gonzalez F, Kirwan JP (2009) Decreased visfatin after exercise training correlates with improved glucose tolerance. Med Sci Sports Exerc 41:1255–1260

Hayashida S, Arimoto A, Kuramoto Y, Kozako T, Honda SI, Shimeno H, Soeda S (2010) Fasting promotes the expression of SIRT1, an NAD(+)-dependent protein deacetylase, via activation of PPARalpha in mice. Mol Cell Biochem 339:285–292

Hofsø D, Ueland T, Hager H, Jenssen T, Bollerslev J, Godang K, Aukrust P, Røislien J, Hjelmesaeth J (2009) Inflammatory mediators in morbidly obese subjects: associations with glucose abnormalities and changes after oral glucose. Eur J Endocrinol 161:451–458

Hug C, Lodish HF (2005) Medicine. Visfatin: a new adipokine. Science 307:366–367

Imai S, Armstrong CM, Kaeberlein M, Guarente L (2000) Transcriptional silencing and longevity protein Sir2 is an NAD-dependent histone deacetylase. Nature 403:795–800

Imai S, Kiess W (2009) Therapeutic potential of SIRT1 and NAMPT-mediated NAD biosynthesis in type 2 diabetes. Front Biosci 14:2983–2995

Jia SH, Li Y, Parodo J, Kapus A, Fan L, Rotstein OD, Marshall JC (2004) Pre-B cell colony-enhancing factor inhibits neutrophil apoptosis in experimental inflammation and clinical sepsis. J Clin Invest 113:1318–1327

Jin H, Jiang B, Tang J, Lu W, Wang W, Zhou L, Shang W, Li F, Ma Q, Yang Y, Chen M (2008) Serum visfatin concentrations in obese adolescents and its correlation with age and high-density lipoprotein cholesterol. Diabetes Res Clin Pract 79:412–418

Kadoglou NP, Sailer N, Moumtzouoglou A, Kapelouzou A, Tsanikidis H, Vitta I, Karkos C, Karayannacos PE, Gerasimidis T, Liapis CD (2010) Visfatin (Nampt) and ghrelin as novel markers of carotid atherosclerosis in patients with type 2 diabetes. Exp Clin Endocrinol Diabetes 118:75–80

Kaeberlein M, McVey M, Guarente L (1999) The SIR2/3/4 complex and SIR2 alone promote longevity in *Saccharomyces cerevisiae* by two different mechanisms. Genes Dev 13:2570–2580

Kendal-Wright CE, Hubbard D, Bryant-Greenwood GD (2008) Chronic stretching of amniotic epithelial cells increases pre-B cell colony-enhancing factor (PBEF/Visfatin) expression and protects them from apoptosis. Placenta 29:255–265

Khan JA, Tao X, Tong L (2006) Molecular basis for the inhibition of human NMPRTase, a novel target for anticancer agents. Nat Struct Mol Biol 13:582–588

Kim MK, Lee JH, Kim H, Park SJ, Kim SH, Kang GB, Lee YS, Kim JB, Kim KK, Suh SW, Eom SH (2006) Crystal structure of visfatin/pre-B cell colony-enhancing factor 1/nicotinamide phosphoribosyltransferase, free and in complex with the anti-cancer agent FK-866. J Mol Biol 362:66–77

Körner A, Garten A, Blüher M, Tauscher R, Kratzsch J, Kiess W (2007) Molecular characteristics of serum visfatin and differential detection by immunoassays. J Clin Endocrinol Metab 92:4783–4791

Laudes M, Oberhauser F, Schulte DM, Freude S, Bilkovski R, Mauer J, Rappl G, Abken H, Hahn M, Schulz O, Krone W (2010) Visfatin/PBEF/Nampt and Resistin expressions in circulating blood monocytes are differentially related to obesity and type 2 diabetes in humans. Horm Metab Res 42:268–273

Leclercq IA, Da Silva MA, Schroyen B, Van Hul N, Geerts A (2007) Insulin resistance in hepatocyte sand sinusoidal liver cells: mechanisms and consequences. J Hepatol 47:142–156

Lee HC (2001) Physiological functions of cyclic ADP-ribose and NAADP as calcium messengers. Annu Rev Pharmacol Toxicol 41:317–345

Li Y, Zhang Y, Dorweiler B, Cui D, Wang T, Woo CW, Brunkan CS, Wolberger C, Imai SI, Tabas I (2008) Extracellular Nampt promotes macrophage survival via a non-enzymatic interleukin-6/STAT3 signaling mechanism. J Biol Chem 283:34833–34843

Lin SJ, Guarente L (2003) Nicotinamide adenine dinucleotide, a metabolic regulator of transcription, longevity and disease. Curr Opin Cell Biol 15:241–246

Luk T, Malam Z, Marshall JC (2008) Pre-B cell colony-enhancing factor (PBEF)/visfatin: a novel mediator of innate immunity. J Leukoc Biol 83:804–816

Magni G, Amici A, Emanuelli M, Ruggieri S (1999) Enzymology of NAD+ synthesis. Adv Enzymol Relat Areas Mol Biol 73:135–182

Malavazos AE, Ermetici F, Cereda E, Coman C, Locati M, Morricone L, Corsi MM, Ambrosi B (2008) Epicardial fat thickness: relationship with plasma visfatin and plasminogen activator inhibitor-1 levels in visceral obesity. Nutr Metab Cardiovasc Dis 18:523–530

Martin P, Shea RJ, Mulks MH (2001) Identification of a plasmid-encoded gene from *Haemophilus ducreyi* which confers NAD independence. J Bacteriol 183:1168–1174

Ménissier de Murcia J, Ricoul M, Tartier L, Niedergang C, Huber A, Dantzer F, Schreiber V, Amé JC, Dierich A, LeMeur M, Sabatier L, Chambon P, de Murcia G (2003) Functional interaction between PARP-1 and PARP-2 in chromosome stability and embryonic development in mouse. EMBO J 22:2255–2263

Milne JC, Lambert PD, Schenk S, Carney DP, Smith JJ, Gagne DJ, Jin L, Boss O, Perni RB, Vu CB, Bernis JE, Xie R, Disch JS, Ng PY, Nunes JJ, Lynch AV, Yang H, Galonek H, Israelian K, Choy W, Iffland A, Lavu S, Medvedik O, Sinclair DA, Olefsky JM, Jirousek MR, Elliot PJ,

Westphal CH (2007) Small molecule activators of Sirt1 as therapeutics for the treatment of type 2 diabetes. Nature 450:712–716

Mirzaei K, Hossein-Nezhad A, Javad Hosseinzadeh-Attar M, Jafari N, Najmafshar A, Mohammadzadeh N, Larijani B (2009) Visfatin genotype may modify the insulin resistance and lipid profile in type 2 diabetes patients. Minerva Endocrinol 34:273–279

Moller N, Gormsen L, Fuglsang J, Gjedsted J (2003) Effects of ageing on insulin secretion and action. Horm Res 60:102–104

Moschen AR, Kaser A, Enrich B, Mosheimer B, Theurl M, Niederegger H, Tilg H (2007) Visfatin, an adipocytokine with antiinflammatory and immunomodulating properties. J Immunol 178:1748–1758

Moschen AR, Molnar C, Wolf AM, Weiss H, Graziadei I, Kaser S, Ebenbichler CF, Stadlmann S, Moser PL, Tilg H (2009) Effects of weight loss induced by bariatric surgery on hepatic adipocytokine expression. J Hepatol 51:765–777

Moynihan KA, Grimm AA, Plueger MM, Bernal-Mizrachi E, Ford E, Cras-Méneur C, Permutt MA, Imai S (2005) Increased dosage of mammalian Sir2 in pancreatic beta cells enhances glucose-stimulated insulin secretion in mice. Cell Metab 2:80–82

Nüsken KD, Nüsken E, Petrasch M, Rauh M, Dötsch J (2007) Preanalytical influences on the measurement of visfatin by enzyme immuno assay. Clin Chim Acta 382:154–156

Nüsken KD, Petrasch M, Rauh M, Stöhr W, Nüsken E, Schneider H, Dötsch J (2009) Active visfatin is elevated in serum of maintenance haemodialysis patients and correlates inversely with circulating HDL cholesterol. Nephrol Dial Transplant 24:2832–2838

Ognjanovic S, Bryant-Greenwood GD (2002) Pre-B-cell colony-enhancing factor, a novel cytokine of human fetal membranes. Am J Obstet Gynecol 187:1051–1058

Ognjanovic S, Ku TL, Bryant-Greenwood GD (2005) Pre-B-cell colony-enhancing factor is a secreted cytokine-like protein from the human amniotic epithelium. Am J Obstet Gynecol 193:273–282

Pagano C, Pilon C, Olivieri M, Mason P, Fabris R, Serra R, Milan G, Rossato M, Federspil G, Vettor R (2006) Reduced plasma visfatin/pre-B cell colony-enhancing factor in obesity is not related to insulin resistance in humans. J Clin Endocrinol Metab 91:3165–3170

Peng XD, Xie H, Zhao Q, Wu XP, Sun ZQ, Liao EY (2008) Relationships between serum adiponectin, leptin, resistin, visfatin levels and bone mineral density, and bone biochemical markers in Chinese men. Clin Chim Acta 387(1–2):31–35

Pilz S, Mangge H, Obermayer-Pietsch B, März W (2007) Visfatin/pre-B-cell colony-enhancing factor: a protein with various suggested functions. J Endocrinol Invest 30:138–144

Poirier P, Giles TD, Bray GA, Hong Y, Stern JS, Pi-Sunyer FX, Eckel RH (2006) Obesity and cardiovascular disease: pathophysiology, evaluation and effect of weight loss. Circulation 113:898–918

Preiss J, Handler P (1957) Enzymatic synthesis of nicotinamide mononucleotide. J Biol Chem 225:759–770

Ramsey KM, Mills KF, Satoh A, Imai S (2008) Age-associated loss of Sirt1-mediated enhancement of glucose-stimulated insulin secretion in beta cell-specific Sirt1-overexpressing (BESTO) mice. Aging Cell 7:78–88

Retnakaran R, Youn BS, Liu Y, Hanley AJG, Lee NS, Park JW, Song ES, Vu V, Kim W, Tungtrongchitr R, Havel PJ, Swarbrick MM, Shaw C, Sweeney G (2008) Correlation of circulating full-lenght visfatin (PBEF/Nampt) with metabolic parameters in subjects with and without diabetes: a cross-sectional study. Clin Endocrinol (Oxf) 69:885–893

Revollo JR, Grimm AA, Imai S (2004) The NAD biosynthesis pathway mediated by nicotinamide phosphoribosyltransferase regulates Sir2 activity in mammalian cells. J Biol Chem 279:50754–50763

Revollo JR, Grimm AA, Imai S (2007a) The regulation of nicotinamide adenine dinucleotide biosynthesis by Nampt/PBEF/visfatin in mammals. Curr Opin Gastroenterol 23:164–170

Revollo JR, Körner A, Mills KF, Satoh A, Wang T, Garten A, Dasgupta B, Sasaki Y, Wolberger C, Townsend RR, Milbrandt J, Kiess W, Imai S (2007b) Nampt/PBEF/Visfatin regulates insulin secretion in beta cells as a systemic NAD biosynthetic enzyme. Cell Metab 6:363–375

Rodgers JT, Lerin C, Haas W, Gygi SP, Spiegelman BM, Puigserver P (2005) Nutrient control of glucose homeostasis through a complex of PGC-1alpha and Sirt1. Nature 434:113–118

Rogina B, Helfand SL (2004) Sir2 mediates longevity in the fly through a pathway related to calorie restriction. Proc Natl Acad Sci USA 101:15998–16003

Romacho T, Azcutia V, Vázquez-Bella M, Matesanz N, Cercas E, Nevado J, Carraro R, Rodrí-guez-Mañas L, Sánchez-Ferrer CF, Peiró C (2009) Extracellular PBEF/NAMPT/visfatin activates pro-inflammatory signalling in human vascular smooth muscle cells through nicotin-amide phosphoribosyltransferase activity. Diabetologia 52:2455–2463

Rongvaux A, Andris F, Van Gool F, Leo O (2003) Reconstructing eukaryotic NAD metabolism. Bioessays 25:683–690

Rongvaux A, Shea RJ, Mulks MH, Gigot D, Urbain J, Leo O, Andris F (2002) Pre-B-cell colony enhancing factor, whose expression is upregulated in activated lymphocytes, is a nicotinamide phosphoribosyltransferase, a cytosolic enzyme involved in NAD biosynthesis. Eur J Immunol 32:3225–3234

Samal B, Sun Y, Stearns G, Xie C, Suggs S, McNiece I (1994) Cloning and characterization of the cDNA encoding a novel human pre-B-cell colony-enhancing factor. Mol Cell Biol 14:1431–1437

Schenk S, Saberi M, Olefsky JM (2008) Insulin sensitivity: modulation by nutrients and inflam-mation. J Clin Invest 118:2992–3002

Schwer B, Verdin E (2008) Conserved metabolic regulatory functions of sirtuins. Cell Metab 7:104–112

Seo JA, Jang ES, Kim BG, Ryu OH, Kim HY, Lee KW, Kim SG, Choi KM, Baik SH, Choi DS, Kim NH (2008) Plasma visfatin levels are positively associated with circulating interleukin-6 in apparently healthy Korean women. Diabetes Res Clin Pract 79:108–111

Sethi JK (2007) Is PBEF/visfatin/Nampt an authentic adipokine relevant to the metabolic syn-drome?. Curr Hypertens Rep 9:33–38

Skokowa J, Lan D, Thakur BK, Wang F, Gupta K, Cario G, Brechlin AM, Schambach A, Hinrichsen L, Meyer G, Gaestel M, Stanulla M, Tong Q, Welte K (2009) NAMPT is essential for the G-CSF-induced myeloid differentiation via a NAD(+)-sirtuin-1-dependent pathway. Nat Med 15:151–158

Sommer G, Kralisch S, Kloting N, Kamprad M, Schrock K, Kratzsch J, Tonjes A, Lossner U, Bluher M, Stumvoll M, Fasshauer M (2010) Visfatin is a positive regulator of MCP-1 in human adipocytes in vitro and in mice in vivo. Obesity (Silver Spring) 18:1486–1492

Song HK, Lee MH, Kim BK, Park YG, Ko GJ, Kang YS, Han JY, Han SY, Han KH, Kim HK, Cha DR (2008) Visfatin: a new player in mesangial cell physiology and diabetic nephropathy. Am J Physiol Renal Physiol 295:F1485–F1494

Sowers JR (2003) Obesity as a cardiovascular risk factor. Am J Med 115:37S–41S

Stephens JM, Vidal-Puig A (2006) An update on visfatin/pre-B cell colony-enhancing factor, an ubiquitously expressed, illusive cytokine that is regulated in obesity. Curr Opin Lipidol 17:128–131

Sun Q, Li L, Li R, Yang M, Liu H, Nowicki MJ, Zong H, Xu J, Yang G (2009) Overexpression of visfatin/PBEF/Nampt alters whole-body insulin sensitivity and lipid profile in rats. Ann Med 41:311–320

Szamatowicz J, Kužmicki M, Telejko B, Zonenberg A, Nikolajuk A, Kretowski A, Górska M (2009) Serum visfatin concentration is elevated in pregnant women irrespectively of the presence of gestational diabetes. Ginekol Pol 80:14–18

Tanaka M, Nozaki M, Fukuhara A, Segawa K, Aoki N, Matsuda M, Komuro R, Shimomura I (2007) Visfatin is released from 3T3-L1 adipocytes via a non-classical pathway. Biochem Biophys Res Commun 359:194–201

Telejko B, Kuzmicki M, Zonenberg A, Szamatowicz J, Wawrusiewicz-Kurylonek N, Nikolajuk A, Kretowski A, Gorska M (2009) Visfatin in gestational diabetes: serum level and mRNA expression in fat and placental tissue. Diabetes Res Clin Pract 84:68–75

Tissenbaum HA, Guarente L (2001) Increased dosage of a sir-2 gene extends lifespan in Caenor-habditis elegans. Nature 410:227–230

Toruner F, Altinova AE, Bukan AE, Akbay E, Ersoy R, Arslan M (2009) Plasma visfatin concentrations in subjects with type 1 diabetes mellitus. Horm Res 72:33–37

Unluturk U, Harmanci A, Yildiz BO, Bayraktar M (2010) Dynamics of Nampt/visfatin and high molecular weight adiponectin in response to oral glucose load in obese and lean women. Clin Endocrinol (Oxf) 72:469–474

van der Veer E, Nong Z, O'Neil C, Urquhart B, Freeman D, Pickering JG (2005) Pre-B-cell colony-enhancing factor regulates NAD+-dependent protein deacetylase activity and promotes vascular smooth muscle cell maturation. Circ Res 97:25–34

Wang P, van Greevenbroek MM, Bouwman FG, Brouwers MC, van der Kallen CJ, Smit E, Keijer J, Mariman EC (2007) The circulating PBEF/NAMPT/visfatin level is associated with a beneficial blood lipid profile. Pflugers Arch 454:971–976

Wang T, Zhang X, Bheda P, Revollo JR, Imai S, Wolberger C (2006) Structure of Nampt/PBEF/ visfatin, a mammalian NAD+ biosynthetic enzyme. Nat Struct Mol Biol 13:661–662

Westphal CH, Dipp MA, Guarente L (2008) A therapeutic role of sirtuins in diseases of aging? Trends Biochem Sci 32:555–560

Xie H, Tang SY, Luo XH, Huang J, Cui RR, Yuan LQ, Zhou HD, Wu XP, Liao EY (2007) Insulin-like effects of visfatin on human osteoblasts. Calcif Tissue Int 80:201–210

Xu LG, Wu M, Hu J, Zhai Z, Shu HB (2002) Identification of downstream genes up-regulated by the tumor necrosis factor family member TALL-1. J Leukoc Biol 72:410–416

Yang H, Lavu S, Sinclair DA (2006) Nampt/PBEF/Visfatin: a regulator of mammalian health and longevity? Exp Gerontol 41:718–726

Yildiz BO, Bozdag G, Otegen U, Harmanci A, Boynukalin K, Vural Z, Kirazli S, Yarali H (2010) Visfatin and retinol-binding protein 4 concentrations in lean, glucose-tolerant women with PCOS. Reprod Biomed Online 20:150–155

Yilmaz MI, Saglam M, Qureshi AR, Carrero JJ, Caglar K, Eyileten T, Sonmez A, Cakir E, Oguz Y, Vural A, Yenicesu M, Stenvinkel P, Lindholm B, Axelsson J (2008) Endothelial dysfunction in type-2 diabetics with early diabetic nephropathy is associated with low circulating adiponectin. Nephrol Dial Transplant 23:1621–1627

Inhibition of Ganglioside Biosynthesis as a Novel Therapeutic Approach in Insulin Resistance

Jin-ichi Inokuchi

Contents

Abstract A new concept "Life style-related diseases, such as type 2 diabetes, are a membrane microdomain disorder caused by aberrant expression of gangliosides" has arisen. By examining this working hypothesis, we demonstrate the molecular pathogenesis of type 2 diabetes and insulin resistance focusing on the interaction between insulin receptor and gangliosides in microdomains and propose the new therapeutic strategy "membrane microdomain ortho-signaling therapy".

Keywords Ganglioside · GM3 synthase · GM3 · Insulin receptor · Insulin resistance · Membrane microdomains (lipid rafts) · Microdomain disorder · Microdomain ortho-signaling therapy

J.-i. Inokuchi
Division of Glycopathology and CREST, Japan Science and Technology Agency, Institute of Molecular Biomembranes and Glycobiology, Tohoku Pharmaceutical University, 4-4-1, komat-sushima, Aoba-ku, Sendai 981-8558, Miyagi, Japan
e-mail: jin@tohoku-pharm.ac.jp

M. Schwanstecher (ed.), *Diabetes - Perspectives in Drug Therapy*,
Handbook of Experimental Pharmacology 203,
DOI 10.1007/978-3-642-17214-4_8, © Springer-Verlag Berlin Heidelberg 2011

1 Introduction

Caveolae are a subset of membrane microdomains (lipid raft) particularly abundant in adipocytes. Critical dependence of the insulin metabolic signal transduction on caveolae/microdomains in adipocytes has been demonstrated. These microdomains can be biochemically isolated with their detergent insolubility and were designated as detergent-resistant microdomains (DRMs). Gangliosides are known as structurally and functionally important components in microdomains. We demonstrated that increased GM3 expression was accompanied in the state of insulin resistance in mouse 3T3-L1 adipocytes induced by TNFα and in the adipose tissues of obese/diabetic rodent models such as Zucker *fa/fa* rats and *ob/ob* mice (Tagami et al. 2002). We examined the effect of TNFα on the composition and function of DRMs in adipocytes and demonstrated that increased GM3 levels result in the elimination of insulin receptor (IR) from the DRM, while caveolin and flotillin remain in the DRMs, leading to the inhibition of insulin's metabolic signaling (Kabayama et al. 2005). These findings are further supported by the report that mice lacking GM3 synthase exhibit enhanced insulin signaling (Yamashita et al. 2003). To gain insight into molecular mechanisms behind interactions of IR, caveolin-1 (Cav1), and GM3 in adipocytes, we have performed immunoprecipitations, cross-linking studies of IR and GM3, and live cell studies using fluorescence recovery after photobleaching (FRAP) technique. We found that (1) IR forms complexes with Cav1 and GM3 independently; (2) in GM3-enriched membranes, the mobility of IR is increased by dissociation of the IR–Cav1 interaction; (3) the lysine residue localized just above the transmembrane domain of the IR β-subunit is essential for the interaction of IR with GM3. Since insulin metabolic signal transduction in adipocytes is known to be critically dependent on caveolae, we propose a new pathological feature of insulin resistance in adipocytes caused by dissociation of the IR–Cav1 complex by the interactions of IR with GM3 in microdomains (Kabayama et al. 2007).

2 Ganglioside GM3 Is an Inducer of Insulin Resistance

Insulin elicits a wide variety of biological activities, which can be globally categorized into metabolic and mitogenic actions. The binding of insulin to IR activates IR internal tyrosine kinase activity. The activated tyrosine-phosphorylated IR was able to recruit and phosphorylate adaptor proteins such as insulin receptor substrate (IRS). The phosphorylated IRS activates PI3 kinase, resulting in the translocation of glucose transporter 4 (GLUT-4) to plasma membrane to facilitate glucose uptake. This IR–IRS–PI3 kinase signaling cascade is the representative metabolic pathway of insulin. On the other hand, the mitogenic pathway in insulin signaling initiates phosphorylation of Shc by the activated IR and then activates Ras-MAPK signaling.

When mouse adipocytes were cultured in low concentrations of TNFα, which do not cause generalized suppression of adipocyte gene expression including IRS-1 and GLUT-4, interference of insulin action by TNFα occurred (Guo and Donner 1996). This requires prolonged treatment (at least 72 h), unlike many acute effects of this cytokine. The slowness of the effect suggests that TNFα induces the synthesis of an inhibitor that is the actual effector. We demonstrated that the state of insulin resistance in adipocytes treated with 0.1 nM TNFα was accompanied by a progressive increase in cell surface GM3. This was reflected by increases in cellular GM3 content, *GM3 synthase* activity, and GM3 synthase mRNA content, indicating that TNFα upregulates GM3 synthesis at the transcriptional level in cultured adipocytes (Tagami et al. 2002). To elucidate whether the increased GM3 in 3T3-L1 adipocytes treated with TNFα is involved in insulin resistance, we used an inhibitor of glucosylceramide synthase, D-*threo*-1-phenyl-2-decanoylamino-3-morpholino-1-propanol (D-PDMP) (Inokuchi and Radin 1987), to deplete cellular glycosphingolipids derived from glucosylceramide. D-PDMP proved the ability to counteract TNF-induced increase of GM3 content in adipocytes and completely normalize the TNF-induced defect in tyrosine phosphorylation of IRS-1 in response to insulin stimulation (Fig. 1a). These findings are supported by the observation that knockout mice lacking GM3 synthase exhibits enhancement of insulin signaling (Yamashita et al. 2003).

   Hotamisligil et al. reported that treatment of adipocytes with TNFα induces an increase in the serine phosphorylation of IRS-1 (Hotamisligil et al. 1993).

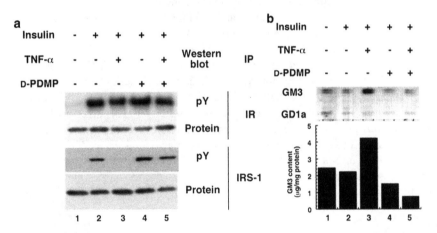

Fig. 1 TNFα increases the expression of GM3 and prevention of GM3 synthesis reverses TNFα-induced suppression of insulin signaling in adipocytes. (**a**) 3T3-L1 adipocytes were cultured in maintenance medium without (lanes 1, 2, and 4) or with (lanes 3 and 5) 0.1 nM TNFα for 96 h and in order to deplete GM3, 20 μM D-PDMP was also included (lanes 4 and 5). Before insulin stimulation (100 nM for 3 min), cells were starved in serum-free media containing 0.5% bovine serum albumin in the absence or presence of TNFα and D-PDMP as above for 8 h. Proteins in cell lysates were immunoprecipitated with antiserum to IR and IRS-1, fractionated by SDS-PAGE, and transferred to Immobilon-P. Western blot was then proved with antiphosphotyrosine monoclonal antibody, stripped, and reproved with antiserum to IR and IRS-1. (**b**) 3T3-L1 adipocytes were incubated in the absence or presence of TNFα and D-PDMP as in (**a**) and the ganglioside fraction was visualized by resorcinol staining on HPTLC

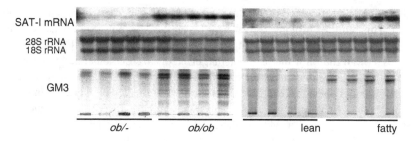

Fig. 2 Increased GM3 synthase mRNA in adipose tissue of typical rodent models of insulin resistance. Northern blot analysis of GM3 synthase mRNA was performed using total mRNA from adipose tissues of *ob/ob* mice and Zucker *fa/fa* rats and their lean counterparts

This phosphorylation is an important event since immunoprecipitated IRS-1, which has been serine phosphorylated in response to TNFα, is a direct inhibitor of IR tyrosine kinase activity. We have shown that TNFα-induced serine phosphorylation of IRS-1 in adipocytes was completely suppressed by inhibition of GM3 biosynthesis with D-PDMP treatment, suggesting that the elevated GM3 synthesis induced by TNFα caused the upregulation of serine phosphorylation of IRS-1 (Tagami et al. 2002) (Fig. 1b). Since TNF-induced serine phosphorylation of IRS-1 may occur through the activation of a variety of kinases including protein kinase C, c-Jun NH2-terminal kinase, p44/42 kinase, and PI3 kinase, it is important to identify the actual kinase(s) activated by endogenous GM3.

It was shown that adipose tissues of the obese–diabetic *db/db, ob/ob*, KK-A^y mice, and the Zucker *fa/fa* rat produced significant levels of TNFα (Hotamisligil et al. 1993). Much less expression was seen in adipose tissues obtained from the lean control animals. Interestingly, these obese–diabetic animals did not show evidence of altered expression of other cytokines, such as IL-1 or IFNγ (Hotamisligil et al. 1993; Hotamisligil and Spiegelman 1994). Thus, we were interested in measuring the expression of GM3 synthase mRNA in the epididymal fat of Zucker *fa/fa* rats and *ob/ob* mice. Northern blot analysis of GM3 synthase mRNA contents in the adipose tissues from these two typical models of insulin resistance exhibited significantly high levels compared to their lean counterparts (Fig. 2) (Tagami et al. 2002).

3 Caveolae Microdomains and Insulin Signaling

Caveolae are a subset of membrane microdomains particularly abundant in adipocytes (Fan et al. 1983; Parpal et al. 2001). Critical dependence of the insulin metabolic signal transduction on caveolae/microdomains in adipocytes has been demonstrated (Bickel. 2002; Cohen et al. 2003a). Disruption of microdomains by cholesterol extraction with methyl-β-cyclodextrin resulted in progressive inhibition of tyrosine phosphorylation of IRS-1 and activation of glucose transport in response to insulin, although autophosphorylation of IR and activation of MAP kinase were not impaired (Parpal et al. 2001). Similarities between these cell culture results and the findings in many cases of clinical insulin resistance (Virkamaki et al. 1999) suggest a potential role for microdomains in the pathogenesis of this disorder.

Couet et al. demonstrated the presence of a caveolin-binding motif (fXXXXfXXf [f: an aromatic amino acid, X: any amino acid]) in the β-subunit of IRs that could bind to the scaffold domain of caveolin (Couet et al. 1997). Moreover, mutation of this motif resulted in the inhibition of insulin signaling (Nystrom et al. 1999). Indeed, mutations of the IR β-subunit have been found in type 2 diabetic patients (Imamura et al. 1994). Lisanti's laboratory reported that Cav1-null mice developed insulin resistance when placed on a high fat diet (Cohen et al. 2003b). Interestingly, insulin signaling, as measured by IR phosphorylation and its downstream targets, was selectively decreased in the adipocytes of these animals, while signaling in both muscle and liver cells was normal. This signaling defect was attributed to a 90% decrease in IR protein content in the adipocytes, with no changes in mRNA levels, indicating that Cav1 serves to stabilize the IR protein (Cohen et al. 2003a, b). These studies clearly indicate the critical importance of the interaction between caveolin and IR in executing successful insulin signaling in adipocytes.

Although the direct interaction between Cav1 and IR has been shown as described above, studies of the presence of IRs in DRM have provided conflicting data (Gustavsson et al. 1999; Iwanishi et al. 1993; Kimura et al. 2002; Mastick and Saltiel 1997; Muller et al. 2001). Saltiel and colleagues found that insulin stimulation of 3T3-L1 adipocytes was associated with tyrosine phosphorylation of Cav1 (Mastick et al. 1995). However, since only trace levels of IR were recovered in the caveolae microdomains in assays with a buffer of 1% Triton X-100, they speculated on the presence of intermediate molecule(s) bridging IR and caveolin (Mastick and Saltiel 1997). Gustavsson et al. also observed the dissociation of IRs from caveolin-containing DRM after treatments of 0.3 and 0.1% Triton X-100 (Gustavsson et al. 1999). It has been reported that a comparison of protein and lipid contents of DRM prepared with a variety of detergents indicated considerable differences in their ability to selectively solubilize membrane proteins and to enrich sphingolipids and cholesterol over glycerophospholipids and that Triton X-100 was the most reliable detergent (Schuck et al. 2003). Therefore, we performed a flotation assay with a wide range of Triton X-100 concentrations to identify the protein of interest, which might weakly associate with DRM. In an assay system containing less than 0.08% Triton X-100, we were able to show that in normal adipocytes IRs can localize to DRM. Thus, by employing low detergent concentrations we were able to demonstrate, for the first time, the presence of IR in DRM (Kabayama et al. 2005). As summarized in Table 1, there is strong evidence suggesting that the localization of IRs in caveolae microdomains is essential for metabolic signaling of insulin.

Table 1 Localization of insulin receptor in caveolae microdomains is essential for the metabolic signaling of insulin

Function	Evidence	Reference
Direct biding of IR and caveolin-1	IR has caveolin-binding domain	Couet et al. (1997)
	Coimmunoprecipitation of IR and caveolin	Nystrom et al. (1999)
Colocalization of IR and caveolin-1	IR and caveolin in light-density fractions by sucrose density floatation assay	Kabayama et al. (2005)
	Fluorescence microscope	Gustavsson et al. (1999)
	Electron microscope	Karlsson et al. (2004), Foti et al. (2007)
Signaling	Stimulation of caveolin-1 tyrosine phosphorylation by insulin	Mastick et al. (1995), Kimura et al. (2002)
	Caveolin-deficient mice show insulin resistance due to accelerated degradation of IR in adipose tissue	Cohen et al. (2003b), Capozza et al. (2005)
	Cholesterol depletion disrupts caveolae and metabolic signaling of insulin	Parpal et al. (2001), Gustavsson et al. (1996)
	Increased GM3 eliminates IR from DRM and inhibits IR–IRS-1 signaling	Kabayama et al. (2005), Kabayama et al. (2007)

4 Insulin Resistance as a Membrane Microdomain Disorder

In a state of insulin resistance induced in adipocytes by TNFα, we presented evidence that the transformation to a resistant state may depend on increased ganglioside GM3 biosynthesis following upregulated GM3 synthase gene expression. Additionally, GM3 may function as an inhibitor of insulin signaling during chronic exposure to TNFα (Tagami et al. 2002). Since GSL, including GM3, is an important component of DRM/caveolae, we have pursued the possibility that increased GM3 levels in DRM confer insulin resistance upon TNFα-treated adipocytes. We examined the effect of TNFα on the composition and function of DRM in adipocytes and demonstrated that increased GM3 levels result in the elimination of IRs from the DRM, while raft marker proteins such as caveolin and flotillin remain in the DRM (Kabayama et al. 2005). Although the localization of IRs to DRM may be maintained by the association with Cav1 as mentioned above, the excess accumulation of GM3 in the DRM may weaken IR–caveolin interaction. Therefore, to examine interactions among IR, Cav1, and GM3 in 3T3-L1 adipocytes, we initially performed coimmunoprecipitation assays. Cav1 has a scaffolding domain to which IR and other functional transmembrane proteins bind through a caveolin-binding domain in their cytoplasmic region (Couet et al. 1997; Nystrom et al. 1999). As expected from another study (Nystrom et al. 1999), IR was coprecipitated with Cav1 (Fig. 3a). GM3 was coprecipitated with IR but not with Cav1 (Fig. 3b upper panel). In addition, IR but not Cav1 was coprecipitated with GM3 (Fig. 3b, lower panel). Thus, IR can bind both Cav1 and GM3, but there is no interaction between GM3 and Cav1, suggesting that IR can form distinct complexes with each of

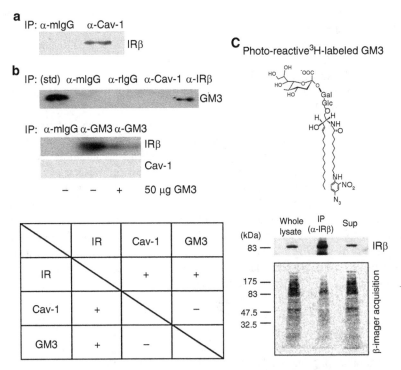

Fig. 3 The insulin receptor forms distinct complexes with Cav1 and GM3 in adipocytes. (**a**) Coimmunoprecipitation assay of Cav1 and IR. Postnuclear supernatants (PNS) of whole cell lysates were immunoprecipitated with an anti-Cav1 antibody or anti-mouse IgG (−), and the precipitates were subjected to SDS-PAGE followed by immunoblotting with an anti-IRβ antibody. (**b**) GM3 associates with IR but not with Cav1. *Upper panel*: PNS were immunoprecipitated with an anti-Cav1 antibody, an anti-IRβ antibody, or an anti-mouse or anti-rat IgG (−). The precipitates were subjected to TLC followed by immunostaining with the anti-GM3 antibody M2590. *Lower panel*: Immunoprecipitation was performed with the anti-GM3 antibody DH2, in the presence or absence of 50 μg GM3, or with anti-mouse IgG (−). The precipitates were then subjected to SDS-PAGE followed by immunoblotting with an anti-IRβ or anti-Cav1 antibody. (**c**) Cross-linking assay of GM3 and IR. Adipocytes were treated with photoactivatable [3H]-labeled GM3, washed, and then irradiated. Cells were then lysed and subjected to immunoprecipitation with an anti-IRβ antibody. PNS, anti-IRβ immunoprecipitates (IP), and the supernatant from the immunoprecipitation (Sup) were subjected to SDS-PAGE, followed by immunoblotting with an anti-IRβ antibody and autoradiography (Kabayama et al. 2007)

them. The association between IR and the anti-GM3 antibody was abolished by the presence of GM3, confirming the specific binding ability of the anti-GM3 antibody to GM3 in the immunoprecipitation medium (Fig. 3b, lower panel).

We next examined GM3–protein interactions occurring within the plasma membrane of living cells by performing a cross-linking assay using a photoactivatable radioactive derivative of GM3 (Fig. 3c). Adipocytes were preincubated with [3H] GM3(N₃) and then irradiated to induce cross-linking of GM3. Target proteins were then separated by SDS-PAGE and visualized by autoradiography. A broad range of radioactivity reflecting GM3–protein complexes could be detected from 80 to 200 kDa, suggesting a close association between GM3 and a variety of cell surface proteins, including IR. Moreover, a specific radioactive band corresponding to the 90-kDa IR β-subunit was immunoprecipitated with anti-IRβ antibodies, confirming the direct association of GM3 and IR. Therefore, we found that IR forms complexes with Cav1 and GM3 independently in 3T3-L1 adipocytes (Kabayama et al. 2007).

Lipids are asymmetrically distributed in the outer and inner leaflets of plasma membranes. In typical mammalian cells, most acidic phospholipids are located in the inner leaflet, and only acidic glycosphingolipids such as sulfatides and ganglio-sides are in the outer. The binding of proteins to lipid membranes is often mediated by electrostatic interactions between the proteins' basic domains and acidic lipids. Gangliosides, which bear sialic acid residues, exist ubiquitously in the outer leaflet of the vertebrate plasma membrane. GM3 is the most abundant ganglioside and the primary ganglioside found in adipocytes (Ohashi 1979). Glycosphingolipids, includ-ing gangliosides, share a common minimum energy conformational structure in which the oligosaccharide chain is oriented at a defined angle to the axis of the ceramide (Hakomori 2002). In addition, GM3 spontaneously forms clusters with its own saturated fatty acyl chains, regardless of any repulsion between the negatively charged units in the sugar chains (Sonnino et al. 2007). Thus, GM3 clusters with other cell surface gangliosides generate a negatively charged environment just above the plasma membrane. Conversely, IR has a sequence in its transmembrane domain, homologous among mammals, which allows presentation of the basic amino acid lysine (IR944) just above transmembrane domain. Therefore, during lateral diffu-sion, an electrostatic interaction between the lysine residue at IR944 and the GM3 cluster could occur due to their proximity on the plasma membrane (Fig.4a).

We previously developed GM3-reconstituted cells by stably transfecting the GM3 synthase (SAT-I) gene into GM3-deficient cells (Uemura et al. 2003) (Fig. 4b, left panel). Using the FRAP technique, we examined the mobility of IR in the plasma membranes of GM3-reconstituted (GM3 (+)) cells and mock (GM3 (−)) cells expressing equal levels of Cav1 (Fig. 4b, right panel inset). The mobility of IR-GFP expressed in the GM3 (+) cells was statistically (10%) higher than that in the GM3 (−) cells (Fig. 4b, right panel), providing further evidence that GM3 is able to enhance IR mobility by dissociating the Cav1 and IR complex in living cells.

The binding between IR and Cav1 has been studied in detail (Nystrom et al. 1999). To similarly analyze interactions between IR and GM3, we constructed several mutants of IR in which the lysine at IR944 was replaced with the basic amino acid, arginine, or with the neutral amino acid valine, serine, or glutamine (Fig. 4c). The fluorescence recovery of IR(K944G), IR(K944S), and IR(K944V) 100 s after bleaching was decreased by 10% compared to those of IR(WT) and IR(K944R) in GM3 (+) cells (Fig. 4c, lower left panel). However, in GM3 (−) cells, no such difference in the mobility between IR(wt) and IR(K944S) was observed (Fig. 4c, lower right panel). This demonstrates that the lysine in the

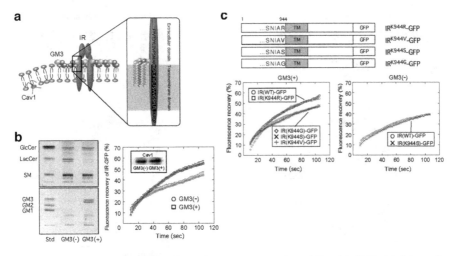

Fig. 4 The lysine residue IR944 is essential for the interaction of IR with GM3. (**a**) Schematic representation of the proposed interaction of the lysine residue at IR944, which is located just above the transmembrane domain, and GM3 at the cell surface. (**b**) Enhanced mobility of IR in GM3-enriched membrane. *Left panel*: Glycosphingolipid (GSL) analysis of GM3-reconstituted cells (GM3 (+)) and mock cells (GM3 (−)). GSLs extracted from these cells, corresponding to 1 mg of cellular protein, were separated on HPTLC plates and stained with resorcinol–HCl reagent, to visualize gangliosides, or with cupric acetate–phosphoric acid reagent for neutral GSLs. *Right panel*: FRAP analyses. Fluorescence recovery of IR-GFP in GM3 (−) and GM3 (+) cells expressing equal levels of Cav1(*inset*). (**c**) Specificity of the interaction between lysine at IR944 and GM3 by FRAP analyses. *Upper panel*: Schematic structure of IR-GFP mutants in which the lysine at IR944 is replaced with basic and neutral amino acids. *Lower panel*: Fluorescence recovery of IR-GFP mutants in GM3(+) and GM3(−) cells

wild type is essential for its binding to GM3 due to its basic charge (Kabayama et al. 2007).

Here, we propose a mechanism behind the shift of IR from the caveolae to the glycosphingolipid-enriched domain (GEM) in adipocytes during a state of insulin resistance. Figure 5 shows a schematic representation of raft/microdomains comprising caveolae and noncaveolae rafts such as GEM. Caveolae and GEM reportedly can be separated by an anti-Cav1 antibody (Iwabuchi et al. 1998). IR may be constitutively resident in caveolae via its binding to the scaffolding domain of Cav1 through the caveolin-binding domain in its cytoplasmic region. Binding of IR and Cav1 is necessary for successful insulin metabolic signaling (Table 1). In adipocytes, the localization of IR in the caveolae is interrupted by elevated levels of the endogenous ganglioside GM3 during a state of insulin resistance induced by inflammatory response (e.g. TNFα) (Tagami et al. 2002). By live cell studies using FRAP techniques we have proven a mechanism, at least in part, in which the dissociation of the IR–Cav1 complex is caused by the interaction of a lysine residue, located just above the transmembrane domain in IR β-subunit, and the increased GM3 clustered at the cell surface (Kabayama et al. 2007) (Fig. 5).

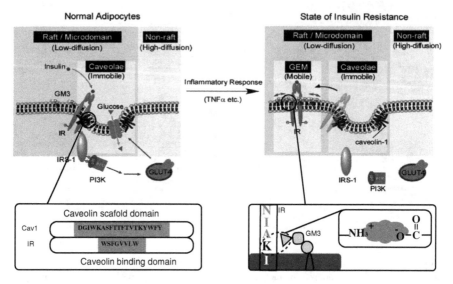

Fig. 5 Proposed mechanism behind the shift of insulin receptors from the caveolae to the glyco-sphingolipid-enriched microdomains (GEM) in adipocytes during a state of insulin resistance. A schematic representation of raft/microdomains comprising caveolae and noncaveolae rafts such as GEM. Caveolae and GEM reportedly can be separated by an anti-Cav1 antibody. IR may be constitutively resident in caveolae via its binding to the scaffolding domain of Cav1 through the caveolin-binding domain in its cytoplasmic region. Binding of IR and Cav1 is necessary for successful insulin metabolic signaling (Table 1). In adipocytes, the localization of IR in the caveolae is interrupted by elevated levels of the endogenous ganglioside GM3 during a state of insulin resistance induced by TNFα (Kabayama et al. 2005). This study has proved a mechanism, at least in part, in which the dissociation of the IR/Cav1 complex is caused by the interaction of a lysine residue at IR944, located just above the transmembrane domain, and the increased GM3 clustered at the cell surface

5   Serum GM3 Levels as a New Biomarker of Metabolic Syndrome

GM3 is the major ganglioside present in serum and is known to be associated with serum lipoproteins (Senn et al. 1989). However, there have been no studies examining a relationship between serum GM3 levels and diabetes or abdominal obesity. So, we investigated the relationship between serum GM3 levels and adiposity indices, as well as between serum GM3 levels and metabolic risk variables (Sato et al. 2008). Serum GM3 levels were higher in hyperglycemic patients (1.4-fold), hyperlipidemic patients (1.4-fold), and hyperglycemic patients with hyperlipidemia (1.6-fold) than in normal subjects. In addition, serum GM3 levels were significantly increased in type 2 diabetic patients with severe obesity (visceral fat area >200 cm^2 and BMI >30). The GM3 level was positively correlated with LDL-c (0.403, $p = 0.012$) in type 2 diabetes mellitus, but not affected by blood pressure. In addition, the high levels of small dense LDL (>10 mg/dl) were associated with the elevation of GM3. Serum GM3 levels were affected by glucose and lipid metabolism abnormalities and by visceral obesity. Interestingly, increased small dense LDL was reportedly associated with the development of atherosclerosis (Austin et al. 1988; de Graaf et al. 1991; Tribble et al. 1995), and GM3 has been detected in atherosclerotic lesions (Bobryshev et al. 2001, 1997). Thus, our findings provide evidence that GM3 may be a useful marker for the management of metabolic syndrome including insulin resistance, as well as for the early diagnosis of atherosclerosis.

6   A Possible Therapeutic Intervention of Metabolic Syndrome by Inhibiting Ganglioside Synthesis

Critical involvement of ganglioside GM3 in insulin resistance and metabolic syndrome including type 2 diabetes has now become evident based on the following key observations: (1) TNFα increases the expression of GM3 in adipocytes and the TNFα-induced insulin resistance is prevented by treatment with a glucosylceramide synthase inhibitor, D-PDMP, and decreases GM3 contents (Tagami et al. 2002); (2) GM3 contents increase in the adipose tissue of Zucker *fa/fa* rats and *ob/ob* mice, which are typical rodent models of obesity (Tagami et al. 2002) and diet-induced obesity (unpublished observation); (3) insulin sensitivity is enhanced in GM3 synthase knockout mice (Yamashita et al. 2003); (4) the accumulation of GM3 in insulin resistance results in dissociation of the IR from caveolae (Kabayama et al. 2005); (5) dissociation of the IR from caveolae is caused by electrostatic interaction between GM3 and the lysine residue (Lys-944) located just above the transmembrane of the IR(Kabayama et al. 2007); and (6) treatment with glucosylceramide synthase inhibitors significantly improved insulin sensitivity and glucose homeostasis in rodent models of obesity (Aerts et al. 2007; Zhao et al. 2007). Taken together, a new therapeutic intervention by inhibiting GM3 biosynthesis can be proposed for the treatment of metabolic syndrome including type 2 diabetes.

7   Concluding Remarks

A growing body of evidence implicates glycosphingolipids including gangliosides in the pathogenesis of insulin resistance. We demonstrated that in 3T3-L1 adipocytes in a state of TNF-induced insulin resistance, the inhibition of insulin metabolic signaling was associated with an accumulation of the ganglioside GM3, and, moreover, the pharmacological inhibition of GM3 biosynthesis by the glucosylceramide synthase inhibitor D-PDMP resulted in the nearly complete recovery of TNFα-induced suppression of insulin signaling, suggesting a new target for therapy against insulin resistance and type 2 diabetes (Tagami et al. 2002). Recently, an

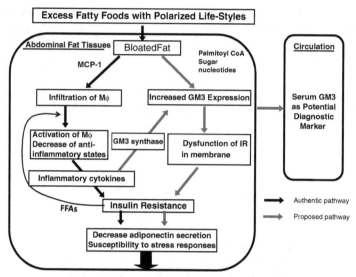

Fig. 6 Insulin resistance through increased GM3 levels in metabolic syndrome. MCP-1: mono-cyte chemotactic protein-1; Mφ: macrophage

improved PDMP analogue (Zhao et al. 2007) and another type of glucosylceramide synthase inhibitor (Aerts et al. 2007) were proven to have therapeutic value by oral administration in diabetic rodent models. In addition, our data substantiate a rationale for designing novel therapies against type 2 diabetes and related diseases based on inhibition of ganglioside biosynthesis. Figure 6 represents a synopsis of the proposed role of the aberrant expression of GM3 in metabolic syndrome, type 2 diabetes, and atherosclerosis.

Acknowledgments The author thank all my collaborators concerned with the studies described in this review article, who are listed as the co-authors of our papers.

References

Aerts JM, Ottenhoff R, Powlson AS, Grefhorst A, van Eijk M, Dubbelhuis PF, Aten J, Kuipers F, Serlie MJ, Wennekes T, Sethi JK, O'Rahilly S, Overkleeft HS (2007) Pharmacological inhibition of glucosylceramide synthase enhances insulin sensitivity. Diabetes 56:1341–1349

Austin MA, Breslow JL, Hennekens CH, Buring JE, Willett WC, Krauss RM (1988) Low-density lipoprotein subclass patterns and risk of myocardial infarction. JAMA 260:1917–1921

Bickel PE (2002) Lipid rafts and insulin signaling. Am J Physiol Endocrinol Metab 282:E1–E10

Bobryshev YV, Lord RS, Golovanova NK, Gracheva EV, Zvezdina ND, Sadovskaya VL, Prokazova NV (1997) Incorporation and localisation of ganglioside GM3 in human intimal atherosclerotic lesions. Biochim Biophys Acta 1361:287–294

Bobryshev YV, Lord RS, Golovanova NK, Gracheva EV, Zvezdina ND, Prokazova NV (2001) Phenotype determination of anti-GM3 positive cells in atherosclerotic lesions of the human aorta. Hypothetical role of ganglioside GM3 in foam cell formation. Biochim Biophys Acta 1535:87–99

Capozza F, Cohen AW, Cheung MW, Sotgia F, Schubert W, Battista M, Lee H, Frank PG, Lisanti MP (2005) Muscle-specific interaction of caveolin isoforms: differential complex formation between caveolins in fibroblastic vs. muscle cells. Am J Physiol Cell Physiol 288:C677–C691

Cohen AW, Combs TP, Scherer PE, Lisanti MP (2003a) Role of caveolin and caveolae in insulin signaling and diabetes. Am J Physiol Endocrinol Metab 285:E1151–E1160

Cohen AW, Razani B, Wang XB, Combs TP, Williams TM, Scherer PE, Lisanti MP (2003b) Caveolin-1-deficient mice show insulin resistance and defective insulin receptor protein expression in adipose tissue. Am J Physiol Cell Physiol 285:C222–C235

Couet J, Li S, Okamoto T, Ikezu T, Lisanti MP (1997) Identification of peptide and protein ligands for the caveolin-scaffolding domain. Implications for the interaction of caveolin with caveolae-associated proteins. J Biol Chem 272:6525–6533

de Graaf J, Hak-Lemmers HL, Hectors MP, Demacker PN, Hendriks JC, Stalenhoef AF (1991) Enhanced susceptibility to in vitro oxidation of the dense low density lipoprotein subfraction in healthy subjects. Arterioscler Thromb 11:298–306

Fan JY, Carpentier JL, van Obberghen E, Grunfeld C, Gorden P, Orci L (1983) Morphological changes of the 3T3-L1 fibroblast plasma membrane upon differentiation to the adipocyte form. J Cell Sci 61:219–230

Foti M, Porcheron G, Fournier M, Maeder C, Carpentier JL (2007) The neck of caveolae is a distinct plasma membrane subdomain that concentrates insulin receptors in 3T3-L1 adipocytes. Proc Natl Acad Sci USA 104:1242–1247

Guo D, Donner DB (1996) Tumor necrosis factor promotes phosphorylation and binding of insulin receptor substrate 1 to phosphatidylinositol 3-kinase in 3T3-L1 adipocytes. J Biol Chem 271:615–618

Gustavsson J, Parpal S, Strålfors P (1996) Insulin-stimulated glucose uptake involves the transition of glucose transporters to a caveolae-rich fraction within the plasma membrane: implications for type II diabetes. Mol Med 2:367–372

Gustavsson J, Parpal S, Karlsson M, Ramsing C, Thorn H, Borg M, Lindroth M, Peterson KH, Magnusson KE, Strålfors P (1999) Localization of the insulin receptor in caveolae of adipocyte plasma membrane. FASEB J 13:1961–1971

Hakomori SI (2002) Inaugural article: the glycosynapse. Proc Natl Acad Sci USA 99:225–232

Hotamisligil GS, Spiegelman BM (1994) Tumor necrosis factor alpha: a key component of the obesity-diabetes link. Diabetes 43:1271–1278

Hotamisligil GS, Shargill NS, Spiegelman BM (1993) Adipose expression of tumor necrosis factor-alpha: direct role in obesity-linked insulin resistance. Science 259:87–91

Imamura T, Takata Y, Sasaoka T, Takada Y, Morioka H, Haruta T, Sawa T, Iwanishi M, Hu YG, Suzuki Y (1994) Two naturally occurring mutations in the kinase domain of insulin receptor accelerate degradation of the insulin receptor and impair the kinase activity. J Biol Chem 269:31019–31027

Inokuchi J, Radin N (1987) Preparation of the active isomer of 1-phenyl-2-decanoylamino-3-morpholino-1-propanol, inhibitor of murine glucocerebroside synthetase. J Lipid Res 28:565–571

Iwabuchi K, Handa K, Hakomori S (1998) Separation of "glycosphingolipid signaling domain" from caveolin-containing membrane fraction in mouse melanoma B16 cells and its role in cell adhesion coupled with signaling. J Biol Chem 273:33766–33773

Iwanishi M, Haruta T, Takata Y, Ishibashi O, Sasaoka T, Egawa K, Imamura T, Naitou K, Itazu T, Kobayashi M (1993) A mutation (Trp1193–>Leu1193) in the tyrosine kinase domain of the insulin receptor associated with type A syndrome of insulin resistance. Diabetologia 36:414–422

Kabayama K, Sato T, Kitamura F, Uemura S, Kang BW, Igarashi Y, Inokuchi J (2005) TNFalpha-induced insulin resistance in adipocytes as a membrane microdomain disorder: involvement of ganglioside GM3. Glycobiology 15:21–29

Kabayama K, Sato T, Saito K, Loberto N, Prinetti A, Sonnino S, Kinjo M, Igarashi Y, Inokuchi J (2007) Dissociation of the insulin receptor and caveolin-1 complex by ganglioside GM3 in the state of insulin resistance. Proc Natl Acad Sci USA 104:13678–13683

Karlsson M, Thorn H, Danielsson A, Stenkula KG, Ost A, Gustavsson J, Nystrom FH, Strålfors P (2004) Colocalization of insulin receptor and insulin receptor substrate-1 to caveolae in primary human adipocytes. Cholesterol depletion blocks insulin signalling for metabolic and mitogenic control. Eur J Biochem 271:2471–2479

Kimura A, Mora S, Shigematsu S, Pessin JE, Saltiel AR (2002) The insulin receptor catalyzes the tyrosine phosphorylation of caveolin-1. J Biol Chem 277:30153–30158

Mastick CC, Saltiel AR (1997) Insulin-stimulated tyrosine phosphorylation of caveolin is specific for the differentiated adipocyte phenotype in 3T3-L1 cells. J Biol Chem 272:20706–20714

Mastick CC, Brady MJ, Saltiel AR (1995) Insulin stimulates the tyrosine phosphorylation of caveolin. J Cell Biol 129:1523–1531

Muller G, Jung C, Wied S, Welte S, Jordan H, Frick W (2001) Redistribution of glycolipid raft domain components induces insulin-mimetic signaling in rat adipocytes. Mol Cell Biol 21:4553–4567

Nystrom FH, Chen H, Cong LN, Li Y, Quon MJ (1999) Caveolin-1 interacts with the insulin receptor and can differentially modulate insulin signaling in transfected Cos-7 cells and rat adipose cells. Mol Endocrinol 13:2013–2024

Ohashi M (1979) A comparison of the ganglioside distributions of fat tissues in various animals by two-dimensional thin layer chromatography. Lipids 14:52–57

Parpal S, Karlsson M, Thorn H, Strålfors P (2001) Cholesterol depletion disrupts caveolae and insulin receptor signaling for metabolic control via insulin receptor substrate-1, but not for mitogen-activated protein kinase control. J Biol Chem 276:9670–9678

Sato T, Nihei Y, Nagafuku M, Tagami S, Chin R, Kawamura M, Miyazaki S, Suzuki M, Sugahara S, Takahashi Y, Saito A, Igarashi Y, Inokuchi J (2008) Circulating levels of ganglioside GM3 in metabolic syndrome: a pilot study. Obes Res Clin Pract 2:231–238

Schuck S, Honsho M, Ekroos K, Shevchenko A, Simons K (2003) Resistance of cell membranes to different detergents. Proc Natl Acad Sci USA 100:5795–5800

Senn HJ, Orth M, Fitzke E, Wieland H, Gerok W (1989) Gangliosides in normal human serum. Concentration, pattern and transport by lipoproteins. Eur J Biochem 181:657–662

Sonnino S, Mauri L, Chigorno V, Prinetti A (2007) Gangliosides as components of lipid membrane domains. Glycobiology 17:1R–13R

Tagami S, Inokuchi J, Kabayama K, Yoshimura H, Kitamura F, Uemura S, Ogawa C, Ishii A, Saito M, Ohtsuka Y, Sakaue S, Igarashi Y (2002) Ganglioside GM3 participates in the pathological conditions of insulin resistance. J Biol Chem 277:3085–3092

Tribble DL, Krauss RM, Lansberg MG, Thiel PM, van den Berg JJ (1995) Greater oxidative susceptibility of the surface monolayer in small dense LDL may contribute to differences in copper-induced oxidation among LDL density subfractions. J Lipid Res 36:662–671

Uemura S, Kabayama K, Noguchi M, Igarashi Y, Inokuchi J (2003) Sialylation and sulfation of lactosylceramide distinctly regulate anchorage-independent growth, apoptosis, and gene expression in 3LL Lewis lung carcinoma cells. Glycobiology 13:207–216

Virkamaki A, Ueki K, Kahn CR (1999) Protein-protein interaction in insulin signaling and the molecular mechanisms of insulin resistance. J Clin Invest 103:931–943

Yamashita T, Hashiramoto A, Haluzik M, Mizukami H, Beck S, Norton A, Kono M, Tsuji S, Daniotti JL, Werth N, Sandhoff R, Sandhoff K, Proia RL (2003) Enhanced insulin sensitivity in mice lacking ganglioside GM3. Proc Natl Acad Sci USA 100:3445–3449

Zhao H, Przybylska M, Wu IH, Zhang J, Siegel C, Komarnitsky S, Yew NS, Cheng SH (2007) Inhibiting glycosphingolipid synthesis improves glycemic control and insulin sensitivity in animal models of type 2 diabetes. Diabetes 56:1210–1218

Overcoming Insulin Resistance with Ciliary Neurotrophic Factor

Tamara L. Allen, Vance B. Matthews, and Mark A. Febbraio

Contents

Abstract The incidence of obesity and related co-morbidities such as insulin resistance, dyslipidemia and hypertension are increasing at an alarming rate worldwide. Current interventions seem ineffective to halt this progression. With the failure of leptin as an anti-obesity therapeutic, ciliary neurotrophic factor (CNTF) has proven efficacious in models of obesity and leptin resistance, where leptin proved ineffective. CNTF is a gp130 ligand that has been found to act centrally and peripherally to promote weight loss and insulin sensitivity in both human and rodent models. Future research into novel gp130 ligands may offer new candidates for obesity-related drug therapy.

Keywords Ciliary Neurotrophic Factor · gp130 Ligands · Insulin Resistance · Leptin · Obesity

T.L. Allen, V.B. Matthews, and M.A. Febbraio (✉)
Cellular and Molecular Metabolism Laboratory, Baker IDI Heart and Diabetes Institute, P.O. Box 6492, St Kilda Road Central, 8008, VIC, Australia
e-mail: mark.febbraio@bakeridi.edu.au

M. Schwanstecher (ed.), *Diabetes - Perspectives in Drug Therapy*,
Handbook of Experimental Pharmacology 203,
DOI 10.1007/978-3-642-17214-4_9, © Springer-Verlag Berlin Heidelberg 2011

1   Introduction

Obesity is rapidly reaching epidemic proportions in many areas of the world. An estimated 1.6 billion people worldwide are now classified as overweight, with 246 million people with diabetes and an additional 308 million with impaired glucose tolerance (WHO 2006). Once considered diseases of age and affluence, the prevalence of obesity and diabetes has increased dramatically in both developing countries and the younger population (Mascie-Taylor and Karim 2003). Obesity is associated with dyslipidemia, elevated blood pressure, insulin resistance and progressive β cell failure leading to type 2 diabetes. In fact, in western countries, 90% of type 2 diabetes cases are due to increased weight (James et al. 2003). As a result, the financial and social costs of obesity and diabetes are profound. Advances in treating obesity will reduce the occurrence and progression of associated diseases such as insulin resistance and type 2 diabetes. Current therapies for obesity are limited at best and as such, there is a requirement for new therapeutics. Compounds that could combine weight loss and reduce insulin resistance would have a significant clinical impact on these often co-existing diseases. Ciliary neurotrophic factor (CNTF) and other gp130 ligands offer a unique class of potential therapeutics for use in obesity and associated diseases.

2   Current Therapeutics for Obesity and Type 2 Diabetes

There is evidence that a dramatic reduction in the incidence of type 2 diabetes and the prevention of an even larger number of cases are possible with the implementation of a healthy diet with increased activity (Knowler et al. 2002). Unfortunately, it is becoming increasingly evident that the vast population is unable to adhere to these lifestyle changes and thus require additional interventions. There are several classes of therapeutics currently available for the treatment of type 2 diabetes: (1) secretagogues (sulfonylureas and non-sulfonylureas), (2) sensitisers (biguanides, TZDs), (3) α-glucosidase inhibitors and (4) incretin-based agents (Fonseca and Kulkarni 2008; Penfornis et al. 2008). Secretagogues stimulate the pancreas to secrete insulin, thus allowing the liver to suppress gluconeogenesis and increase glucose uptake into the muscle. Biguanides (metformin) function not only by suppressing hepatic gluconeogenesis and glycogenolysis but also increase insulin sensitivity. Thiazolidinediones (TZDs, e.g. rosiglitazone, pioglitazone) act by re-sensitising peripheral tissues to the action of insulin. The α-glucosidase inhibitors (acarbose and miglitol) reduce the uptake of carbohydrates from the small intestine. Finally, incretin-based agents include dipeptidyl peptidase IV (DPP-4) inhibitors (e.g. sitagliptin, vildagliptin) and glucagon-like peptide-1 (GLP-1) analogues (e.g. exenatide, liraglutide). DPP-4 inhibitors attenuate the break down of endogenously secreted incretins (GLP-1 and GIP), while GLP-1 analogues increase circulating GLP-1 levels to supraphysiological levels by virtue of being DPP-4 resistant. Both compounds lead to increased insulin secretion and decreased glucagon secretion. Many of these compounds are used in combination with each other and with exogenous insulin. The use of many of these compounds, however, is associated with several side effects including hypoglycemia, weight gain and congestive heart failure (TZDs).

Bariatric surgery has been particularly efficacious in reducing or eliminating type 2 diabetes in addition to weight loss (Cummings et al. 2008). However, surgical intervention is not a wide scale resolution for an escalating epidemic, particularly considering patients are usually not considered candidates until they reach a BMI of 35 or greater. Clearly, early intervention is far more desirable in order to reduce morbidity and mortality associated with obesity and type 2 diabetes. There are currently two therapeutics available to treat obesity (Padwal and Majumdar 2007). Orlistat (Xenical) is a gastric and pancreatic lipase inhibitor designed to reduce the uptake of fats from the gut. Sibutramine is a monoamine-reuptake inhibitor that works predominantly to increase satiety. Both the compounds induced modest weight loss but studies have been plagued with lack of compliance and evidence of the effects on long-term morbidity and mortality. Furthermore, several adverse effects have been reported including gastrointestinal effects (Orlistat) and increases in pulse rate and blood pressure (Sibutramine). In light of this, there is a requirement to develop additional therapeutics suitable for long-term use in obesity. Ideally, a compound that could concomitantly induce weight loss and insulin sensitisation would offer a dual pronged attack on obesity and its co-morbidities.

3   Leptin: A Flash in the Pan

Leptin was once heralded as the breakthrough for obesity. It was discovered as a molecule that was secreted from adipose tissue and activated pathways in the brain that affected energy balance and feeding (Elmquist et al. 1998; Friedman and Halaas 1998; Schwartz et al. 2000; Zhang et al. 1994). Since then, leptin has also been attributed to numerous peripheral metabolic effects including glucose uptake and fatty acid oxidation in skeletal muscle (Minokoshi et al. 2002; Muoio et al. 1997; Steinberg et al. 2003; Watt et al. 2006a). Unfortunately, the discovery that obese people had already elevated levels of circulating leptin, indicative of leptin resistance, quickly ended any hope that exogenous leptin could be used therapeutically (Friedman and Halaas 1998). This leptin resistance is thought to arise via several mechanisms. The first involves the reduced transportation of leptin across the blood–brain barrier, since leptin retains its weight loss effects when administered centrally to a mouse model of diet-induced obesity displaying peripheral leptin resistance (Van Heek et al. 1997). This is further supported by findings that the efficiency of leptin transport into the cerebrospinal fluid was reduced in obese people with high levels of plasma leptin (Schwartz et al. 1996). In addition to this mechanism, the induction of SOCS3 (suppressor of cytokine signalling 3) expression is also thought to contribute to leptin resistance. Leptin receptor signalling leads to activation of the JAK/STAT pathway and transcription of the STAT-dependent gene, SOCS3. SOCS3 then binds via its src-homology-2 (SH-2) domain to the phosphorylation sites on the active leptin receptor and associated JAK, thereby inhibiting further receptor activation (Bjorbaek et al. 2000). Both obese humans and rodents (Bjorbaek et al. 1998; Peralta et al. 2002) have been found to overexpress SOCS leading to the belief that this negative feedback may contribute to a reduction in leptin receptor activity and manifest as leptin resistance. Further to this, haploinsufficient SOCS3 mice exhibit an increased sensitivity to leptin and show resistance to high fat diet induced obesity (Howard et al. 2004; Mori et al. 2004).

4 Leptin Receptor Versus gp130 Receptor Signalling

The leptin receptor (ObR) was first identified in 1995 (Tartaglia et al. 1995). Splice variants were discovered shortly thereafter and the LRβ (ObRb) or the long form of the receptor was identified as the receptor disrupted in the genetically obese and diabetic *db/db* mice. The leptin receptor is a member of the cytokine receptor superfamily and is most closely related to the gp130 receptor, the major signal transduction receptor used by the IL-6 family of cytokines (Lee et al. 1996). Both the leptin receptor and the gp130 receptor function as dimers and signal predominantly through the Janus kinase/signal transducer and activator of transcription (JAK/STAT) pathway. Leptin also activates the MAPK, PI3K, AMPK and mTOR pathways (Cota et al. 2008; Hegyi et al. 2004; Maroni et al. 2005; Minokoshi et al. 2002; Niswender et al. 2004). The gp130 receptor similarly activates the RAS/MAPK, IRS/PI3K, AMPK as well as mTOR pathways (Boulton et al. 1994; Carey et al. 2006; Cota et al. 2008; Daeipour et al. 1993; Watt et al. 2006a; Yokogami et al. 2000).

The intracellular domain of the leptin receptor has two tyrosine residues (Tyr985, Tyr1138), which upon phosphorylation by JAK leads to recruitment of SHP-1 and STAT3, respectively, thereby mediating the activation of the MAPK, PI3K and STAT pathways (Banks et al. 2000). Receptor activation leads to expression of SOCS3 which binds to phosphorylated JAK as well as tyrosine residue 985 on the leptin receptor mediating the inhibition of STAT3 signalling (Bjorbaek et al. 1999, 2000) (Fig. 1a). Further to this, SOCS3 has also been found to target JAK proteins for proteasomal degradation (Kamura et al. 1998; Zhang et al. 1999). The inhibition of gp130 receptor signalling occurs in an analogous manner. In the case of IL-6, the human gp130 receptors contain five tyrosine residues (tyr 759, 767, 814, 905, 915) that are phosphorylated upon receptor activation. Similar to the leptin receptor, SOCS3 can inhibit signalling by binding to the SHP-2 binding tyrosine residue 759 as well as JAK. The gp130 receptor has four STAT3 binding sites distal to tyrosine residue 759, three more than that found in the leptin receptor (Fig. 1b). The existence of these extra STAT3 binding sites may explain why gp130 ligands can overcome SOCS3 inhibition of signalling, while leptin cannot.

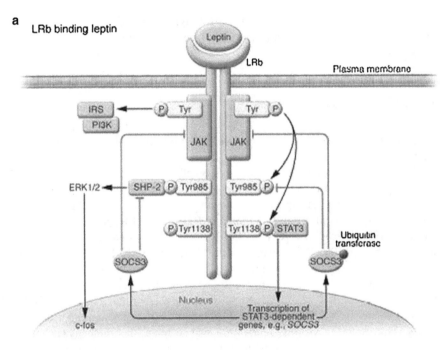

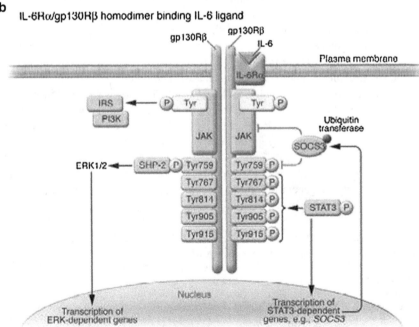

Fig. 1 *Leptin signalling versus gp130 signalling.* Leptin receptor signalling (**a**) occurs through dimerisation of two leptin receptors (LRb). This results in the phosphorylation and activation of JAK, and two tyrosine residues (tyr985, tyr1138) in the leptin receptor that leads to activation of

These findings may further explain why gp130 ligands (e.g. CNTF) are still efficacious in reducing weight in leptin-resistant models of obesity.

5 gp130 Receptor Ligands

The IL-6 family of cytokines includes interleukin-6 (IL-6), interleukin-11 (IL-11),
leukaemia inhibitory factor (LIF), oncostatin M (OSM), cardiotrophin-1 (CT-1),
cardiotrophin-like cytokine (CLC) and neuropoietin (NP). These cytokines function
by binding to their respective α receptors and inducing the dimerisation of the two β
receptors responsible for signal transduction (Fig. 2). For some cytokines, this
dimerisation consists of a gp130 homodimer (IL-6, IL-11) while others consist of
a gp130/LIFR heterodimer (CNTF, CT-1, LIF). OSM is capable of signalling
through both a gp130/OSMR and gp130/LIFR heterodimer. Unlike, other members,
OSM and LIF do not require an α receptor. CLC is a unique IL-6 family cytokine in
that it is capable of forming a dimer with two different α receptors, cytokine-like
receptor (CLF) (Elson et al. 2000) and sCNTFRα (Plun-Favreau et al. 2001).
Incapable of being secreted, CLC must form a dimer with either receptor intracellularly in order to facilitate its secretion. Once secreted, this dimer is capable of
activating the gp130/LIFR heterodimer. Lastly, neuropoietin (NP) is the newest
member of the IL-6 family of cytokines (Derouet et al. 2004). Recently, NP, CLC
and CNTF have all been found to share the same cytokine binding site on CNTFRα,
supporting previous evidence that they are all alternative activators of the CNTFR/
gp130/LIFR complex (Rousseau et al. 2008).

It is the structural homology rather than sequence homology that classifies these
cytokines as gp130 cytokines, all sharing a 4-helical bundle structure (Bazan 1991).
This helical structure contains conserved motifs that are responsible for the binding
to certain components of the receptor complex. These motifs are found in discon-
tinuous modules and have been successfully swapped to create novel, chimeric
cytokines (Kallen et al. 1999). Not only did this finding highlight how gp130
cytokines interacted with their respective receptor components, but it has also led
to the possibility of generating novel cytokines that may offer novel therapeutic
value (discussed further in Sect. 7).

Fig. 1 (continued) down stream pathways (MAPK, PI3K, STAT3). Activated STAT3 initiates the
transcription of STAT3 dependent genes including SOCS3. SOCS3 is then capable of interacting
with tyr985 of the leptin receptor and JAK proteins to inhibit further signalling. (**b**) Gp130
signalling is similar to that of leptin. For example, IL-6 binds its α receptor, IL-6R, which then
induces the dimerisation of a gp130 homodimer. Receptor activation leads to phosphorylation of
JAK and five tyrosine residues (tyr759, 767, 814, 905, 915) in the gp130 receptor and activation of
down stream pathways (PI3K, MAPK, STAT3). SOCS3 can similarly inhibit gp130 receptor
signalling. However, unlike leptin, gp130 ligands can overcome the SOCS3 inhibition in obesity.
The reason for this is still unclear but may involve the existence of four STAT3 binding sites on the
gp130 receptor, three more than the leptin receptor. Taken from (Febbraio 2007)

Fig. 2 *IL-6 Family of Cytokines*. The IL-6 family of cytokines function by binding to their respective α receptors inducing dimerisation of the β receptors. IL-6 and IL-11 bind their receptors (IL-6R and IL-11R, respectively) and signal through a gp130 homodimer. LIF and OSM do not require an α receptor and signal through a LIFR/gp130 heterodimer. OSM can also use a OSMR/gp130 heterodimer. CNTF, NP and CLC/CLF all utilise the CNTFRα and signal through the LIFR/gp130 heterodimer. CT-1 also signals through the LIFR/gp130 heterodimer. Adapted from Heinrich et al. (2003)

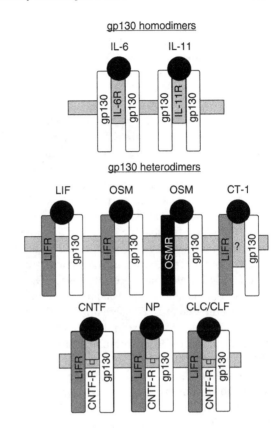

6   CNTF

CNTF was originally discovered as a factor that supported the survival of parasympathetic neurons of the chick ciliary ganglion in vitro (Adler et al. 1979). Subsequent studies have recognised its ability to promote cell survival or differentiation in a number of neuronal and glial cell types (Sleeman et al. 2000). CNTF is a 200 amino acid, 23 kDa protein that lacks a signal sequence peptide. As a result, CNTF is not regarded as a secreted protein. Despite this, CNTF has been found in the serum in both healthy and diseased states (Vergara and Ramirez 2004). This has led to the postulation that it may be secreted by a pathway similar to that proposed for IL-1β or the non-classical pathway described for chick CNTF (Andrei et al. 1999; Reiness et al. 2001; Rubartelli et al. 1990).

The CNTF receptor (CNTFRα) is most highly expressed not only in neural tissue (Davis et al. 1991) but is also detected in skeletal muscle, kidney, liver, testis, lung, bone marrow and adrenal gland (Bellido et al. 1996; Davis et al. 1991; Ip et al. 1992; MacLennan et al. 1994). The receptor is approximately 70 kDa and is attached to the membrane via a glycosyl–phosphatidyl–inositol link. This link makes it possible for the receptor to be cleaved by phospholipase C to release the soluble form of the CNTFR. This soluble CNTFR is then able to bind to CNTF and activate tissues that otherwise might be refractory to the effects of CNTF due to lack of receptor expression (Davis et al. 1993). Once CNTF binds the CNTFR, it initiates dimerisation of the β receptors, LIFR and gp130, leading to receptor tyrosine phosphorylation. Of critical biological significance, it has also been noted that CNTF can use the IL-6R as an alternative α receptor and maintain signalling through a LIFR/gp130 heterodimer (Fig. 3). The predominant pathway activated by CNTF is the JAK/STAT pathway. Activated STAT3 dimerise, move to the nucleus and induce transcription of target genes such as SOCS3, *c-fos*, *tis-11*, cyclooxgenase-2 and fibrinogen (Helgren et al. 1994; Kelly et al. 2004; Nesbitt et al. 1993). Other molecules activated by the CNTF include ERK1/2, PI3K, Raf-1, STAT1, GRB2, SHC, pp120, phospholipase Cγ and PTP1D (Boulton et al. 1994).

Crucial residues in CNTF have been found to alter the potency of human CNTF. A substitution of glutamine 63 for an arginine and cysteine 17 for an alanine, in addition to the deletion of 15 amino acids in the N-terminal end of the protein result in increased potency, stability and solubility compared to the parent molecule (Di Marco et al. 1996; Saggio et al. 1995). The resulting recombinant protein was termed Axokine ($CNTF_{Ax15}$) and has been subsequently used in vivo and in vitro to establish the role of CNTF signalling in obesity and metabolism (discussed later).

With escalating interest in the role of CNTF in obesity, it was then questioned whether mutations or polymorphisms in the *cntf* gene might contribute to obesity. A G–A null mutation in the first exon of the *cntf* gene, leading to absence of protein, was not associated with early onset obesity (Munzberg et al. 1998). However, a later study showed this mutation was associated with 10 kg increase in body mass in men but not in women (O'Dell et al. 2002). Three novel polymorphisms were identified in the CNTF receptor gene but only one, a C–T substitution at position 174, was

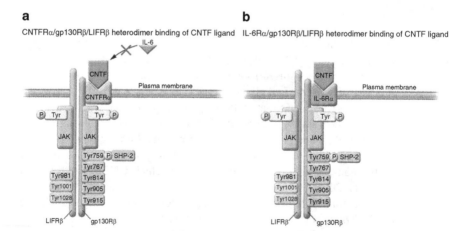

Fig. 3 *CNTF can alternatively use the IL-6R.* (**a**) CNTF binds the CNTFR and signals through the LIFR/gp130 heterodimer. (**b**) Alternatively, CNTF can also bind to the IL-6R and signal only through the LIFR/gp130 receptor. In contrast, IL-6 cannot bind the CNTFR and signal through the LIFR/gp130 heterodimer. Taken from (Febbraio 2007)

associated with greater body mass and BMI (Roth et al. 2003). These findings are clearly different from those for leptin and its receptor. Leptin and leptin receptor mutations are known to cause severe obesity in both humans and rodents (Montague et al. 1997; Tartaglia et al. 1995; Zhang et al. 1994). Genetic ablation of CNTF in mice causes no obvious phenotype, except for loss of motor neurons in old age, leading to muscle weakness (Masu et al. 1993). This, of course, can be explained by the existence of an additional ligand for the CNTFR (CLC). However, loss of the CNTFR leads to death postnatally due to an inability to suckle and significant loss of motor neurons in the brain stem and spinal cord motor nuclei (DeChiara et al. 1995).

7 Metabolic Effects of CNTF

CNTF was first discovered to have a potential role in energy balance during a trial of recombinant human CNTF (rhCNTF) for the treatment of amyotrophic lateral sclerosis (ALS). Obese patients receiving rhCNTF noted decreased food intake and involuntary weight loss (ACTS 1996; Miller et al. 1996). This finding was supported by Ettinger et al. (2003) in a randomised dose-ranging study of CNTF$_{Ax15}$ revealing it was effective at inducing greater weight loss in obese patients than placebo. These observations created significant interest in the feasibility of CNTF in the treatment of obesity. Subsequent studies highlighted the effectiveness of CNTF at reducing weight in both genetic (*ob/ob*, *db/db*) and diet-induced models of obesity, the latter being in a state of leptin resistance (Bluher et al. 2004; Gloaguen et al. 1997; Lambert et al. 2001; Liu et al. 2007a; Sleeman et al. 2003; Watt et al. 2006a). Initial concern was raised that the weight loss effects might be attributed to a cachectic-like effect, as observed for other cytokines, e.g. IL-1. However, while lower doses of CNTF$_{Ax15}$ resulted in significant weight loss, it neither activated similar neural pathways as IL-1, nor did it result in taste aversion, muscle wasting or increased circulating corticosteriods, as noted for IL-1 (Lambert et al. 2001). In contrast, high doses of CNTF$_{Ax15}$ did lead to increased taste aversion, corticosterone levels and muscle loss (Henderson et al. 1994; Lambert et al. 2001; Martin et al. 1996; Matthys and Billiau 1997). In addition to weight loss, CNTF$_{Ax15}$ administration improved hyperinsulinemia, hyperglycemia and hyperlipidemia associated with obesity (Bluher et al. 2004; Gloaguen et al. 1997; Lambert et al. 2001; Sleeman et al. 2003; Watt et al. 2006a). The mechanism by which CNTF$_{Ax15}$ achieves these endpoints has been extensively investigated and involves both central and peripheral effects. A summary of in vivo metabolic effects of CNTF is presented in Table 1.

7.1 Central Effects of CNTF

CNTF was first surmised to act in a similar fashion to leptin when it was discovered that it activated the area of the hypothalamus involved with appetite and energy

Table 1 Metabolic effects of rhCNTF or $CNTF_{Ax15}$ in vivo

Phenotype following treatment	Study
Induced marked weight loss in human and rodent populations	ACTS (1996), Cota et al. (2008), Ettinger et al. (2003), Lambert et al. (2001), Liu et al. (2007a, b), Sleeman et al. (2003), Steinberg et al. (2006), Watt et al. (2006a)
Maintained a decreased body weight after cessation of CNTF treatment	Bluher et al. (2004), Ettinger et al. (2003), Sleeman et al. (2003)
Promoted hypophagia	Bluher et al. (2004), Cota et al. (2008), Gloaguen et al. (1997), Lambert et al. (2001), Steinberg et al. (2006), Watt et al. (2006a)
Reversed hyperinsulinemia and promoted insulin sensitivity in rodent models of obesity/ or infused with lipid	Bluher et al. (2004), Gloaguen et al. (1997), Lambert et al. (2001), Watt et al. (2006a, b)
Acts on hypothalamic neurons to suppress AMPK and alters expression of orexigenic/anorexigenic transcripts	Janoschek et al. (2006), Kokoeva et al. (2005), Lambert et al. (2001), Steinberg et al. (2006), Xu et al. (1998), Ziotopoulou et al. (2000)
Promoted activation of AMPK-dependent fatty acid oxidation in skeletal muscle	Watt et al. (2006a)
Decreased the build up of lipid in skeletal muscle/ and or liver and the activation of serine kinase cascades	Sleeman et al. (2003), Watt et al. (2006a, b)
Increased glucose uptake in skeletal muscle and insulin signalling in skeletal muscle and liver	Sleeman et al. (2003), Watt et al. (2006a, b)
Improved liver function and metabolic rate in *db/db* mice	Bluher et al. (2004), Sleeman et al. (2003)
Increased mitochondrial complex 4 activity, UCP1, NRF-1 and TFam expression in brown adipose tissue	Bluher et al. (2004), Liu et al. (2007a, b)
Enhanced circulating adiponectin levels	Bluher et al. (2008)

balance similarly to leptin (Gloaguen et al. 1997; Lambert et al. 2001). The receptors for leptin and CNTF were then found co-expressed in the arcuate nucleus (ARC) and paraventricular nuclei (PVN) of the hypothalamus and treatment with both the compounds led to a rapid phosphorylation of STAT3 and induction of the *c-fos* and *tis-11* target genes (Gloaguen et al. 1997; Lambert et al. 2001). Furthermore, $CNTF_{Ax15}$ treatment inhibited the increase in neuropeptide Y (NPY), agouti-related protein (AGRP), gamma-aminobutyric acid (GABA) and pCREB (cAMP response element binding protein) in response to food restriction and fasting (Kalra et al. 1998; Lambert et al. 2001; Pu et al. 2000; Xu et al. 1998). The NPY, AGRP and GABA proteins are potent orexigenic proteins whose expression increases rapidly in the arcuate and paraventricular nuclei in response to food deprivation (Brady et al. 1990; Sahu et al. 1988). Similarly, the phosphorylation of CREB increases in the PVN upon food deprivation. In addition, $CNTF_{Ax15}$ was found to have reduced efficacy in stimulating *c-fos* expression and reducing weight in animals where the gp130 receptor was specifically knocked out in pro-opiomelanocortin

(POMC) neurons (Janoschek et al. 2006). POMC neurons are a major site of leptin action in the brain and are involved in energy balance (Gropp et al. 2005). These findings demonstrate that CNTF elicits effects on hypothalamic neurons to suppress food intake as well as the stress signals associated with food deprivation. CNTF administration in both humans and rodents has noted reduced rebound weight gain compared to placebo controls after cessation of treatment and food restriction (Bluher et al. 2004; Ettinger et al. 2003; Gloaguen et al. 1997; Lambert et al. 2001). It has been proposed that changes in the hypothalamic expression of appetite-stimulating peptides during fasting or food deprivation are associated with a memory of missed calories and mediate post restriction binge eating and rebound weight gain (Lambert et al. 2001). The finding that CNTF reduces the induction of such signals may explain the absence of binge eating and abrogated weight gain after treatment had ceased. In addition, the delay in weight gain after CNTF treatment can also be attributed to neurogenesis. Kokoeva et al. (2005) discovered centrally administered CNTF led to cell proliferation in the hypothalamus of mice. Co-administration of a mitotic blocker had no effect on the short-term weight reduction elicited by CNTF but completely abrogated the long-term CNTF weight reduction effects. Upon cessation of treatment, mice treated with the mitotic blocker immediately gained weight and returned to the weight of vehicle-treated animals within 20 days. Subsequent to this study, additional evidences cast some doubt as to whether CNTF treatment in the aforementioned study resulted in the proliferation of bona fide neurons or rather promoted the survival of immature pre-existing neurons (Vogel 2005). While this effect is advantageous to weight reduction, the long-term effects of this "neurogenesis" or neuronal survival are not known and may not be a desirable effect of a long-term obesity therapy.

More recent research has revealed that both leptin and CNTF inhibit AMPK in these areas of the brain (Andersson et al. 2004; Minokoshi et al. 2004; Steinberg et al. 2006). Intracerebroventricular (icv) administration of CNTF increased STAT3 phosphorylation while reducing the phosphorylation and activity AMPKα2 as well as the phosphorylation of ACC in the ARC (Steinberg et al. 2006). When mice were fed a high-fat diet for 12 weeks before ip (intraperitoneal) administration of either leptin or CNTF$_{Ax15}$, only CNTF$_{Ax15}$ maintained its inhibition of AMPKα2 activity in the ARC and phosphorylation of ACC in ARC and PVN. AMPK is a central mediator in the hypothalamic control of energy balance. Dominant negative AMPK expression in the hypothalamus reduces food intake and body weight in mice while expression of constitutively active AMPK leads to increased food intake and body weight. Accordingly, these changes in AMPK lead to alterations in the expression of orexigenic genes, NPY and AGRP, in ARC depending on fed state (Minokoshi et al. 2004). Recent work has identified the ACC/malonyl-CoA/CPT-1 and mTOR signalling pathways as possible downstream effectors of AMPK in the hypothalamus (Cota et al. 2006, 2008; Obici et al. 2003). Interestingly, it has been found that AMPK is activated in astrocytes in the striatum and this activation is accompanied by increases in phosphorylated ACC (pACC), and the oxidation of fatty acids and ketone bodies. This is thought to be a protective function of astrocytes to prevent damage to neurons and glial cells in face of metabolic insults such as prolonged palmitate exposure (Escartin et al. 2007). The reason why AMPK is inhibited in the specific hypothalamic neurons while simultaneously being activated in different parts of the CNS and the periphery by CNTF (discussed later) still remains under debate.

7.2 Peripheral Metabolic Effects of CNTF

Original thoughts were that the effectiveness of CNTF, and indeed leptin, in reducing weight was exerted solely through actions on the feeding centres of the brain. However, extensive research has highlighted numerous positive metabolic effects of CNTF in peripheral tissues.

Numerous studies have been conducted using multiple models of obesity to establish the peripheral effects of $CNTF_{Ax15}$ in vivo. Sleeman et al. (2003) noted that *db/db* mice lost more weight, had increased metabolic rate, energy expenditure and improved glucose tolerance, fasting glucose, fasting insulin, serum non-esterified fatty acids and triglyceride levels in $CNTF_{Ax15}$ treated animals than pair-fed (caloric restricted) control animals. Furthermore, $CNTF_{Ax15}$ administration reduced hepatic lipid deposition and improved liver function. Analysis of hepatic gene expression revealed that $CNTF_{Ax15}$ reduced transcript levels of stearoyl-CoA desaturase (SCD-1), involved in lipid synthesis, and increased carnitine-palmitoyl transferase (CPT-1), involved in fatty acid oxidation. These changes coincided with increases in hepatic insulin sensitivity as visualised by improved IRS-1 phosphorylation, p85/IRS-1 association and AKT phosphorylation.

Bluher et al. (2004) conducted a similar study using diet-induced obese mice and the brown adipose deficient UCP-1 DTA rodent model of obesity. Similar to previous studies, they noted weight loss, primarily due to fat mass loss and improved glucose tolerance and insulin sensitivity in diet-induced obese mice treated with $CNTF_{Ax15}$ compared to pair-fed control animals. No changes were seen in energy expenditure during the treatment period, but authors noted an increase in energy expenditure in the formerly $CNTF_{Ax15}$ treated animals in the days following cessation of treatment. The respiratory exchange ratio (RER) was also reduced during and after treatment. In line with this apparent increase in thermogenesis, UCP-1 expression was found to be elevated in the brown adipose tissue of these mice. To ascertain the involvement of brown adipose tissue thermogenesis in the mechanism of $CNTF_{Ax15}$ action, UCP-1 DTA mice were treated with $CNTF_{Ax15}$. These mice similarly lost weight with treatment but not beyond that achieved in pair-fed animals. The authors concluded that while the central effects of $CNTF_{Ax15}$ are maintained in UCP-1 DTA mice, the peripheral effects are slightly abrogated, possibly through a loss of UCP-1/brown adipose tissue mediated thermogenesis.

In another study of *db/db* mice, human recombinant CNTF was also found to upregulate the expression of UCP-1 in brown adipose tissue (Liu et al. 2007b).

This study further revealed that the transcript levels of nuclear respiratory factor-1 (NRF-1) and mitochondrial transcription factor A (TFam), thought to be involved in mitochondrial biogenesis, were similarly upregulated. In keeping with this, there were increased levels of cytochrome c and activity of mitochondrial complex IV in the brown adipose tissue of treated animals.

In vitro analysis is also critical in dissecting the direct effects of CNTF without interference of cross talk from other tissues and the effects of pair feeding. Ott et al. (2002) established a direct role for CNTF in signalling in brown adipose tissue with treatment leading to the phosphorylation of STAT3, p42/44 MAP kinase, AKT and p70 S6 kinase as well as enhanced β adrenergic induction of UCP-1 expression. In an additional study, Ott et al. (2004) also noted suppression in leptin secretion from cultured brown adipocytes after chronic treatment with CNTF, an effect not attributed to impaired adipocyte differentiation. They also found that acute treatment of CNTF reduced leptin expression in both brown and white adipocytes. Further analysis revealed this decrease in leptin expression was abolished with co-treatment with a PI3K inhibitor, while JAK2, MAPK and PKA inhibitors had no effect. This supplies evidence that CNTF can reduce leptin levels directly without the complication noted in in vivo studies of reduced leptin levels as a result of marked fat mass loss.

An additional pitfall of attempting to discriminate the tissue effects of CNTF in vivo is that studies do not always exclusively determine whether the effects are purely peripheral or whether there is some degree of mediation through the central nervous system. To establish that CNTF does have centrally independent peripheral effects, Watt et al. (2006a) administered $CNTF_{Ax15}$ both ip and icv and ascertained the activation of STAT3 and AMPK in muscles. Increases in pSTAT3, pAMPK and the activity of AMPKα1 and α2 in red gastrocnemius muscle were only noted with ip administration establishing that these effects were not mediated through CNTF effects in the CNS. Further to this, expression of PGC-1α, CPT-1 and UCP-3 were all increased after ip treatment, with icv treatment inducing no changes. This study went on to show that $CNTF_{Ax15}$ mediated activation of AMPK and phosphorylation of ACC in muscle led to significant increases in fatty acid oxidation. This increase in muscle fatty acid oxidation is specifically mediated through AMPK since transduction of L6 myotubes with a dominant negative AMPK virus was capable of abolishing this $CNTF_{Ax15}$ mediated increase in fatty acid oxidation. In mice treated for 7 days with $CNTF_{Ax15}$, muscle triacylglycerol, diacylglycerol and ceramide were decreased compared to pair-fed controls. Insulin signalling was similarly restored in muscle from $CNTF_{Ax15}$ treated high-fat fed mice compared to high-fat fed controls, as displayed by an increase in pAKT and pIRS-1, in addition to muscle glucose uptake.

Of immense interest in the study by Watt et al. (2006a) is the finding that the CNTF-mediated AMPK activation in cultured muscle cells was not dependent on the phosphorylation of STAT3 since a dominant negative STAT3 did not affect AMPK activation. However, SIMM mice (signalling-module mutation), in which the C terminal part of the gp130 receptor containing the STAT3 binding sites has been deleted, are refractory to the effects of CNTF to activate AMPK in muscle. This established that whilst STAT3 activation is not required for $CNTF_{Ax15}$

mediated AMPK activation, a section of the C terminal component of the gp130 receptor is. Furthermore, it was noted that CNTF maintained its ability to phosphorylate AMPK and ACC in human primary muscle cells transduced with an adenovirus over-expressing SOCS3, while leptin's effects were abolished by SOCS3.

Further expanding on the effects of $CNTF_{Ax15}$ in muscle, a study into the short-term effects of $CNTF_{Ax15}$ treatment on insulin action in muscle and liver during lipid oversupply was conducted (Watt et al. 2006b). Rats were subjected to a 2-h lipid infusion with or without pre-treatment with $CNTF_{Ax15}$ followed by a hyper-insulinemic–euglycemic clamp. Animals pre-treated with CNTF were protected from the lipid-induced reductions in systemic insulin sensitivity (decreased glucose disposal rate and suppression of hepatic glucose production). Furthermore, $CNTF_{Ax15}$ reduced the accumulation of triacylglycerols and ceramides as well as the activation of mixed lineage kinase 3 (MLK3) and c-jun N-terminal kinase 1 (JNK1) in both muscle and liver. Insulin signalling was also restored in both liver and muscle. With obesity now considered a state of inflammation, this finding establishes a role for CNTF in reducing the markers of inflammation in insulin responsive peripheral tissues.

One of the most striking effects of CNTF treatment is the rapid and significant loss of fat mass. While this is an obvious effect from a decrease in the food intake, it cannot be dismissed that CNTF may also have direct effects on white adipose tissue functioning. Crowe et al. (2008) established that diet-induced obese mice treated with $CNTF_{Ax15}$ exhibited decreased triglyceride content but an increased number of adipocytes compared to pair fed controls. This increase in adipocyte numbers was accompanied with an increase in the expression of PPARγ and C/EBPα. Lipogenesis and lipolysis were decreased while fatty acid oxidation increased in the epididymal fat pads of these animals. Cultured 3T3-L1 adipocytes treated with $CNTF_{Ax15}$ were also analysed and similarly found to have increased fatty acid oxidation. This increase in oxidation coincided with an increase in the phosphorylation of AMPK and ACC. Chronic CNTF treatment also increased mitochondrial number, protein expression of cytochrome c, ATP synthase-α and OxPhos complex II as well as NRF-1, PPARγ coactivator-1α (PGC-1α) and CPT-1.

Further analysis of white adipose tissue in vitro noted that while 3T3-L1 preadipocytes were highly responsive to CNTF as measured by the phosphorylation of STAT3, p42/44 MAPK and AKT, mature adipocytes lost their responsiveness to CNTF with reduced pSTAT and no increase of pAKT. This was attributed to decreased protein levels of CNTFRα and LIFR during adipocyte differentiation (gp130 levels remained the same) (Zvonic et al. 2003). This begs the question of how CNTF is having profound effects in white adipose tissue if the receptor expression is low. There is always an issue with disparity between the expression of genes in cultured cell lines versus tissue. Indeed it has been found that the CNTFRα is expressed in adipose tissue (Zvonic et al. 2003). An alternative is also that CNTF effects may be mediated through the IL-6Rα (Fig. 3). Schuster et al. (2003) discovered that in BAF/3 cells only expressing gp130, LIFR and IL-6R, CNTF was capable of eliciting a proliferative response, albeit not as robustly as

IL-6. In BAF/3 cells expressing the identical compliment of receptors, Watt et al. (2006a) demonstrated that $CNTF_{Ax15}$ also increased the pSTAT3 and pAMPK, although not quite as dramatically as IL-6. This alternative usage of the IL-6R by CNTF has also been suggested as the mechanism for increases in fibrinogen expression in CNTF-treated primary rat heptocytes (Nesbitt et al. 1993).

In summary, the metabolic effects of CNTF appear twofold, first in the hypothalamus to reduce appetite and secondly through effects on multiple insulin-sensitive peripheral tissues. It can improve insulin sensitivity in muscle, liver and fat with increased fat oxidation in muscle and fat. It increases energy expenditure and thermogenesis partly through events in brown adipose tissue, in addition to remodelling white adipose tissue. Lastly, it activates the expression of genes and signalling pathways involved in insulin signalling, fat utilisation and mitochondrial activity. Collectively, these changes are accountable for the robust anti-obesogenic and anti-diabetic actions of CNTF.

8   Future Directions for gp130 Ligands

Despite the promising effects CNTF has on obesity and insulin resistance, phase II studies discovered between 45% and 87% of patients treated with $CNTF_{Ax15}$ developed antibodies (Ettinger et al. 2003). Furthermore, in phase III clinical trials, half of the $CNTF_{Ax15}$ treated patients developed neutralising antibodies after 12 weeks that limited further weight loss (Duff and Baile 2003). CNTF, unlike other cytokines, is not found in circulation, or found at very low levels; thus it can be expected that high levels of exogenous protein could initiate an immune response. Another contributing factor could be that whilst the CNTFR is highly expressed in the CNS, much lower levels are found in peripheral tissues leading to the requirement of higher doses to elicit a robust response.

The question remains as to whether the gp130 family of cytokines still offers hope of anti-obesity therapeutics. The finding that this family of cytokines possesses the same three-dimensional helical structures and contains discrete motifs that are responsible for specific receptor component binding has led to the generation of novel chimeric cytokines (Kallen et al. 1999). In theory, this method could potentially create cytokines that have specific signalling abilities by utilising a unique combination of receptors. Such a cytokine would ideally have low immunogenicity, utilise highly expressed receptors and initiate favourable metabolic effects. Further work in this area will determine whether such an approach will supply the novel candidate therapeutics that are urgently sought for the treatment of obesity and type 2 diabetes.

Acknowledgements The authors wish to thank Hi Le for assistance in producing this manuscript, the support of the National Health and Medical Research Council (NHMRC), the Australian Research Council and the Diabetes Australia Research Trust. VBM is supported in part by a Baker IDI Heart and Diabetes Institute Early Career Scientist Grant. MAF is supported by a Principal Research Fellowship from the NHMRC.

References

ACTS (1996) A double-blind placebo-controlled clinical trial of subcutaneous recombinant human ciliary neurotrophic factor (rHCNTF) in amyotrophic lateral sclerosis. ALS CNTF Treatment Study Group. Neurology 46(5):1244–1249

Adler R, Landa KB, Manthorpe M, Varon S (1979) Cholinergic neuronotrophic factors: intraocular distribution of trophic activity for ciliary neurons. Science 204:1434–1436

Andersson U, Filipsson K, Abbott CR, Woods A, Smith K, Bloom SR, Carling D, Small CJ (2004) AMP-activated protein kinase plays a role in the control of food intake. J Biol Chem 279:12005–12008

Andrei C, Dazzi C, Lotti L, Torrisi MR, Chimini G, Rubartelli A (1999) The secretory route of the leaderless protein interleukin 1beta involves exocytosis of endolysosome-related vesicles. Mol Biol Cell 10:1463–1475

Banks AS, Davis SM, Bates SH, Myers MG Jr (2000) Activation of downstream signals by the long form of the leptin receptor. J Biol Chem 275:14563–14572

Bazan JF (1991) Neuropoietic cytokines in the hematopoietic fold. Neuron 7:197–208

Bellido T, Stahl N, Farruggella TJ, Borba V, Yancopoulos GD, Manolagas SC (1996) Detection of receptors for interleukin-6, interleukin-11, leukemia inhibitory factor, oncostatin M, and ciliary neurotrophic factor in bone marrow stromal/osteoblastic cells. J Clin Invest 97:431–437

Bjorbaek C, Elmquist JK, Frantz JD, Shoelson SE, Flier JS (1998) Identification of SOCS-3 as a potential mediator of central leptin resistance. Mol Cell 1:619–625

Bjorbaek C, El-Haschimi K, Frantz JD, Flier JS (1999) The role of SOCS-3 in leptin signaling and leptin resistance. J Biol Chem 274:30059–30065

Bjorbaek C, Lavery HJ, Bates SH, Olson RK, Davis SM, Flier JS, Myers MG Jr (2000) SOCS3 mediates feedback inhibition of the leptin receptor via Tyr985. J Biol Chem 275:40649–40657

Bluher S, Moschos S, Bullen J Jr, Kokkotou E, Maratos-Flier E, Wiegand SJ, Sleeman MW, Mantzoros CS (2004) Ciliary neurotrophic factorAx15 alters energy homeostasis, decreases body weight, and improves metabolic control in diet-induced obese and UCP1-DTA mice. Diabetes 53:2787–2796

Bluher S, Bullen J, Mantzoros CS (2008) Altered levels of adiponectin and adiponectin receptors may underlie the effect of ciliary neurotrophic factor (CNTF) to enhance insulin sensitivity in diet-induced obese mice. Horm Metab Res 40:225–227

Boulton TG, Stahl N, Yancopoulos GD (1994) Ciliary neurotrophic factor/leukemia inhibitory factor/interleukin 6/oncostatin M family of cytokines induces tyrosine phosphorylation of a common set of proteins overlapping those induced by other cytokines and growth factors. J Biol Chem 269:11648–11655

Brady LS, Smith MA, Gold PW, Herkenham M (1990) Altered expression of hypothalamic neuropeptide mRNAs in food-restricted and food-deprived rats. Neuroendocrinology 52:441–447

Carey AL, Steinberg GR, Macaulay SL, Thomas WG, Holmes AG, Ramm G, Prelovsek O, Hohnen-Behrens C, Watt MJ, James DE, Kemp BE, Pedersen BK, Febbraio MA (2006) Interleukin-6 increases insulin-stimulated glucose disposal in humans and glucose uptake and fatty acid oxidation in vitro via AMP-activated protein kinase. Diabetes 55:2688–2697

Cota D, Proulx K, Smith KA, Kozma SC, Thomas G, Woods SC, Seeley RJ (2006) Hypothalamic mTOR signaling regulates food intake. Science 312:927–930

Cota D, Matter EK, Woods SC, Seeley RJ (2008) The role of hypothalamic mammalian target of rapamycin complex 1 signaling in diet-induced obesity. J Neurosci 28:7202–7208

Crowe S, Turpin SM, Ke F, Kemp BE, Watt MJ (2008) Metabolic remodeling in adipocytes promotes ciliary neurotrophic factor-mediated fat loss in obesity. Endocrinology 149:2546–2556

Cummings S, Apovian CM, Khaodhiar L (2008) Obesity surgery: evidence for diabetes prevention/management. J Am Diet Assoc 108:S40–S44

Daeipour M, Kumar G, Amaral MC, Nel AE (1993) Recombinant IL-6 activates p42 and p44 mitogen-activated protein kinases in the IL-6 responsive B cell line, AF-10. J Immunol 150:4743–4753

Davis S, Aldrich TH, Valenzuela DM, Wong VV, Furth ME, Squinto SP, Yancopoulos GD (1991) The receptor for ciliary neurotrophic factor. Science 253:59–63

Davis S, Aldrich TH, Ip NY, Stahl N, Scherer S, Farruggella T, DiStefano PS, Curtis R, Panayotatos N, Gascan H et al (1993) Released form of CNTF receptor alpha component as a soluble mediator of CNTF responses. Science 259:1736–1739

DeChiara TM, Vejsada R, Poueymirou WT, Acheson A, Suri C, Conover JC, Friedman B, McClain J, Pan L, Stahl N, Ip NY, Yancopoulos GD (1995) Mice lacking the CNTF receptor, unlike mice lacking CNTF, exhibit profound motor neuron deficits at birth. Cell 83:313–322

Derouet D, Rousseau F, Alfonsi F, Froger J, Hermann J, Barbier F, Perret D, Diveu C, Guillet C, Preisser L, Dumont A, Barbado M, Morel A, deLapeyriere O, Gascan H, Chevalier S (2004) Neuropoietin, a new IL-6-related cytokine signaling through the ciliary neurotrophic factor receptor. Proc Natl Acad Sci USA 101:4827–4832

Di Marco A, Gloaguen I, Graziani R, Paonessa G, Saggio I, Hudson KR, Laufer R (1996) Identification of ciliary neurotrophic factor (CNTF) residues essential for leukemia inhibitory factor receptor binding and generation of CNTF receptor antagonists. Proc Natl Acad Sci USA 93:9247–9252

Duff E, Baile CA (2003) Ciliary neurotrophic factor: a role in obesity? Nutr Rev 61:423–426

Elmquist JK, Maratos-Flier E, Saper CB, Flier JS (1998) Unraveling the central nervous system pathways underlying responses to leptin. Nat Neurosci 1:445–450

Elson GC, Lelievre E, Guillet C, Chevalier S, Plun-Favreau H, Froger J, Suard I, de Coignac AB, Delneste Y, Bonnefoy JY, Gauchat JF, Gascan H (2000) CLF associates with CLC to form a functional heteromeric ligand for the CNTF receptor complex. Nat Neurosci 3:867–872

Escartin C, Pierre K, Colin A, Brouillet E, Delzescaux T, Guillermier M, Dhenain M, Deglon N, Hantraye P, Pellerin L, Bonvento G (2007) Activation of astrocytes by CNTF induces metabolic plasticity and increases resistance to metabolic insults. J Neurosci 27:7094–7104

Ettinger MP, Littlejohn TW, Schwartz SL, Weiss SR, McIlwain HH, Heymsfield SB, Bray GA, Roberts WG, Heyman ER, Stambler N, Heshka S, Vicary C, Guler HP (2003) Recombinant variant of ciliary neurotrophic factor for weight loss in obese adults: a randomized, dose-ranging study. JAMA 289:1826–1832

Febbraio MA (2007) gp130 receptor ligands as potential therapeutic targets for obesity. J Clin Invest 117:841–849

Fonseca VA, Kulkarni KD (2008) Management of type 2 diabetes: oral agents, insulin, and injectables. J Am Diet Assoc 108:S29–S33

Friedman JM, Halaas JL (1998) Leptin and the regulation of body weight in mammals. Nature 395:763–770

Gloaguen I, Costa P, Demartis A, Lazzaro D, Di Marco A, Graziani R, Paonessa G, Chen F, Rosenblum CI, Van der Ploeg LH, Cortese R, Ciliberto G, Laufer R (1997) Ciliary neurotrophic factor corrects obesity and diabetes associated with leptin deficiency and resistance. Proc Natl Acad Sci USA 94:6456–6461

Gropp E, Shanabrough M, Borok E, Xu AW, Janoschek R, Buch T, Plum L, Balthasar N, Hampel B, Waisman A, Barsh GS, Horvath TL, Bruning JC (2005) Agouti-related peptide-expressing neurons are mandatory for feeding. Nat Neurosci 8:1289–1291

Hegyi K, Fulop K, Kovacs K, Toth S, Falus A (2004) Leptin-induced signal transduction pathways. Cell Biol Int 28:159–169

Heinrich PC, Behrmann I, Haan S, Hermanns HM, Muller-Newen G, Schaper F (2003) Principles of interleukin (IL)-6-type cytokine signalling and its regulation. Biochem J 374:1–20

Helgren ME, Squinto SP, Davis HL, Parry DJ, Boulton TG, Heck CS, Zhu Y, Yancopoulos GD, Lindsay RM, DiStefano PS (1994) Trophic effect of ciliary neurotrophic factor on denervated skeletal muscle. Cell 76:493–504

Henderson JT, Seniuk NA, Richardson PM, Gauldie J, Roder JC (1994) Systemic administration of ciliary neurotrophic factor induces cachexia in rodents. J Clin Invest 93:2632–2638

Howard JK, Cave BJ, Oksanen LJ, Tzameli I, Bjorbaek C, Flier JS (2004) Enhanced leptin sensitivity and attenuation of diet-induced obesity in mice with haploinsufficiency of Socs3. Nat Med 10:734–738

Ip NY, Nye SH, Boulton TG, Davis S, Taga T, Li Y, Birren SJ, Yasukawa K, Kishimoto T, Anderson DJ et al (1992) CNTF and LIF act on neuronal cells via shared signaling pathways that involve the IL-6 signal transducing receptor component gp130. Cell 69:1121–1132

James W, Jackson-Leach R, Mhurdu C, Kalamara E, Shayeghi M, Rigby N, Nishida C, Rodgers A, Ezzati M, Lopez A, Rodgers A, Murray C (2003) Overweight and obesity. In: Comparative quantification of health risks: global and regional burden of disease attributable to selected major risk factors. WHO, Geneva

Janoschek R, Plum L, Koch L, Munzberg H, Diano S, Shanabrough M, Muller W, Horvath TL, Bruning JC (2006) gp130 signaling in proopiomelanocortin neurons mediates the acute anorectic response to centrally applied ciliary neurotrophic factor. Proc Natl Acad Sci USA 103:10707–10712

Kallen KJ, Grotzinger J, Lelievre E, Vollmer P, Aasland D, Renne C, Mullberg J, Myer zum Buschenfelde KH, Gascan H, Rose-John S (1999) Receptor recognition sites of cytokines are organized as exchangeable modules. Transfer of the leukemia inhibitory factor receptor-binding site from ciliary neurotrophic factor to interleukin-6. J Biol Chem 274:11859–11867

Kalra SP, Xu B, Dube MG, Moldawer LL, Martin D, Kalra PS (1998) Leptin and ciliary neurotropic factor (CNTF) inhibit fasting-induced suppression of luteinizing hormone release in rats: role of neuropeptide Y. Neurosci Lett 240:45–49

Kamura T, Sato S, Haque D, Liu L, Kaelin WG Jr, Conaway RC, Conaway JW (1998) The Elongin BC complex interacts with the conserved SOCS-box motif present in members of the SOCS, ras, WD-40 repeat, and ankyrin repeat families. Genes Dev 12:3872–3881

Kelly JF, Elias CF, Lee CE, Ahima RS, Seeley RJ, Bjorbaek C, Oka T, Saper CB, Flier JS, Elmquist JK (2004) Ciliary neurotrophic factor and leptin induce distinct patterns of immediate early gene expression in the brain. Diabetes 53:911–920

Knowler WC, Barrett-Connor E, Fowler SE, Hamman RF, Lachin JM, Walker EA, Nathan DM (2002) Reduction in the incidence of type 2 diabetes with lifestyle intervention or metformin. N Engl J Med 346:393–403

Kokoeva MV, Yin H, Flier JS (2005) Neurogenesis in the hypothalamus of adult mice: potential role in energy balance. Science 310:679–683

Lambert PD, Anderson KD, Sleeman MW, Wong V, Tan J, Hijarunguru A, Corcoran TL, Murray JD, Thabet KE, Yancopoulos GD, Wiegand SJ (2001) Ciliary neurotrophic factor activates leptin-like pathways and reduces body fat, without cachexia or rebound weight gain, even in leptin-resistant obesity. Proc Natl Acad Sci USA 98:4652–4657

Lee GH, Proenca R, Montez JM, Carroll KM, Darvishzadeh JG, Lee JI, Friedman JM (1996) Abnormal splicing of the leptin receptor in diabetic mice. Nature 379:632–635

Liu QS, Gao M, Zhu SY, Li SJ, Zhang L, Wang QJ, Du GH (2007a) The novel mechanism of recombinant human ciliary neurotrophic factor on the anti-diabetes activity. Basic Clin Pharmacol Toxicol 101:78–84

Liu QS, Wang QJ, Du GH, Zhu SY, Gao M, Zhang L, Zhu JM, Cao JF (2007b) Recombinant human ciliary neurotrophic factor reduces weight partly by regulating nuclear respiratory factor 1 and mitochondrial transcription factor A. Eur J Pharmacol 563:77–82

MacLennan AJ, Gaskin AA, Lado DC (1994) CNTF receptor alpha mRNA expression in rodent cell lines and developing rat. Brain Res Mol Brain Res 25:251–256

Maroni P, Bendinelli P, Piccoletti R (2005) Intracellular signal transduction pathways induced by leptin in C2C12 cells. Cell Biol Int 29:542–550

Martin D, Merkel E, Tucker KK, McManaman JL, Albert D, Relton J, Russell DA (1996) Cachectic effect of ciliary neurotrophic factor on innervated skeletal muscle. Am J Physiol 271:R1422–R1428

Mascie-Taylor CG, Karim E (2003) The burden of chronic disease. Science 302:1921–1922

Masu Y, Wolf E, Holtmann B, Sendtner M, Brem G, Thoenen H (1993) Disruption of the CNTF gene results in motor neuron degeneration. Nature 365:27–32

Matthys P, Billiau A (1997) Cytokines and cachexia. Nutrition 13:763–770

Miller RG, Petajan JH, Bryan WW, Armon C, Barohn RJ, Goodpasture JC, Hoagland RJ, Parry GJ, Ross MA, Stromatt SC (1996) A placebo-controlled trial of recombinant human ciliary neurotrophic (rhCNTF) factor in amyotrophic lateral sclerosis. rhCNTF ALS Study Group (ACTS). Ann Neurol 39:256–260

Minokoshi Y, Kim YB, Peroni OD, Fryer LG, Muller C, Carling D, Kahn BB (2002) Leptin stimulates fatty-acid oxidation by activating AMP-activated protein kinase. Nature 415:339–343

Minokoshi Y, Alquier T, Furukawa N, Kim YB, Lee A, Xue B, Mu J, Foufelle F, Ferre P, Birnbaum MJ, Stuck BJ, Kahn BB (2004) AMP-kinase regulates food intake by responding to hormonal and nutrient signals in the hypothalamus. Nature 428:569–574

Montague CT, Farooqi IS, Whitehead JP, Soos MA, Rau H, Wareham NJ, Sewter CP, Digby JE, Mohammed SN, Hurst JA, Cheetham CH, Earley AR, Barnett AH, Prins JB, O'Rahilly S (1997) Congenital leptin deficiency is associated with severe early-onset obesity in humans. Nature 387:903–908

Mori H, Hanada R, Hanada T, Aki D, Mashima R, Nishinakamura H, Torisu T, Chien KR, Yasukawa H, Yoshimura A (2004) Socs3 deficiency in the brain elevates leptin sensitivity and confers resistance to diet-induced obesity. Nat Med 10:739–743

Munzberg H, Tafel J, Busing B, Hinney A, Ziegler A, Mayer H, Siegfried W, Matthaei S, Greten H, Hebebrand J, Hamann A (1998) Screening for variability in the ciliary neurotrophic factor (CNTF) gene: no evidence for association with human obesity. Exp Clin Endocrinol Diabetes 106:108–112

Muoio DM, Dohm GL, Fiedorek FT Jr, Tapscott EB, Coleman RA (1997) Leptin directly alters lipid partitioning in skeletal muscle. Diabetes 46:1360–1363

Nesbitt JE, Fuentes NL, Fuller GM (1993) Ciliary neurotrophic factor regulates fibrinogen gene expression in hepatocytes by binding to the interleukin-6 receptor. Biochem Biophys Res Commun 190:544–550

Niswender KD, Baskin DG, Schwartz MW (2004) Insulin and its evolving partnership with leptin in the hypothalamic control of energy homeostasis. Trends Endocrinol Metab 15:362–369

O'Dell SD, Syddall HE, Sayer AA, Cooper C, Fall CH, Dennison EM, Phillips DI, Gaunt TR, Briggs PJ, Day IN (2002) Null mutation in human ciliary neurotrophic factor gene confers higher body mass index in males. Eur J Hum Genet 10:749–752

Obici S, Feng Z, Arduini A, Conti R, Rossetti L (2003) Inhibition of hypothalamic carnitine palmitoyltransferase-1 decreases food intake and glucose production. Nat Med 9:756–761

Ott V, Fasshauer M, Dalski A, Klein HH, Klein J (2002) Direct effects of ciliary neurotrophic factor on brown adipocytes: evidence for a role in peripheral regulation of energy homeostasis. J Endocrinol 173:R1–R8

Ott V, Fasshauer M, Meier B, Dalski A, Kraus D, Gettys TW, Perwitz N, Klein J (2004) Ciliary neurotrophic factor influences endocrine adipocyte function: inhibition of leptin via PI 3-kinase. Mol Cell Endocrinol 224:21–27

Padwal RS, Majumdar SR (2007) Drug treatments for obesity: orlistat, sibutramine, and rimonabant. Lancet 369:71–77

Penfornis A, Borot S, Raccah D (2008) Therapeutic approach of type 2 diabetes mellitus with GLP-1 based therapies. Diabetes Metab 34(Suppl 2):S78–S90

Peralta S, Carrascosa JM, Gallardo N, Ros M, Arribas C (2002) Ageing increases SOCS-3 expression in rat hypothalamus: effects of food restriction. Biochem Biophys Res Commun 296:425–428

Plun-Favreau H, Elson G, Chabbert M, Froger J, deLapeyriere O, Lelievre E, Guillet C, Hermann J, Gauchat JF, Gascan H, Chevalier S (2001) The ciliary neurotrophic factor receptor alpha component induces the secretion of and is required for functional responses to cardiotrophin-like cytokine. EMBO J 20:1692–1703

Pu S, Dhillon H, Moldawer LL, Kalra PS, Kalra SP (2000) Neuropeptide Y counteracts the anorectic and weight reducing effects of ciliary neurotropic factor. J Neuroendocrinol 12:827–832

Reiness CG, Seppa MJ, Dion DM, Sweeney S, Foster DN, Nishi R (2001) Chick ciliary neurotrophic factor is secreted via a nonclassical pathway. Mol Cell Neurosci 17:931–944

Roth SM, Metter EJ, Lee MR, Hurley BF, Ferrell RE (2003) C174T polymorphism in the CNTF receptor gene is associated with fat-free mass in men and women. J Appl Physiol 95: 1425–1430

Rousseau F, Chevalier S, Guillet C, Ravon E, Diveu C, Froger J, Barbier F, Grimaud L, Gascan H (2008) Ciliary neurotrophic factor, cardiotrophin-like cytokine and neuropoietin share a conserved binding site on the ciliary neurotrophic factor receptor alpha chain. J Biol Chem 283 (44):30341–30350

Rubartelli A, Cozzolino F, Talio M, Sitia R (1990) A novel secretory pathway for interleukin-1 beta, a protein lacking a signal sequence. EMBO J 9:1503–1510

Saggio I, Gloaguen I, Poiana G, Laufer R (1995) CNTF variants with increased biological potency and receptor selectivity define a functional site of receptor interaction. EMBO J 14:3045–3054

Sahu A, Kalra PS, Kalra SP (1988) Food deprivation and ingestion induce reciprocal changes in neuropeptide Y concentrations in the paraventricular nucleus. Peptides 9:83–86

Schuster B, Kovaleva M, Sun Y, Regenhard P, Matthews V, Grotzinger J, Rose-John S, Kallen KJ (2003) Signaling of human ciliary neurotrophic factor (CNTF) revisited. The interleukin-6 receptor can serve as an alpha-receptor for CTNF. J Biol Chem 278:9528–9535

Schwartz MW, Peskind E, Raskind M, Boyko EJ, Porte D Jr (1996) Cerebrospinal fluid leptin levels: relationship to plasma levels and to adiposity in humans. Nat Med 2:589–593

Schwartz MW, Woods SC, Porte D Jr, Seeley RJ, Baskin DG (2000) Central nervous system control of food intake. Nature 404:661–671

Sleeman MW, Anderson KD, Lambert PD, Yancopoulos GD, Wiegand SJ (2000) The ciliary neurotrophic factor and its receptor, CNTFR alpha. Pharm Acta Helv 74:265–272

Sleeman MW, Garcia K, Liu R, Murray JD, Malinova L, Moncrieffe M, Yancopoulos GD, Wiegand SJ (2003) Ciliary neurotrophic factor improves diabetic parameters and hepatic steatosis and increases basal metabolic rate in db/db mice. Proc Natl Acad Sci USA 100:14297–14302

Steinberg GR, Rush JW, Dyck DJ (2003) AMPK expression and phosphorylation are increased in rodent muscle after chronic leptin treatment. Am J Physiol Endocrinol Metab 284:E648–E654

Steinberg GR, Watt MJ, Fam BC, Proietto J, Andrikopoulos S, Allen AM, Febbraio MA, Kemp BE (2006) Ciliary neurotrophic factor suppresses hypothalamic AMP-kinase signaling in leptin-resistant obese mice. Endocrinology 147:3906–3914

Tartaglia LA, Dembski M, Weng X, Deng N, Culpepper J, Devos R, Richards GJ, Campfield LA, Clark FT, Deeds J, Muir C, Sanker S, Moriarty A, Moore KJ, Smutko JS, Mays GG, Wool EA, Monroe CA, Tepper RI (1995) Identification and expression cloning of a leptin receptor, OB-R. Cell 83:1263–1271

Van Heek M, Compton DS, France CF, Tedesco RP, Fawzi AB, Graziano MP, Sybertz EJ, Strader CD, Davis HR Jr (1997) Diet-induced obese mice develop peripheral, but not central, resistance to leptin. J Clin Invest 99:385–390

Vergara C, Ramirez B (2004) CNTF, a pleiotropic cytokine: emphasis on its myotrophic role. Brain Res Brain Res Rev 47:161–173

Vogel G (2005) Neuroscience. Does brain cell growth drive weight loss? Science 310:602

Watt MJ, Dzamko N, Thomas WG, Rose-John S, Ernst M, Carling D, Kemp BE, Febbraio MA, Steinberg GR (2006a) CNTF reverses obesity-induced insulin resistance by activating skeletal muscle AMPK. Nat Med 12:541–548

Watt MJ, Hevener A, Lancaster GI, Febbraio MA (2006b) Ciliary neurotrophic factor prevents acute lipid-induced insulin resistance by attenuating ceramide accumulation and phosphorylation of c-Jun N-terminal kinase in peripheral tissues. Endocrinology 147:2077–2085

WHO (2006) http://www.who.int/mediacentre/factsheets/fs311/en/index.html. World Health Organisation, Geneva

Xu B, Dube MG, Kalra PS, Farmerie WG, Kaibara A, Moldawer LL, Martin D, Kalra SP (1998) Anorectic effects of the cytokine, ciliary neurotropic factor, are mediated by hypothalamic neuropeptide Y: comparison with leptin. Endocrinology 139:466–473

Yokogami K, Wakisaka S, Avruch J, Reeves SA (2000) Serine phosphorylation and maximal activation of STAT3 during CNTF signaling is mediated by the rapamycin target mTOR. Curr Biol 10:47–50

Zhang Y, Proenca R, Maffei M, Barone M, Leopold L, Friedman JM (1994) Positional cloning of the mouse obese gene and its human homologue. Nature 372:425–432

Zhang JG, Farley A, Nicholson SE, Willson TA, Zugaro LM, Simpson RJ, Moritz RL, Cary D, Richardson R, Hausmann G, Kile BJ, Kent SB, Alexander WS, Metcalf D, Hilton DJ, Nicola NA, Baca M (1999) The conserved SOCS box motif in suppressors of cytokine signaling binds to elongins B and C and may couple bound proteins to proteasomal degradation. Proc Natl Acad Sci USA 96:2071–2076

Ziotopoulou M, Erani DM, Hileman SM, Bjorbaek C, Mantzoros CS (2000) Unlike leptin, ciliary neurotrophic factor does not reverse the starvation-induced changes of serum corticosterone and hypothalamic neuropeptide levels but induces expression of hypothalamic inhibitors of leptin signaling. Diabetes 49:1890–1896

Zvonic S, Cornelius P, Stewart WC, Mynatt RL, Stephens JM (2003) The regulation and activation of ciliary neurotrophic factor signaling proteins in adipocytes. J Biol Chem 278:2228–2235

Thermogenesis and Related Metabolic Targets in Anti-Diabetic Therapy

Jonathan R.S. Arch

Contents

J.R.S. Arch
Clore Laboratory, University of Buckingham, Buckingham MK18 1EG, UK

M. Schwanstecher (ed.), *Diabetes - Perspectives in Drug Therapy*,
Handbook of Experimental Pharmacology 203,
DOI 10.1007/978-3-642-17214-4_10, © Springer-Verlag Berlin Heidelberg 2011

Abstract Exercise, together with a low-energy diet, is the first-line treatment for type 2 diabetes. Exercise improves insulin sensitivity by increasing the number or function of muscle mitochondria and the capacity for aerobic metabolism, all of which are low in many insulin-resistant subjects. Cannabinoid 1-receptor antagonists and β-adrenoceptor agonists improve insulin sensitivity in humans and promote fat oxidation in rodents independently of reduced food intake. Current drugs for the treatment of diabetes are not, however, noted for their ability to increase fat oxidation, although the thiazolidinediones increase the capacity for fat oxidation in skeletal muscle, whilst paradoxically increasing weight gain.

There are a number of targets for anti-diabetic drugs that may improve insulin sensitivity by increasing the capacity for fat oxidation. Their mechanisms of action are linked, notably through AMP-activated protein kinase, adiponectin, and the sympathetic nervous system. If ligands for these targets have obvious acute thermogenic activity, it is often because they increase sympathetic activity. This promotes fuel mobilisation, as well as fuel oxidation. When thermogenesis is not obvious, researchers often argue that it has occurred by using the inappropriate device of treating animals for days or weeks until there is weight (mainly fat) loss and then expressing energy expenditure relative to body weight. In reality, thermogenesis may have occurred, but it is too small to detect, and this device distracts us from really appreciating why insulin sensitivity has improved. This is that by increasing fatty acid oxidation more than fatty acid supply, drugs lower the concentrations of fatty acid metabolites that cause insulin resistance. Insulin sensitivity improves long before any anti-obesity effect can be detected.

Keywords Drug discovery · Exercise · Fatty acid oxidation · Insulin sensitivity · Thermogenesis

1   Introduction

Why might we treat type 2 diabetes with a drug that increases energy expenditure (thermogenesis)? Perhaps, because exercise is beneficial in type 2 diabetes. Perhaps because compounds that increase energy expenditure – or at least aerobic capacity – have been shown to be efficacious in models of type 2 diabetes, even though they may not have become drugs. Perhaps, because this approach also addresses obesity

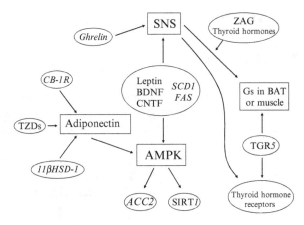

Fig. 1 Mechanistic links between targets for insulin-sensitising anti-diabetic drugs. The multiple linkages of the sympathetic nervous system, AMPK, and adiponectin are emphasised. The linkages do not indicate that targets influence insulin sensitivity exclusively through the targets to which they are linked. Targets that require inhibitors or antagonists are shown in *italics*. Leptin, CNTF, BDNF, SCD1, and FAS are shown in one oval for ease of presentation. Abbreviations: *11βHSD* 11β-hydroxysteroid dehydrogenase-1; *ACC2* acetyl-CoA caboxylase-2; *AICAR* 5-aminoimidazole-4-carboxamide-1-β-D-ribofuranoside (but in the literature on purine metabolism synonymous with ZMP – see Sect. 6.1); *AMPK* AMP-activated protein kinase; *BDNF* brain-derived neurotrophic factor; *CB-1R* cannabinoid 1-receptor; *CNTF* ciliary neurotrophic factor; *FAS* fatty acid synthase; *FBPase* fructose-1,6-bisphosphatase; *Gs* Gs protein (activated by many G-protein-coupled receptors); *PGC-1α* peroxisome proliferator-activated receptor gamma coactivator 1-α; *PPAR* peroxisome proliferator-activated receptor; *SCD1* stearoyl-CoA desaturases-1; *SIRT1* sirtuin1; *TZD* thiazolidinedione; *ZAG* Zn-α₂-glycoprotein; *ZMP* 5-amino-4-imidazolecarboxamide ribonucleotide or 5-aminoimidazole-4-carboxamide-1-β-D-ribofuranotide

and other features of the metabolic syndrome. And perhaps, being a novel approach, a thermogenic drug would increase therapeutic options and synergise with other drugs.

This chapter addresses these arguments and then goes on to review targets for thermogenic drugs. Many of these targets are linked mechanistically, as illustrated in Fig. 1. Some of these targets are covered in a recent review focused on obesity (Tseng et al. 2010) and others are the subject of entire chapters in this book, but the focus of this chapter is on whether ligands for these targets might increase insulin sensitivity by increasing thermogenesis or at least the capacity for thermogenesis. A recurring theme is that targets associated with improved insulin sensitivity are generally also associated with increased fatty acid oxidation. The chapter concludes by considering some general issues for the discovery and development of such drugs, in particular whether thermogenesis can easily be detected in the absence of increased sympathetic activity or another stimulus to fuel supply.

2 Rationale: Why Target Thermogenesis?

2.1 Obesity and Diabetes

The prevalence of type 2 diabetes is greatly increased in obese subjects (Chan et al. 1994). This is both because obesity is associated with insulin resistance and because β-cells are damaged by exposure to elevated lipid and glucose levels (Rutter and Parton 2008). Visceral obesity (intra-abdominal adipose tissue) is the major culprit. Indeed, some studies suggest that subcutaneous fat may even be protective against markers of cardiometabolic risk (Buemann et al. 2005; Livingston 2006; Hocking et al. 2008), though others give a more complex picture (Janiszewski et al. 2008; Frederiksen et al. 2009; Porter et al. 2009).

2.2 Exercise in the Treatment of Diabetes

Low-calorie diets and exercise are usually the first line of treatment not only for obesity but also for type 2 diabetic patients (Hayes and Kriska 2008). Exercise of sufficient intensity generally improves insulin sensitivity (Assah et al. 2008; O'Gorman and Krook 2008; Qi et al. 2008; Slentz et al. 2009), though it is not effective in all patients (Teran-Garcia et al. 2005; Burns et al. 2007), partly owing to interactions with genetic variation (Qi et al. 2008). Exercise must be regular for its benefits to be persistent (Hawley and Lessard 2008). Combined aerobic and resistance training may be of great value (Zanuso et al. 2010). Some studies suggest that exercise can improve insulin sensitivity without causing weight loss (Cox et al. 2004; Holloszy 2005; Bo et al. 2008). This might be expected if exercise increases muscle mass or the oxidative capacity of skeletal muscle but, surprisingly, in one study fat loss elicited by exercise resulted in no greater improvement in glucose tolerance and insulin action than similar fat loss elicited by caloric restriction (Weiss et al. 2006).

Improved insulin sensitivity appears to protect the β-cell. This reduces the risk of developing diabetes, because β-cell failure is ultimately responsible for declining blood glucose control in diabetes. Thus, lifestyle modifications that involve a large exercise component and thiazolidinedione drugs, both of which improve insulin sensitivity, have a continuing effect over 4 years on the incidence of diabetes in subjects at risk of developing the disease. By contrast, clinical trials on the prevention of diabetes by metformin and acarbose, which have less effect on insulin sensitivity, show maximal effects after only 2 years (Buchanan 2007).

Drugs that mimic the effects of exercise should therefore be of benefit in most type 2 diabetic patients. The efficacy of such a drug relative to one that reduces energy intake may be debated, but since the mechanisms by which increased energy expenditure and reduced intake improve glucose homeostasis are not identical, a thermogenic drug would expand therapeutic options beyond those of anorectic drugs.

2.3 Exercise in the Treatment of the Metabolic Syndrome

Stimulation of thermogenesis may provide a means of treating multiple components of the metabolic syndrome. There is some scepticism as to whether the various definitions of the metabolic syndrome identify patients who are at risk of cardiovascular disease any better than if risk markers are considered individually (Taslim and Tai 2009). Nevertheless, an anti-diabetic drug that also reduces other markers of cardiovascular disease risk should be more valuable than one that does not, and the need for polypharmacy to treat multiple components of the syndrome would be reduced.

Many definitions of the metabolic syndrome include insulin resistance, abdominal obesity (abdominal subcutaneous as well as visceral adipose tissue, as assessed by waist circumference), high plasma triglycerides, low high-density lipoprotein cholesterol, and hypertension (Alberti et al. 2009). These features of the metabolic syndrome are all altered favourably by exercise (Hagberg et al. 2000; Duncan 2006; Orozco et al. 2008). Moreover, adipose tissue inflammation, which may be a key link in the metabolic syndrome (Yudkin 2007), is reduced by exercise in mice (Bradley et al. 2008).

Some individual studies have found that exercise preferentially reduces visceral rather than subcutaneous fat or that it can cause loss of visceral fat without weight loss (Thomas et al. 2000; Fujimoto et al. 2007; Ohkawara et al. 2007). This supports the argument that exercise, or drugs that elicit the metabolic effects of exercise, offers the best treatment for the metabolic syndrome. Systematic reviews of the literature suggest, however, that all weight loss interventions preferentially reduce visceral fat when weight loss is low, whilst greater weight loss increases the proportion of subcutaneous fat lost (Chaston and Dixon 2008; Hall and Hallgreen 2008; Hallgreen and Hall 2008). A systematic review of the literature also provided no evidence that exercise was any better than other interventions at causing loss of fat rather than lean tissue (Chaston and Dixon 2008). This is somewhat surprising: studies in which diet and diet plus exercise have been compared directly do demonstrate that exercise promotes loss of fat rather than lean tissue (Janssen et al. 2002).

2.4 Mitochondrial Function and Capacity for Fat Oxidation in Diabetes

Exercise increases the number of mitochondria in skeletal muscle (Hawley and Holloszy 2009); some studies have also found improved mitochondrial function (Hood et al. 2006; Menshikova et al. 2007; Schrauwen and Hesselink 2008). Mitochondrial function or number is low in the skeletal muscle of many insulin-resistant or type 2 diabetic subjects (Abdul-Ghani and DeFronzo 2008; Schiff et al. 2009). Muscle mitochondrial content is also low in healthy subjects with a family history of diabetes (Ukropcova et al. 2007) and such subjects have lower insulin sensitivity and benefit more from exercise than controls (Barwell et al. 2008).

Thus, thermogenic drugs that increase mitochondrial number or function in skeletal muscle may correct an underlying defect in insulin-resistant type 2 diabetic subjects.

Reduced mitochondrial content in skeletal muscle has been linked to the feature of type 2 diabetes known as metabolic inflexibility (Ukropcova et al. 2007). This is an impaired ability to switch between fatty acid and glucose fuels in response to insulin, fasting, feeding or the food quotient of the diet. The evidence for this link, and that metabolic inflexibility is responsible for insulin resistance, has been questioned (Galgani et al. 2008). Exercise training has, however, been found to improve both mitochondrial function and metabolic flexibility, as well as insulin resistance (Meex et al. 2010).

2.5 Lessons from β-Adrenoceptor Agonists

β-Adrenoceptor agonists of various selectivities for β_1-, β_2- and β_3-adrenoceptors stimulate energy expenditure in both humans and rodents (Arch 2008). For the last 25 years, there has been a particular interest in the potential of β_3-adrenoceptor agonists in the treatment of obesity. These compounds cause the loss of fat (but not lean tissue) in obese rodents, and they do not affect the cardiovascular system to the same degree as β_1- or β_2-adrenoceptor agonists. Moreover, β_3-adrenoceptor agonists are exquisitely effective at improving insulin sensitivity in obese rodents, including models of type 2 diabetes. Difficulties in finding orally bioavailable agonists that are selective for the human, rather than the rodent, β_3-adrenoceptor, together with a lesser role for β_3-adrenoceptors in the regulation of energy balance in humans, have thwarted the development of β_3-adrenoceptor agonists as drugs for the treatment of either obesity or diabetes. Nevertheless, stimulation of β_2- or β_3-adrenoceptors does enhance insulin action in humans. As in rodents, insulin sensitisation occurs with dosing regimens that are insufficient to elicit weight loss (Arch 2002; Arch 2008). Interest in β_3-adrenoceptor agonists or similar approaches has been revived by new evidence that brown adipose tissue is present in adult humans (van Marken Lichtenbelt et al. 2009) and activated by exposure to cold. Moreover, its amount and activity is reduced in obesity (Cypess et al. 2009). This might encourage new approaches to be taken to the development of β_3-adrenoceptor agonists.

The reason why β-adrenoceptor agonists are so effective at improving insulin sensitivity may be that, like exercise training (O'Gorman and Krook 2008), they promote fat oxidation. Most studies have found that β-adrenoceptor agonists stimulate fat oxidation in preference to carbohydrate oxidation, and inhibition of fat oxidation prevents thermogenesis in response to β_3-adrenoceptor agonists in rodents (Wilson et al. 1986; Arch 2008). Moreover, inhibition of fat mobilisation (lipolysis) using anti-lipolytic drugs in rodents, dogs, or humans (Havel et al. 1964; Eaton et al. 1965; Kennedy and Ellis 1969; Mjos 1971; Lafrance et al. 1979; Schiffelers et al. 1998) or genetic methods in mice (Grujic et al. 1997; Gavrilova et al. 2000) reduces, sometimes markedly, acute increases in energy expenditure in response to β-adrenoceptor agonists. β-Adrenoceptor agonists promote glycogenolysis as well as lipolysis, but fat is oxidised in preference to carbohydrate through mechanisms first described by Randle and his co-workers (Bebernitz and Schuster 2002).

Stimulation of lipolysis is unlikely to be the whole explanation for why β_3-adrenoceptor agonists promote fat oxidation, however. Prevention of lipid mobilisation reduces the acute thermogenic response to β_3-adrenoceptor agonists, but this does not mean that they have no effect on fat oxidation or fail to improve insulin sensitivity. When β_3-adrenoceptor agonists stimulate lipolysis, they exacerbate insulin resistance acutely because fatty acids inhibit glucose utilisation, but in the longer term it seems that β_3-adrenoceptor agonists improve insulin sensitivity by promoting fatty acid oxidation more than fatty acid supply. Thus, β_3-adrenoceptor agonists acutely raise plasma non-esterified fatty acid levels, but after repeated administration any small rise in levels after each dose is rapidly followed by a lowering of fatty acid levels below those in control animals, until the next dose is administered (Virtanen et al. 1997; Liu et al. 1998; Sugimoto et al. 2005). Desensitisation of the acute lipolytic affect may be partly responsible for the diminishing rise in fatty acid levels after each dose. In this author's unpublished work, however, there was no diminution in the rise in glycerol levels, suggesting that lipolysis was not diminished and the diminishing rise in fatty acid levels must have been because their utilisation was increased. In any event, reduced lipolysis does not explain why repeated administration of β_3-adrenoceptor agonists lowers fatty acid levels after any small elevation in levels after each dose has subsided: there must be a stimulation of fatty acid oxidation that outweighs any effect on fatty acid supply. β_3-Adrenoceptor agonists have not been shown to affect mitochondria in skeletal muscle, but they increase mitochondrial function and number in brown and white adipose tissue, which may go some way to explaining their effect on fatty acid oxidation (Granneman et al. 2005).

Fat oxidation driven by unrestrained lipolysis will raise the concentrations of lipid metabolites, whereas fat oxidation driven by mitochondrial mechanisms downstream of these metabolites is likely to lower their concentrations. Diacylglycerol and possibly ceramide and fatty acyl-CoA (Kraegen and Cooney 2008) inhibit insulin signalling at the level of insulin receptor substrates (Morino et al. 2006; Yu et al. 2002) or protein kinase B/Akt (Schmitz-Peiffer et al. 1999), through activation of protein kinase $C\theta$ or ε, inhibitor κB kinase (IKKβ), or c-Jun N-terminal kinase (JNK) (Montecucco et al. 2008). Lowering the concentration of these metabolites by stimulating their oxidation might explain why β_3-adrenoceptor agonists increase insulin sensitivity, at least in situations in which there is insulin resistance. Stimulation of fatty acid oxidation should lower the concentrations of fatty acid metabolites rapidly because their pool sizes are small. It will take much longer to drain the large store of triglyceride from white adipose tissue for oxidation elsewhere. Hence, this hypothesis may explain why β_3-adrenoceptor agonists reduce insulin resistance in rodents, monkeys, and humans long before weight loss is detected. There appears, however, to be only one paper that has addressed the effect of β_3-adrenoceptor agonists on lipid metabolite levels. This showed that a β_3-adrenoceptor agonist reduced the concentration of diacylglycerol in skeletal muscle of obese rats (Darimont et al. 2004).

Anorectic drugs must also promote fat oxidation once glycogen stores have been depleted. This may be why restriction of energy intake sensitises the body to insulin more than if the individual was "naturally" leaner (Wing and Phelan 2005). As discussed above (Sect. 2.2), it remains to be seen whether thermogenic drugs will be more effective than anorectic drugs as insulin sensitisers.

3   Current and Recent Drugs

3.1   Diabetes Drugs

Current therapies for type 2 diabetes do not in general stimulate thermogenesis. Insulin, sulphonylureas and thiazolidinediones in fact promote weight gain. In rodents, thiazolidinediones increase the capacity for thermogenesis in response to sympathetic activity, but they decrease the activity of the sympathetic nervous system (Festuccia et al. 2008). The glucagon-like peptide-1 receptor agonists exenetide and liraglutide and the modified amylin peptide pramlintide cause weight loss, but this is because they reduce energy intake rather than because they increase energy expenditure (Edwards et al. 2001; Harder et al. 2004; Pratley 2008; Smith et al. 2008). This seems somewhat surprising in the case of pramlintide because a number of studies have shown that amylin increases energy expenditure in rodents, apparently by increasing sympathetic activity (Osaka et al. 2008). At its clinical dose in humans, however, pramlintide does not appear to increase sympathetic activity (Hoogwerf et al. 2008), energy expenditure, or indeed insulin sensitivity. The α-glucosidase inhibitor acarbose (probably also miglitol and voglibose) causes a little weight loss, but it does this by reducing energy absorption rather than increasing energy expenditure (Van de Laar et al. 2006). The dipeptidylpeptidase IV inhibitors sitagliptin, vildagliptin, and others in development do not affect body weight (see Gallwitz 2011; Mikhail 2008). However, treatment of type 2 diabetic patients for 7 days with vildagliptin enhanced lipid mobilisation and oxidation, possibly due to sympathetic activation (Boschmann et al. 2009).

Metformin may cause a slight reduction in body weight (Golay 2008). Its mechanism of action in diabetes seems to involve activation of AMP-activated protein kinase (AMPK) (Hardie 2008), which is also believed to play a role in the effect of exercise on insulin sensitivity (Hawley and Lessard 2008; O'Gorman and Krook 2008) and is a target for thermogenic drugs (see Sect. 6.1). One might therefore expect metformin to increase energy expenditure, but the effect of metformin on body weight seems to be due to reduced food intake rather than increased energy expenditure (Keates and Bailey 1993; Perriello et al. 1994; Chong et al. 1995; Paolisso et al. 1998). However, there is evidence that metformin increases fat oxidation or decreases lipogenesis, at least transiently (Avignon et al. 2000; Cool et al. 2006; Braun et al. 2008). Metformin (like other AMPK activators – Sect. 6.1) may have a thermogenic effect that is too small to detect (see Sect. 7).

3.2 Obesity Drugs

In contrast with drugs for diabetes, two drugs that have been used recently in the treatment of obesity increase energy expenditure.

3.2.1 Sibutramine

The noradrenaline and serotonin reuptake inhibitor sibutramine has been withdrawn in Europe, the USA and other countries. In addition to reducing food intake, some, though not all, studies (Addy et al. 2008) show that it increases energy expenditure by increasing sympathetic activity. This has the adverse consequence of raising blood pressure and it increases the risk of nonfatal myocardial infarction and nonfatal stroke in patients who have a history of cardiovascular disease (James et al. 2010). It should not be prescribed for such patients. In patients with good weight loss, which tends to reduce blood pressure, this tendency may be countered. Whether increased energy expenditure contributes significantly to the effect of sibutramine on body weight is unclear (Hansen et al. 1998; Finer 2002). Sibutramine improves insulin sensitivity (Hung et al. 2005), but it does not appear to have significant metabolic benefits beyond those expected from weight loss, in contrast to what might be expected for a sympathomimetic thermogenic drug (Arch 2008).

3.2.2 Rimonabant

The cannabinoid 1-receptor antagonist rimonabant was withdrawn from the European market in 2008, never having been approved in the USA. In contrast with sibutramine, rimonabant improved lipid profiles and glucose homeostasis in humans (as in rodents) beyond what might be expected from the weight loss that it caused (Hollander 2007; Scheen 2008). About half of these metabolic benefits appeared to be independent of weight loss.

The improvement in lipid profile in humans in response to rimonabant was suggested to be due to an elevation in plasma adiponectin concentration that was independent of weight loss (Despres et al. 2005). Rimonabant directly stimulates adiponectin secretion from cultured adipocytes (Bensaid et al. 2003), and adiponectin stimulates fat oxidation and improves insulin sensitivity in skeletal muscle. Rimonabant may also stimulate fat oxidation in skeletal muscle and other tissues independently of its effect on adiponectin secretion (Lafontan et al. 2007). In support of an adiponectin-independent effect, rimonabant ameliorated insulin resistance in Lep^{ob}/Lep^{ob} mice that also lacked the adiponectin gene, though its effect was less than when the adiponectin gene was present (Watanabe et al. 2009). In diet-induced obese mice, absence of the adiponectin gene did not prevent rimonabant from causing weight loss, but it did prevent rimonabant from causing a statistically significant improvement in insulin sensitivity (Migrenne et al. 2009). Rimonabant increased mitochondrial biogenesis in murine white adipocytes by inducing endothelial nitric oxide synthase (Tedesco et al. 2008) and it activated AMPK in human myotubes (Cavuoto et al. 2007), suggesting molecular links for adiponectin-independent effects.

The beneficial effects of rimonabant on lipid metabolism and insulin sensitivity in rats required it to be administered peripherally, whereas both central and peripheral administration reduced food intake (Nogueiras et al. 2008). However, peripheral administration of the non-CNS-penetrant cannabinoid 1-receptor anta-gonist LH-21 to obese Zucker (fa/fa) rats reduced food intake, but surprisingly did not improve hypertriglyceridaemia or hypercholesterolaemia (Pavon et al. 2008). The anorectic effect of LH-21 is consistent with evidence that rimonabant can reduce food intake by acting at peripheral receptors (Gomez et al. 2002), though this is disputed by authors of a report that shows that the anorectic effect of rimonabant does not require intact gut vagal or sympathetic afferents (Madsen et al. 2009). An alternative, but speculative, explanation of the failure of LH-21 to affect metabolism may be that it is a neutral antagonist of the cannabinoid 1-receptor, whereas rimonabant is an inverse agonist (Pavon et al. 2008). Another poorly brain-penetrant cannabinoid 1-receptor antagonist, AM6545, improved the metabolic profile of obese mice, but it did not block various behavioural effects elicited by a cannabinoid agonist. The improvement in metabolic profile depended on the presence of cannabinoid 1-receptors in the liver (Tam et al. 2010). Consistent with this, mice that lack cannabinoid 1-receptors in liver but not in other organs are resistant to diet-induced steatosis, dyslipidaemia, and insulin resistance, but not to obesity (Osei-Hyiaman et al. 2005). However, AM6545 reduced food intake in mice that lack the cannabinoid 1-receptor (Cluny et al. 2010), so its value as a tool compound is dubious. A recent report on a derivative of rimonabant that penetrates the brain poorly suggests that brain penetration is required for anti-obesity activity, but this does not exclude the possibility that such a compound would be useful for diabetes (Son et al. 2010). Thus, considerable interest remains in the possibility that peripherally acting cannabinoid 1-receptor antagonists might be useful in the treatment of diabetes, even though the data on different antagonists are not entirely consistent (Bermudez-Silva et al. 2010).

It is unclear whether the beneficial metabolic effects of rimonabant in humans are due to stimulation of energy expenditure. There is no published evidence that rimonabant affects energy expenditure in humans, but a pharmacologically similar compound, taranabant, increased energy expenditure in one study (Addy et al. 2008). Rimonabant has been shown to increase energy expenditure in rats (Herling et al. 2008; Kunz et al. 2008). In obese rats, the thermogenic effect of rimonabant was due to increased fat oxidation; carbohydrate oxidation decreased. Moreover, increased energy expenditure made a greater contribution than decreased energy intake to the anti-obesity effect of rimonabant in rats (Herling et al. 2008). A cannabinoid 1-receptor antagonist that had been in clinical development, AVE1625, also rapidly raised energy expenditure and fat oxidation and suppressed carbohydrate oxidation in rats (Herling et al. 2007).

The effects of rimonabant and AVE1625 on energy expenditure and fuel utilisation in rats appear within minutes of their administration. Since the method of calculation of energy expenditure and fat oxidation used in these studies probably took no account of the time needed for oxygen and carbon dioxide in a respiratory chamber to respond to changes in their rates of utilisation and production by the rat (Arch et al. 2006), the effects of the drugs may be even more rapid than they appear. One mechanism that might cause such rapid effects on energy expenditure and fat oxidation is sympathetic activation. Cannabinoid 1-receptor antagonism has been shown both to disinhibit transmitter release at peripheral sympathetic nerves (Marsicano and Lutz 2006; Mnich et al. 2010) and to activate sympathetic activity centrally causing brown adipose tissue activation (Verty et al. 2009). Agonism inhibits sympathetic activity by a central mechanism (Niederhoffer et al. 2003). Consistent with a general activation of sympathetic activity, whether peripherally or centrally mediated, rimonabant increased cardiac contractility and blood pressure in hypertensive rats (Batkai et al. 2004).

Stimulation of adiponectin secretion might explain why rimonabant causes a rapid increase in energy expenditure and fat oxidation. Rimonabant raises the expression of adiponectin mRNA within 30 min in cultured adipocytes (Bensaid et al. 2003). Adiponectin cannot then act through changes in protein expression: this would take too long. A more likely link is, once again, sympathetic activation, because adiponectin rapidly activates sympathetic activity in brown adipose tissue (Masaki et al. 2003). But whether adiponectin secretion and activation of sympathetic nerves by adiponectin are in combination rapid enough to explain the rapid effect of adiponectin on energy expenditure and fat oxidation is questionable. On balance, it seems most likely that sympathetic activation is responsible for the rapid effects of rimonabant on energy expenditure and fat oxidation in rodents, but adiponectin secretion may be partly responsible for the weight loss-independent effects of rimonabant on metabolism, especially in humans.

Indeed, whether the rapid, possibly sympathetically mediated, effects of rimonabant and other cannabinoid 1-receptor antagonists on energy expenditure in rodents are of any relevance to the metabolic effects of rimonabant in humans is doubtful. Rimonabant decreased hypertension in obese patients (Ruilope et al. 2008), whereas it increased blood pressure in hypertensive rats (Batkai et al. 2004), possibly reflecting a species difference in the activation of sympathetic activity. Blockade by rimonabant of liver, muscle, and adipocyte cannabinoid 1-receptors may be sufficient to account for its weight-independent metabolic benefits in humans (Kunos et al. 2009). Through such mechanisms, rimonabant may increase the capacity for fat oxidation without having an easily detectable effect on energy expenditure (see Sect. 7).

In conclusion, rimonabant elicits metabolic benefits in humans and rodents that are beyond what would be expected from the weight loss that it causes. It has been shown in rodents but not humans that it increases energy expenditure through a sympathetically mediated mechanism (or mechanisms) and elicits weight loss beyond what would be expected from reduced food intake.

4 Targets in Hormonal Systems

4.1 Sympathetic Nervous System

Sympathomimetic drugs can increase resting metabolic rate in humans by about 30% (Schiffelers et al. 2000). This is far less than in rodents where an increase of the order of two- to threefold for a mouse at thermoneutrality is possible (Wilson et al. 1984; Wernstedt et al. 2006; Feldmann et al. 2009). Nevertheless, sympathomimetic drugs, such as β-adrenoceptor agonists, have been shown to improve insulin sensitivity markedly in humans (Mitchell et al. 1989; Smith et al. 1990).

There is some evidence that components of the sympathetic nervous system, including the component that regulates metabolism and thermogenesis, are subject to differential central regulation (Terao et al. 1994; Morrison 2001; van den Hoek et al. 2008). However, most centrally acting sympathomimetic agents, like sibutramine (Sect. 3.2.1), have cardiovascular side effects. Even leptin (see below) has limited selectivity, despite its primary role being to ensure that adipose tissue fat stores are adequate (Haynes et al. 1997; Rahmouni and Morgan 2007). Peripheral pre- or post-synaptic receptors of the sympathetic nervous system (possibly illustrated by the cannabinoid 1-receptor or the β_3-adrenoceptor) may offer more potential for selective thermogenic drugs.

4.2 Zn-α_2-Glycoprotein

Lipid-mobilising factor (LMF)/Zn-α_2-glycoprotein (ZAG), which is secreted by tumours and adipocytes (Hale et al. 2001; Bao et al. 2005), may act at $G\alpha_s$-coupled receptors. It has been argued that it stimulates the β_3-adrenoceptor because it raised cyclic AMP levels in cells transfected with the human β_3-adrenoceptor and this effect was attenuated by as little as 1 nM SR59230A (Russell et al. 2002), a β_3-adrenoceptor antagonist. It is surprising, however, that its efficacy relative to isoprenaline should be higher in human than in murine white adipocytes (Hirai et al. 1998), since β_3-adrenoceptors have much lower lipolytic efficacy in human than in murine adipocytes (Sennitt et al. 1998). It is also surprising that SR59230A was quite so potent in the light of reported affinities for the human cloned β_3-adrenoceptor (Arch 2000). Other effects of ZAG have been attenuated by 10 μM SR59230A (Sanders and Tisdale 2004a, b). Moreover, the binding of [^{125}I]-LMF to CHO cells that expressed β_3-adrenoceptors was reduced by 10 μM SR59230A (Russell et al. 2002). As little as 0.1 μM SR59230A should antagonise rodent β_1-as well as β_3-adrenoceptors (Manara et al. 1996). Crucially, obesity and decreased lipolysis in ZAG-deficient mice could not be corrected using β_3-adrenoceptor agonists (Rolli et al. 2007). ZAG increases the expression of Gs_α (Islam-Ali et al. 2001). This property explains its biological activities better than direct stimulation of β_3-adrenoceptors. It is important to know whether ZAG raises the concentration of cyclic AMP in cells transfected with Gs-coupled receptors other than the β_3-adrenoceptor.

Nevertheless, whatever the receptor for ZAG is, it stimulates fatty acid oxidation and exhibits anti-diabetic activity in mice (Russell and Tisdale 2002, 2010). It increased gastrocnemius weight almost threefold in ob/ob mice within 5 days (Russell and Tisdale 2010), resembling the β_2-adrenoceptor agonist clenbuterol. Its absence caused obesity (Rolli et al. 2007). Moreover, ZAG (also known as AZGP1) mRNA expression is lower in tissue from obese compared to lean humans (Dahlman et al. 2005; Marrades et al. 2008). Its thermogenic activity and its potential as an anti-diabetic agent merits further investigation.

4.3 Thyroid Hormones

Thyroid hormones were used in the treatment of obesity as long ago as the 1890s. Despite increasing energy expenditure, the complex metabolic effects of thyroid hormones cause insulin resistance and impaired glucose tolerance. Thyroid hormones also cause cardiac stimulation, loss of skeletal muscle, bone wasting, fatigue, and CNS effects (Crunkhorn and Patti 2008).

Recent interest has focused on selective stimulants of thyroid hormone receptor-β. Thyroid hormone receptor-β is poorly expressed in heart and skeletal muscle compared to thyroid hormone receptor-α. Consequently, selective stimulants of thyroid hormone receptor-β cause less cardiac stimulation, muscle wasting, and bone loss than non-selective stimulants of both receptors (Grover et al. 2007; Villicev et al. 2007; Ribeiro 2008). There is a particular interest in thyroid hormone receptor-β agonists that are selectively taken up by the liver (GC-1, KB2115) or are extracted by and activated in the liver (MB07811). Compared to triiodothyronine, GC-1 may also differentially activate thyroid response elements and thereby activate a different set of genes (Baxter and Webb 2009).

Thyroid hormone receptor-β agonists were originally seen as potential treatments for obesity. Currently, the main interest is in the treatment of dyslipidaemia, reflecting the focus on liver-selective compounds (Baxter and Webb 2009), but thyroid hormone receptor-β agonists have also been shown to improve insulin sensitivity in rodent models of obesity and diabetes (Bryzgalova et al. 2008). This would be expected from the ability of these compounds to increase energy expenditure and mitochondrial function and reduce body fat and hepatic steatosis. However, interest in diabetes and obesity indications may have dimmed with reports that GC-1 was less thermogenic than triiodothyronine in hypothyroid rats (Venditti et al. 2010) and that another compound, GC-24, had less effect on weight gain in dietary obese compared to normal mice (Castillo et al. 2010). Moreover, stimulation of gluconeogenesis would not be beneficial (Baxter and Webb 2009).

4.4 TGR5: A Bile Acid Receptor

Activation of the Gs-protein-coupled bile acid receptor TGR5 may be a way of activating the thyroid hormone system in tissues, such as adipose tissue, where it is expressed. I will argue that whilst this is true, it may not be essential for the thermogenic effects of TGR5 agonists.

The potential of TGR5 as a target for drugs for the treatment of obesity and diabetes was highlighted by the finding that diet-induced obesity in mice was ameliorated by supplementing their diet with cholic acid. Glucose tolerance also improved. Cholic acid also activates the FXRα receptor, but this did not appear to mediate the anti-obesity effect because it was not reproduced by the synthetic FXRα agonist GW4064 (Watanabe et al. 2006). The reduction in body weight gain elicited by cholic acid was achieved without a reduction in food intake, pointing to it having a thermogenic effect. However, thermogenesis was only demonstrated after 4 months and only by expressing oxygen consumption relative to body weight: demonstration of an acute effect or an increase in energy expenditure per animal would have been more convincing (see Sect. 7). Nevertheless, the effects of cholic acid on a mitochondrial structure and gene expression in brown adipose tissue support the view that it increased the capacity for thermogenesis (Watanabe et al. 2006).

One of the genes whose expression was increased in brown adipose was type 2 iodothyronine deiodinase (D2), which activates thyroxine by converting it into triiodothyronine. Cholic acid failed to affect diet-induced thermogenesis in mice that lacked D2, leading the authors to conclude that its mechanism of action depends primarily on activation of D2 (Watanabe et al. 2006). It might, however, be that D2 plays a permissive and amplifying role in the response to TGR5 agonists. Thus, TGR5 is coupled to Gs and elevation of cyclic AMP. β_3-Adrenoceptor receptors are also coupled to Gs and agonists activate brown adipose tissue thermogenesis acutely. In the absence of a functional thyroid hormone system, β_3-adrenoceptor agonists have a greatly reduced effect (Rubio et al. 1995; Golozoubova et al. 2004), but it is not believed that β_3-adrenoceptor agonists stimulate thermogenesis by activating D2. Nevertheless, β_3- and other β-adrenoceptor agonists increase the activity of D2 (Hofer et al. 2000) and this amplifies their effects.

A taurine conjugate of cholic acid activated D2 in human skeletal muscle myoblasts (Watanabe et al. 2006), suggesting that TGR5 agonists may increase thermogenesis in skeletal muscle as well as in brown adipose tissue. Moreover, bile acids, acting via TGR5, and the TGR5 agonist INT-777 stimulate glucagon-like peptide-1 secretion by enteroendocrine cells (Thomas et al. 2009). This effect is unrelated to thermogenesis, but adds to the therapeutic potential of TGR5 agonists for the treatment of diabetes. High expression of TGR5 in monocytes and macrophages may also be useful by resulting in suppression of inflammation (Kawamata et al. 2003).

4.5   Glucocorticoids and 11β-Hydroxysteroid Dehydrogenase-1

In contrast with thyroid hormones, the glucocorticoid system has usually been seen as offering targets for anti-diabetic rather than anti-obesity drugs. In the context of this chapter, it is appropriate to ask whether this approach is in part due to an effect on thermogenesis. The glucocorticoid system clearly has a profound effect on energy balance, as evidenced by visceral obesity in Cushing's syndrome and by the ability of adrenalectomy to prevent all forms of rodent obesity (Bray 2000).

Simply inhibiting cortisol production or blocking the glucocorticoid receptor does not offer a viable approach to the treatment of metabolic disease because removal of feedback inhibition to the brain results in increased corticotropin releasing hormone (CRH) and adrenocorticotropic hormone (ACTH) secretion. One approach to this problem has been to make compounds that are localised to the liver. The anti-diabetic activity of such compounds is not associated with marked anti-obesity activity (Zinker et al. 2007).

A more popular approach has been to inhibit the activity of the enzyme 11β-hydroxysteroid dehydrogenase-1 (11βHSD-1). This is the subject of a chapter by Wang (2011). In tissues, 11βHSD-1 catalyses the conversion of cortisone, which is inactive, to cortisol, which is active. (In rodents it catalyses the conversion of 11-dehydrocorticosterone to corticosterone.) Provided the effects of inhibitors of 11βHSD-1 on cortisol concentration are limited to the tissues in which they act, they should not increase CRH and ACTH secretion. One tissue that must be excluded from this statement is the hypothalamus because inhibition of 11βHSD-1 will reduce exposure of the hypothalamic glucocorticoid receptor to cortisol, thereby increasing CRH production. In addition, inhibition of the liver enzyme does not have a tissue-limited effect: it increases the plasma cortisol concentration. Moreover, inhibition of the liver enzyme provides limited metabolic benefit (Livingstone and Walker 2003). On the other hand, inhibition of the adipose tissue, especially intra-abdominal adipose tissue 11βHSD-1 seems to be of great value (Berthiaume et al. 2007).

Inhibitors of 11βHSD-1 improve insulin sensitivity in mice partly because they reduce obesity. The anti-obesity effect is due to not only reduced food intake (Alberts et al. 2003; Hermanowski-Vosatka et al. 2005), but also prevention of the decrease in energy expenditure that would normally be associated with reduced energy intake (Wang 2006). It is clear that 11βHSD-1 influences energy expenditure in mice because mice that lack this enzyme are protected from diet-induced obesity despite consuming more calories (Morton et al. 2004). Direct evidence of raised energy expenditure is limited, however, to data for oxygen consumption expressed relative to body weight (Kershaw et al. 2005).

How inhibition or lack of 11βHSD-1 increases energy expenditure is unclear. The inhibitor BVT116429 raised the plasma concentration of adiponectin in KKAy (Sundbom et al. 2008) and this would be expected to promote fat oxidation (see Sect. 4.10). However, BVT2733 from the same company (Biovitrum), which maintains energy expenditure in mice despite reducing food intake (Wang 2006), did not raise adiponectin levels (Sundbom et al. 2008). BVT116429 has been shown to inhibit 11βHSD-1 activity in adipose tissue taken from inhibitor-treated mice (Johansson et al. 2008). One might predict that BVT116429 would have more effect than BVT2733 in adipose tissue, perhaps explaining their differential effects on adiponectin levels, but such a difference has not been reported. It is noticeable that acute effects of 11βHSD-1 inhibitors on energy expenditure have not been reported.

The most advanced 11βHSD-1 inhibitor in clinical trials appears to be INCB13739. After 12 weeks of treatment, it reduced A1C by 0.6% in patients with inadequately controlled diabetes (Rosenstock et al. 2010).

4.6 Leptin

Leptin is a 146 amino acid peptide that is released mainly from adipocytes. It signals to the brain whether fat stores are adequate. In normal rodents, leptin not only decreases food intake, but it also prevents energy expenditure falling in animals that are below thermoneutrality and have a restricted energy intake (Doring et al. 1998). The latter effect is partly due to increased sympathetic outflow (Arch 2008; Asensio et al. 2008). Other mechanisms are activation of AMPK, mediated by leptin receptors in skeletal muscle (Minokoshi et al. 2002) and increased locomotor activity (Choi et al. 2008), possibly mediated by STAT3 activation (Mesaros et al. 2008). Leptin- or leptin receptor-deficient rodents are insulin resistant as well as obese, and administration of leptin to leptin-deficient mice restores insulin sensitivity. Leptin also improves insulin sensitivity in normal rodents and in mouse models of type 2 diabetes that have normal or slightly elevated leptin levels, but it has little or no effect when leptin levels are already greatly increased (Toyoshima et al. 2005; Kusakabe et al. 2009).

There is no evidence that leptin increases energy expenditure in most humans; neither does it reduce food intake in most humans (Heymsfield et al. 1999; Proietto and Thorburn 2003). However, it does reduce food intake in rare leptin-deficient subjects, and in contrast to what happens in subjects on a reduced energy diet, basal metabolic rate does not fall as the weight is lost (Farooqi and O'Rahilly 2004).

Leptin is effective in the treatment of insulin resistance and dyslipidaemia in lipodystrophic patients (including HIV-infected patients), who have low plasma leptin concentrations (Chong et al. 2010). Obese humans, however, generally have raised plasma leptin levels and are considered to be resistant to leptin. Endogenous leptin cannot be totally ineffective, or else we would all be as food-obsessed and grossly obese as untreated leptin-deficient individuals (Farooqi et al. 2007), and in fact mean weight loss over 24 weeks was 5.8 kg more in obese subjects treated with 0.3 mg/kg body weight per day leptin than in placebo-treated subjects (Heymsfield et al. 1999). However, the blood levels of leptin achieved at this dose were more than 20-fold above baseline (Heymsfield et al. 1999) or what others have reported for subjects of similar BMI (Kennedy et al. 1997). Immune responses at the injection site and the cost of leptin may have precluded it being used as a treatment for obesity.

Various approaches are being taken to improve sensitivity to leptin. One is to develop small molecular weight leptin mimetics (Vaillancourt et al. 2001; Maneuf et al. 2004; Mirshamsi et al. 2007). This may bypass the mechanism that transports leptin into the brain, which is one possible cause of leptin resistance (Banks 2008). Another approach, being progressed by Amylin Pharmaceuticals, is to treat subjects with a combination of metreleptin (a modified leptin) and pramlintide (a modified form of amylin), which is already marketed for the treatment of diabetes. This combination caused more weight loss in obese humans than either peptide alone (Roth et al. 2008). It is intriguing, however, that weight loss with pramlintide alone was similar to weight loss with metreleptin alone, because many believe that pramlintide but not leptin causes weight loss. There was no placebo group (who would have been given a 40% energy-deficient diet), so it is not possible to say whether both compounds were effective when used alone (and so their effects were additive rather than synergistic) or both were ineffective when used alone. A third approach may be to combine leptin with chemical chaperones, such as 4-phenyl butyric acid and tauroursodeoxycholic acid, which are used clinically in other diseases. When mice were pre-treated with these compounds, they became as much as tenfold more sensitive to leptin and lost weight, even when fed on a high fat diet (Ozcan et al. 2009).

If such approaches can overcome leptin resistance, there remains the question of whether leptin has a thermogenic or metabolic, rather than just an anorectic, effect in humans. Whatever the mechanism, there is some evidence that leptin increases insulin sensitivity in humans, even when endogenous leptin levels are normal, as it clearly does in rodents (Toyoshima et al. 2005; Arch 2008). In humans, a high plasma leptin concentration usually correlates with an adverse metabolic profile (as they do in diet-induced obese rodents) and an increased risk of type 2 diabetes. However, when the influence of obesity as a confounder is taken into account, high leptin levels are associated with a decreased risk of diabetes (Schmidt et al. 2006; Arch 2007).

4.7   Fibroblast Growth Factor 21

Fibroblast growth factor 21 (FGF21) and its receptors are recent additions to pharmaceutical company targets for anti-diabetic and anti-obesity drugs. FGF21 is a member of the FGF19 subfamily of fibroblast growth factors. It is secreted primarily from the liver and promotes hepatic lipid oxidation. Both stimulation and inhibition of lipolysis in adipocytes have been described. Administration of FGF21 to diet-induced obese mice, mice with a dysfunctional leptin system, or diabetic rhesus monkeys reduces body weight and improves insulin sensitivity (Beenken and Mohammadi 2009; Xu et al. 2009). FGF21 clearly increases energy expenditure: its thermogenic effect has a fairly rapid onset, it is apparent when energy expenditure is expressed per mouse, rather than relative to body weight, and, in some studies, it has stimulated food intake as well as energy expenditure without affecting body weight or body composition (Xu et al. 2009; Sarruf et al. 2010).

Like leptin, the plasma concentration of FGF21 is elevated in obese rodents and humans, and FGF21 signalling was reduced in liver and fat of diet-induced obese mice (Fisher et al. 2010). Whether FGF21, like leptin, proves to be ineffective in obese humans, or whether, like insulin, it retains useful efficacy, remains to be determined.

4.8   Ghrelin

Ghrelin is a 28 amino acid peptide produced mainly in the stomach, but also in other gastrointestinal tissues, the hypothalamus (Nakazato et al. 2001), pancreatic islet α and β cells (Sun et al. 2007), and other tissues (Soares and Leite-Moreira 2008). Its active form is generally believed to be acylated with a medium chain fatty acid, especially octanoic acid, but des-acyl ghrelin may also have a physiological role (Gauna et al. 2007; Inhoff et al. 2009; Rodriguez et al. 2009). By stimulating the growth hormone secretagogue type 1a receptor in the hypothalamus, acyl ghrelin stimulates energy intake (Wren et al. 2001). It seems, however, that ghrelin is not a meal initiation signal, but rather a signal that prepares skeletal muscle and possibly liver to store instead of oxidising fat. Thus, plasma levels of acyl ghrelin are not elevated following prolonged fasting, but only when dietary lipids provide medium chain fatty acids for acylation of des-acyl ghrelin (Kirchner et al. 2009).

Some, but not all, studies also find that ghrelin decreases energy expenditure (Theander-Carrillo et al. 2006; Strassburg et al. 2008; Kirchner et al. 2009; Salome et al. 2009). A single dose of a catalytic antibody that hydrolysed the octanoyl moiety of ghrelin increased energy expenditure within 1 h of its administration (Mayorov et al. 2008). Ghrelin-induced weight gain was undetectable in mice that lacked all three β-adrenoceptors (Theander-Carrillo et al. 2006) and ghrelin suppressed noradrenaline release in brown adipose tissue (Mano-Otagiri et al. 2010), suggesting that decreased energy expenditure was a consequence of decreased sympathetic activity. Both stimulatory and inhibitory effects of ghrelin on insulin secretion have been reported, with the majority demonstrating inhibition of secretion in rodents and humans (Sun et al. 2007).

These effects of ghrelin suggest that antagonists of its receptor, or inhibition of ghrelin O-acyltransferase, which activates ghrelin by octanoylation (Gualillo et al. 2008), should have potential for the treatment of diabetes. In the context of this chapter, however, the question is whether any benefit is due to thermogenesis.

There are conflicting reports concerning whether ghrelin-deficient mice are protected from diet-induced obesity and insulin resistance (Sun et al. 2003; Wortley et al. 2005). In a study in which absence of ghrelin did not reduce obesity in Lep^{ob}/Lep^{ob} mice, blood glucose was reduced and glucose tolerance improved, but this was associated with increased insulin levels and so provided no evidence of increased insulin sensitivity (Sun et al. 2006). Studies on mice that lack the ghrelin receptor (Zigman et al. 2005; Longo et al. 2008), especially mice that also lack ghrelin (Pfluger et al. 2008), have provided better evidence of improved insulin sensitivity associated with increased energy expenditure. Motor activity was increased in some of these studies.

Small molecular weight ghrelin receptor antagonists that ameliorate diet-induced obesity and improve glucose tolerance have been described (Rudolph et al. 2007). Improved glucose tolerance following a single dose of one such compound appeared, however, to be due to stimulation of insulin secretion, whilst the effect on body weight with repeated dosing was primarily due to reduced food intake (Esler et al. 2007). It has been suggested that an inverse agonist might be used to reduce constitutive ghrelin receptor activity between meals, but a recent report that links lack of constitutive activity in a mutant receptor with obesity in humans suggests that a neutral antagonist would be more beneficial or at least that the inverse antagonist must (as inverse agonists usually do) have antagonist activity as well (Holst and Schwartz 2006).

Thus, ghrelin antagonists have potential for the treatment of diabetes, but it is not clear whether any benefit is due to stimulation of energy expenditure.

4.9 Ciliary and Brain-Derived Neurotrophic Factors

Ciliary neurotrophic factor (CNTF) is a member of the interleukin-6 family of cytokines, which is produced by astrocytes following brain injury. It is the subject of a chapter by Allen et al. (2011). It caused weight loss in a clinical trial in amyotrophic lateral sclerosis (motor neurone disease) (ALS 1996). The variant $CNTF_{Ax15}$ developed under the name Axokine® reduced body weight in a phase II clinical trial (Ettinger et al. 2003), but it performed poorly in a phase III trial, partly due to the formation of neutralising antibodies (Matthews and Febbraio 2008). In rodents, CNTF and Axokine increased energy expenditure and caused weight loss that was not all due to reduced food intake (Lambert et al. 2001; Bluher et al. 2004). Moreover, they improve insulin sensitivity and diabetes in insulin-resistant and diabetic mice (Sleeman et al. 2003; Watt et al. 2006b). One study, however, showed no greater effects on obesity or insulin sensitivity than those elicited by pair feeding (Cui et al. 2010).

CNTF binds both to the CNTF receptor and to the IL-6 receptor. Both receptors then heterodimerise with the leukaemia inhibitory factor receptor and the glycoprotein 130 receptor. This results in activation of the Janus Kinase/signal transducer and activator of transcription (JAK/STAT) pathway (Matthews and Febbraio 2008). This is a key pathway activated by leptin. However, CNTF corrects obesity and diabetes in animal models, such as diet-induced obesity, in which leptin is ineffective (Gloaguen et al. 1997; Lambert et al. 2001). The reason for this may be that CNTF can overcome inhibition of JAK/STAT signalling by suppressor of cytokine signalling-3 (SOCS-3) because the gp130 receptor has four STAT3 binding sites that must be nullified, whereas the long form of the leptin receptor (Rb) has only one such site (Matthews and Febbraio 2008). It may also be relevant that plasma leptin levels are raised in obesity leading to leptin resistance, but hypothalamic CNTF levels are reduced (Vacher et al. 2008).

The CNTF receptor and the leptin Rb receptor are co-localised in the hypothalamus, suggesting that the thermogenic effect of CNTF, like that of leptin, may be partly due to activation of the sympathetic nervous system. CNTF also acts directly on skeletal muscle and brown adipocytes to increase fatty acid oxidation, the capacity for thermogenesis, energy expenditure, and insulin action (Ott et al. 2002; Watt et al. 2006a, b). It increases fatty acid oxidation in skeletal muscle by a mechanism that requires activation of AMPK (Watt et al. 2006a). This is similar to the direct effect of leptin in skeletal muscle.

Because CNTF activates the IL-6 as well as the CNTF receptor, Matthews and Febbraio (2008) have suggested that a suitable approach to the treatment of obesity-related metabolic disease would be to design a CNTF-like molecule that has greater affinity for the IL-6 than the CNTF receptor and specifically targets peripheral tissues, such as skeletal muscle and adipose tissue, in which IL-6 receptor is more abundant than the CNTF receptors.

Brain-derived neurotrophic factor (BDNF) is small dimeric protein that is expressed in a number of organs in addition to brain. It is similar to CNTF in that it stimulates energy expenditure (Tsuchida et al. 2001) and has anti-obesity activity in animal models that are resistant to leptin (Nakagawa et al. 2003). Plasma levels are, like those of CNTF but unlike those of leptin, low in obese or type 2 diabetic subjects (Krabbe et al. 2007; Araya et al. 2008). BDNF mRNA and protein increase in skeletal muscle after exercise. Muscle BDNF seems to act mostly in an autocrine or paracrine fashion because it is elevated for at least 24 h after exercise, whereas serum levels increase for only 2 h. Serum BDNF may also be derived from platelets (Matthews et al. 2009). The thermogenic effect of BDNF has been attributed to a centrally mediated stimulation of sympathetic activity (Nonomura et al. 2001) but, like CNTF, it also acts directly on skeletal muscle to stimulate fatty acid oxidation via an AMPK-dependent mechanism (Matthews et al. 2009). This raises the possibility that a peripherally acting mimetic might be of value in the treatment of diabetes.

4.10   Adiponectin

Adiponectin is a 30 kDa protein secreted by adipocytes, which circulates at high concentrations (0.5–30 µg/ml) in plasma. Large adipocytes secrete less adiponectin than small adipocytes, and plasma adiponectin concentrations are generally lower in obese and insulin-resistant subjects than in lean and insulin-sensitive subjects (Kadowaki and Yamauchi 2005; Shetty et al. 2009). Adiponectin circulates as a variety of multimeric complexes. Until recently, it was believed that the 12- to 36-mer high molecular weight form correlates best with insulin sensitivity, but this is now disputed (Almeda-Valdes et al. 2010; Elisha et al. 2010). Various polymorphisms of the adiponectin gene are associated with the metabolic syndrome (Shetty et al. 2009).

Administration of full-length adiponectin improves insulin sensitivity in mouse models of type 2 diabetes (Kadowaki and Yamauchi 2005). A globular form of adiponectin, produced by its proteolytic cleavage, reduced weight gain in diet-induced obese mice without decreasing energy intake. Fatty acid oxidation increased in skeletal muscle (Fruebis et al. 2001). However, others have reported that high fat feeding leads to resistance to the stimulation by adiponectin of fatty acid oxidation in skeletal muscle (Mullen et al. 2009). Overexpression of adipo-nectin in Lep^{ob}/Lep^{ob} mice improved glucose homeostasis. Subcutaneous adipose tissue was greatly increased, suggesting that, like the thiazolidinedione drugs (see below), adiponectin promotes the storage of fat where it can do little harm (Shetty et al. 2009). Both central and (in Lep^{ob}/Lep^{ob} mice) systemic administration of adiponectin increased oxygen consumption relative to body weight, but not per animal, after 3 days. More significantly, adiponectin increased oxygen consumption in Lep^{ob}/Lep^{ob} mice after only 3 h, at which time body weight had presumably changed little. Moreover, body temperature increased (Qi et al. 2004).

The thiazolidinedione drugs increase plasma adiponectin concentrations (Coletta et al. 2009), even though they promote, rather than reduce, adiposity. This may be partly explained by the fact that these drugs increase adiponectin secretion from omental but not subcutaneous adipocytes (Motoshima et al. 2002), whereas it is subcutaneous fat stores that increase (Yang and Smith 2007). Indeed, as indicated above, the increased secretion of adiponectin may play a role in the hyperplasia of subcutaneous adipocytes (Shetty et al. 2009). A further paradox is that the insulin-sensitising effect of the thiazolidinedione drugs is partly due to increased secretion of adiponectin (Kubota et al. 2006; Banga et al. 2009; Shetty et al. 2009), but these drugs do not increase whole-body energy expenditure. They do, however, increase the expression of genes involved in mitochondrial function and fat oxidation in skeletal muscle (Coletta et al. 2009). This would be expected to result in improved insulin sensitivity in muscle – the dominant influence on whole-body insulin sensitivity. In other words, fat balance in skeletal muscle is more important than whole-body energy balance in determining whole-body insulin sensitivity.

Adiponectin signals via at least three receptors, AdipoR1 (PAQR1; progestin-adipoQ receptor1) and AdipoR2 (PAQR2), which are most highly expressed in muscle and liver, respectively (Yamauchi et al. 2003), and the more recently described receptor PAQR3 (Gonez et al. 2008). In liver, AdipoR1 is more tightly linked to activation of AMPK, whereas AdipoR2 is more tightly linked to activation of peroxisome proliferator-activated receptor α (PPARα). Knockout of either receptor appeared to cause insulin resistance (Yamauchi et al. 2007). However, in another study, AdipoR1 knockout mice had increased adiposity associated with decreased glucose tolerance, physical activity, and energy expenditure, whereas the opposite phenotype was found in AdipoR2 knockout mice (Bjursell et al. 2007).

Although adiponectin receptors, especially AdipoR1, are potential targets for anti-diabetic drugs, no directly acting, non-peptide agonists have been identified. It is unlikely that adiponectin itself could be used as a drug because of the quantities required, the need to control the proportions of its various forms and its short half-life (Shetty et al. 2009). It might perhaps be possible to use a modified form of adiponectin. Osmotin, a PR-5 family of plant defence protein, seems to activate AMPK via adiponectin receptors (Narasimhan et al. 2005; Yamauchi and Kadowaki 2008). The patent literature suggests that there is interest in osmotin and in drugs that increase adiponectin secretion or receptor expression, rather than in small molecular weight agonists.

5 Targets in Lipid Metabolism

Inhibitors of fatty acid and triglyceride synthesis often alter energy balance by promoting fat oxidation, though some also inhibit food intake. The molecular mechanism that links inhibition of lipid synthesis with stimulation of fat oxidation is best established for inhibitors of acetyl-CoA carboxylase (ACC).

5.1 Acetyl-CoA Carboxylase

ACC produces malonyl-CoA from acetyl-CoA. ACC1 is a cytosolic enzyme and is the predominant ACC in tissues that have a high capacity for fatty acid synthesis. ACC2 is associated with mitochondria and is the predominant ACC in tissues that have a high capacity for fatty acid oxidation. It therefore appears that the malonyl-CoA produced by ACC1 is used by fatty acid synthase to initiate the building of fatty acid chains, whereas malonyl-CoA produced by ACC2 acts primarily to prevent the transfer of fatty acids into mitochondria by carnitine palmitoyl transferase-1 (Wakil and Abu-Elheiga 2009). Some workers disagree with this strict distinction, however (Harada et al. 2007).

ACC2 knockout mice are lean despite consuming more food than wild-type mice. They are protected from diet-induced obesity and insulin resistance (Abu-Elheiga et al. 2003; Choi et al. 2007). Knocking out ACC1 in all tissues of the mouse was lethal to the embryo (Abu-Elheiga et al. 2005). Knocking out ACC1 in liver only, which was not lethal, did not prevent lipogenesis in liver in one study because ACC2 was upregulated and took over the role of ACC1 in lipogenesis (Harada et al. 2007). In another study, lipogenesis did decrease in mice fed on a fat-free diet despite upregulation of ACC2 and lipogenic enzymes (Mao et al. 2006). Chronic suppression of ACC1 in a β-cell line impaired insulin secretion (Ronnebaum et al. 2008). ACC2 therefore appears to be a better drug target than ACC1.

Selective inhibitors of ACC2 have been claimed in patents (Corbett and Harwood 2007; Corbett 2009), but only non-selective inhibitors, including the anti-fungal compound soraphen, have been described in the scientific literature. These non-selective inhibitors inhibit fatty acid synthesis and promote fatty acid oxidation, but increased energy expenditure has not been detected (Harwood et al. 2003; Schreurs et al. 2009).

AMPK phosphorylates and thereby inactivates ACC2, and so drugs that activate AMPK (see below) may have similar thermogenic and anti-diabetic effects to ACC2 inhibitors. With either approach, however, it may be important that a drug does not penetrate the hypothalamus because elevation of AMPK activity or the concentration of malonyl-CoA in the hypothalamus is associated with increased energy intake and, in the case of elevated malonyl-CoA, decreased fatty acid oxidation (Kola 2008; Lane et al. 2008).

5.2   Fatty Acid Synthase

The fatty acid synthase inhibitors cerulenin and C75 were originally claimed to inhibit feeding by inhibiting the hypothalamic enzyme and increasing the concentration of malonyl-CoA (Loftus et al. 2000). Subsequent studies showed that C75 also promotes fat oxidation, partly by directly activating carnitine O-palmitoyl transferase-1 (CPT-1) in the periphery (Thupari et al. 2002; Rohrbach et al. 2005; Tu et al. 2005). Cerulenin does not activate CPT-1 directly, but it increased sympathetic activity 3–5 h after intraperitoneal injection and this led to increased CPT-1 activity (Aja et al. 2008). There is surprisingly little information to support the potential of fatty acid synthase inhibitors as anti-diabetic agents, however.

5.3   Stearoyl-CoA Desaturase-1

Stearoyl-CoA desaturases (SCDs) introduce a double bond into palmitoyl- and stearoyl-CoA at the Δ9 position to form palmitoyl- and oleoyl-CoA, respectively. There are at least four SCD isoenzymes in mice and two (SCD1 and SCD5) in humans. Of these, SCD1 has generated by far the most interest as a target for inhibitors that might be used in the treatment of diabetes and obesity (Flowers and Ntambi 2008; Popeijus et al. 2008).

The interest in SCD1 inhibitors arose from the discovery that SCD1 knockout mice have reduced adiposity, increased insulin sensitivity, and are resistant to diet-induced obesity. This resistance is due to increased energy expenditure (Cohen et al. 2002; Ntambi et al. 2002). Recently, the SCD1 inhibitor has been reported to improve insulin sensitivity in rat models of insulin resistance (Issandou et al. 2009). Various molecular mechanisms have been proposed to link inhibition of SCD1 with thermogenesis and fatty acid oxidation. These include inhibition of ACC2 by saturated fatty acyl-CoAs, which are more potent inhibitors than unsaturated fatty acyl-CoAs (Cohen et al. 2002), and activation of AMPK, which also inhibits ACC2. It was suggested that activation of AMPK might be due to the mice being more active (Dobrzyn et al. 2004). The possibility that SCD1 knockout mice are more active and have increased energy expenditure due to increased sympathetic activity should be explored, because Lou/C rats, which have low SCD1 activity, are hyperactive and have raised sympathetic activity (Perrin et al. 2003; Soulage et al. 2008).

There is evidence that SCD1 is associated with acetyl-CoA:diacylglycerol acyltransferase (DGAT) 2, which together with DGAT1 (see Sect. 5.4 below) catalyses the final step of triglyceride synthesis. Local desaturation of fatty acids at their site of esterification may promote triglyceride synthesis (Man et al. 2006). Prevention of triglyceride synthesis might promote fatty acid oxidation. The concentration of ceramide is reduced in skeletal muscle of SCD1 knockout mice (Dobrzyn et al. 2005), providing a possible link between increased fatty acid oxidation and improved insulin sensitivity. Intriguingly, deficiency of SCD1 did not improve insulin sensitivity in Lep^{ob}/Lep^{ob} despite causing these mice to have a lower body weight than "normal" Lep^{ob}/Lep^{ob} mice (Miyazaki et al. 2009).

There may be a more prosaic explanation for increased energy expenditure in SCD1 knockout mice. The validity of SCD1 as a drug target has been questioned by a report that SCD1 knockout mice are driven to increase their energy expenditure because disruption of their epidermal lipid barrier causes them to lose heat rapidly (Binczek et al. 2007). They are able to maintain their body temperature by increasing their sympathetic activity when they are maintained at normal animal house temperatures, but at 4°C they become hypothermic (Lee et al. 2004). Skin-specific deletion of SCD1 is enough to increase energy expenditure, cause intolerance to cold, and protect mice from diet-induced obesity (Sampath et al. 2009).

There are other issues as well. SCD1 protects pancreatic β-cells from lipoapoptosis (Busch et al. 2005) and β-cell loss is hastened in BTBR Lep^{ob}/Lep^{ob} mice if they lack SCD1 (Flowers et al. 2007). SCD1 protected the rat L6 muscle cell line from fatty acid-induced insulin resistance (Pinnamaneni et al. 2006), so inhibiting the enzyme does not seem logical. In humans, in contrast with mice, metabolic disease tends to be associated with decreased SCD1 mRNA (Popeijus et al. 2008). Inhibition of SCD1 can promote atherosclerosis (Brown and Rudel 2010). Lastly, there have been descriptions at meetings of skin and eye lesions in animals treated with SCD1 inhibitors, though this may not be a problem with all SCD1 inhibitors (Uto et al. 2010).

There may be potential for SCD1 inhibitors that are directed towards the liver. Liver-specific deficiency of SCD1 protected mice from high carbohydrate (but not high fat) diet-induced obesity and hepatic steatosis. Decreased monounsaturated fatty acid production prevented the upregulation of lipogenic genes in these mice (Flowers and Ntambi 2009). Treatment of rodents with SCD1 antisense oligonucleotides, which reach the liver but not other organs, including the skin or β-cells, lowered plasma glucose and insulin and reversed hepatic insulin resistance (Jiang et al. 2005; Gutierrez-Juarez et al. 2006). Moreover, SCD1 antisense oligonucleotides increased fatty acid oxidation in mouse hepatocytes (Jiang et al. 2005). Liver-selective small molecular weight SCD1 inhibitors have been described (Koltun et al. 2009), but it remains to be seen whether they have potential in the treatment of diabetes.

5.4 Acetyl-CoA:Diacylglycerol Acyltransferase (DGAT)

The final step of triglyceride synthesis is catalysed in rodents by the enzymes DGAT1 and DGAT2. In higher mammals, including humans, monoacylglycerol acyltransferase 3 (MGAT3) is also able to perform this role. MGAT3 and DGAT2 belong to the same protein family, but MGAT3 is not found in rodents (Zammit et al. 2008).

Interest in DGAT1 as a target for drugs arose from the finding that DGAT1 knockout mice are resistant to diet-induced obesity and insulin resistance (Smith et al. 2000; Chen et al. 2002a). The male mice had increased energy expenditure (Wang et al. 2007), although in the early papers this was only demonstrated when energy expenditure was expressed relative to body weight (Smith et al. 2000; Chen et al. 2003). The DGAT1 knockout mice displayed increased physical activity and capacity for fatty acid oxidation (increased uncoupling protein-1 expression) in brown adipose and other tissues. If the capacity for fatty oxidation is increased in other tissues, this might explain why the concentration of diacylglycerol – the substrate of DGAT – paradoxically tended to be low in white adipose tissue, skeletal muscle, and especially liver of DGAT1 knockout mice (Yu and Ginsberg 2004; Chen and Farese 2005).

Surprisingly, overexpression of DGAT1 in white adipose tissue of C57Bl/6 mice resulted in a greater susceptibility to diet-induced obesity but not to impaired glucose tolerance (Chen et al. 2002b). However, a similar modification to FVB mice did not cause obesity but did cause insulin resistance (Chen et al. 2005). This may be because FVB mice are resistant to adipose tissue expansion. It could be that increased triglyceride synthesis in fat reduced uptake of VLDL and the triglyceride was redistributed to liver.

Even more surprisingly, overexpression of DGAT1 specifically in muscle protected mice from high fat diet-induced insulin resistance. This was consistent, however, with the unexpected finding that muscle diacylglycerol and ceramide contents were decreased (Liu et al. 2007). The most likely explanation for these findings is that fatty acids that enter the muscle are rapidly converted to triglycerides, lowering levels of fatty acid metabolites, such as diacylglycerol and ceramide, that cause insulin resistance. This is analogous to what happens in the muscles of athletes (Liu et al. 2007). It is interesting that when DGAT1 knockout mice were fed on a chow diet, their glucose tolerance was no better than that of wild-type mice (Wang et al. 2007). This might be because there was a limited capacity to store fatty acids as triglycerides, a condition analogous to, but less extreme than, lipodystrophy. These findings show that genetic manipulation of DGAT1 does not make the case for DGAT1 inhibitors as treatments for diabetes as clearly as at first thought.

It has been suggested that DGAT1 plays a greater role than DGAT2 in triacylglycerol synthesis in the lumen of the endoplasmic reticulum. Thus, selective inhibition of DGAT1 might inhibit VLDL secretion and cause hepatic steatosis. On the other hand, a similar mechanism operating in enterocytes might delay the rate at which chylomicrons appear in the plasma following a meal (Zammit et al. 2008). This might lead to a reduction in liver triglycerides.

Reports have begun to appear on DGAT1 inhibitors, but their emphasis has been on reduced weight gain and serum and liver triglyceride concentrations, rather than insulin sensitivity (Zhao et al. 2008; Birch et al. 2010; King et al. 2010). One inhibitor reduced body weight in diet-induced obese mice without reducing food intake, but it was not demonstrated that it increased locomotor activity or energy expenditure (Zhao et al. 2008). This raises the possibility that raised energy expenditure and locomotor activity in the DGAT1 knockout mouse are due to the absence of DGAT1 in the brain. Thus, the value of DGAT1 inhibitors in the treatment of diabetes has yet to be established in rodents.

6 Other Intracellular Targets

6.1 AMP-Activated Protein Kinase

AMPK is a heterotrimeric enzyme, potentially formed by 12 different combinations of monomers, which generally promotes metabolic pathways involved in energy provision, such as glucose uptake and fatty acid oxidation, whilst inhibiting those involved in energy storage, such as fatty acid, triglyceride, and cholesterol synthesis. It does this both by acutely phosphorylating key enzymes, such as ACC2 (whose activity it inhibits), and by altering gene expression. Chronic AMPK activation promotes mitochondrial biogenesis, and AMPK plays a role in mitochondrial biogenesis in response to exercise. Downstream mechanisms that promote mitochondrial biogenesis include both acute activation, as a consequence of enhanced SIRT1 activation (Canto et al. 2009), and increased expression of the transcription factor peroxisome proliferator-activated receptor gamma coactivator 1-alpha (PGC-1α; see Sect. 6.3). The large and rapidly expanding literature on the role that AMPK plays in energy balance and glucose homeostasis is summarised in many excellent recent reviews (Hegarty et al. 2009; Viollet et al. 2009a, b; Zhang et al. 2009; Zhou et al. 2009; Fogarty and Hardie 2010). AMPK is the subject of a chapter by Violett and Andreelli (2011).

The role of AMPK in promoting energy providing pathways is mostly consistent with it being a suitable target for anti-obesity and anti-diabetic targets. This, however, depends upon the fuels mobilised also being used. Generalised activation of AMPK in the hypothalamus promotes energy supply in that it promotes feeding (Zhang et al. 2009), but this is the opposite of what is needed for obesity and diabetes. It is especially important for diabetic patients that when glucose is mobilised it is used. Activation of AMPK in skeletal muscle inhibits glycogen synthase (Hegarty et al. 2009) and in dogs the AMPK activator 5-aminoimidazole-4-carboxamide-1-β-D-ribofuranoside (AICAR) stimulates hepatic glucose output (Camacho et al. 2005; Pencek et al. 2005). Whilst its effects on glycogen synthesis and degradation tend to promote hepatic glucose output, AMPK suppresses gluco-neogenesis (Viollet et al. 2009b). A caution is that many of these statements are based on studies using AICAR, which may act by mechanisms other than activation of AMPK (see below).

One effect of AMPK activation that does not promote energy provision is that it inhibits lipolysis in adipose tissue (Daval et al. 2006), but this is beneficial for an anti-diabetic drug. On the other hand, activation of AMPK in β-cells inhibits glucose-stimulated insulin secretion. It has been suggested that this might help to protect the β-cell in the long term (Viollet et al. 2009b), but there is evidence that activation of AMPK might actually be toxic to β-cells in some situations (Riboulet-Chavey et al. 2008; Ryu et al. 2009).

A number of the other targets linked to AMPK have already been described in this chapter. Activation of AMPK not only inhibits ACC2 acutely, but also reduces the expression of both ACC2 and fatty acid synthase. AMPK also activates SIRT1 (see Sect. 6.3). Metformin acts primarily by activating AMPK, probably by raising the concentration of AMP. Leptin, sympathetic activation, and adiponectin increase AMPK activity in skeletal muscle and liver (Minokoshi et al. 2002; Yamauchi et al. 2002, 2007). In addition, the thiazolidinedione PPARγ agonists increase AMPK activity in skeletal muscle, partly by promoting adiponectin secretion and partly acting directly (LeBrasseur et al. 2006). The mechanisms by which these agents increase AMPK activity are not established. It is possible that perturbation of mitochondrial function with consequent elevation of the concentration of AMP, rather than PPARγ agonism, mediates the effect of the thiazolidinediones (Brunmair et al. 2004). The effects of adiponectin and leptin on glucose homeosta-sis are partly dependent on activation of AMPK (Hegarty et al. 2003). The benefi-cial effects of metformin appear to be primarily dependent on AMPK (Hardie 2008) and its upstream kinase LKB1 (Shaw et al. 2005), though part of its action may be independent of AMPK (Saeedi et al. 2008). Elevation of the concentration of AMP may be a consequence of inhibition of complex I of the mitochondrial respiratory chain (Owen et al. 2000).

A number of strains of genetically modified mice have been created in which the α_1- or α_2-catalytic or γ-AMP binding subunits of AMPK have been inactivated or constitutively activated (Viollet et al. 2009a). Of note are mice that express an inactive α_2-subunit in muscle. These mice developed impaired whole-body glucose tolerance and skeletal muscle insulin resistance (Fujii et al. 2008). Liver-specific deletion of the α_2-subunit resulted in mild hyperglycaemia and glucose intolerance (Andreelli et al. 2006), whilst adenovirus-mediated expression of a constitutively active form of the α_2-subunit protected mice from diabetes (Foretz et al. 2005).

Various compounds have been described that activate AMPK directly or are metabolised to compounds that activate AMPK directly. The best known is AICAR, which improves glucose homeostasis in various rodent models of diabetes (Hegarty et al. 2009). The abbreviation AICAR is confusing because it is used in the literature on purine synthesis to describe the ribotide, but in the literature on metabolism to describe the riboside. The ribotide (5-amino-4-imidazolecarboxa-mide ribonucleotide or 5-aminoimidazole-4-carboxamide-1-β-D-ribofuranotide) is called ZMP in the literature on metabolism. The riboside is taken up by cells and phosphorylated to the ribotide. One problem with ZMP is that it mimics actions of AMP other than activation of AMPK. In particular, it stimulates glycogen phosphorylase, which exacerbates blood glucose control, and inhibits fructose-1,6-bisphosphatase (FBPase), which inhibits gluconeogenesis and improves blood glucose control (Zhou et al. 2009). There may also be protein kinases other than AMPK that are activated by AICAR because it increased ACC2 phosphorylation and fatty acid oxidation in mice that expressed inactive AMPK in skeletal muscle (Dzamko et al. 2008). Another problem, apparently ignored by other authors, is that by increasing the cellular ZMP concentration, AICAR (the riboside) may increase the extracellular concentration of adenosine. This is how methotrexate is believed to reduce inflammation in rheumatoid arthritis (Cronstein 2005). Adenosine can both increase and decrease insulin sensitivity, probably depending upon whether it acts via A1 or A2 receptors (Budohoski et al. 1984; Derave and Hespel 1999; Thong et al. 2007).

The thienopyridone A769662 activates AMPK directly but does not activate glycogen phosphorylase or inhibit FBPase. However, its utility may be compromised by it activating glucose uptake in skeletal muscle through a phosphatidylinositol 3-kinase-dependent pathway, independent of AMPK activation (Treebak et al. 2009). It also inhibits Na^+/K^+-ATPase (Benziane et al. 2009). Like AICAR, it reduced the blood glucose concentration in Lep^{ob}/Lep^{ob} mice, but it is not clear whether any of this effect was independent of its effect on body weight (Cool et al. 2006). Other activators of AMPK have been described, but most of these are not direct activators and many may activate AMPK by raising the tissue AMP concentration (Zhou et al. 2009). The polyphenol resveratrol, which is found in the skin of grapes, causes weight loss and improves glucose homeostasis in wild-type but not in AMPKα knockout mice, but how it activates AMPK is unclear. It was claimed that it increases energy expenditure, but energy expenditure was expressed relative to body weight, possibly after many weeks of treatment, and it does not appear to have affected energy expenditure per animal (Um et al. 2010). Resveratrol has a number of possible mechanisms of action (Pirola and Frojdo 2008). Nootkatone, a constituent of grapefruit, may activate AMPK by activating upstream AMPK kinases (Murase et al. 2010). A single dose of nootkatone caused a rapid increase in energy expenditure in mice (reminiscent of activators of the sympathetic nervous system). It is noticeable, by contrast, that acute thermogenic effects of directly acting AMPK activators have not been reported, but both A769662 and metformin reduced the respiratory exchange ratio (RER) of rats for about 3 h, after which the ratio increased slightly for about 3 h (Cool et al. 2006). The reduction in RER implies depletion of fat due to increased fat oxidation or decreased lipogenesis.

Ultimately, the main problem with direct activators of AMPK is that AMPK has many roles, extending beyond major metabolic pathways. It may be that drugs that act at specific isoforms of the α-subunit or other sophisticated strategies might reduce unwanted side effects. Drugs targeted to skeletal muscle, or more realistically, the liver are most likely to succeed (Zhang et al. 2009; Zhou et al. 2009).

6.2 Peroxisome Proliferator-Activated Receptor β/δ

A number of transcription factors have been identified whose activation or over-expression promotes mitochondrial biogenesis. Skeletal muscle develops more type 1 (oxidative) fibres, and adipose tissue develops more adipocytes that have the features of brown adipocytes, such as expression of uncoupling protein-1. The best known of these transcription factors is PGC-1α. Most of them do not have natural small molecule regulators, however, and they have not been shown to be good targets for drug.

PPARβ/δ, on the other hand, is a transcription factor that is also a hormone receptor and it can be activated by small molecules. Overexpression of PPARβ/δ in skeletal muscle or adipose tissue increases the capacity of these tissues for fatty acid oxidation and protects mice from diet-induced obesity and glucose intolerance (Wang et al. 2003, 2004). Conversely, PPARβ/δ knockout mice are more susceptible than wild-type mice to diet-induced glucose intolerance (Lee et al. 2006) and in one study (Wang et al. 2003), though not others (Peters et al. 2000; Lee et al. 2006), obesity.

The best described selective PPARδ agonists are GW501516 and its close structural analogue GW042 (also known as GW610742 or GW610742X). GW501516 has been taken as far as phase II clinical studies but is no longer being developed (Billin 2008). Some caution must be taken when interpreting rodent data on these compounds because, although they are of the order of 1,000-fold selective as agonists of human PPARβ/δ compared to PPARα or PPARγ, they are less selective (about 50-fold in the case of GW501516) for murine PPARβ/δ (Oliver et al. 2001; Sznaidman et al. 2003). This may explain why oral administration of both compounds causes hepatomegaly in mice, a known effect of PPARα agonists (Tanaka et al. 2003; Harrington et al. 2007; Faiola et al. 2008). Moreover, GW501516 activated AMPK and glucose uptake in human primary skeletal muscle cells by a PPARβ/δ- and PPARα-independent mechanism (Kramer et al. 2007). Interestingly, the only evidence that GW501516 increases energy expenditure is that oxygen consumption relative to body weight was decreased after 35–39 days treatment, but body weight was reduced at this time. There was a trend to reduced food intake, but this was not statistically significant (Tanaka et al. 2003), so it is likely that weight loss was partly due to increased energy expenditure, even though this was not marked. In my colleagues' unpublished work (D Hislop and M A Cawthorne), GW501516 (10 mg/kg, po for 2 weeks) did not increase energy expenditure per animal detectably in Lep^{ob}/Lep^{ob} mice, but it improved oral glucose tolerance. Similar results were obtained with a compound that is more selective for murine PPARβ/δ, but it was also found that the mice were sensitised to the thermogenic effect of a β$_3$-adrenoceptor agonist (R A Ngala et al, unpublished). Published work also shows that GW501516 improves glucose tolerance or lowers blood glucose and plasma insulin in various animal models of diabetes and insulin resistance (Tanaka et al. 2003; Lee et al. 2006). GW501516 reduced fasting glucose and insulin in a phase II clinical study slightly, but these effects were not statistically significant (Billin 2008).

At least two other PPARβ/δ agonists remain in clinical development according to company Web sites. One of these is the Metabolex compound MBX-8025, which improved insulin sensitivity in a phase II study. There was a trend to decreased weight consumption and body fat. The other is the Kalypsys compound KD-3010, which has completed a phase 1b study. Much of the focus with PPARβ/δ agonists has been on dyslipidaemia (Billin 2008).

Finally, it should be noted that regulatory authorities regard PPARβ/δ agonists as potential carcinogens and the hurdles to their development are considerable (Billin 2008).

6.3   Sirtuin1

The sirtuins, of which there are at least seven mammalian homologues, are histone deacetylases that are involved in gene silencing (Finkel et al. 2009). Sirtuin1 (SIRT1) is a possible target for drugs for the treatment of diabetes. SIRT1 deacetylates not only histones but also various transcription factors, including PGC-1α. Deacetylation activates PGC-1α and promotes mitochondrial biogenesis (Liang et al. 2009). Some of the effects of AMPK activation, in particular its effect on mitochondrial biogenesis, may be mediated by SIRT1, because by promoting the oxidation of NADH and increasing the NAD^+ concentration, AMPK enhances SIRT1 activity (Canto et al. 2009).

The current focus is on activators rather than inhibitors of SIRT1 (see below). This is consistent with a report that moderate generalised overexpression of SIRT1 improved glucose tolerance in diet-induced obese and $LepR^{db}/LepR^{db}$ mice. Adiponectin levels increased and, possibly in consequence, hepatic glucose production decreased (Banks et al. 2008). Body weight, body composition, food intake, and oxygen consumption were largely unchanged. Body weight and body composition were also unchanged by overexpression of SIT1 in normal, chow-fed mice, but both food intake and oxygen consumption were decreased. This illustrates how energy turnover can change without there being any alteration in body weight.

In contrast with this report on overexpression of SIRT1, two studies on the knockdown of SIRT1 specifically in liver suggest that inhibitors of SIRT1 should be of value in the treatment of diabetes. Both studies found that knockdown of hepatic SIRT1 decreased fasting blood glucose and improved whole-body insulin sensitivity (Rodgers and Puigserver 2007; Erion et al. 2009). In the first of these studies, SIRT1 was also overexpressed in liver; this reduced glucose tolerance, but the effect was prevented if PGC-1α was overexpressed. The key to understanding these various findings may be that SIRT1 by activating PGC-1α promotes hepatic gluconeogenesis (Rodgers and Puigserver 2007). Thus, the effects of SIRT1 in liver (increased gluconeogenesis) and in skeletal muscle (increased capacity for fatty acid oxidation) may have opposing effects on glucose homeostasis.

Further support for developing activators rather than inhibitors is that activation of SIRT1 in β-cells enhances glucose-stimulated insulin secretion and β-cell survival (Liang et al. 2009). In addition, low SIRT1 mRNA in adipose tissue was associated with impaired stimulation of energy expenditure by insulin, low expression of mitochondrial genes in adipose tissue, and low levels of mitochondrial DNA in skeletal muscle (Rutanen et al. 2010).

Two reports have described three compounds, claimed to be SIRT1 activators that, at high dose levels, improve insulin sensitivity in animal models of insulin resistance. The authors of the first report were mostly from Sitris Pharmaceuticals (Milne et al. 2007). The second report focuses on one of the compounds (SRT1720) described in the first report; only two of the ten authors were from Sitris (Feige et al. 2008). The first report finds increased "mitochondrial capacity" in SIRT1 activator-treated mice, but no effects on body weight. The second, by contrast, describes prevention of diet-induced obesity by SRT1720 without reduced food intake. Energy expenditure was increased relative to body weight, but since all the weight lost appears to have been fat, this is not an appropriate way to express the data (see Sect. 7). Other effects of the SIRT1 activator used in the second study were increased numbers of type 1 fibres in gastrocnemius (but not soleus) muscle, increased running endurance (as found with PPARδ activation), and increased capacity for fat oxidation in brown adipocytes. The expression of type 2 deiodinase and other genes for proteins that control energy expenditure was increased in brown adipose tissue, but surprisingly expression of uncoupling protein-1 mRNA was not increased (Feige et al. 2008). The relevance of these reports to SIRT1 has, however, been contested; the identification of these compounds as SIRT1 activators appears to have been an artefact of the assay method. The compounds interact with multiple receptors, enzymes, transporters, and ion channels, and in any event SRT1720 neither lowered plasma glucose, nor improved mitochondrial capacity in mice fed on a high fat diet. Resveratrol (see Sect. 6.1), which has also been claimed to activate SIRT1, was also shown not to act directly on the enzyme (Pacholec et al. 2010). Nevertheless, resveratrol-stimulated glucose uptake in L6 muscle cells was prevented both by the sirtuin inhibitor nicotinamide and by the AMPK inhibitor Compound C (Breen et al. 2008).

7 Perspectives and Implications for Drug Discovery and Development

Figure 1 illustrates how most of the targets that have been described are linked, especially to adiponectin, AMPK, and the sympathetic nervous system. Another link, and a recurring theme of this chapter, is that metabolic targets for insulin-sensitising drugs are generally associated with an increased capacity for fat oxidation in skeletal muscle and liver. This is not fat oxidation driven (pushed) by an increased supply of fatty acids, but rather an increased capacity to draw (pull) fat into tissues for oxidation. One possible consequence is that the plasma non-esterified fatty acid concentration may fall. A more likely consequence is that, in those tissues in which the capacity for fatty acid oxidation is increased, the concentrations of lipid metabolites that cause insulin resistance fall and glucose uptake becomes more sensitive to insulin. It is not essential that the rate of fatty acid oxidation increases significantly; what does matter is that the "pull" for fatty acid oxidation exceeds the "push" of fatty acid supply, so that the concentration of these key lipid metabolites fall.

7.1 Detection of Thermogenesis

In fact, the target that has given us the drugs that are best known as "insulin sensitisers", PPARγ, has not given us drugs that increase whole-body fat oxidation and thermogenesis. Rosiglitazone and pioglitazone increase fat stores, but crucially, this fat is safely stored away in subcutaneous adipocytes and does not satisfy the demands of skeletal muscle and (especially in rodents) brown adipose tissue, in which the capacity for fat oxidation is increased (Benton et al. 2008; Festuccia et al. 2008). These tissues respond by becoming more sensitive to insulin and using glucose instead of fat. Metformin has a weaker insulin-sensitising effect than the thiazolidinediones and its effects are transient (Pavo et al. 2003; Karlsson et al. 2005; Basu et al. 2008). It is not obviously thermogenic, but there is some evidence that it increases whole-body fat oxidation, though only transiently (Avignon et al. 2000; Cool et al. 2006; Braun et al. 2008). Genetic modification of mice so that β_3-adrenoceptors were expressed in brown but not white adipocytes prevented obvious thermogenesis in response to a β_3-adrenoceptor agonist, but it was not stated whether the β_3-adrenoceptor agonist could still lower respiratory quotient or (with repeated administration in diet-induced obesity) reduce insulin sensitivity (Grujic et al. 1997). I suggest that these effects would still have occurred.

All this means that non-anorectic insulin-sensitising drugs need not be obviously thermogenic, nor need genetically modified mice that model the potential of the drug have markedly higher energy expenditure than their wild-type counterparts. Mice that are treated with a prototype drug or genetically modified may be leaner, but it may be very difficult to demonstrate that their energy expenditure is raised. For example, the loss of 3 g of fat, which should be easily detectable, corresponds to the loss of 27 kcal. This is about 5% of what a mouse eats in 5 weeks, and a 5% increase in energy expenditure is not easy to detect. Speakman has made this point more eloquently (Speakman 2010). Some genetically modified mice, such as ACC2 or 11βHSD knockout mice, are lean despite raised energy intake, leaving no doubt that fat loss must be due to raised energy expenditure, even if this is not obvious from direct measurement of energy expenditure.

In those cases in which tool compounds cause an acute and easily detectable increase in energy expenditure, this is usually because sympathetic activity is raised (or mimicked in the case of β-adrenoceptor agonists). The reason why sympathomimetic drugs have such a clear thermogenic effect is probably that they mobilise fuels as well as promote their combustion. Energy expenditure subsides to or below the control level between doses of the compound, or energy intake increases. If one of these things did not happen, the animal would soon lose all its fat. Thus, obese mice lose more lipid after 28 days of treatment with some β_3-adrenoceptor agonists than their lean littermates have in their entire bodies (Arch and Ainsworth 1983). The lean animals respond less to β_3-adrenoceptor agonists and when the dose is high they increase their food intake (Arch et al. 1984; Clapham et al. 2001).

7.2 Manipulation of Energy Expenditure Data

The problem of detecting small increases in energy expenditure sufficient to cause weight loss does not stop researchers from claiming that they can detect raised energy expenditure, even when sympathetic activity or energy intake is not raised. The device that they use is to wait until the tool compound or the genetic modification has caused weight loss and then express energy expenditure relative to body weight or body weight$^{0.75}$. If energy expenditure for humans were adjusted in the same way, we would have to conclude that most obese humans have a low metabolic rate, which nobody believes. Body weight$^{0.75}$ is a term that was found to normalise energy turnover between species but has never been validated as a means of normalising energy turnover between lean and obese members of the same species (Arch et al. 2006). Almost invariably the leaner animals have increased energy expenditure relative to body weight or body weight$^{0.75}$. This is hardly surprising: fat contributes about a fifth or less of what lean tissue contributes to energy expenditure. Amazingly, the same correction is not applied to energy intake. If it was, one would find that in most cases energy intake relative to body weight or body weight$^{0.75}$ is also reduced in the leaner animals. Relative to body weight$^{0.75}$, it is not just energy expenditure that is similar in mice and elephants – energy intake is as well.

Various workers have objected to the use of this device and have suggested alternative approaches (Himms-Hagen 1997; Packard and Boardman 1999; Toth 2001; Arch et al. 2006; Butler and Kozak 2010). Authors, referees, and editors appear, however, to prefer to "turn a blind eye" to this issue. This author recommends that manipulated energy expenditure data are treated sceptically unless the mean body weights of the groups being compared are very similar. It seems reasonable to normalise energy expenditure relative to body weight within a group in order to reduce within group variance, but before comparing between groups the energy expenditure of each mouse within a group should then be converted back to whole animal energy expenditure as if it had the body weight of the average for the group. It does not seem reasonable, however, to normalise relative to body weight for the purpose of comparing groups of different body weights. Normalisation relative to lean body mass, if measured, is more acceptable because basal metabolic rate is largely determined by lean body mass, but even this is not ideal if the intercept of the relationship between energy expenditure and lean body mass is not zero (Speakman 2010).

7.3   Translation from Rodents to Humans

If it can be shown without manipulation of data that whole-body energy expenditure is raised in rodents, what are the implications for humans? First, it is possible that if thermogenesis in rodents is due to increased sympathetic activity, it will not be apparent in humans (as in the case of rimonabant). Rodents have a far higher capacity for thermogenesis than humans due to their greater amount of brown adipose tissue. They are usually studied below thermoneutrality so that effects of compounds or genetic modification on heat loss alter their requirements for thermogenesis. This may be the main cause of thermogenesis in SCD1 knockout mice (see Sect. 5.3). Humans, by contrast, usually live and are studied near thermoneutrality. Secondly, if, as in the case of sibutramine, sympathetic activity is raised in humans, there is the risk that blood pressure will be raised.

And if the weight loss occurs in rodents, whether or not due to detectable thermogenesis, what are the implications for humans? Should a drug that causes weight loss in rodents be developed initially for the treatment of obesity – the traditional view of thermogenic drugs? Probably not. Two arguments suggest that type 2 diabetes should be the first indication. First, exercise and some thermogenic agents (notably β_3-adrenoceptor agonists) are capable of improving insulin sensitivity without eliciting weight loss. Stimulation of fatty acid oxidation rapidly lowers the concentrations of fatty acid metabolites that cause insulin resistance, but it takes far longer to burn off large stores of lipid in white adipose tissue. Efficacy is more likely to be achieved in diabetes than obesity and proof of concept trials can be shorter. Secondly, the regulatory requirements for the development of anti-diabetic drugs are less stringent than those for the development of anti-obesity drugs. Phase III clinical studies are shorter and the benefits of efficacy easier to justify. Moreover, the European Medicines Agency still requires that anti-obesity drugs cannot subsequently be approved for the treatment of diabetes without first ignoring any benefit that is a consequence of weight loss. This is a perverse argument in view of the fact that the effects of some diabetes drugs (exenatide, pramlintide and possibly metformin) on markers of blood glucose control are partly due to weight loss. Fortunately, the US Food and Drug Authority no longer takes the same position. Nevertheless, it may be a safer strategy to register a drug for the treatment of diabetes and claim weight loss as a benefit than to register it for obesity and then attempt to have it approved for diabetes.

So in conclusion, the message of this chapter is that drugs that increase the capacity for fatty acid oxidation, especially in skeletal muscle, may improve whole-body insulin sensitivity. Do not be surprised if it is impossible to detect an acute increase in energy expenditure, or a chronic increase in energy expenditure per animal or human, or relative to lean body mass. And if others report an acute increase in energy expenditure, check whether this is due to an increase in sympathetic activity and whether blood pressure is also raised.

Acknowledgements I thank Professors Mike Cawthorne and Paul Trayhurn for their comments on the manuscript.

References

Abdul-Ghani MA, DeFronzo RA (2008) Mitochondrial dysfunction, insulin resistance, and type 2 diabetes mellitus. Curr Diab Rep 8:173–178

Abu-Elheiga L, Oh W, Kordari P, Wakil SJ (2003) Acetyl-CoA carboxylase 2 mutant mice are protected against obesity and diabetes induced by high-fat/high-carbohydrate diets. Proc Natl Acad Sci USA 100:10207–10212

Abu-Elheiga L, Matzuk MM, Kordari P, Oh W, Shaikenov T, Gu Z, Wakil SJ (2005) Mutant mice lacking acetyl-CoA carboxylase 1 are embryonically lethal. Proc Natl Acad Sci USA 102:12011–12016

Addy C, Wright H, Van Laere K, Gantz I, Erondu N, Musser BJ, Lu K, Yuan J, Sanabria-Bohorquez SM, Stoch A, Stevens C, Fong TM, De Lepeleire I, Cilissen C, Cote J, Rosko K, Gendrano IN 3rd, Nguyen AM, Gumbiner B, Rothenberg P, de Hoon J, Bormans G, Depre M, Eng WS, Ravussin E, Klein S, Blundell J, Herman GA, Burns HD, Hargreaves RJ, Wagner J, Gottesdiener K, Amatruda JM, Heymsfield SB (2008) The acyclic CB1R inverse agonist taranabant mediates weight loss by increasing energy expenditure and decreasing caloric intake. Cell Metab 7:68–78

Aja S, Landree LE, Kleman AM, Medghalchi SM, Vadlamudi A, McFadden JM, Aplasca A, Hyun J, Plummer E, Daniels K, Kemm M, Townsend CA, Thupari JN, Kuhajda FP, Moran TH, Ronnett GV (2008) Pharmacological stimulation of brain carnitine palmitoyl-transferase-1 decreases food intake and body weight. Am J Physiol Regul Integr Comp Physiol 294: R352–R361

Alberti KG, Eckel RH, Grundy SM, Zimmet PZ, Cleeman JI, Donato KA, Fruchart JC, James WP, Loria CM, Smith SC Jr (2009) Harmonizing the metabolic syndrome: a joint interim statement of the International Diabetes Federation Task Force on Epidemiology and Prevention; National Heart, Lung, and Blood Institute; American Heart Association; World Heart Federation; International Atherosclerosis Society; and International Association for the Study of Obesity. Circulation 120:1640–1645

Alberts P, Nilsson C, Selen G, Engblom LO, Edling NH, Norling S, Klingstrom G, Larsson C, Forsgren M, Ashkzari M, Nilsson CE, Fiedler M, Bergqvist E, Ohman B, Bjorkstrand E, Abrahmsen LB (2003) Selective inhibition of 11 β-hydroxysteroid dehydrogenase type 1 improves hepatic insulin sensitivity in hyperglycemic mice strains. Endocrinology 144:4755–4762

Allen TL, Matthews VB, Febbraio MB (2011) Overcoming insulin resistance with ciliary neurotrophic factor. Handb Exp Pharmacol. doi:10.1007/978-3-642-17214-4_9

Almeda-Valdes P, Cuevas-Ramos D, Mehta R, Gomez-Perez FJ, Cruz-Bautista I, Arellano-Campos O, Navarrete-Lopez M, Aguilar-Salinas CA (2010) Total and high molecular weight adiponectin have similar utility for the identification of insulin resistance. Cardiovasc Diabetol 9:26

ALS (1996) A double-blind placebo-controlled clinical trial of subcutaneous recombinant human ciliary neurotrophic factor (rHCNTF) in amyotrophic lateral sclerosis. ALS CNTF Treatment Study Group. Neurology 46:1244–1249

Andreelli F, Foretz M, Knauf C, Cani PD, Perrin C, Iglesias MA, Pillot B, Bado A, Tronche F, Mithieux G, Vaulont S, Burcelin R, Viollet B (2006) Liver adenosine monophosphate-activated kinase-alpha2 catalytic subunit is a key target for the control of hepatic glucose production by adiponectin and leptin but not insulin. Endocrinology 147:2432–2441

Araya AV, Orellana X, Espinoza J (2008) Evaluation of the effect of caloric restriction on serum BDNF in overweight and obese subjects: preliminary evidences. Endocrine 33:300–304

Arch JRS (2000) β₃-adrenoreceptor ligands and the pharmacology of the β₃-adrenoreceptor. In: Strosberg A (ed) The β₃-adrenoreceptor. Taylor and Francis, London, pp 48–76

Arch JR (2002) β₃-Adrenoceptor agonists: potential, pitfalls and progress. Eur J Pharmacol 440:99–107

Arch JRS (2007) Comment on: Schmidt MI, Duncan BB, Vigo A et al. (2006) Leptin and incident type 2 diabetes: risk or protection? Diabetologia 50:239–240

Arch JR (2008) The discovery of drugs for obesity, the metabolic effects of leptin and variable receptor pharmacology: perspectives from beta3-adrenoceptor agonists. Naunyn Schmiedebergs Arch Pharmacol 378:225–240

Arch JR, Ainsworth AT (1983) Thermogenic and antiobesity activity of a novel β-adrenoceptor agonist (BRL 26830A) in mice and rats. Am J Clin Nutr 38:549–558

Arch JRS, Ainsworth AT, Ellis RDM, Piercy V, Thody VE, Thurlby PL, Wilson C, Wilson S, Young P (1984) Treatment of obesity with thermogenic β-adrenoceptor agonists: studies on BRL 26830A in rodents. Int J Obes 8(Suppl 1):1–11

Arch JR, Hislop D, Wang SJY, Speakman JR (2006) Some mathematical and technical issues in the measurement and interpretation of open-circuit indirect calorimetry in small animals. Int J Obes 30:1322–1331

Asensio CD, Arsenijevic D, Lehr D, Giacobino J-P, Muzzin P, Rohner-Jeanrenaud F (2008) Effects of leptin on energy metabolism in β-less mice. Int J Obes 32(6):936–942

Assah FK, Brage S, Ekelund U, Wareham NJ (2008) The association of intensity and overall level of physical activity energy expenditure with a marker of insulin resistance. Diabetologia 51:1399–1407

Avignon A, Lapinski H, Rabasa-Lhoret R, Caubel C, Boniface H, Monnier L (2000) Energy metabolism and substrates oxidative patterns in type 2 diabetic patients treated with sulphonylurea alone or in combination with metformin. Diabetes Obes Metab 2:229–235

Banga A, Unal R, Tripathi P, Pokrovskaya I, Owens RJ, Kern PA, Ranganathan G (2009) Adiponectin translation is increased by the PPARgamma agonists pioglitazone and omega-3 fatty acids. Am J Physiol Endocrinol Metab 296:E480–E489

Banks WA (2008) The blood-brain barrier as a cause of obesity. Curr Pharm Des 14:1606–1614

Banks AS, Kon N, Knight C, Matsumoto M, Gutierrez-Juarez R, Rossetti L, Gu W, Accili D (2008) SirT1 gain of function increases energy efficiency and prevents diabetes in mice. Cell Metab 8:333–341

Bao Y, Bing C, Hunter L, Jenkins JR, Wabitsch M, Trayhurn P (2005) Zinc-alpha2-glycoprotein, a lipid mobilizing factor, is expressed and secreted by human (SGBS) adipocytes. FEBS Lett 579:41–47

Barwell ND, Malkova D, Moran CN, Cleland SJ, Packard CJ, Zammit VA, Gill JM (2008) Exercise training has greater effects on insulin sensitivity in daughters of patients with type 2 diabetes than in women with no family history of diabetes. Diabetologia 51:1912–1919

Basu R, Shah P, Basu A, Norby B, Dicke B, Chandramouli V, Cohen O, Landau BR, Rizza RA (2008) Comparison of the effects of pioglitazone and metformin on hepatic and extra-hepatic insulin action in people with type 2 diabetes. Diabetes 57:24–31

Batkai S, Pacher P, Osei-Hyiaman D, Radaeva S, Liu J, Harvey-White J, Offertaler L, Mackie K, Rudd MA, Bukoski RD, Kunos G (2004) Endocannabinoids acting at cannabinoid-1 receptors regulate cardiovascular function in hypertension. Circulation 110:1996–2002

Baxter JD, Webb P (2009) Thyroid hormone mimetics: potential applications in atherosclerosis, obesity and type 2 diabetes. Nat Rev Drug Discov 8:308–320

Bebernitz GR, Schuster HF (2002) The impact of fatty acid oxidation on energy utilization: targets and therapy. Curr Pharm Des 8:1199–1227

Beenken A, Mohammadi M (2009) The FGF family: biology, pathophysiology and therapy. Nat Rev Drug Discov 8:235–253

Bensaid M, Gary-Bobo M, Esclangon A, Maffrand JP, Le Fur G, Oury-Donat F, Soubrie P (2003) The cannabinoid CB1 receptor antagonist SR141716 increases Acrp30 mRNA expression in adipose tissue of obese fa/fa rats and in cultured adipocyte cells. Mol Pharmacol 63:908–914

Benton CR, Holloway GP, Campbell SE, Yoshida Y, Tandon NN, Glatz JF, Luiken JJ, Spriet LL, Bonen A (2008) Rosiglitazone increases fatty acid oxidation and fatty acid translocase (FAT/CD36) but not carnitine palmitoyltransferase I in rat muscle mitochondria. J Physiol 586:1755–1766

Benziane B, Bjornholm M, Lantier L, Viollet B, Zierath JR, Chibalin AV (2009) AMP-activated protein kinase activator A-769662 is an inhibitor of the Na(+)-K(+)-ATPase. Am J Physiol Cell Physiol 297:C1554–C1566

Bermudez-Silva FJ, Viveros MP, McPartland JM, Rodriguez de Fonseca F (2010) The endocannabinoid system, eating behavior and energy homeostasis: the end or a new beginning? Pharmacol Biochem Behav 95:375–382

Berthiaume M, Laplante M, Festuccia W, Gelinas Y, Poulin S, Lalonde J, Joanisse DR, Thieringer R, Deshaies Y (2007) Depot-specific modulation of rat intraabdominal adipose tissue lipid metabolism by pharmacological inhibition of 11beta-hydroxysteroid dehydrogenase type 1. Endocrinology 148:2391–2397

Billin AN (2008) PPAR-beta/delta agonists for Type 2 diabetes and dyslipidemia: an adopted orphan still looking for a home. Expert Opin Invest Drugs 17:1465–1471

Binczek E, Jenke B, Holz B, Gunter RH, Thevis M, Stoffel W (2007) Obesity resistance of the stearoyl-CoA desaturase-deficient (scd1-/-) mouse results from disruption of the epidermal lipid barrier and adaptive thermoregulation. Biol Chem 388:405–418

Birch AM, Buckett LK, Turnbull AV (2010) DGAT1 inhibitors as anti-obesity and anti-diabetic agents. Curr Opin Drug Discov Dev 13:489–496

Bjursell M, Ahnmark A, Bohlooly YM, William-Olsson L, Rhedin M, Peng XR, Ploj K, Gerdin AK, Arnerup G, Elmgren A, Berg AL, Oscarsson J, Linden D (2007) Opposing effects of adiponectin receptors 1 and 2 on energy metabolism. Diabetes 56:583–593

Bluher S, Moschos S, Bullen J Jr, Kokkotou E, Maratos-Flier E, Wiegand SJ, Sleeman MW, Mantzoros CS (2004) Ciliary neurotrophic factorAx15 alters energy homeostasis, decreases body weight, and improves metabolic control in diet-induced obese and UCP1-DTA mice. Diabetes 53:2787–2796

Bo S, Ciccone G, Guidi S, Gambino R, Durazzo M, Gentile L, Cassader M, Cavallo-Perin P, Pagano G (2008) Diet or exercise: what is more effective in preventing or reducing metabolic alterations? Eur J Endocrinol 159(6):685–691

Boschmann M, Engeli S, Dobberstein K, Budziarek P, Strauss A, Boehnke J, Sweep FC, Luft FC, He Y, Foley JE, Jordan J (2009) Dipeptidyl-peptidase-IV inhibition augments postprandial lipid mobilization and oxidation in type 2 diabetic patients. J Clin Endocrinol Metab 94:846–852

Bradley RL, Jeon JY, Liu FF, Maratos-Flier E (2008) Voluntary exercise improves insulin sensitivity and adipose tissue inflammation in diet-induced obese mice. Am J Physiol Endocrinol Metab 295:E586–E594

Braun B, Eze P, Stephens BR, Hagobian TA, Sharoff CG, Chipkin SR, Goldstein B (2008) Impact of metformin on peak aerobic capacity. Appl Physiol Nutr Metab 33:61–67

Bray GA (2000) Reciprocal relation of food intake and sympathetic activity: experimental observations and clinical implications. Int J Obes 24(suppl 2):S8–S17

Breen DM, Sanli T, Giacca A, Tsiani E (2008) Stimulation of muscle cell glucose uptake by resveratrol through sirtuins and AMPK. Biochem Biophys Res Commun 374:117–122

Brown JM, Rudel LL (2010) Stearoyl-coenzyme A desaturase 1 inhibition and the metabolic syndrome: considerations for future drug discovery. Curr Opin Lipidol 21:192–197

Brunmair B, Staniek K, Gras F, Scharf N, Althaym A, Clara R, Roden M, Gnaiger E, Nohl H, Waldhausl W, Furnsinn C (2004) Thiazolidinediones, like metformin, inhibit respiratory complex I: a common mechanism contributing to their antidiabetic actions? Diabetes 53:1052–1059

Bryzgalova G, Effendic S, Khan A, Rehnmark S, Barbounis P, Boulet J, Dong G, Singh R, Shapses S, Malm J, Webb P, Baxter JD, Grover GJ (2008) Anti-obesity, anti-diabetic, and lipid lowering effects of the thyroid receptor beta subtype selective agonist KB-141. J Steroid Biochem Mol Biol 111:262–267

Buchanan TA (2007) (How) can we prevent type 2 diabetes? Diabetes 56:1502–1507

Budohoski L, Challiss RA, Cooney GJ, McManus B, Newsholme EA (1984) Reversal of dietary-induced insulin resistance in muscle of the rat by adenosine deaminase and an adenosine-receptor antagonist. Biochem J 224:327–330

Buemann B, Sorensen TI, Pedersen O, Black E, Holst C, Toubro S, Echwald S, Holst JJ, Rasmussen C, Astrup A (2005) Lower-body fat mass as an independent marker of insulin sensitivity – the role of adiponectin. Int J Obes (Lond) 29:624–631

Burns N, Finucane FM, Hatunic M, Gilman M, Murphy M, Gasparro D, Mari A, Gastaldelli A, Nolan JJ (2007) Early-onset type 2 diabetes in obese white subjects is characterised by a marked defect in beta cell insulin secretion, severe insulin resistance and a lack of response to aerobic exercise training. Diabetologia 50:1500–1508

Busch AK, Gurisik E, Cordery DV, Sudlow M, Denyer GS, Laybutt DR, Hughes WE, Biden TJ (2005) Increased fatty acid desaturation and enhanced expression of stearoyl coenzyme A desaturase protects pancreatic beta-cells from lipoapoptosis. Diabetes 54:2917–2924

Butler AA, Kozak LP (2010) A recurring problem with the analysis of energy expenditure in genetic models expressing lean and obese phenotypes. Diabetes 59:323–329

Camacho RC, Pencek RR, Lacy DB, James FD, Donahue EP, Wasserman DH (2005) Portal venous 5-aminoimidazole-4-carboxamide-1-beta-D-ribofuranoside infusion overcomes hyper-insulinemic suppression of endogenous glucose output. Diabetes 54:373–382

Canto C, Gerhart-Hines Z, Feige JN, Lagouge M, Noriega L, Milne JC, Elliott PJ, Puigserver P, Auwerx J (2009) AMPK regulates energy expenditure by modulating NAD+ metabolism and SIRT1 activity. Nature 458:1056–1060

Castillo M, Freitas BC, Rosene ML, Drigo RA, Grozovsky R, Maciel RM, Patti ME, Ribeiro MO, Bianco AC (2010) Impaired metabolic effects of a thyroid hormone receptor beta-selective agonist in a mouse model of diet-induced obesity. Thyroid 20:545–553

Cavuoto P, McAinch AJ, Hatzinikolas G, Cameron-Smith D, Wittert GA (2007) Effects of cannabinoid receptors on skeletal muscle oxidative pathways. Mol Cell Endocrinol 267:63–69

Chan JM, Rimm EB, Colditz GA, Stampfer MJ, Willett WC (1994) Obesity, fat distribution, and weight gain as risk factors for clinical diabetes in men. Diabetes Care 17:961–969

Chaston TB, Dixon JB (2008) Factors associated with percent change in visceral versus subcutaneous abdominal fat during weight loss: findings from a systematic review. Int J Obes (Lond) 32:619–628

Chen HC, Farese RV Jr (2005) Inhibition of triglyceride synthesis as a treatment strategy for obesity: lessons from DGAT1-deficient mice. Arterioscler Thromb Vasc Biol 25:482–486

Chen HC, Smith SJ, Ladha Z, Jensen DR, Ferreira LD, Pulawa LK, McGuire JG, Pitas RE, Eckel RH, Farese RV Jr (2002a) Increased insulin and leptin sensitivity in mice lacking acyl CoA: diacylglycerol acyltransferase 1. J Clin Invest 109:1049–1055

Chen HC, Stone SJ, Zhou P, Buhman KK, Farese RV Jr (2002b) Dissociation of obesity and impaired glucose disposal in mice overexpressing acyl coenzyme a:diacylglycerol acyltrans-ferase 1 in white adipose tissue. Diabetes 51:3189–3195

Chen HC, Ladha Z, Smith SJ, Farese RV Jr (2003) Analysis of energy expenditure at different ambient temperatures in mice lacking DGAT1. Am J Physiol Endocrinol Metab 284: E213–E218

Chen N, Liu L, Zhang Y, Ginsberg HN, Yu YH (2005) Whole-body insulin resistance in the absence of obesity in FVB mice with overexpression of Dgat1 in adipose tissue. Diabetes 54:3379–3386

Choi CS, Savage DB, Abu-Elheiga L, Liu ZX, Kim S, Kulkarni A, Distefano A, Hwang YJ, Reznick RM, Codella R, Zhang D, Cline GW, Wakil SJ, Shulman GI (2007) Continuous fat oxidation in acetyl-CoA carboxylase 2 knockout mice increases total energy expenditure, reduces fat mass, and improves insulin sensitivity. Proc Natl Acad Sci USA 104:16480–16485

Choi YH, Li C, Hartzell DL, Little DE, Della-Fera MA, Baile CA (2008) ICV leptin effects on spontaneous physical activity and feeding behavior in rats. Behav Brain Res 188:100–108

Chong PK, Jung RT, Rennie MJ, Scrimgeour CM (1995) Energy expenditure in type 2 diabetic patients on metformin and sulphonylurea therapy. Diabet Med 12:401–408

Chong AY, Lupsa BC, Cochran EK, Gorden P (2010) Efficacy of leptin therapy in the different forms of human lipodystrophy. Diabetologia 53:27–35

Clapham JC, Arch JRS, Tadayyon M (2001) Anti-obesity drugs: a critical review of current therapies and future opportunities. Pharmacol Ther 89:81–121

Cluny NL, Vemuri VK, Chambers AP, Limebeer CL, Bedard H, Wood JT, Lutz B, Zimmer A, Parker LA, Makriyannis A, Sharkey KA (2010) A novel peripherally restricted cannabinoid receptor antagonist, AM6545, reduces food intake and body weight, but does not cause malaise, in rodents. Br J Pharmacol 161:629–642

Cohen P, Miyazaki M, Socci ND, Hagge-Greenberg A, Liedtke W, Soukas AA, Sharma R, Hudgins LC, Ntambi JM, Friedman JM (2002) Role for stearoyl-CoA desaturase-1 in leptin-mediated weight loss. Science 297:240–243

Coletta DK, Sriwijitkamol A, Wajcberg E, Tantiwong P, Li M, Prentki M, Madiraju M, Jenkinson CP, Cersosimo E, Musi N, Defronzo RA (2009) Pioglitazone stimulates AMP-activated protein kinase signalling and increases the expression of genes involved in adiponectin signalling, mitochondrial function and fat oxidation in human skeletal muscle in vivo: a randomised trial. Diabetologia 52:723–732

Cool B, Zinker B, Chiou W, Kifle L, Cao N, Perham M, Dickinson R, Adler A, Gagne G, Iyengar R, Zhao G, Marsh K, Kym P, Jung P, Camp HS, Frevert E (2006) Identification and characterization of a small molecule AMPK activator that treats key components of type 2 diabetes and the metabolic syndrome. Cell Metab 3:403–416

Corbett JW (2009) Review of recent acetyl-CoA carboxylase inhibitor patents: mid-2007-2008. Expert Opin Ther Pat 19:943–956

Corbett JW, Harwood JH Jr (2007) Inhibitors of mammalian acetyl-CoA carboxylase. Recent Pat Cardiovasc Drug Discov 2:162–180

Cox KL, Burke V, Morton AR, Beilin LJ, Puddey IB (2004) Independent and additive effects of energy restriction and exercise on glucose and insulin concentrations in sedentary overweight men. Am J Clin Nutr 80:308–316

Cronstein BN (2005) Low-dose methotrexate: a mainstay in the treatment of rheumatoid arthritis. Pharmacol Rev 57:163–172

Crunkhorn S, Patti ME (2008) Links between thyroid hormone action, oxidative metabolism, and diabetes risk? Thyroid 18:227–237

Cui MX, Jiang JF, Zheng RL, Li RL, Wang L, Li WG, Gao MT, Yang LN, Dou JY, Wu YJ (2010) Therapeutic effects of a recombinant mutant of the human ciliary neurotrophic factor in a mouse model of metabolic syndrome. Pharmazie 65:279–283

Cypess AM, Lehman S, Williams G, Tal I, Rodman D, Goldfine AB, Kuo FC, Palmer EL, Tseng YH, Doria A, Kolodny GM, Kahn CR (2009) Identification and importance of brown adipose tissue in adult humans. N Engl J Med 360:1509–1517

Dahlman I, Kaaman M, Olsson T, Tan GD, Bickerton AS, Wahlen K, Andersson J, Nordstrom EA, Blomqvist L, Sjogren A, Forsgren M, Attersand A, Arner P (2005) A unique role of monocyte chemoattractant protein 1 among chemokines in adipose tissue of obese subjects. J Clin Endocrinol Metab 90:5834–5840

Darimont C, Turini M, Epitaux M, Zbinden I, Richelle M, Montell E, Ferrer-Martinez A, Mace K (2004) β3-adrenoceptor agonist prevents alterations of muscle diacylglycerol and adipose tissue phospholipids induced by a cafeteria diet. Nutr Metab (Lond) 1:4

Daval M, Foufelle F, Ferre P (2006) Functions of AMP-activated protein kinase in adipose tissue. J Physiol 574:55–62

Derave W, Hespel P (1999) Role of adenosine in regulating glucose uptake during contractions and hypoxia in rat skeletal muscle. J Physiol 515(Pt 1):255–263

Despres JP, Golay A, Sjostrom L (2005) Effects of rimonabant on metabolic risk factors in overweight patients with dyslipidemia. N Engl J Med 353:2121–2134

Dobrzyn P, Dobrzyn A, Miyazaki M, Cohen P, Asilmaz E, Hardie DG, Friedman JM, Ntambi JM (2004) Stearoyl-CoA desaturase 1 deficiency increases fatty acid oxidation by activating AMP-activated protein kinase in liver. Proc Natl Acad Sci USA 101:6409–6414

Dobrzyn A, Dobrzyn P, Lee SH, Miyazaki M, Cohen P, Asilmaz E, Hardie DG, Friedman JM, Ntambi JM (2005) Stearoyl-CoA desaturase-1 deficiency reduces ceramide synthesis by downregulating serine palmitoyltransferase and increasing beta-oxidation in skeletal muscle. Am J Physiol Endocrinol Metab 288:E599–E607

Doring H, Schwarzer K, Nuesslein-Hildesheim B, Schmidt I (1998) Leptin selectively increases energy expenditure of food-restricted lean mice. Int J Obes Relat Metab Disord 22:83–88

Duncan GE (2006) Exercise, fitness, and cardiovascular disease risk in type 2 diabetes and the metabolic syndrome. Curr Diab Rep 6:29–35

Dzamko N, Schertzer JD, Ryall JG, Steel R, Macaulay SL, Wee S, Chen ZP, Michell BJ, Oakhill JS, Watt MJ, Jorgensen SB, Lynch GS, Kemp BE, Steinberg GR (2008) AMPK-independent pathways regulate skeletal muscle fatty acid oxidation. J Physiol 586:5819–5831

Eaton RP, Steinberg D, Thompson RH (1965) Relationship between free fatty acid turnover and total body oxygen consumption in the euthyroid and hyperthyroid states. J Clin Invest 44:247–260

Edwards CM, Stanley SA, Davis R, Brynes AE, Frost GS, Seal LJ, Ghatei MA, Bloom SR (2001) Exendin-4 reduces fasting and postprandial glucose and decreases energy intake in healthy volunteers. Am J Physiol Endocrinol Metab 281:E155–E161

Elisha B, Ziai S, Karelis AD, Rakel A, Coderre L, Imbeault P, Rabasa-Lhoret R (2010) Similar associations of total adiponectin and high molecular weight adiponectin with cardio-metabolic risk factors in a population of overweight and obese postmenopausal women: a MONET study. Horm Metab Res 42(8):590–594

Erion DM, Yonemitsu S, Nie Y, Nagai Y, Gillum MP, Hsiao JJ, Iwasaki T, Stark R, Weismann D, Yu XX, Murray SF, Bhanot S, Monia BP, Horvath TL, Gao Q, Samuel VT, Shulman GI (2009) SirT1 knockdown in liver decreases basal hepatic glucose production and increases hepatic insulin responsiveness in diabetic rats. Proc Natl Acad Sci USA 106:11288–11293

Esler WP, Rudolph J, Claus TH, Tang W, Barucci N, Brown SE, Bullock W, Daly M, Decarr L, Li Y, Milardo L, Molstad D, Zhu J, Gardell SJ, Livingston JN, Sweet LJ (2007) Small-molecule ghrelin receptor antagonists improve glucose tolerance, suppress appetite, and promote weight loss. Endocrinology 148:5175–5185

Ettinger MP, Littlejohn TW, Schwartz SL, Weiss SR, McIlwain HH, Heymsfield SB, Bray GA, Roberts WG, Heyman ER, Stambler N, Heshka S, Vicary C, Guler HP (2003) Recombinant variant of ciliary neurotrophic factor for weight loss in obese adults: a randomized, dose-ranging study. JAMA 289:1826–1832

Faiola B, Falls JG, Peterson RA, Bordelon NR, Brodie TA, Cummings CA, Romach EH, Miller RT (2008) PPAR alpha, more than PPAR delta, mediates the hepatic and skeletal muscle alterations induced by the PPAR agonist GW0742. Toxicol Sci 105:384–394

Farooqi IS, O'Rahilly S (2004) Monogenic human obesity syndromes. Recent Prog Horm Res 59:409–424

Farooqi IS, Wangensteen T, Collins S, Kimber W, Matarese G, Keogh JM, Lank E, Bottomley B, Lopez-Fernandez J, Ferraz-Amaro I, Dattani MT, Ercan O, Myhre AG, Retterstol L, Stanhope R, Edge JA, McKenzie S, Lessan N, Ghodsi M, De Rosa V, Perna F, Fontana S, Barroso I, Undlien DE, O'Rahilly S (2007) Clinical and molecular genetic spectrum of congenital deficiency of the leptin receptor. N Engl J Med 356:237–247

Feige JN, Lagouge M, Canto C, Strehle A, Houten SM, Milne JC, Lambert PD, Mataki C, Elliott PJ, Auwerx J (2008) Specific SIRT1 activation mimics low energy levels and protects against diet-induced metabolic disorders by enhancing fat oxidation. Cell Metab 8:347–358

Feldmann HM, Golozoubova V, Cannon B, Nedergaard J (2009) UCP1 ablation induces obesity and abolishes diet-induced thermogenesis in mice exempt from thermal stress by living at thermoneutrality. Cell Metab 9:203–209

Festuccia WT, Oztezcan S, Laplante M, Berthiaume M, Michel C, Dohgu S, Denis RG, Brito MN, Brito NA, Miller DS, Banks WA, Bartness TJ, Richard D, Deshaies Y (2008) Peroxisome proliferator-activated receptor-gamma-mediated positive energy balance in the rat is associated with reduced sympathetic drive to adipose tissues and thyroid status. Endocrinology 149:2121–2130

Finer N (2002) Sibutramine: its mode of action and efficacy. Int J Obes Relat Metab Disord 26 (Suppl 4):S29–S33

Finkel T, Deng CX, Mostoslavsky R (2009) Recent progress in the biology and physiology of sirtuins. Nature 460:587–591

Fisher FM, Chui PC, Antonellis PJ, Bina HA, Kharitonenkov A, Flier JS, Maratos-Flier E (2010) Obesity is an FGF21 resistant state. Diabetes 59(11):2781–2789

Flowers MT, Ntambi JM (2008) Role of stearoyl-coenzyme A desaturase in regulating lipid metabolism. Curr Opin Lipidol 19:248–256

Flowers MT, Ntambi JM (2009) Stearoyl-CoA desaturase and its relation to high-carbohydrate diets and obesity. Biochim Biophys Acta 1791:85–91

Flowers JB, Rabaglia ME, Schueler KL, Flowers MT, Lan H, Keller MP, Ntambi JM, Attie AD (2007) Loss of stearoyl-CoA desaturase-1 improves insulin sensitivity in lean mice but worsens diabetes in leptin-deficient obese mice. Diabetes 56:1228–1239

Fogarty S, Hardie DG (2010) Development of protein kinase activators: AMPK as a target in metabolic disorders and cancer. Biochim Biophys Acta 1804:581–591

Foretz M, Ancellin N, Andreelli F, Saintillan Y, Grondin P, Kahn A, Thorens B, Vaulont S, Viollet B (2005) Short-term overexpression of a constitutively active form of AMP-activated protein kinase in the liver leads to mild hypoglycemia and fatty liver. Diabetes 54:1331–1339

Frederiksen L, Nielsen TL, Wraae K, Hagen C, Frystyk J, Flyvbjerg A, Brixen K, Andersen M (2009) Subcutaneous rather than visceral adipose tissue is associated with adiponectin levels and insulin resistance in young men. J Clin Endocrinol Metab 94:4010–4015

Fruebis J, Tsao TS, Javorschi S, Ebbets-Reed D, Erickson MR, Yen FT, Bihain BE, Lodish HF (2001) Proteolytic cleavage product of 30-kDa adipocyte complement-related protein increases fatty acid oxidation in muscle and causes weight loss in mice. Proc Natl Acad Sci USA 98:2005–2010

Fujii N, Ho RC, Manabe Y, Jessen N, Toyoda T, Holland WL, Summers SA, Hirshman MF, Goodyear LJ (2008) Ablation of AMP-activated protein kinase alpha2 activity exacerbates insulin resistance induced by high-fat feeding of mice. Diabetes 57:2958–2966

Fujimoto WY, Jablonski KA, Bray GA, Kriska A, Barrett-Connor E, Haffner S, Hanson R, Hill JO, Hubbard V, Stamm E, Pi-Sunyer FX (2007) Body size and shape changes and the risk of diabetes in the diabetes prevention program. Diabetes 56:1680–1685

Galgani JE, Moro C, Ravussin E (2008) Metabolic flexibility and insulin resistance. Am J Physiol Endocrinol Metab 295:E1009–E1017

Gallwitz B (2011) GLP-1-agonists and dipeptidyl-peptidase IV inhibitors. Handb Exp Pharmacol. doi:10.1007/978-3-642-17214-4_3

Gauna C, Kiewiet RM, Janssen JA, van de Zande B, Delhanty PJ, Ghigo E, Hofland LJ, Themmen AP, van der Lely AJ (2007) Unacylated ghrelin acts as a potent insulin secretagogue in glucose-stimulated conditions. Am J Physiol Endocrinol Metab 293:E697–E704

Gavrilova O, Marcus-Samuels B, Reitman ML (2000) Lack of responses to a β_3-adrenergic agonist in lipoatrophic A-ZIP/F-1 mice. Diabetes 49:1910–1916

Gloaguen I, Costa P, Demartis A, Lazzaro D, Dimarco A, Graziani R, Paonessa G, Chen F, Rosenblum CI, Vanderploeg LT, Cortese R, Ciliberto G, Laufer R (1997) Ciliary neurotrophic factor corrects obesity and diabetes associated with leptin deficiency and resistance. Proc Natl Acad Sci USA 94:6456–6461

Golay A (2008) Metformin and body weight. Int J Obes (Lond) 32:61–72

Golozoubova V, Gullberg H, Matthias A, Cannon B, Vennstrom B, Nedergaard J (2004) Depressed thermogenesis but competent brown adipose tissue recruitment in mice devoid of all hormone-binding thyroid hormone receptors. Mol Endocrinol 18:384–401

Gomez R, Navarro M, Ferrer B, Trigo JM, Bilbao A, Del Arco I, Cippitelli A, Nava F, Piomelli D, Rodriguez de Fonseca F (2002) A peripheral mechanism for CB1 cannabinoid receptor-dependent modulation of feeding. J Neurosci 22:9612–9617

Gonez LJ, Naselli G, Banakh I, Niwa H, Harrison LC (2008) Pancreatic expression and mitochondrial localization of the progestin-adipoQ receptor PAQR10. Mol Med 14:697–704

Granneman JG, Li P, Zhu Z, Lu Y (2005) Metabolic and cellular plasticity in white adipose tissue I: effects of beta3-adrenergic receptor activation. Am J Physiol Endocrinol Metab 289: E608–E616

Grover GJ, Mellstrom K, Malm J (2007) Therapeutic potential for thyroid hormone receptor-beta selective agonists for treating obesity, hyperlipidemia and diabetes. Curr Vasc Pharmacol 5:141–154

Grujic D, Susulic VS, Harper ME, Himms-Hagen J, Cunningham BA, Corkey BE, Lowell BB (1997) β3-adrenergic receptors on white and brown adipocytes mediate β3-selective agonist-induced effects on energy expenditure, insulin secretion, and food intake. A study using transgenic and gene knockout mice. J Biol Chem 272:17686–17693

Gualillo O, Lago F, Dieguez C (2008) Introducing GOAT: a target for obesity and anti-diabetic drugs? Trends Pharmacol Sci 29:398–401

Gutierrez-Juarez R, Pocai A, Mulas C, Ono H, Bhanot S, Monia BP, Rossetti L (2006) Critical role of stearoyl-CoA desaturase-1 (SCD1) in the onset of diet-induced hepatic insulin resistance. J Clin Invest 116:1686–1695

Hagberg JM, Park JJ, Brown MD (2000) The role of exercise training in the treatment of hypertension: an update. Sports Med 30:193–206

Hale LP, Price DT, Sanchez LM, Demark-Wahnefried W, Madden JF (2001) Zinc alpha-2-glycoprotein is expressed by malignant prostatic epithelium and may serve as a potential serum marker for prostate cancer. Clin Cancer Res 7:846–853

Hall KD, Hallgreen CE (2008) Increasing weight loss attenuates the preferential loss of visceral compared with subcutaneous fat: a predicted result of an allometric model. Int J Obes (Lond) 32:722

Hallgreen CE, Hall KD (2008) Allometric relationship between changes of visceral fat and total fat mass. Int J Obes (Lond) 32:845–852

Hansen DL, Toubro S, Stock MJ, Macdonald IA, Astrup A (1998) Thermogenic effects of sibutramine in humans. Am J Clin Nutr 68:1180–1186

Harada N, Oda Z, Hara Y, Fujinami K, Okawa M, Ohbuchi K, Yonemoto M, Ikeda Y, Ohwaki K, Aragane K, Tamai Y, Kusunoki J (2007) Hepatic de novo lipogenesis is present in liver-specific ACC1-deficient mice. Mol Cell Biol 27:1881–1888

Harder H, Nielsen L, Tu DT, Astrup A (2004) The effect of liraglutide, a long-acting glucagon-like peptide 1 derivative, on glycemic control, body composition, and 24-h energy expenditure in patients with type 2 diabetes. Diabetes Care 27:1915–1921

Hardie DG (2008) AMPK: a key regulator of energy balance in the single cell and the whole organism. Int J Obes (Lond) 32(suppl 4): S7–S12

Harrington WW, Britt CS, Wilson JG, Milliken NO, Binz JG, Lobe DC, Oliver WR, Lewis MC, Ignar DM (2007) The effect of PPARα, PPARδ, PPARγ, and PPARpan agonists on body weight, body mass, and serum lipid profiles in diet-induced obese AKR/J mice. PPAR Res 2007:97125

Harwood HJ Jr, Petras SF, Shelly LD, Zaccaro LM, Perry DA, Makowski MR, Hargrove DM, Martin KA, Tracey WR, Chapman JG, Magee WP, Dalvie DK, Soliman VF, Martin WH, Mularski CJ, Eisenbeis SA (2003) Isozyme-nonselective N-substituted bipiperidylcarboxamide acetyl-CoA carboxylase inhibitors reduce tissue malonyl-CoA concentrations, inhibit fatty acid synthesis, and increase fatty acid oxidation in cultured cells and in experimental animals. J Biol Chem 278:37099–37111

Havel RJ, Carlson LA, Ekelund LG, Holmgren A (1964) Studies on the relation between mobilization of free fatty acids and energy metabolism in man: effects of norepinephrine and nicotinic acid. Metabolism 13:1402–1412

Hawley JA, Holloszy JO (2009) Exercise: it's the real thing! Nutr Rev 67:172–178

Hawley JA, Lessard SJ (2008) Exercise training-induced improvements in insulin action. Acta Physiol (Oxf) 192:127–135

Hayes C, Kriska A (2008) Role of physical activity in diabetes management and prevention. J Am Diet Assoc 108:S19–S23

Haynes WG, Morgan DA, Walsh SA, Mark AL, Sivitz WI (1997) Receptor-mediated regional sympathetic nerve activation by leptin. J Clin Invest 100:270–278

Hegarty BD, Furler SM, Ye J, Cooney GJ, Kraegen EW (2003) The role of intramuscular lipid in insulin resistance. Acta Physiol Scand 178:373–383

Hegarty BD, Turner N, Cooney GJ, Kraegen EW (2009) Insulin resistance and fuel homeostasis: the role of AMP-activated protein kinase. Acta Physiol (Oxf) 196:129–145

Herling AW, Gossel M, Haschke G, Stengelin S, Kuhlmann J, Muller G, Schmoll D, Kramer W (2007) CB1 receptor antagonist AVE1625 affects primarily metabolic parameters independently of reduced food intake in Wistar rats. Am J Physiol Endocrinol Metab 293:E826–E832

Herling AW, Kilp S, Elvert R, Haschke G, Kramer W (2008) Increased energy expenditure contributes more to the body weight-reducing effect of rimonabant than reduced food intake in candy-fed wistar rats. Endocrinology 149:2557–2566

Hermanowski-Vosatka A, Balkovec JM, Cheng K, Chen HY, Hernandez M, Koo GC, Le Grand CB, Li Z, Metzger JM, Mundt SS, Noonan H, Nunes CN, Olson SH, Pikounis B, Ren N, Robertson N, Schaeffer JM, Shah K, Springer MS, Strack AM, Strowski M, Wu K, Wu T, Xiao J, Zhang BB, Wright SD, Thieringer R (2005) 11β-HSD1 inhibition ameliorates metabolic syndrome and prevents progression of atherosclerosis in mice. J Exp Med 202:517–527

Heymsfield SB, Greenberg AS, Fujioka K, Dixon RM, Kushner R, Hunt T, Lubina JA, Patane J, Self B, Hunt P, McCamish M (1999) Recombinant leptin for weight loss in obese and lean adults, A randomized, controlled, dose-escalation trial. JAMA 282:1568–1575

Himms-Hagen J (1997) On raising energy expenditure in ob/ob mice. Science 276:1132–1133

Hirai K, Hussey HJ, Barber MD, Price SA, Tisdale MJ (1998) Biological evaluation of a lipid-mobilizing factor isolated from the urine of cancer patients. Cancer Res 58:2359–2365

Hocking SL, Chisholm DJ, James DE (2008) Studies of regional adipose transplantation reveal a unique and beneficial interaction between subcutaneous adipose tissue and the intra-abdominal compartment. Diabetologia 51:900–902

Hofer D, Raices M, Schauenstein K, Porta S, Korsatko W, Hagmuller K, Zaninovich A (2000) The in vivo effects of beta-3-receptor agonist CGP-12177 on thyroxine deiodination in cold-exposed, sympathectomized rat brown fat. Eur J Endocrinol 143:273–277

Hollander P (2007) Endocannabinoid blockade for improving glycemic control and lipids in patients with type 2 diabetes mellitus. Am J Med 120:S18–S28; discussion S29–S32

Holloszy JO (2005) Exercise-induced increase in muscle insulin sensitivity. J Appl Physiol 99:338–343

Holst B, Schwartz TW (2006) Ghrelin receptor mutations–too little height and too much hunger. J Clin Invest 116:637–641

Hood DA, Irrcher I, Ljubicic V, Joseph AM (2006) Coordination of metabolic plasticity in skeletal muscle. J Exp Biol 209:2265–2275

Hoogwerf BJ, Doshi KB, Diab D (2008) Pramlintide, the synthetic analogue of amylin: physiology, pathophysiology, and effects on glycemic control, body weight, and selected biomarkers of vascular risk. Vasc Health Risk Manag 4:355–362

Hung YJ, Chen YC, Pei D, Kuo SW, Hsieh CH, Wu LY, He CT, Lee CH, Fan SC, Sheu WH (2005) Sibutramine improves insulin sensitivity without alteration of serum adiponectin in obese subjects with Type 2 diabetes. Diabet Med 22:1024–1030

Inhoff T, Wiedenmann B, Klapp BF, Monnikes H, Kobelt P (2009) Is desacyl ghrelin a modulator of food intake? Peptides 30:991–994

Islam-Ali B, Khan S, Price SA, Tisdale MJ (2001) Modulation of adipocyte G-protein expression in cancer cachexia by a lipid-mobilizing factor (LMF). Br J Cancer 85:758–763

Issandou M, Bouillot A, Brusq JM, Forest MC, Grillot D, Guillard R, Martin S, Michiels C, Sulpice T, Daugan A (2009) Pharmacological inhibition of stearoyl-CoA desaturase 1 improves insulin sensitivity in insulin-resistant rat models. Eur J Pharmacol 618:28–36

James WP, Caterson ID, Coutinho W, Finer N, Van Gaal LF, Maggioni IP, Torp-Pedersen C, Sharma AM, Shepherd GM, Rode RA, Renz CL (2010) Effect of sibutramine on cardiovascular outcomes in overweight and obese subjects. N Engl J Med 363:905–917

Janiszewski PM, Kuk JL, Ross R (2008) Is the reduction of lower-body subcutaneous adipose tissue associated with elevations in risk factors for diabetes and cardiovascular disease? Diabetologia 51:1475–1482

Janssen I, Fortier A, Hudson R, Ross R (2002) Effects of an energy-restrictive diet with or without exercise on abdominal fat, intermuscular fat, and metabolic risk factors in obese women. Diabetes Care 25:431–438

Jiang G, Li Z, Liu F, Ellsworth K, Dallas-Yang Q, Wu M, Ronan J, Esau C, Murphy C, Szalkowski D, Bergeron R, Doebber T, Zhang BB (2005) Prevention of obesity in mice by antisense oligonucleotide inhibitors of stearoyl-CoA desaturase-1. J Clin Invest 115:1030–1038

Johansson L, Fotsch C, Bartberger MD, Castro VM, Chen M, Emery M, Gustafsson S, Hale C, Hickman D, Homan E, Jordan SR, Komorowski R, Li A, McRae K, Moniz G, Matsumoto G, Orihuela C, Palm G, Veniant M, Wang M, Williams M, Zhang J (2008) 2-amino-1, 3-thiazol-4 (5H)-ones as potent and selective 11beta-hydroxysteroid dehydrogenase type 1 inhibitors: enzyme-ligand co-crystal structure and demonstration of pharmacodynamic effects in C57Bl/6 mice. J Med Chem 51:2933–2943

Kadowaki T, Yamauchi T (2005) Adiponectin and adiponectin receptors. Endocr Rev 26:439–451

Karlsson HK, Hallsten K, Bjornholm M, Tsuchida H, Chibalin AV, Virtanen KA, Heinonen OJ, Lonnqvist F, Nuutila P, Zierath JR (2005) Effects of metformin and rosiglitazone treatment on insulin signaling and glucose uptake in patients with newly diagnosed type 2 diabetes: a randomized controlled study. Diabetes 54:1459–1467

Kawamata Y, Fujii R, Hosoya M, Harada M, Yoshida H, Miwa M, Fukusumi S, Habata Y, Itoh T, Shintani Y, Hinuma S, Fujisawa Y, Fujino M (2003) A G protein-coupled receptor responsive to bile acids. J Biol Chem 278:9435–9440

Keates AC, Bailey CJ (1993) Metformin does not increase energy expenditure of brown fat. Biochem Pharmacol 45:971–973

Kennedy BL, Ellis S (1969) Dissociation of catecholamine-induced calorigenesis from lipolysis and glycogenolysis in intact animals. J Pharmacol Exp Ther 168:137–145

Kennedy A, Gettys TW, Watson P, Wallace P, Ganaway E, Pan Q, Garvey WT (1997) The metabolic significance of leptin in humans: gender-based differences in relationship to adiposity, insulin sensitivity, and energy expenditure. J Clin Endocrinol Metab 82:1293–1300

Kershaw EE, Morton NM, Dhillon H, Ramage L, Seckl JR, Flier JS (2005) Adipocyte-specific glucocorticoid inactivation protects against diet-induced obesity. Diabetes 54:1023–1031

King AJ, Segreti JA, Larson KJ, Souers AJ, Kym PR, Reilly RM, Collins CA, Voorbach MJ, Zhao G, Mittelstadt SW, Cox BF (2010) In vivo efficacy of acyl CoA: diacylglycerol acyltransferase (DGAT) 1 inhibition in rodent models of postprandial hyperlipidemia. Eur J Pharmacol 637:155–161

Kirchner H, Gutierrez JA, Solenberg PJ, Pfluger PT, Czyzyk TA, Willency JA, Schurmann A, Joost HG, Jandacek RJ, Hale JE, Heiman ML, Tschop MH (2009) GOAT links dietary lipids with the endocrine control of energy balance. Nat Med 15:741–745

Kola B (2008) Role of AMP-activated protein kinase in the control of appetite. J Neuroendocrinol 20:942–951

Koltun DO, Vasilevich NI, Parkhill EQ, Glushkov AI, Zilbershtein TM, Mayboroda EI, Boze MA, Cole AG, Henderson I, Zautke NA, Brunn SA, Chu N, Hao J, Mollova N, Leung K, Chisholm JW, Zablocki J (2009) Orally bioavailable, liver-selective stearoyl-CoA desaturase (SCD) inhibitors. Bioorg Med Chem Lett 19:3050–3053

Krabbe KS, Nielsen AR, Krogh-Madsen R, Plomgaard P, Rasmussen P, Erikstrup C, Fischer CP, Lindegaard B, Petersen AM, Taudorf S, Secher NH, Pilegaard H, Bruunsgaard H, Pedersen BK (2007) Brain-derived neurotrophic factor (BDNF) and type 2 diabetes. Diabetologia 50:431–438

Kraegen EW, Cooney GJ (2008) Free fatty acids and skeletal muscle insulin resistance. Curr Opin Lipidol 19:235–241

Kramer DK, Al-Khalili L, Guigas B, Leng Y, Garcia-Roves PM, Krook A (2007) Role of AMP kinase and PPARdelta in the regulation of lipid and glucose metabolism in human skeletal muscle. J Biol Chem 282:19313–19320

Kubota N, Terauchi Y, Kubota T, Kumagai H, Itoh S, Satoh H, Yano W, Ogata H, Tokuyama K, Takamoto I, Mineyama T, Ishikawa M, Moroi M, Sugi K, Yamauchi T, Ueki K, Tobe K, Noda T, Nagai R, Kadowaki T (2006) Pioglitazone ameliorates insulin resistance and diabetes by both adiponectin-dependent and -independent pathways. J Biol Chem 281:8748–8755

Kunos G, Osei-Hyiaman D, Batkai S, Sharkey KA, Makriyannis A (2009) Should peripheral CB (1) cannabinoid receptors be selectively targeted for therapeutic gain? Trends Pharmacol Sci 30:1–7

Kunz I, Meier MK, Bourson A, Fisseha M, Schilling W (2008) Effects of rimonabant, a cannabinoid CB1 receptor ligand, on energy expenditure in lean rats. Int J Obes (Lond) 32:863–870

Kusakabe T, Tanioka H, Ebihara K, Hirata M, Miyamoto L, Miyanaga F, Hige H, Aotani D, Fujisawa T, Masuzaki H, Hosoda K, Nakao K (2009) Beneficial effects of leptin on glycaemic and lipid control in a mouse model of type 2 diabetes with increased adiposity induced by streptozotocin and a high-fat diet. Diabetologia 52:675–683

Lafontan M, Piazza PV, Girard J (2007) Effects of CB1 antagonist on the control of metabolic functions in obese type 2 diabetic patients. Diabetes Metab 33:85–95

Lafrance L, Routhier D, Tetu B, Tetu C (1979) Effects of noradrenaline and nicotinic acid on plasma free fatty acids and oxygen consumption in cold-adapted rats. Can J Physiol Pharmacol 57:725–730

Lambert PD, Anderson KD, Sleeman MW, Wong V, Tan J, Hijarunguru A, Corcoran TL, Murray JD, Thabet KE, Yancopoulos GD, Wiegand SJ (2001) Ciliary neurotrophic factor activates leptin-like pathways and reduces body fat, without cachexia or rebound weight gain, even in leptin-resistant obesity. Proc Natl Acad Sci USA 98:4652–4657

Lane MD, Wolfgang M, Cha SH, Dai Y (2008) Regulation of food intake and energy expenditure by hypothalamic malonyl-CoA. Int J Obes (Lond) 32(suppl 4):S49–S54

LeBrasseur NK, Kelly M, Tsao TS, Farmer SR, Saha AK, Ruderman NB, Tomas E (2006) Thiazolidinediones can rapidly activate AMP-activated protein kinase in mammalian tissues. Am J Physiol Endocrinol Metab 291:E175–E181

Lee SH, Dobrzyn A, Dobrzyn P, Rahman SM, Miyazaki M, Ntambi JM (2004) Lack of stearoyl-CoA desaturase 1 upregulates basal thermogenesis but causes hypothermia in a cold environment. J Lipid Res 45:1674–1682

Lee CH, Olson P, Hevener A, Mehl I, Chong LW, Olefsky JM, Gonzalez FJ, Ham J, Kang H, Peters JM, Evans RM (2006) PPARdelta regulates glucose metabolism and insulin sensitivity. Proc Natl Acad Sci USA 103:3444–3449

Liang F, Kume S, Koya D (2009) SIRT1 and insulin resistance. Nat Rev Endocrinol 5:367–373

Liu X, Perusse F, Bukowiecki LJ (1998) Mechanisms of the antidiabetic effects of the beta 3-adrenergic agonist CL-316243 in obese Zucker-ZDF rats. Am J Physiol 274:R1212–R1219

Liu L, Zhang Y, Chen N, Shi X, Tsang B, Yu YH (2007) Upregulation of myocellular DGAT1 augments triglyceride synthesis in skeletal muscle and protects against fat-induced insulin resistance. J Clin Invest 117:1679–1689

Livingston EH (2006) Lower body subcutaneous fat accumulation and diabetes mellitus risk. Surg Obes Relat Dis 2:362–368

Livingstone DE, Walker BR (2003) Is 11beta-hydroxysteroid dehydrogenase type 1 a therapeutic target? Effects of carbenoxolone in lean and obese Zucker rats. J Pharmacol Exp Ther 305:167–172

Loftus TM, Jaworsky DE, Frehywot GL, Townsend CA, Ronnett GV, Lane MD, Kuhajda FP (2000) Reduced food intake and body weight in mice treated with fatty acid synthase inhibitors. Science 288:2379–2381

Longo KA, Charoenthongtrakul S, Giuliana DJ, Govek EK, McDonagh T, Qi Y, DiStefano PS, Geddes BJ (2008) Improved insulin sensitivity and metabolic flexibility in ghrelin receptor knockout mice. Regul Pept 150:55–61

Madsen AN, Jelsing J, van de Wall EH, Vrang N, Larsen PJ, Schwartz GJ (2009) Rimonabant induced anorexia in rodents is not mediated by vagal or sympathetic gut afferents. Neurosci Lett 449:20–23

Man WC, Miyazaki M, Chu K, Ntambi J (2006) Colocalization of SCD1 and DGAT2: implying preference for endogenous monounsaturated fatty acids in triglyceride synthesis. J Lipid Res 47:1928–1939

Manara L, Badone D, Baroni M, Boccardi G, Cecchi R, Croci T, Giudice A, Guzzi U, Landi M, Le Fur G (1996) Functional identification of rat atypical β-adrenoceptors by the first β₃-selective antagonists, aryloxypropanolaminotetralins. Br J Pharmacol 117:435–442

Maneuf Y, Higginbottom M, Pritchard M, Lione L, Ashford MLJ, Richardson PJ (2004) Small molecule leptin mimetics overcome leptin resistance in obese rats. Fundam Clin Pharmacol 18 (suppl 1):83

Mano-Otagiri A, Iwasaki-Sekino A, Nemoto T, Ohata H, Shuto Y, Nakabayashi H, Sugihara H, Oikawa S, Shibasaki T (2010) Genetic suppression of ghrelin receptors activates brown adipocyte function and decreases fat storage in rats. Regul Pept 160:81–90

Mao J, DeMayo FJ, Li H, Abu-Elheiga L, Gu Z, Shaikenov TE, Kordari P, Chirala SS, Heird WC, Wakil SJ (2006) Liver-specific deletion of acetyl-CoA carboxylase 1 reduces hepatic triglyceride accumulation without affecting glucose homeostasis. Proc Natl Acad Sci USA 103:8552–8557

Marrades MP, Martinez JA, Moreno-Aliaga MJ (2008) ZAG, a lipid mobilizing adipokine, is downregulated in human obesity. J Physiol Biochem 64:61–66

Marsicano G, Lutz B (2006) Neuromodulatory functions of the endocannabinoid system. J Endocrinol Invest 29:27–46

Masaki T, Chiba S, Yasuda T, Tsubone T, Kakuma T, Shimomura I, Funahashi T, Matsuzawa Y, Yoshimatsu H (2003) Peripheral, but not central, administration of adiponectin reduces visceral adiposity and upregulates the expression of uncoupling protein in agouti yellow (Ay/a) obese mice. Diabetes 52:2266–2273

Matthews VB, Febbraio MA (2008) CNTF: a target therapeutic for obesity-related metabolic disease? J Mol Med 86:353–361

Matthews VB, Astrom MB, Chan MH, Bruce CR, Krabbe KS, Prelovsek O, Akerstrom T, Yfanti C, Broholm C, Mortensen OH, Penkowa M, Hojman P, Zankari A, Watt MJ, Bruunsgaard H, Pedersen BK, Febbraio MA (2009) Brain-derived neurotrophic factor is produced by skeletal muscle cells in response to contraction and enhances fat oxidation via activation of AMP-activated protein kinase. Diabetologia 52:1409–1418

Mayorov AV, Amara N, Chang JY, Moss JA, Hixon MS, Ruiz DI, Meijler MM, Zorrilla EP, Janda KD (2008) Catalytic antibody degradation of ghrelin increases whole-body metabolic rate and reduces refeeding in fasting mice. Proc Natl Acad Sci USA 105:17487–17492

Meex RC, Schrauwen-Hinderling VB, Moonen-Kornips E, Schaart G, Mensink M, Phielix E, van de Weijer T, Sels JP, Schrauwen P, Hesselink MK (2010) Restoration of muscle mitochondrial function and metabolic flexibility in type 2 diabetes by exercise training is paralleled by increased myocellular fat storage and improved insulin sensitivity. Diabetes 59:572–579

Menshikova EV, Ritov VB, Ferrell RE, Azuma K, Goodpaster BH, Kelley DE (2007) Characteristics of skeletal muscle mitochondrial biogenesis induced by moderate-intensity exercise and weight loss in obesity. J Appl Physiol 103:21–27

Mesaros A, Koralov SB, Rother E, Wunderlich FT, Ernst MB, Barsh GS, Rajewsky K, Bruning JC (2008) Activation of Stat3 signaling in AgRP neurons promotes locomotor activity. Cell Metab 7:236–248

Migrenne S, Lacombe A, Lefevre AL, Pruniaux MP, Guillot E, Galzin AM, Magnan C (2009) Adiponectin is required to mediate rimonabant-induced improvement of insulin sensitivity but not body weight loss in diet-induced obese mice. Am J Physiol Regul Integr Comp Physiol 296:R929–R935

Mikhail N (2008) Incretin mimetics and dipeptidyl peptidase 4 inhibitors in clinical trials for the treatment of type 2 diabetes. Expert Opin Invest Drugs 17:845–853

Milne JC, Lambert PD, Schenk S, Carney DP, Smith JJ, Gagne DJ, Jin L, Boss O, Perni RB, Vu CB, Bemis JE, Xie R, Disch JS, Ng PY, Nunes JJ, Lynch AV, Yang H, Galonek H, Israelian K, Choy W, Iffland A, Lavu S, Medvedik O, Sinclair DA, Olefsky JM, Jirousek MR, Elliott PJ, Westphal CH (2007) Small molecule activators of SIRT1 as therapeutics for the treatment of type 2 diabetes. Nature 450:712–716

Minokoshi Y, Kim YB, Peroni OD, Fryer LG, Muller C, Carling D, Kahn BB (2002) Leptin stimulates fatty-acid oxidation by activating AMP-activated protein kinase. Nature 415:339–343

Mirshamsi S, Olsson M, Arnelo U, Kinsella JM, Permert J, Ashford ML (2007) BVT.3531 reduces body weight and activates K(ATP) channels in isolated arcuate neurons in rats. Regul Pept 141:19–24

Mitchell TH, Ellis RD, Smith SA, Robb G, Cawthorne MA (1989) Effects of BRL 35135, a β-adrenoceptor agonist with novel selectivity, on glucose tolerance and insulin sensitivity in obese subjects. Int J Obes 13:757–766

Miyazaki M, Sampath H, Liu X, Flowers MT, Chu K, Dobrzyn A, Ntambi JM (2009) Stearoyl-CoA desaturase-1 deficiency attenuates obesity and insulin resistance in leptin-resistant obese mice. Biochem Biophys Res Commun 380:818–822

Mjos OD (1971) Effect of inhibition of lipolysis on myocardial oxygen consumption in the presence of isoproterenol. J Clin Invest 50:1869–1873

Mnich SJ, Hiebsch RR, Huff RM, Muthian S (2010) Anti-inflammatory properties of CB1-receptor antagonist involves beta2 adrenoceptors. J Pharmacol Exp Ther 333:445–453

Montecucco F, Steffens S, Mach F (2008) Insulin resistance: a proinflammatory state mediated by lipid-induced signaling dysfunction and involved in atherosclerotic plaque instability. Mediators Inflamm 2008:767623

Morino K, Petersen KF, Shulman GI (2006) Molecular mechanisms of insulin resistance in humans and their potential links with mitochondrial dysfunction. Diabetes 55 (Suppl 2):S9–S15

Morrison SF (2001) Differential regulation of sympathetic outflows to vasoconstrictor and thermoregulatory effectors. Ann NY Acad Sci 940:286–298

Morton NM, Paterson JM, Masuzaki H, Holmes MC, Staels B, Fievet C, Walker BR, Flier JS, Mullins JJ, Seckl JR (2004) Novel adipose tissue-mediated resistance to diet-induced visceral obesity in 11 β-hydroxysteroid dehydrogenase type 1-deficient mice. Diabetes 53:931–938

Motoshima H, Wu X, Sinha MK, Hardy VE, Rosato EL, Barbot DJ, Rosato FE, Goldstein BJ (2002) Differential regulation of adiponectin secretion from cultured human omental and subcutaneous adipocytes: effects of insulin and rosiglitazone. J Clin Endocrinol Metab 87:5662–5667

Mullen KL, Pritchard J, Ritchie I, Snook LA, Chabowski A, Bonen A, Wright D, Dyck DJ (2009) Adiponectin resistance precedes the accumulation of skeletal muscle lipids and insulin resistance in high-fat-fed rats. Am J Physiol Regul Integr Comp Physiol 296:R243–R251

Murase T, Misawa K, Haramizu S, Minegishi Y, Hase T (2010) Nootkatone, a characteristic constituent of grapefruit, stimulates energy metabolism and prevents diet-induced obesity by activating AMPK. Am J Physiol Endocrinol Metab 299:E266–E275

Nakagawa T, Ogawa Y, Ebihara K, Yamanaka M, Tsuchida A, Taiji M, Noguchi H, Nakao K (2003) Anti-obesity and anti-diabetic effects of brain-derived neurotrophic factor in rodent models of leptin resistance. Int J Obes Relat Metab Disord 27:557–565

Nakazato M, Murakami N, Date Y, Kojima M, Matsuo H, Kangawa K, Matsukura S (2001) A role for ghrelin in the central regulation of feeding. Nature 409:194–198

Narasimhan ML, Coca MA, Jin J, Yamauchi T, Ito Y, Kadowaki T, Kim KK, Pardo JM, Damsz B, Hasegawa PM, Yun DJ, Bressan RA (2005) Osmotin is a homolog of mammalian adiponectin and controls apoptosis in yeast through a homolog of mammalian adiponectin receptor. Mol Cell 17:171–180

Niederhoffer N, Schmid K, Szabo B (2003) The peripheral sympathetic nervous system is the major target of cannabinoids in eliciting cardiovascular depression. Naunyn Schmiedebergs Arch Pharmacol 367:434–443

Nogueiras R, Veyrat-Durebex C, Suchanek PM, Klein M, Tschop J, Caldwell C, Woods SC, Wittmann G, Watanabe M, Liposits Z, Fekete C, Reizes O, Rohner-Jeanrenaud F, Tschop MH (2008) Peripheral, but not central, CB1 antagonism provides food intake independent metabolic benefits in diet-induced obese rats. Diabetes 57(11):2977–2991

Nonomura T, Tsuchida A, Ono-Kishino M, Nakagawa T, Taiji M, Noguchi H (2001) Brain-derived neurotrophic factor regulates energy expenditure through the central nervous system in obese diabetic mice. Int J Exp Diabetes Res 2:201–209

Ntambi JM, Miyazaki M, Stoehr JP, Lan H, Kendziorski CM, Yandell BS, Song Y, Cohen P, Friedman JM, Attie AD (2002) Loss of stearoyl-CoA desaturase-1 function protects mice against adiposity. Proc Natl Acad Sci USA 99:11482–11486

O'Gorman DJ, Krook A (2008) Exercise and the treatment of diabetes and obesity. Endocrinol Metab Clin North Am 37:887–903

Ohkawara K, Tanaka S, Miyachi M, Ishikawa-Takata K, Tabata I (2007) A dose-response relation between aerobic exercise and visceral fat reduction: systematic review of clinical trials. Int J Obes (Lond) 31:1786–1797

Oliver WR Jr, Shenk JL, Snaith MR, Russell CS, Plunket KD, Bodkin NL, Lewis MC, Winegar DA, Sznaidman ML, Lambert MH, Xu HE, Sternbach DD, Kliewer SA, Hansen BC, Willson TM (2001) A selective peroxisome proliferator-activated receptor delta agonist promotes reverse cholesterol transport. Proc Natl Acad Sci USA 98:5306–5311

Orozco LJ, Buchleitner AM, Gimenez-Perez G, Roque IFM, Richter B, Mauricio D (2008) Exercise or exercise and diet for preventing type 2 diabetes mellitus. Cochrane Database Syst Rev:CD003054

Osaka T, Tsukamoto A, Koyama Y, Inoue S (2008) Central and peripheral administration of amylin induces energy expenditure in anesthetized rats. Peptides 29:1028–1035

Osei-Hyiaman D, DePetrillo M, Pacher P, Liu J, Radaeva S, Batkai S, Harvey-White J, Mackie K, Offertaler L, Wang L, Kunos G (2005) Endocannabinoid activation at hepatic CB1 receptors stimulates fatty acid synthesis and contributes to diet-induced obesity. J Clin Invest 115:1298–1305

Ott V, Fasshauer M, Dalski A, Klein HH, Klein J (2002) Direct effects of ciliary neurotrophic factor on brown adipocytes: evidence for a role in peripheral regulation of energy homeostasis. J Endocrinol 173:R1–R8

Owen MR, Doran E, Halestrap AP (2000) Evidence that metformin exerts its anti-diabetic effects through inhibition of complex 1 of the mitochondrial respiratory chain. Biochem J 348 (Pt 3):607–614

Ozcan L, Ergin AS, Lu A, Chung J, Sarkar S, Nie D, Myers MG Jr, Ozcan U (2009) Endoplasmic reticulum stress plays a central role in development of leptin resistance. Cell Metab 9:35–51

Pacholec M, Bleasdale JE, Chrunyk B, Cunningham D, Flynn D, Garofalo RS, Griffith D, Griffor M, Loulakis P, Pabst B, Qiu X, Stockman B, Thanabal V, Varghese A, Ward J, Withka J, Ahn K (2010) SRT1720, SRT2183, SRT1460, and resveratrol are not direct activators of SIRT1. J Biol Chem 285:8340–8351

Packard GC, Boardman TJ (1999) The use of percentages and size-specific indices to normalise physiological data for variation in body size: wasted time wasted effort? Comp Biochem Physiol A 122:37–44

Paolisso G, Amato L, Eccellente R, Gambardella A, Tagliamonte MR, Varricchio G, Carella C, Giugliano D, D'Onofrio F (1998) Effect of metformin on food intake in obese subjects. Eur J Clin Invest 28:441–446

Pavo I, Jermendy G, Varkonyi TT, Kerenyi Z, Gyimesi A, Shoustov S, Shestakova M, Herz M, Johns D, Schluchter BJ, Festa A, Tan MH (2003) Effect of pioglitazone compared with metformin on glycemic control and indicators of insulin sensitivity in recently diagnosed patients with type 2 diabetes. J Clin Endocrinol Metab 88:1637–1645

Pavon FJ, Serrano A, Perez-Valero V, Jagerovic N, Hernandez-Folgado L, Bermudez-Silva FJ, Macias M, Goya P, de Fonseca FR (2008) Central versus peripheral antagonism of cannabinoid CB1 receptor in obesity: effects of LH-21, a peripherally acting neutral cannabinoid receptor antagonist, in Zucker rats. J Neuroendocrinol 20(Suppl 1):116–123

Pencek RR, Shearer J, Camacho RC, James FD, Lacy DB, Fueger PT, Donahue EP, Snead W, Wasserman DH (2005) 5-Aminoimidazole-4-carboxamide-1-beta-D-ribofuranoside causes acute hepatic insulin resistance in vivo. Diabetes 54:355–360

Perriello G, Misericordia P, Volpi E, Santucci A, Santucci C, Ferrannini E, Ventura MM, Santeusanio F, Brunetti P, Bolli GB (1994) Acute antihyperglycemic mechanisms of metformin in NIDDM. Evidence for suppression of lipid oxidation and hepatic glucose production. Diabetes 43:920–928

Perrin D, Mamet J, Geloen A, Morel G, Dalmaz Y, Pequignot JM (2003) Sympathetic and brain monoaminergic regulation of energy balance in obesity-resistant rats (Lou/C). Auton Neurosci 109:1–9

Peters JM, Lee SS, Li W, Ward JM, Gavrilova O, Everett C, Reitman ML, Hudson LD, Gonzalez FJ (2000) Growth, adipose, brain, and skin alterations resulting from targeted disruption of the mouse peroxisome proliferator-activated receptor beta(delta). Mol Cell Biol 20:5119–5128

Pfluger PT, Kirchner H, Gunnel S, Schrott B, Perez-Tilve D, Fu S, Benoit SC, Horvath T, Joost HG, Wortley KE, Sleeman MW, Tschop MH (2008) Simultaneous deletion of ghrelin and its receptor increases motor activity and energy expenditure. Am J Physiol Gastrointest Liver Physiol 294:G610–G618

Pinnamaneni SK, Southgate RJ, Febbraio MA, Watt MJ (2006) Stearoyl CoA desaturase 1 is elevated in obesity but protects against fatty acid-induced skeletal muscle insulin resistance in vitro. Diabetologia 49:3027–3037

Pirola L, Frojdo S (2008) Resveratrol: one molecule, many targets. IUBMB Life 60:323–332

Popeijus HE, Saris WH, Mensink RP (2008) Role of stearoyl-CoA desaturases in obesity and the metabolic syndrome. Int J Obes (Lond) 32:1076–1082

Porter SA, Massaro JM, Hoffmann U, Vasan RS, O'Donnel CJ, Fox CS (2009) Abdominal subcutaneous adipose tissue: a protective fat depot? Diabetes Care 32:1068–1075

Pratley RE (2008) Overview of glucagon-like peptide-1 analogs and dipeptidyl peptidase-4 inhibitors for type 2 diabetes. Medscape J Med 10:171

Proietto J, Thorburn AW (2003) The therapeutic potential of leptin. Expert Opin Invest Drugs 12:373–378

Qi Y, Takahashi N, Hileman SM, Patel HR, Berg AH, Pajvani UB, Scherer PE, Ahima RS (2004) Adiponectin acts in the brain to decrease body weight. Nat Med 10:524–529

Qi L, Hu FB, Hu G (2008) Genes, environment, and interactions in prevention of type 2 diabetes: a focus on physical activity and lifestyle changes. Curr Mol Med 8:519–532

Rahmouni K, Morgan DA (2007) Hypothalamic arcuate nucleus mediates the sympathetic and arterial pressure responses to leptin. Hypertension 49:647–652

Ribeiro MO (2008) Effects of thyroid hormone analogs on lipid metabolism and thermogenesis. Thyroid 18:197–203

Riboulet-Chavey A, Diraison F, Siew LK, Wong FS, Rutter GA (2008) Inhibition of AMP-activated protein kinase protects pancreatic beta-cells from cytokine-mediated apoptosis and CD8+ T-cell-induced cytotoxicity. Diabetes 57:415–423

Rodgers JT, Puigserver P (2007) Fasting-dependent glucose and lipid metabolic response through hepatic sirtuin 1. Proc Natl Acad Sci USA 104:12861–12866

Rodriguez A, Gomez-Ambrosi J, Catalan V, Gil MJ, Becerril S, Sainz N, Silva C, Salvador J, Colina I, Fruhbeck G (2009) Acylated and desacyl ghrelin stimulate lipid accumulation in human visceral adipocytes. Int J Obes (Lond) 33:541–552

Rohrbach KW, Han S, Gan J, O'Tanyi EJ, Zhang H, Chi CL, Taub R, Largent BL, Cheng D (2005) Disconnection between the early onset anorectic effects by C75 and hypothalamic fatty acid synthase inhibition in rodents. Eur J Pharmacol 511:31–41

Rolli V, Radosavljevic M, Astier V, Macquin C, Castan-Laurell I, Visentin V, Guigne C, Carpene C, Valet P, Gilfillan S, Bahram S (2007) Lipolysis is altered in MHC class I zinc-alpha(2)-glycoprotein deficient mice. FEBS Lett 581:394–400

Ronnebaum SM, Joseph JW, Ilkayeva O, Burgess SC, Lu D, Becker TC, Sherry AD, Newgard CB (2008) Chronic suppression of acetyl-CoA carboxylase 1 in beta-cells impairs insulin secretion via inhibition of glucose rather than lipid metabolism. J Biol Chem 283:14248–14256

Rosenstock J, Banarer S, Fonseca VA, Inzucchi SE, Sun W, Yao W, Hollis G, Flores R, Levy R, Williams WV, Seckl JR, Huber R (2010) The 11-beta-hydroxysteroid dehydrogenase type 1 inhibitor INCB13739 improves hyperglycemia in patients with type 2 diabetes inadequately controlled by metformin monotherapy. Diabetes Care 33:1516–1522

Roth JD, Roland BL, Cole RL, Trevaskis JL, Weyer C, Koda JE, Anderson CM, Parkes DG, Baron AD (2008) Leptin responsiveness restored by amylin agonism in diet-induced obesity: evidence from nonclinical and clinical studies. Proc Natl Acad Sci USA 105:7257–7262

Rubio A, Raasmaja A, Maia AL, Kim KR, Silva JE (1995) Effects of thyroid hormone on norepinephrine signaling in brown adipose tissue. I. Beta 1- and beta 2-adrenergic receptors and cyclic adenosine $3'$, $5'$-monophosphate generation. Endocrinology 136:3267–3276

Rudolph J, Esler WP, O'Connor S, Coish PD, Wickens PL, Brands M, Bierer DE, Bloomquist BT, Bondar G, Chen L, Chuang CY, Claus TH, Fathi Z, Fu W, Khire UR, Kristie JA, Liu XG, Lowe DB, McClure AC, Michels M, Ortiz AA, Ramsden PD, Schoenleber RW, Shelekhin TE, Vakalopoulos A, Tang W, Wang L, Yi L, Gardell SJ, Livingston JN, Sweet LJ, Bullock WH (2007) Quinazolinone derivatives as orally available ghrelin receptor antagonists for the treatment of diabetes and obesity. J Med Chem 50:5202–5216

Ruilope LM, Despres JP, Scheen A, Pi-Sunyer X, Mancia G, Zanchetti A, Van Gaal L (2008) Effect of rimonabant on blood pressure in overweight/obese patients with/without co-morbidities: analysis of pooled RIO study results. J Hypertens 26:357–367

Russell ST, Tisdale MJ (2002) Effect of a tumour-derived lipid-mobilising factor on glucose and lipid metabolism in vivo. Br J Cancer 87:580–584

Russell ST, Tisdale MJ (2010) Antidiabetic properties of zinc-alpha2-glycoprotein in ob/ob mice. Endocrinology 151:948–957

Russell ST, Hirai K, Tisdale MJ (2002) Role of β_3-adrenergic receptors in the action of a tumour lipid mobilizing factor. Br J Cancer 86:424–428

Rutanen J, Yaluri N, Modi S, Pihlajamaki J, Vanttinen M, Itkonen P, Kainulainen S, Yamamoto H, Lagouge M, Sinclair DA, Elliott P, Westphal C, Auwerx J, Laakso M (2010) SIRT1 mRNA expression may be associated with energy expenditure and insulin sensitivity. Diabetes 59:829–835

Rutter GA, Parton LE (2008) The beta-cell in type 2 diabetes and in obesity. Front Horm Res 36:118–134

Ryu GR, Lee MK, Lee E, Ko SH, Ahn YB, Kim JW, Yoon KH, Song KH (2009) Activation of AMP-activated protein kinase mediates acute and severe hypoxic injury to pancreatic beta cells. Biochem Biophys Res Commun 386:356–362

Saeedi R, Parsons HL, Wambolt RB, Paulson K, Sharma V, Dyck JR, Brownsey RW, Allard MF (2008) Metabolic actions of metformin in the heart can occur by AMPK-independent mechanisms. Am J Physiol Heart Circ Physiol 294:H2497–H2506

Salome N, Hansson C, Taube M, Gustafsson-Ericson L, Egecioglu E, Karlsson-Lindahl L, Fehrentz JA, Martinez J, Perrissoud D, Dickson SL (2009) On the central mechanism underlying ghrelin's chronic pro-obesity effects in rats: new insights from studies exploiting a potent ghrelin receptor (GHS-R1A) antagonist. J Neuroendocrinol 21(9):777–85

Sampath H, Flowers MT, Liu X, Paton CM, Sullivan R, Chu K, Zhao M, Ntambi JM (2009) Skin-specific deletion of stearoyl-CoA desaturase-1 alters skin lipid composition and protects mice from high-fat diet-induced obesity. J Biol Chem 284(30):19961–73

Sanders PM, Tisdale MJ (2004a) Effect of zinc-alpha2-glycoprotein (ZAG) on expression of uncoupling proteins in skeletal muscle and adipose tissue. Cancer Lett 212:71–81

Sanders PM, Tisdale MJ (2004b) Role of lipid-mobilising factor (LMF) in protecting tumour cells from oxidative damage. Br J Cancer 90:1274–1278

Sarruf DA, Thaler JP, Morton GJ, German J, Fischer JD, Ogimoto K, Schwartz MW (2010) Fibroblast growth factor 21 action in the brain increases energy expenditure and insulin sensitivity in obese rats. Diabetes 59:1817–1824

Scheen AJ (2008) CB1 receptor blockade and its impact on cardiometabolic risk factors: overview of the RIO programme with rimonabant. J Neuroendocrinol 20(Suppl 1):139–146

Schiff M, Loublier S, Coulibaly A, Benit P, de Baulny HO, Rustin P (2009) Mitochondria and diabetes mellitus: untangling a conflictive relationship? J Inherit Metab Dis 32:684–698

Schiffelers SL, Brouwer EM, Saris WH, van Baak MA (1998) Inhibition of lipolysis reduces β1-adrenoceptor-mediated thermogenesis in man. Metabolism 47:1462–1467

Schiffelers SL, Blaak EE, Saris WH, van Baak MA (2000) In vivo β3-adrenergic stimulation of human thermogenesis and lipid use. Clin Pharmacol Ther 67:558–566

Schmidt MI, Duncan BB, Vigo A, Pankow JS, Couper D, Ballantyne CM, Hoogeveen RC, Heiss G (2006) Leptin and incident type 2 diabetes: risk or protection? Diabetologia 49:2086–2096

Schmitz-Peiffer C, Craig DL, Biden TJ (1999) Ceramide generation is sufficient to account for the inhibition of the insulin-stimulated PKB pathway in C2C12 skeletal muscle cells pretreated with palmitate. J Biol Chem 274:24202–24210

Schrauwen P, Hesselink MK (2008) Reduced tricarboxylic acid cycle flux in type 2 diabetes mellitus? Diabetologia 51:1694–1697

Schreurs M, van Dijk TH, Gerding A, Havinga R, Reijngoud DJ, Kuipers F (2009) Soraphen, an inhibitor of the acetyl-CoA carboxylase system, improves peripheral insulin sensitivity in mice fed a high-fat diet. Diabetes Obes Metab 11(10):987–91

Sennitt MV, Kaumann AJ, Molenaar P, Beeley LJ, Young PW, Kelly J, Chapman H, Henson SM, Berge JM, Dean DK, Kotecha NR, Morgan HK, Rami HK, Ward RW, Thompson M, Wilson S, Smith SA, Cawthorne MA, Stock MJ, Arch JR (1998) The contribution of classical (β$_{1/2}$-) and atypical β-adrenoceptors to the stimulation of human white adipocyte lipolysis and right atrial appendage contraction by novel β$_3$-adrenoceptor agonists of differing selectivities. J Pharmacol Exp Ther 285:1084–1095

Shaw RJ, Lamia KA, Vasquez D, Koo SH, Bardeesy N, Depinho RA, Montminy M, Cantley LC (2005) The kinase LKB1 mediates glucose homeostasis in liver and therapeutic effects of metformin. Science 310:1642–1646

Shetty S, Kusminski CM, Scherer PE (2009) Adiponectin in health and disease: evaluation of adiponectin-targeted drug development strategies. Trends Pharmacol Sci 30:234–239

Sleeman MW, Garcia K, Liu R, Murray JD, Malinova L, Moncrieffe M, Yancopoulos GD, Wiegand SJ (2003) Ciliary neurotrophic factor improves diabetic parameters and hepatic steatosis and increases basal metabolic rate in db/db mice. Proc Natl Acad Sci USA 100:14297–14302

Slentz CA, Houmard JA, Kraus WE (2009) Exercise, abdominal obesity, skeletal muscle, and metabolic risk: evidence for a dose response. Obesity (Silver Spring) 17(suppl 3):S27–S33

Smith SA, Sennitt MV, Cawthorne MA (1990) BRL 35135: an orally active antihyperglycaemic agent with weight reducing effects. In: Bailey CJ, Flatt PR (eds) New antidiabetic drugs. Smith-Gordon, London, pp 177–189

Smith SJ, Cases S, Jensen DR, Chen HC, Sande E, Tow B, Sanan DA, Raber J, Eckel RH, Farese RV Jr (2000) Obesity resistance and multiple mechanisms of triglyceride synthesis in mice lacking Dgat. Nat Genet 25:87–90

Smith SR, Aronne LJ, Burns CM, Kesty NC, Halseth AE, Weyer C (2008) Sustained weight loss following 12-month pramlintide treatment as an adjunct to lifestyle intervention in obesity. Diabetes Care 31:1816–1823

Soares JB, Leite-Moreira AF (2008) Ghrelin, des-acyl ghrelin and obestatin: three pieces of the same puzzle. Peptides 29:1255–1270

Son MH, Kim HD, Chae YN, Kim MK, Shin CY, Ahn GJ, Choi SH, Yang EK, Park KJ, Chae HW, Moon HS, Kim SH, Shin YG, Yoon SH (2010) Peripherally acting CB1-receptor antagonist: the relative importance of central and peripheral CB1 receptors in adiposity control. Int J Obes (Lond) 34:547–556

Soulage C, Zarrouki B, Soares AF, Lagarde M, Geloen A (2008) Lou/C obesity-resistant rat exhibits hyperactivity, hypermetabolism, alterations in white adipose tissue cellularity, and lipid tissue profiles. Endocrinology 149:615–625

Speakman JR (2010) FTO effect on energy demand versus food intake. Nature 464:E1; discussion E2

Strassburg S, Anker SD, Castaneda TR, Burget L, Perez-Tilve D, Pfluger PT, Nogueiras R, Halem H, Dong JZ, Culler MD, Datta R, Tschop MH (2008) Long-term effects of ghrelin and ghrelin receptor agonists on energy balance in rats. Am J Physiol Endocrinol Metab 295:E78–E84

Sugimoto T, Ogawa W, Kasuga M, Yokoyama Y (2005) Chronic effects of AJ-9677 on energy expenditure and energy source utilization in rats. Eur J Pharmacol 519:135–145

Sun Y, Ahmed S, Smith RG (2003) Deletion of ghrelin impairs neither growth nor appetite. Mol Cell Biol 23:7973–7981

Sun Y, Asnicar M, Saha PK, Chan L, Smith RG (2006) Ablation of ghrelin improves the diabetic but not obese phenotype of ob/ob mice. Cell Metab 3:379–386

Sun Y, Asnicar M, Smith RG (2007) Central and peripheral roles of ghrelin on glucose homeostasis. Neuroendocrinology 86:215–228

Sundbom M, Kaiser C, Bjorkstrand E, Castro VM, Larsson C, Selen G, Nyhem CS, James SR (2008) Inhibition of 11betaHSD1 with the S-phenylethylaminothiazolone BVT116429 increases adiponectin concentrations and improves glucose homeostasis in diabetic KKAy mice. BMC Pharmacol 8:3

Sznaidman ML, Haffner CD, Maloney PR, Fivush A, Chao E, Goreham D, Sierra ML, LeGrumelec C, Xu HE, Montana VG, Lambert MH, Willson TM, Oliver WR Jr, Sternbach DD (2003) Novel selective small molecule agonists for peroxisome proliferator-activated receptor delta (PPARdelta)–synthesis and biological activity. Bioorg Med Chem Lett 13:1517–1521

Tam J, Vemuri VK, Liu J, Batkai S, Mukhopadhyay B, Godlewski G, Osei-Hyiaman D, Ohnuma S, Ambudkar SV, Pickel J, Makriyannis A, Kunos G (2010) Peripheral CB1 cannabinoid receptor blockade improves cardiometabolic risk in mouse models of obesity. J Clin Invest 120:2953–2966

Tanaka T, Yamamoto J, Iwasaki S, Asaba H, Hamura H, Ikeda Y, Watanabe M, Magoori K, Ioka RX, Tachibana K, Watanabe Y, Uchiyama Y, Sumi K, Iguchi H, Ito S, Doi T, Hamakubo T, Naito M, Auwerx J, Yanagisawa M, Kodama T, Sakai J (2003) Activation of peroxisome proliferator-activated receptor delta induces fatty acid β-oxidation in skeletal muscle and attenuates metabolic syndrome. Proc Natl Acad Sci USA 100:15924–15929

Taslim S, Tai ES (2009) The relevance of the metabolic syndrome. Ann Acad Med Singapore 38:29–35

Tedesco L, Valerio A, Cervino C, Cardile A, Pagano C, Vettor R, Pasquali R, Carruba MO, Marsicano G, Lutz B, Pagotto U, Nisoli E (2008) Cannabinoid type 1 receptor blockade promotes mitochondrial biogenesis through endothelial nitric oxide synthase expression in white adipocytes. Diabetes 57:2028–2036

Teran-Garcia M, Rankinen T, Koza RA, Rao DC, Bouchard C (2005) Endurance training-induced changes in insulin sensitivity and gene expression. Am J Physiol Endocrinol Metab 288: E1168–E1178

Terao A, Oikawa M, Saito M (1994) Tissue-specific increase in norepinephrine turnover by central interleukin-1, but not by interleukin-6, in rats. Am J Physiol 266:R400–R404

Theander-Carrillo C, Wiedmer P, Cettour-Rose P, Nogueiras R, Perez-Tilve D, Pfluger P, Castaneda TR, Muzzin P, Schurmann A, Szanto I, Tschop MH, Rohner-Jeanrenaud F (2006) Ghrelin action in the brain controls adipocyte metabolism. J Clin Invest 116:1983–1993

Thomas EL, Brynes AE, McCarthy J, Goldstone AP, Hajnal JV, Saeed N, Frost G, Bell JD (2000) Preferential loss of visceral fat following aerobic exercise, measured by magnetic resonance imaging. Lipids 35:769–776

Thomas C, Gioiello A, Noriega L, Strehle A, Oury J, Rizzo G, Macchiarulo A, Yamamoto H, Mataki C, Pruzanski M, Pellicciari R, Auwerx J, Schoonjans K (2009) TGR5-mediated bile acid sensing controls glucose homeostasis. Cell Metab 10:167–177

Thong FS, Lally JS, Dyck DJ, Greer F, Bonen A, Graham TE (2007) Activation of the A1 adenosine receptor increases insulin-stimulated glucose transport in isolated rat soleus muscle. Appl Physiol Nutr Metab 32:701–710

Thupari JN, Landree LE, Ronnett GV, Kuhajda FP (2002) C75 increases peripheral energy utilization and fatty acid oxidation in diet-induced obesity. Proc Natl Acad Sci USA 99:9498–9502

Toth MJ (2001) Comparing energy expenditure data among individuals differing in body size and composition: statistical and physiological considerations. Curr Opin Clin Nutr Metab Care 4:391–397

Toyoshima Y, Gavrilova O, Yakar S, Jou W, Pack S, Asghar Z, Wheeler MB, LeRoith D (2005) Leptin improves insulin resistance and hyperglycemia in a mouse model of type 2 diabetes. Endocrinology 146:4024–4035

Treebak JT, Birk JB, Hansen BF, Olsen GS, Wojtaszewski JF (2009) A-769662 activates AMPK beta1-containing complexes but induces glucose uptake through a PI3-kinase-dependent pathway in mouse skeletal muscle. Am J Physiol Cell Physiol 297:C1041–C1052

Tseng YH, Cypess AM, Kahn CR (2010) Cellular bioenergetics as a target for obesity therapy. Nat Rev Drug Discov 9:465–482

Tsuchida A, Nonomura T, Ono-Kishino M, Nakagawa T, Taiji M, Noguchi H (2001) Acute effects of brain-derived neurotrophic factor on energy expenditure in obese diabetic mice. Int J Obes Relat Metab Disord 25:1286–1293

Tu Y, Thupari JN, Kim EK, Pinn ML, Moran TH, Ronnett GV, Kuhajda FP (2005) C75 alters central and peripheral gene expression to reduce food intake and increase energy expenditure. Endocrinology 146:486–493

Ukropcova B, Sereda O, de Jonge L, Bogacka I, Nguyen T, Xie H, Bray GA, Smith SR (2007) Family history of diabetes links impaired substrate switching and reduced mitochondrial content in skeletal muscle. Diabetes 56:720–727

Um JH, Park SJ, Kang H, Yang S, Foretz M, McBurney MW, Kim MK, Viollet B, Chung JH (2010) AMP-activated protein kinase-deficient mice are resistant to the metabolic effects of resveratrol. Diabetes 59:554–563

Uto Y, Kiyotsuka Y, Ueno Y, Miyazawa Y, Kurata H, Ogata T, Deguchi T, Yamada M, Watanabe N, Konishi M, Kurikawa N, Takagi T, Wakimoto S, Kono K, Ohsumi J (2010) Novel spiropiperidine-based stearoyl-CoA desaturase-1 inhibitors: Identification of 1'-{6-[5-(pyridin-3-ylmethyl)-1, 3, 4-oxadiazol-2-yl]pyridazin-3-yl}-5-(trif luoromethyl)-3, 4-dihydrospiro[chromene-2, 4'-piperidine]. Bioorg Med Chem Lett 20:746–754

Vacher CM, Crepin D, Aubourg A, Couvreur O, Bailleux V, Nicolas V, Ferezou J, Gripois D, Gertler A, Taouis M (2008) A putative physiological role of hypothalamic CNTF in the control of energy homeostasis. FEBS Lett 582:3832–3838

Vaillancourt VA, Larsen SD, Tanis SP, Burr JE, Connell MA, Cudahy MM, Evans BR, Fisher PV, May PD, Meglasson MD, Robinson DD, Stevens FC, Tucker JA, Vidmar TJ, Yu JH (2001) Synthesis and biological activity of aminoguanidine and diaminoguanidine analogues of the antidiabetic/antiobesity agent 3-guanidinopropionic acid. J Med Chem 44:1231–1248

Van de Laar FA, Lucassen PL, Akkermans RP, Van de Lisdonk EH, De Grauw WJ (2006) Alpha-glucosidase inhibitors for people with impaired glucose tolerance or impaired fasting blood glucose. Cochrane Database Syst Rev:CD005061

van den Hoek AM, van Heijningen C, Schroder-van der Elst JP, Ouwens DM, Havekes LM, Romijn JA, Kalsbeek A, Pijl H (2008) Intracerebroventricular administration of neuropeptide Y induces hepatic insulin resistance via sympathetic innervation. Diabetes 57:2304–2310

van Marken Lichtenbelt WD, Vanhommerig JW, Smulders NM, Drossaerts JM, Kemerink GJ, Bouvy ND, Schrauwen P, Teule GJ (2009) Cold-activated brown adipose tissue in healthy men. N Engl J Med 360:1500–1508

Venditti P, Chiellini G, Di Stefano L, Napolitano G, Zucchi R, Columbano A, Scanlan TS, Di Meo S (2010) The TRbeta-selective agonist, GC-1, stimulates mitochondrial oxidative processes to a lesser extent than triiodothyronine. J Endocrinol 205:279–289

Verty AN, Allen AM, Oldfield BJ (2009) The effects of rimonabant on brown adipose tissue in rat: implications for energy expenditure. Obesity (Silver Spring) 17:254–261

Villicev CM, Freitas FR, Aoki MS, Taffarel C, Scanlan TS, Moriscot AS, Ribeiro MO, Bianco AC, Gouveia CH (2007) Thyroid hormone receptor beta-specific agonist GC-1 increases energy expenditure and prevents fat-mass accumulation in rats. J Endocrinol 193:21–29

Violett B, Andreelli F (2011) AMP-activated protein kinase and metabolic control. Handb Exp Pharmacol. doi:10.1007/978-3-642-17214-4_13

Viollet B, Athea Y, Mounier R, Guigas B, Zarrinpashneh E, Horman S, Lantier L, Hebrard S, Devin-Leclerc J, Beauloye C, Foretz M, Andreelli F, Ventura-Clapier R, Bertrand L (2009a) AMPK: Lessons from transgenic and knockout animals. Front Biosci 14:19–44

Viollet B, Lantier L, Devin-Leclerc J, Hebrard S, Amouyal C, Mounier R, Foretz M, Andreelli F (2009b) Targeting the AMPK pathway for the treatment of Type 2 diabetes. Front Biosci 14:3380–3400

Virtanen KA, Rouru J, Hanninen V, Savontaus E, Rouvari T, Teirmaa T, Koulu M, Huupponen R (1997) Chronic treatment with BRL 35135 potentiates the action of insulin on lipid metabolism. Eur J Pharmacol 332:215–218

Wakil SJ, Abu-Elheiga LA (2009) Fatty acid metabolism: target for metabolic syndrome. J Lipid Res 50(Suppl):S138–S143

Wang M (2006) Inhibitors of 11β-hydroxysteroid dehydrogenase type 1 for the treatment of metabolic syndrome. Curr Opin Investig Drugs 7:319–323

Wang M (2011) Inhibitors of 11β-hydroxysteroid dehydrogenase type 1 in antidiabetic therapy. Handb Exp Pharmacol. doi:10.1007/978-3-642-17214-4_6

Wang YX, Lee CH, Tiep S, Yu RT, Ham J, Kang H, Evans RM (2003) Peroxisome-proliferator-activated receptor δ activates fat metabolism to prevent obesity. Cell 113:159–170

Wang YX, Zhang CL, Yu RT, Cho HK, Nelson MC, Bayuga-Ocampo CR, Ham J, Kang H, Evans RM (2004) Regulation of muscle fiber type and running endurance by PPARdelta. PLoS Biol 2:e294

Wang SJ, Cornick C, O'Dowd J, Cawthorne MA, Arch JR (2007) Improved glucose tolerance in acyl CoA:diacylglycerol acyltransferase 1-null mice is dependent on diet. Lipids Health Dis 6:2

Watanabe M, Houten SM, Mataki C, Christoffolete MA, Kim BW, Sato H, Messaddeq N, Harney JW, Ezaki O, Kodama T, Schoonjans K, Bianco AC, Auwerx J (2006) Bile acids induce energy expenditure by promoting intracellular thyroid hormone activation. Nature 439:484–489

Watanabe T, Kubota N, Ohsugi M, Kubota T, Takamoto I, Iwabu M, Awazawa M, Katsuyama H, Hasegawa C, Tokuyama K, Moroi M, Sugi K, Yamauchi T, Noda T, Nagai R, Terauchi Y, Tobe K, Ueki K, Kadowaki T (2009) Rimonabant ameliorates insulin resistance via both adiponectin-dependent and adiponectin-independent pathways. J Biol Chem 284:1803–1812

Watt MJ, Dzamko N, Thomas WG, Rose-John S, Ernst M, Carling D, Kemp BE, Febbraio MA, Steinberg GR (2006a) CNTF reverses obesity-induced insulin resistance by activating skeletal muscle AMPK. Nat Med 12:541–548

Watt MJ, Hevener A, Lancaster GI, Febbraio MA (2006b) Ciliary neurotrophic factor prevents acute lipid-induced insulin resistance by attenuating ceramide accumulation and phosphorylation of c-Jun N-terminal kinase in peripheral tissues. Endocrinology 147:2077–2085

Weiss EP, Racette SB, Villareal DT, Fontana L, Steger-May K, Schechtman KB, Klein S, Holloszy JO (2006) Improvements in glucose tolerance and insulin action induced by increasing energy expenditure or decreasing energy intake: a randomized controlled trial. Am J Clin Nutr 84:1033–1042

Wernstedt I, Edgley A, Berndtsson A, Faldt J, Bergstrom G, Wallenius V, Jansson JO (2006) Reduced stress- and cold-induced increase in energy expenditure in interleukin-6-deficient mice. Am J Physiol Regul Integr Comp Physiol 291:R551–R557

Wilson S, Arch JR, Thurlby PL (1984) Genetically obese C57BL/6 ob/ob mice respond normally to sympathomimetic compounds. Life Sci 35:1301–1309

Wilson S, Thurlby PL, Arch JR (1986) Substrate supply for thermogenesis induced by the β-adrenoceptor agonist BRL 26830A. Can J Physiol Pharmacol 65:113–119

Wing RR, Phelan S (2005) Long-term weight loss maintenance. Am J Clin Nutr 82:222S–225S

Wortley KE, del Rincon JP, Murray JD, Garcia K, Iida K, Thorner MO, Sleeman MW (2005) Absence of ghrelin protects against early-onset obesity. J Clin Invest 115:3573–3578

Wren AM, Small CJ, Abbott CR, Dhillo WS, Seal LJ, Cohen MA, Batterham RL, Taheri S, Stanley SA, Ghatei MA, Bloom SR (2001) Ghrelin causes hyperphagia and obesity in rats. Diabetes 50:2540–2547

Xu J, Lloyd DJ, Hale C, Stanislaus S, Chen M, Sivits G, Vonderfecht S, Hecht R, Li YS, Lindberg RA, Chen JL, Jung DY, Zhang Z, Ko HJ, Kim JK, Veniant MM (2009) Fibroblast growth factor 21 reverses hepatic steatosis, increases energy expenditure, and improves insulin sensitivity in diet-induced obese mice. Diabetes 58:250–259

Yamauchi T, Kadowaki T (2008) Physiological and pathophysiological roles of adiponectin and adiponectin receptors in the integrated regulation of metabolic and cardiovascular diseases. Int J Obes (Lond) 32(suppl 7):S13–S18

Yamauchi T, Kamon J, Minokoshi Y, Ito Y, Waki H, Uchida S, Yamashita S, Noda M, Kita S, Ueki K, Eto K, Akanuma Y, Froguel P, Foufelle F, Ferre P, Carling D, Kimura S, Nagai R, Kahn BB, Kadowaki T (2002) Adiponectin stimulates glucose utilization and fatty-acid oxidation by activating AMP-activated protein kinase. Nat Med 8:1288–1295

Yamauchi T, Kamon J, Ito Y, Tsuchida A, Yokomizo T, Kita S, Sugiyama T, Miyagishi M, Hara K, Tsunoda M, Murakami K, Ohteki T, Uchida S, Takekawa S, Waki H, Tsuno NH, Shibata Y, Terauchi Y, Froguel P, Tobe K, Koyasu S, Taira K, Kitamura T, Shimizu T, Nagai R, Kadowaki T (2003) Cloning of adiponectin receptors that mediate antidiabetic metabolic effects. Nature 423:762–769

Yamauchi T, Nio Y, Maki T, Kobayashi M, Takazawa T, Iwabu M, Okada-Iwabu M, Kawamoto S, Kubota N, Kubota T, Ito Y, Kamon J, Tsuchida A, Kumagai K, Kozono H, Hada Y, Ogata H, Tokuyama K, Tsunoda M, Ide T, Murakami K, Awazawa M, Takamoto I, Froguel P, Hara K, Tobe K, Nagai R, Ueki K, Kadowaki T (2007) Targeted disruption of AdipoR1 and AdipoR2 causes abrogation of adiponectin binding and metabolic actions. Nat Med 13:332–339

Yang X, Smith U (2007) Adipose tissue distribution and risk of metabolic disease: does thiazo-lidinedione-induced adipose tissue redistribution provide a clue to the answer? Diabetologia 50:1127–1139

Yu C, Chen Y, Cline GW, Zhang D, Zong H, Wang Y, Bergeron R, Kim JK, Cushman SW, Cooney GJ, Atcheson B, White MF, Kraegen EW, Shulman GI (2002) Mechanism by which fatty acids inhibit insulin activation of insulin receptor substrate-1 (IRS-1)-associated phosphatidylinositol 3-kinase activity in muscle. J Biol Chem 277:50230–50236

Yu YH, Ginsberg HN (2004) The role of acyl-CoA:diacylglycerol acyltransferase (DGAT) in energy metabolism. Ann Med 36:252–261

Yudkin JS (2007) Insulin resistance and the metabolic syndrome–or the pitfalls of epidemiology. Diabetologia 50:1576–1586

Zammit VA, Buckett LK, Turnbull AV, Wure H, Proven A (2008) Diacylglycerol acyltransferases: potential roles as pharmacological targets. Pharmacol Ther 118:295–302

Zanuso S, Jimenez A, Pugliese G, Corigliano G, Balducci S (2010) Exercise for the management of type 2 diabetes: a review of the evidence. Acta Diabetol 47:15–22

Zhang BB, Zhou G, Li C (2009) AMPK: an emerging drug target for diabetes and the metabolic syndrome. Cell Metab 9:407–416

Zhao G, Souers AJ, Voorbach M, Falls HD, Droz B, Brodjian S, Lau YY, Iyengar RR, Gao J, Judd AS, Wagaw SH, Ravn MM, Engstrom KM, Lynch JK, Mulhern MM, Freeman J, Dayton BD, Wang X, Grihalde N, Fry D, Beno DW, Marsh KC, Su Z, Diaz GJ, Collins CA, Sham H, Reilly RM, Brune ME, Kym PR (2008) Validation of diacyl glycerolacyltransferase i as a novel target for the treatment of obesity and dyslipidemia using a potent and selective small molecule inhibitor. J Med Chem 51(3):380–3

Zhou G, Sebhat IK, Zhang BB (2009) AMPK activators–potential therapeutics for metabolic and other diseases. Acta Physiol (Oxf) 196:175–190

Zigman JM, Nakano Y, Coppari R, Balthasar N, Marcus JN, Lee CE, Jones JE, Deysher AE, Waxman AR, White RD, Williams TD, Lachey JL, Seeley RJ, Lowell BB, Elmquist JK (2005) Mice lacking ghrelin receptors resist the development of diet-induced obesity. J Clin Invest 115:3564–3572

Zinker B, Mika A, Nguyen P, Wilcox D, Ohman L, von Geldern TW, Opgenorth T, Jacobson P (2007) Liver-selective glucocorticoid receptor antagonism decreases glucose production and increases glucose disposal, ameliorating insulin resistance. Metabolism 56:380–387

Interleukin-Targeted Therapy for Metabolic Syndrome and Type 2 Diabetes

Kathrin Maedler, Gitanjali Dharmadhikari, Desiree M. Schumann, and Joachim Størling

Contents

Abstract Interleukin-1β (IL-1β) is a key regulator of the body's inflammatory response and is produced after infection, injury, and an antigenic challenge. Cloned in 1984, the single polypeptide IL-1β has been shown to exert numerous biological effects. It plays a role in various diseases, including autoimmune diseases such as rheumatoid arthritis, inflammatory bowel diseases, and Type 1 diabetes, as well as in diseases associated with metabolic syndrome such as atherosclerosis, chronic heart failure, and Type 2 diabetes. The macrophage is the primary source of IL-1β, but epidermal, epithelial, lymphoid, and vascular tissues also synthesize IL-1. Recently, IL-1β production and secretion have also been reported from pancreatic islets. Insulin-producing β-cells within the pancreatic islets are specifically prone to

K. Maedler (✉) and G. Dharmadhikari
Centre for Biomolecular Interactions Bremen, University of Bremen, Leobener Straße NW2, Room B2080, mailbox 330440, 28334, 28359 Bremen, Germany
e-mail: kmaedler@uni-bremen.de

D.M. Schumann
Boehringer-Ingelheim, Cardiometabolic Diseases Research, Biberach, Germany

J. Størling
Hagedorn Research Institute, Gentofte, Denmark

M. Schwanstecher (ed.), *Diabetes - Perspectives in Drug Therapy*,
Handbook of Experimental Pharmacology 203,
DOI 10.1007/978-3-642-17214-4_11, © Springer-Verlag Berlin Heidelberg 2011

IL-β-induced destruction and loss of function. Macrophage-derived IL-1β production in insulin-sensitive organs leads to the progression of inflammation and
induction of insulin resistance in obesity. This chapter explains the mechanisms
involved in the inflammatory response during diabetes progression with specific
attention to the IL-1β signal effects influencing insulin action and insulin secretion.
We highlight recent clinical studies, rodent and in vitro experiments with isolated
islets using IL-1β as a potential target for the therapy of Type 2 diabetes.

Keywords β-Cell · IL-1β · Diabetes · Inflammation · Obesity · Interleukin-1
receptor antagonist

1 Introduction: The IL-1 Family

Twenty-five years ago, IL-1β was cloned in the lab of Charles Dinarello (Auron
et al. 1984). Meanwhile, 11 ligands and 10 receptors of the IL-1 family have been
discovered. The proinflammatory and agonistic ligands are IL-1α, IL-1β, IL-18,
FIL-1ε, IL-1H2, IL-1ε, and IL-33; and the anti-inflammatory and antagonistic
ligands are IL-1Ra, FIL-1δ, IL-1H4, and IL-1Hy2 (Dinarello 2009). IL-1α, IL-1β,
and IL-1Ra bind to IL-1R1; IL-1β and the IL-1β precursors bind to IL-1R2; IL-33
binds to IL-1R4; IL-18 and IL-1H4 to IL-1R5; FIL-1ε, IL-1H2, and IL-1ε to
IL-1R6; and IL-1R8, IL-1R9 and TIR8 remain orphan receptors (Boraschi and
Tagliabue 2006). IL-1β is mainly produced by activated macrophages. Production
and secretion of IL-1β have been linked not only to various autoimmune and
autoinflammatory diseases, but also to metabolic dysregulation (Dinarello 2009).
Signaling pathways of IL-1β have been shown to result in impaired insulin secretion and action (Maedler et al. 2009). Clearly, other cytokines and chemokines are
involved in the inflammatory responses; however, this chapter focuses on the
possibility of blocking only IL-1β as a target for improving glycemia in T2DM.

A recent paper showing that genetic variation in the IL-1 gene family is
associated with hyperglycemia and insulin resistance provides another proof for
the involvement of IL-1β in the pathogenesis of diabetes (Luotola et al. 2009).

2 IL-1β Links Obesity and Diabetes

Chronic subclinical inflammation is present in obesity, insulin resistance, and
T2DM. The diseases related to metabolic syndrome are characterized by abnormal
cytokine production, including elevated circulating IL-1β, increased acute-phase
proteins, e.g., CRP (Koenig et al. 2006), and activation of inflammatory signaling
pathways (Wellen and Hotamisligil 2005).

Proinflammatory cytokines can cause insulin resistance in adipose tissue, skeletal muscle, and liver by inhibiting insulin signal transduction. The sources of

cytokines in insulin-resistant states are the insulin target tissues themselves, primarily fat and liver, but to a larger extent the activated tissue resident macrophages (de Luca and Olefsky 2008).

While macrophage infiltration in adipose and brain tissue has been shown in many studies (Schenk et al. 2008), increased islet macrophage infiltration has only recently been observed in pancreatic sections from patients with T2DM (Ehses et al. 2007; Richardson et al. 2009) and in T2DM animal models, such as the GK rat (Homo-Delarche et al. 2006), the HFD and *db/db* mouse (Ehses et al. 2007), and the hyperglycemic Cohen diabetic rat (Weksler-Zangen et al. 2008). While IL-1β signals induce destruction and impaired insulin secretion in the β-cells, insulin signaling is disturbed in the insulin target tissues (Fig. 1).

Insulin receptor signaling is complex. To summarize shortly, signaling downstream of the insulin receptor involves phosphorylation of IRS1/2 and the activation of the PI3K–AKT pathway (responsible for insulin action on glucose uptake) and the Ras-mitogen-activated protein kinase (MAPK) pathway (responsible for suppression of gluconeogenesis, reviewed in (Taniguchi et al. 2006)). Due to inflammation, IRS1 can be alternatively phosphorylated on serine 307, which leads to downstream activation of the NF-κB pathway, phosphorylation of C-jun N-terminal kinase 1 (JNK1), and activation of the JNK/AP-1 pathway and thus disturbed insulin signaling. Furthermore, IL-1β induces suppressor of cytokine signaling (SOCS), which leads to degradation of insulin receptor substrate (IRS) proteins (Rui et al. 2002).

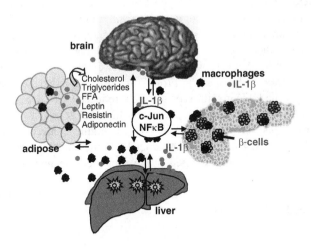

Fig. 1 The inflammatory axis in metabolic diseases and interplay between macrophage-derived IL-1β and its action in adipose tissue, brain, pancreas, and liver. Macrophages migrate into insulin-sensitive organs and produce proinflammatory signals, which change the cell fate. In adipose tissue, this leads to increased production of cholesterol, triglycerides, cytokines, and the adipokines lepin and resistin, while adiponectin is decreased. Insulin sensitivity is impaired and glucose uptake disturbed. Mediated through intracellular signaling cascades, NF-κB and c-Jun are activated and insulin resistance in the liver and brain and impaired insulin secretion in the β-cells develop [adapted from Maedler et al. (2009)]

2.1 IL-1β in Adipocytes

Infiltration of macrophages in adipose tissue is tightly correlated with obesity in mice and humans (Weisberg et al. 2003; Xu et al. 2003). Important modulators of inflammation are the adipocytokines, i.e., leptin, resistin, and adiponectin, which play a central role in the regulation of insulin resistance and β-cell function (Koerner et al. 2005; Tilg and Moschen 2006).

In obesity, not only circulating free fatty acids (FFA) and lipids but also leptin and resistin are increased; whereas adiponectin, which is known to prevent inflammation (Tilg and Moschen 2006) and is negatively correlated with insulin resistance, is decreased (Rasouli and Kern 2008). Leptin has been shown to exert pro- as well as anti-inflammatory properties, probably dependent on its dose and exposure time. While in vivo, leptin overexpression normalizes glycemia in the diabetic NOD mice as well as in STZ- and alloxan-induced diabetes (Yu et al. 2008), chronic leptin incubation in vitro leads to impaired β-cell function and survival (Maedler et al. 2004; Roduit and Thorens 1997; Seufert et al. 1999). Leptin has been shown to manipulate levels of IL-1β and IL-1Ra. While leptin acutely induces IL-1Ra expression in islets and monocytes (Gabay et al. 2001; Maedler et al. 2004), there is a chronic reduction of IL-1Ra and induction of IL-1β secretion.

IL-1Ra expression is increased in white adipose tissue in obese individuals with increased circulating FFA and lipids (Juge-Aubry et al. 2003). In contrast, daily IL-1Ra injections in HFD-fed mice normalize circulating FFA, lipids, as well as adipokines. Although the percentage of macrophages in a given adipose tissue depot is positively correlated with adiposity and adipocyte size (Weisberg et al. 2003), the normalization of lipids and adipokines by IL-1Ra seems to be independent of fat mass, since IL-1Ra treatment neither influences fat mass nor adipocyte size. In contrast, mRNA levels of the inflammatory cytokines IL-1β and TNF-α, the macrophage marker F4/80, and the proinflammatory macrophage marker CD11c are increased by the HFD in wild-type mice but reduced by IL-1Ra overexpression (Sauter et al. 2008). Interestingly, specifically the marker of the "classically activated" macrophages M1 (Lumeng et al. 2007) is highly induced by the HFD and normalized by IL-1Ra. Thus, the HFD-induced proinflammatory state of adipocytes may be the reason for the increased adipokines (resistin and leptin) and lipid production.

Undoubtedly, the effect of IL-1Ra on adipocyte-derived factors plays a protective role at the level of the β-cell.

2.2 IL-1β in the Liver

The bone marrow-derived macrophage cells in the liver are the Kupffer cells. Kupffer cells secrete cytokines, among them IL-1β, NO, and free radicals, which could, per se, induce β-cell failure (Barshes et al. 2005). This is specifically deleterious in the environment of transplanted islets in the liver. Cytokines (IL-1β, IFN-γ, and TNF-α) are particularly elevated after islet transplantation (Bottino et al. 1998), and liver tissue macrophages participate in cell injury and graft failure (Kaufman et al. 1990, 1994). Strategies to inhibit IL-1β-induced β-cell failure, e.g., by salicylate treatment of the islets (Tran et al. 2002; Zeender et al. 2004) may therefore improve graft survival.

Similar to the role of macrophages in obese adipose tissue, secretion of IL-1β by the Kupffer cells could be central to hepatic insulin resistance in obesity. Cytokine-induced JNK phosphorylation and activation of the NF-κB pathway are indicative of insulin resistance in the liver, e.g., depletion of JNK in myeloid cells (including Kupffer cells) in mice leads to HFD-induced hepatic steatosis without an increase in inflammatory markers in the liver and no development of insulin resistance (Solinas et al. 2007). Furthermore, hepatocyte-specific inhibition of NF-κB (Cai et al. 2005) or of IKK-β (Arkan et al. 2005) in myeloid cells improves hepatic insulin sensitivity. These studies show that independent of obesity, the inflammatory status in the liver primarily regulates insulin sensitivity.

2.3   IL-1β in the Brain

In the healthy brain, members of the IL-1 family are expressed at low or undetectable levels (Allan et al. 2005). During neuro-inflammation, IL-1β is dramatically upregulated by various local and systemic brain insults including ischemia, trauma, hypoxia, and neurotoxic inflammatory stimuli (Allan et al. 2005).

IL-1β in the brain is produced primarily by microglia, which also express caspase-1 (Touzani et al. 1999). To a lesser extent, astrocytes, oligodendroglia, neurons, cerebrovascular cells, and circulating immune cells after infiltrating the brain under inflammatory conditions produce IL-1β (Rothwell and Luheshi 2000).

IL-1β has a number of diverse actions in the CNS to modify feeding behavior, fever (Dinarello and Wolff 1982), central pain modulation (Wolf et al. 2003), stress responses (Goshen et al. 2003) memory (Schneider et al. 1998), and neuroendocrine responses, mainly through actions in the hypothalamus (Sims and Dower 1994).

There is evidence of a hypothalamic control of insulin sensitivity, which is disturbed when elevated levels of proinflammatory cytokines are circulating. Studies in mice show that HFD promotes hypothalamic resistance to the main anorexigenic hormones, leptin and insulin, leading to the progressive loss of the balance between food intake and thermogenesis and, therefore, resulting in body mass gain (De Souza et al. 2005; Milanski et al. 2009; Munzberg et al. 2004). HFD feeding of rats resulted in hypothalamic induction of IL-1β, TNF-α, IL-6, and IL-10. Activation of the toll-like receptor 4 signaling induces local cytokine expression in the hypothalamus and promotes endoplasmic reticulum stress and insulin resistance (Milanski et al. 2009).

The structural and metabolic damage found in Alzheimer's disease is in part due to sustained elevation of IL-1β (Holden and Mooney 1995; Vandenabeele and

Fiers 1991; Zuliani et al. 2007). It upregulates expression of β-amyloid precursor protein (β-APP) and stimulates the processing of β-APP, resulting in amyloido-genic fragments in neurons (Goldgaber et al. 1989). Similarly, the β-APP deposits found in the Alzheimer brain share the same molecular structure as the amylin oligomer deposits found in the pancreatic β-cells in T2DM and are equally neuro-toxic (Haataja et al. 2008). On the basis of the observations in the human islet amyloid polypeptide transgenic rat (Butler et al. 2004), there is evidence that IL-1β is expressed within the islets after the induction of severe hyperglycemia (unpub-lished observation), indicating that IL-1β expression can only be observed at high glucose levels. It remains to be elucidated if the toxicity of amylin oligomers on the β-cell involves IL-1β signals.

Possibly, the activation of cytokine-induced proinflammatory pathways (e.g., JNK) plays a major role in the modulation of neurodegeneration (Borsello and Forloni 2007). In line with this hypothesis, JNKs are negatively regulating insulin sensitivity in the obese state.

Four different pathways are shown in the brain as a consequence of diet-induced activation of inflammatory signaling: (1) induction of suppressor of cytokine signaling-3 (SOCS-3) expression (Howard et al. 2004), (2) activation of c-Jun N-terminal kinase (JNK) and I-kappa kinase (IKK) (De Souza et al. 2005), (3) induction of protein tyrosine phosphatase 1B (PTP1B) (Bence et al. 2006), and (4) activation of TLR4 signaling (Milanski et al. 2009). Thus, obesity and HFD induce activation of proinflammatory pathways in the brain, which may directly develop insulin resistance and lead to diminished glucose regulation by the insulin target tissues.

3   IL-1β Signaling in the β-Cell

Only when the β-cell compensates for the higher insulin demand during insulin resistance, normoglycemia can be maintained. A relative insulin deficiency leads to diabetes. From numerous in vitro studies from isolated islets and β-cell lines, we know that the β-cell is especially sensitive to cytokines. Consequently, circulating cytokines are likely to rapidly affect β-cell function and survival.

Soon after the cloning of IL-1β, Mandrup-Poulsen and colleagues observed that IL-1β impairs β-cell function (Mandrup-Poulsen et al. 1985, 1986). In addition to impaired insulin secretion, IL-1β was found to induce β-cell death, which was potentiated by the cytokines IFN-γ and TNF-α (Eizirik 1988; Pukel et al. 1988). In the pancreatic islet, IL-1R1 is present in the β-cells (Deyerle et al. 1992) and not in the α-cells (Scarim et al. 1997), and thus the β-cells are a target for IL-1a, IL-1β, and IL-1Ra.

Surprisingly, IL-1R1 is highly expressed in the β-cell; more than tenfold higher expression of IL-1RI mRNA was observed in isolated islets than in total pancreas, which is attributed to the expression in the β-cell. Furthermore, β-cell IL-1R1 expression levels are higher than in any other tissue. (Boni-Schnetzler et al.

2009), which may explain the high sensitivity of the β-cell to IL-1. Blocking IL-1β with specific IL-1β-neutralizing antibodies protected from the cytotoxic effects induced by activated mononuclear cell conditioned medium (Bendtzen et al. 1986), indicating that IL-1β may play an important role in the molecular mechanisms underlying autoimmune β-cell destruction.

Since then, IL-1β signaling and the underlying mechanisms of IL-1β-induced β-cell destruction have been investigated. Importantly, IL-1β induces its own and the expression of other cytokines, e.g., IL-2, -3, -6, and interferons (Dinarello 1988). In turn, cells that produce IL-1β also respond to IL-1β (Warner et al. 1987). IL-1β initiates signal transduction by binding to IL-1R1 in the β-cell. This leads to docking of the IL-1RAcP to the IL-1/IL-1R1 complex, which is followed by recruitment of the adaptor protein MyD88. IRAK-4, Tollip, and IRAK-1 are then recruited, allowing IRAK-1 to activate TRAF6, which in turn triggers activation of TAK1. TAK1 is able to stimulate two main pathways: the IKK–NF-κB pathway and the mitogen-activated/stress-activated protein kinase (MAPK/SAPK) pathway (Frobose et al. 2006). In addition to TAK1, MEKK1 seems to participate in the activation of both NF-κB and SAPK in β-cells (Mokhtari et al. 2008). Phosphorylation of I-κB, a cytosolic inhibitor of *NF-κB*, by IKK leads to I-κB degradation and NF-κB translocation to the nucleus, thus regulating the transcription of many target genes, such as iNOS expression and NO production, a toxic reactive radical. Consistently, interfering with NF-κB activation decreases IL-1β-induced β-cell death (Giannoukakis et al. 2000; Kim et al. 2007).

IL-1β can also activate protein kinase C delta, which leads to β-cell apoptosis presumably through iNOS expression (Carpenter et al. 2001, 2002). Notably, IL-1β induces *Fas* expression on β-cells (Augstein et al. 2003; Stassi et al. 1995), increasing their sensitivity to FasL and accelerating apoptosis via cleavage of downstream caspases [see Fig. 2 and reviewed in Donath et al. (2003)]. A distal consequence of IL-1β signaling in β-cells is the induction of endoplasmic reticulum (ER) stress. IL-1β depletes ER Ca^{2+}, leading to ER stress and induction of several ER stress markers including CHOP. The induction of ER stress by IL-1β can be prevented by inhibition of iNOS, suggesting that NO mediates ER stress (Cardozo et al. 2005). This is consistent with the notion that a chemical NO donor causes ER Ca^{2+} depletion and ER stress (Oyadomari et al. 2002). What is currently unclear is the importance of ER stress in IL-1β-induced β-cell impairment. Studies addressing the role of ER stress-induced CHOP so far indicate that ER stress and CHOP do not contribute to cytokine-induced β-cell death (Akerfeldt et al. 2008). Thus, while there is little doubt that ER stress is induced in β-cells by IL-1β, it is uncertain whether ER stress contributes to apoptosis or whether it may simply be a secondary effect and thus only plays a minor role, if any, in IL-1β-mediated apoptosis.

The MAPK/SAPK pathways consist of ERK1/2, p38, and JNK1/2, all of which are activated by IL-1β in β-cells (Larsen et al. 1998; Welsh 1996). Using both pharmacological and molecular inhibitor approaches, NF-κB, ERK1/2, p38, and JNK1/2 have been demonstrated to be involved in IL-1β-induced β-cell apoptosis (Abdelli et al. 2007; Bonny et al. 2001; Larsen et al. 1998; Pavlovic et al. 2000; Saldeen et al. 2001).

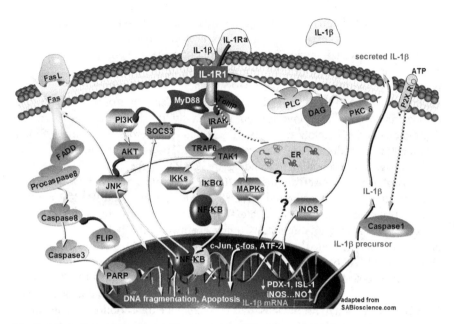

Fig. 2 Mechanisms of IL-1β signaling in the β-cell. Details are described in the text [adapted from Maedler et al. (2009)]

Another target of IL-1β signaling in β-cells is the survival kinase pathway PI3K–Akt. IL-1β reduces both PI3K (Emanuelli et al. 2004) and Akt (Storling et al. 2005) activation. Since Akt is a negative regulator of JNK/SAPK in β-cells (Aikin et al. 2004), reduced Akt signaling may allow increased and sustained proapoptotic JNK activation.

In general, signal transduction initiated by a ligand binding to membrane receptors leads to activation or induction of negative feedback mechanisms to ensure only transient signaling. This is also true for signal transduction evoked by proinflammatory cytokines such as IL-1β. IL-1β induces expression of SOCS-3 in β-cells (Emanuelli et al. 2004; Karlsen et al. 2001). SOCS-3 is a member of a family of proteins that function to terminate cytokine signaling, thereby constituting a negative feedback loop (Ronn et al. 2007). Although IL-1β induces SOCS-3 expression in β-cells, this induction seems to be insufficient to completely terminate IL-1β signal transduction, since prolonged NF-κB and MAPK/SAPK signaling is observed in β-cells exposed to IL-1β (Aikin et al. 2004; Larsen et al. 1998; Ortis et al. 2006). Putatively, either the amount of SOCS-3 induced by IL-1β in β-cells is too low to effectively block signaling or the kinetics of SOCS-3 induction by IL-1β may be abnormally slow in β-cells. In any case, forced SOCS-3 overexpression effectively inhibits IL-1β signaling at the level of TRAF6, leading to dampening of both the NF-κB and MAPK/SAPK pathways, thus protecting against apoptosis (Frobose et al. 2006; Ronn et al. 2008). Interestingly, IL-1β-induced endogenous SOCS-3 targets insulin signaling in the β-cell by associating with the insulin receptor (IR), thereby preventing activation of IRS and PI3K (Emanuelli et al. 2004). By this mechanism, SOCS-3 induction is likely to contribute to IL-1β-induced desensitization of insulin signaling, which is important for optimal β-cell function. One may speculate whether IL-1β-induced SOCS-3 expression is preferentially directed toward IR signals while leaving the IL-1β signaling cascade unaffected. The IL-1β signaling pathways are shown in Fig. 2.

4 IL-1β Secretion

The primary sources of IL-1β are blood monocytes, tissue macrophages, and dendritic cells. B lymphocytes and NK cells also produce IL-1β (Dinarello 2009). The release of the leaderless cytokine, IL-1β, cannot be initiated through the Golgi apparatus. Inactive pro-IL-1β precursor accumulates in the cytosol and is processed by caspase-1 (also named Interleukin-converting enzyme, ICE) into the mature secreted IL-1β. The maturation occurs in a large multiprotein complex. ATP activates the $P2X_7$ receptor, which forms a pore in response to ligand stimulation and regulates cell permeability and cytokine release (Narcisse et al. 2005).

Resident islet macrophages are fundamental in the development of autoimmune diabetes (Arnush et al. 1998; Lacy 1994) and it is postulated that IL-1β secreted from such intra-islet macrophages results in β-cell destruction (Arnush et al. 1998). Recent studies show that the β-cells themselves are able to secrete IL-1β, which is induced by double-stranded RNA, a mechanism by which viral infection may mediate β-cell damage (Heitmeier et al. 2001) by elevated glucose concentrations (Boni-Schnetzler et al. 2008; Maedler et al. 2002) and by free fatty acids (Boni-Schnetzler et al. 2009).

A recent study shows that glucose-induced IL-1β secretion involves Caspase-1 activation mediated by the NALP3 inflammasome. The inflammasone is activated by bacterial toxins and endogenous stress signals (e.g., ATP and β-amyloid) through the formation of reactive oxygen species (Schroder et al. 2010; Zhou et al. 2009). Glucose-induced IL-1β secretion is prevented in $NALP3^{-/-}$ mice, indicating that IL-1β is generated through glucose-induced ROS production and oxidative stress (Zhou et al. 2009). The thioredoxin (TRX)-interacting protein (TXNIP), which has been linked to insulin resistance (Parikh et al. 2007), functions as an activator of NALP3. In line with this data, another recent study shows that TXNIP is highly increased by elevated glucose in β-cells and that TXNIP-deficient islets are protected against glucose toxicity (Chen et al. 2009).

Despite the high expression of IL-1R1 in β-cells, expression of the NALP3 inflammasone components NALP3, ASC, and Caspase-1 show relatively low expression levels (Zhou et al. 2009), which may explain the modest release of IL-1β from islets.

Upregulation of the Fas receptor plays a central role in the mediation of β-cell death (Cnop et al. 2005; Donath et al. 2005). IL-1β rapidly induces Fas upregulation, whereas glucose only induces Fas in chronic conditions (Elouil et al. 2005).

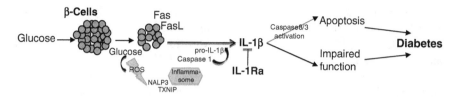

Fig. 3 Dual role of glucose on β-cell turnover. Stimulation of β-cells with glucose induces insulin secretion and β-cell proliferation. In contrast, chronic glucose exposure leads to upregulation of the Fas receptor and ligation with FasL to caspase activation, apoptosis, and impaired function, which contributes to β-cell failure in diabetes. Under such conditions, IL-1β is produced and secreted by the β-cell. This is mediated through ROS-induced induction of the NALP3 inflammasome, which activates Caspase-1 and maturation of active IL-1β from pro-IL-1β. Preincubation of the islets with the naturally occurring IL-1 antagonist interleukin-1 receptor antagonist (IL-1Ra) inhibits glucose-induced apoptosis and improves β-cell function and could therefore be a valuable tool for diabetes therapy

The dual role of glucose on β-cell turnover is illustrated in Fig. 3. While glucose promotes insulin secretion and β-cell survival in the short term, chronic glucose induces Fas upregulation, IL-1β secretion, which leads downstream to caspase cleavage, β-cell death, and loss of insulin secretion.

In two animal models, *Psammomys obesus* and Goto-Kakizaki (GK) rat, pancreatic β-cells express IL-1β under hyperglycemic conditions (Maedler et al. 2002; Mine et al. 2004). In *P. obesus*, normalizing hyperglycemia with phlorizin, an inhibitor of the renal tubular glucose reuptake, inhibited intra-islet IL-1β expression (Maedler et al. 2002). In contrast, Jorns et al. found no IL-1β expression within the islets (Jorns et al. 2006). IL-1β production by islet cells was confirmed in several studies (Boni-Schnetzler et al. 2008; Venieratos et al. 2010; Welsh et al. 2005; Zhou et al. 2009). While glucose-induced IL-1β mRNA production was not found in human islets that had been preincubated in suspension for 3–5 days (Welsh et al. 2005), Boni-Schnetzler et al. show that glucose response in islets is negatively correlated with basal IL-1β expression levels (Boni-Schnetzler et al. 2008). These studies show that IL-β may also mediate β-cell destruction in Type 2 diabetes [T2DM, reviewed in Donath et al. (2005)]. It is tempting to suggest IL-1β as a target for the treatment of diabetes. However, whether changes in circulating cytokines are physiologically relevant in the face of locally produced inflammatory mediators remains unknown.

5 Blocking IL-1β Signals Protects the β-Cell

As described above, IL-1β has been shown to impair insulin release, to induce Fas expression, thus enabling Fas-triggered apoptosis in rodent and human islets (Corbett et al. 1993; Giannoukakis et al. 1999, 2000; Loweth et al. 1998, 2000; Maedler et al. 2001; Mandrup-Poulsen et al. 1985, 1986, 1993; Rabinovitch et al.

1990; Stassi et al. 1997), and to share similarities with glucose-induced apoptosis (see Fig. 3). In parallel to the essential role of glucose in mediating insulin secretion and proliferation, a low concentration of IL-1β also stimulates insulin release and proliferation in rat and human islets (Maedler et al. 2006; Schumann et al. 2005; Spinas et al. 1986, 1987, 1988). The beneficial IL-1β effects seem to be partly mediated by the increased secretion of the naturally occurring anti-inflammatory cytokine and antagonist of IL-1α and IL-1β, the interleukin-1 receptor antagonist (IL-1Ra). Since it was discovered in 1987 (Dinarello 2000; Seckinger et al. 1987a, b), four forms of IL-1Ra have been described, of which three are intracellular proteins (icIL-1Ra I, II and III) and one is secreted (sIL-1Ra) (Arend and Guthridge 2000). Similar to IL-1β, IL-1Ra binds to type 1 and 2 IL-1 receptors but lacks a second binding domain. Therefore, IL-1Ra does not recruit the IL-1 receptor accessory protein, the second component of the receptor complex.

Endogenous production and secretion of IL-1Ra limits inflammation and tissue damage (Dinarello 2009). In vivo, exogenous IL-1Ra counteracts low-dose streptozotocin-induced diabetes (Sandberg et al. 1994) and autoimmune diabetes (Nicoletti et al. 1994) and promotes graft survival (Nicoletti et al. 1994; Sandberg et al. 1997; Stoffels et al. 2002; Tellez et al. 2007) and islet survival after transplantation (Satoh et al. 2007).

We have recently shown that IL-1Ra is secreted from the β-cell and expressed in β-cell granules (Maedler et al. 2004). IL-1Ra protects cultured human islets from the deleterious effects of glucose (Maedler et al. 2002) as well as IL-1β (Mandrup-Poulsen et al. 1993; Sandberg et al. 1997, 1993; Stoffels et al. 2002; Tellez et al. 2005). Inhibition of IL-1Ra with small interfering RNAs or long-term treatment with leptin leads to β-cell apoptosis and impaired function, which may provide a further link between obesity and diabetes.

The definite secretion and regulation mechanisms of IL-1Ra are unknown. Like IL-1β, IL-1Ra may also be secreted by a leaderless pathway via activation of the P2X$_7$ receptor (Glas et al. 2009; Wilson et al. 2004). In pancreatic islets from obese individuals, P2X$_7$ receptors are highly expressed and these receptors were almost undetectable in T2DM (Glas et al. 2009). In accordance with the P2X$_7$ receptor expression levels, increased IL-1Ra serum levels correlate with obesity and insulin resistance (Abbatecola et al. 2004; Meier et al. 2002; Ruotsalainen et al. 2006; Salmenniemi et al. 2004), but IL-1Ra is decreased in T2DM (Marculescu et al. 2002). Recent results from the Whitehall Study show that IL-1Ra levels are increased before the onset of T2DM (Herder et al. 2008), which are consistent with findings in mice fed with a high fat/high sucrose diet (HFD). IL-1Ra levels were increased after 4 and 8 weeks of diet together with an increase in β-cell mass and body weight. Serum concentrations of IL-1Ra are influenced by adipose tissue, which is a major source of IL-1Ra (Juge-Aubry et al. 2003). After 16 weeks, when the HFD-fed mice displayed glucose intolerance and β-cell apoptosis, IL-1Ra levels were lower than in the normal diet-fed mice. Mice deficient for the P2X$_7$ receptor were unable to compensatorily increase β-cell mass in response to the HFD feeding and had no adaptive increase in IL-1Ra levels (Glas et al. 2009).

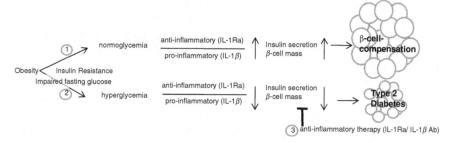

Fig. 4 Our hypothetical model illustrating the consequence of obesity on the development of Type 2 diabetes. (1) When IL-1Ra is highly expressed in the β-cell and the IL-1Ra/IL-1β balance is toward the protective IL-1Ra, β-cell mass and insulin secretion increase. The β-cell is able to adapt to a situation of higher insulin demand. (2) On the other hand, decreased β-cell expression of IL-1Ra, together with hyperglycemia-induced β-cell production of IL-1β, shifts the balance toward the proapoptotic IL-1β, leading to decreased β-cell mass, impaired β-cell function, and increased β-cell apoptosis. Glucose levels can no longer be regulated. This results in a vicious cycle and Type 2 diabetes develops. (3) But overexpression of IL-1Ra could reverse the process and protect from hyperglycemia-induced β-cell apoptosis [adapted from Maedler et al. (2009)]

The increased IL-1Ra could be an attempt of the body to counteract the deleterious effects of IL-1β and to preserve β-cell survival, insulin secretion, and insulin sensitivity. It is hypothesized that IL-1Ra could have an additional metabolic effect that leads to insulin resistance. However, when we treated mice daily for 12 weeks with IL-1Ra, we did not observe changes in insulin sensitivity at any time point (Sauter et al. 2008).

Whether serum IL-1Ra levels would explain the progression of diabetes in obese individuals and whether serum IL-1Ra affects IL-1Ra expression in the β-cell is not known. We hypothesize that a decreased β-cell IL-1Ra expression could trigger the progression from obesity to diabetes and high IL-1Ra expression could possibly protect the β-cell and enable it to adapt to conditions of higher insulin demand; this is illustrated in the cartoon shown in Fig. 4.

5.1 Lessons from IL-1 Mouse Models

Having shown the deleterious effects of IL-1β on the β-cell, one would hypothesize that the IL-1β-knockout mouse would be the ideal model for improved β-cell survival and function. Conversely, IL-1β-KO mice show impaired glucose tolerance, decreased β-cell mass, and decreased expression of β-cell transcription factors (e.g., PDX-1 and Pax-4) (Maedler et al. 2006), indicating that IL-1β has a dual role in the β-cell and activated pathways, e.g., FLIP, Fas, and NF-κB might be needed for insulin secretion and survival (Maedler et al. 2006; Liadis et al. 2007; Schumann et al. 2007). In line with these data, Caspase-8-knockout (Liadis et al. 2007) and Fas-deficient mice (Schumann et al. 2007) show impaired glucose tolerance. NF-κB is for a long time known to be responsible for IL-1β-induced β-cell destruction (Flodstrom et al. 1996). In contrast, NF-κB also induces activation of the antiapoptotic gene A20, which protects against cell death (Liuwantara et al. 2006) and promotes insulin secretion (Hammar et al. 2004). β-Cell-specific NF-κB depletion accelerates diabetes in the NOD mouse (Kim et al. 2007).

Despite their basally impaired glucose tolerance, IL-1β-KO mice are protected against the diabetogenic effects of the HFD as well as against glucotoxicity (Maedler et al. 2006), which supports the concept that IL-β mediates nutrient-induced β-cell dysfunction during the development of T2DM.

In NOD mice, IL-1R deficiency slows but does not prevent diabetes progression (Thomas et al. 2004), and caspase-1 (interleukin-converting enzyme) deficiency has no effect on diabetes progression (Schott et al. 2004), although both IL-1R subtype 1 and caspase-1 are highly expressed in islets from wild-type NOD mice (Jafarian-Tehrani et al. 1995). It is possible that pathways other than IL-1β signals are involved in diabetes in NOD mice since it was shown that IL-10 promotes diabetes in NOD mice independent of Fas, perforin, TNFR 1, and TNFR 2 (Balasa et al. 2000).

5.2   Blocking IL-1β Signals In Vivo Inhibits Diabetes Progression

Recently, the hypothesis that blocking IL-1β as a successful strategy for the therapy of T2DM has been proved by several studies. Daily injection of IL-1Ra in mice fed an HFD improved glycemia, glucose-stimulated insulin secretion, and survival (Sauter et al. 2008), reduced hyperglycaemia, and reversed the islet inflammatory phenotype in the GK rat (Ehses et al. 2008). Treatment with an IL-1β antibody also improved glycemic control in diet-induced obesity in mice (Owyang et al. 2010; Osborn et al. 2008).

Importantly, results from a recent clinical study in patients with T2DM showed that IL-1Ra improved glycemic control and β-cell function (Larsen et al. 2007). After 13 weeks of treatment, C-peptide secretion was increased and inflammatory markers, e.g., interleukin-6 and C-reactive protein were reduced in the IL-1Ra group. HbA1c was significantly lower in the IL-1Ra compared to the placebo group, which correlated with the body surface area in the IL-1Ra group. The dose of 100 mg IL-1Ra was given daily to the patients without weight adjustment. Currently, ongoing trials that include dose adjustment to the body weight may result in better glycemic control in the higher body surface area group. The effect of interleukin-1 antagonism on β-cell function is currently tested in patients with recent onset of T1DM (Pickersgill and Mandrup-Poulsen 2009). Both IL-1Ra and anti-IL-1β antibody Xoma 052 do not completely block IL-1β signaling. While IL-1Ra is a competitive antagonist to IL-1β, XOMA 052 has a novel mechanism of action that reduces IL-1β activity by 40- to 50-fold rather than completely blocking it (Donath et al. 2008; Owyang et al. 2010). Given the dual role of IL-1β on β-cell survival and insulin secretion, this may be an important characteristic of both drugs.

As shown by these recent studies, blocking IL-1β signaling may be a powerful new treatment for T2DM, which does not rely on replacing insulin exogenously but acts at the level of the β-cell to improve β-cell survival and to improve endogenous insulin secretion and action. Moreover, blocking IL-1β may also improve insulin sensitivity. Further studies will be necessary to clarify the contradiction of IL-1Ra's modulation of insulin sensitivity and the impact of IL-β on β-cell survival in T2DM.

Acknowledgments This work was supported by the German Research Foundation (DFG, Emmy Noether Programm, MA4172/1-1), the Juvenile Diabetes Research Foundation (JDRF 26-2008-861), and the European Foundation for the Study of Diabetes (EFSD)/Merck Sharp & Dohme (MSD) European Studies on Beta Cell Function and Survival. The authors thank the members of the Islet Biology laboratory in Bremen for critical discussion.

References

Abbatecola AM, Ferrucci L, Grella R, Bandinelli S, Bonafe M, Barbieri M, Corsi AM, Lauretani F, Franceschi C, Paolisso G (2004) Diverse effect of inflammatory markers on insulin resistance and insulin-resistance syndrome in the elderly. J Am Geriatr Soc 52(3):399–404

Abdelli S, Abderrahmani A, Hering BJ, Beckmann JS, Bonny C (2007) The c-jun n-terminal kinase jnk participates in cytokine- and isolation stress-induced rat pancreatic islet apoptosis. Diabetologia 50(8):1660–1669

Aikin R, Maysinger D, Rosenberg L (2004) Cross-talk between phosphatidylinositol 3-kinase/akt and c-jun nh2-terminal kinase mediates survival of isolated human islets. Endocrinology 145(10):4522–4531

Akerfeldt MC, Howes J, Chan JY, Stevens VA, Boubenna N, McGuire HM, King C, Biden TJ, Laybutt DR (2008) Cytokine-induced beta-cell death is independent of endoplasmic reticulum stress signaling. Diabetes 57(11):3034–3044

Allan SM, Tyrrell PJ, Rothwell NJ (2005) Interleukin-1 and neuronal injury. Nat Rev Immunol 5(8):629–640. doi:nri1664 [pii] 10.1038/nri1664

Arend WP, Guthridge CJ (2000) Biological role of interleukin 1 receptor antagonist isoforms. Ann Rheum Dis 59(Suppl 1):i60–i64

Arkan MC, Hevener AL, Greten FR, Maeda S, Li ZW, Long JM, Wynshaw-Boris A, Poli G, Olefsky J, Karin M (2005) Ikk-beta links inflammation to obesity-induced insulin resistance. Nat Med 11(2):191–198

Arnush M, Heitmeier MR, Scarim AL, Marino MH, Manning PT (1998) Corbett JA Il-1 produced and released endogenously within human islets inhibits beta cell function. J Clin Invest 102(3):516–526, (3):26

Augstein P, Dunger A, Heinke P, Wachlin G, Berg S, Hehmke B, Salzsieder E (2003) Prevention of autoimmune diabetes in nod mice by troglitazone is associated with modulation of icam-1 expression on pancreatic islet cells and ifn-gamma expression in splenic t cells. Biochem Biophys Res Commun 304(2):378–384

Auron PE, Webb AC, Rosenwasser LJ, Mucci SF, Rich A, Wolff SM, Dinarello CA (1984) Nucleotide sequence of human monocyte interleukin 1 precursor cdna. Proc Natl Acad Sci USA 81(24):7907–7911

Balasa B, La Cava A, Van Gunst K, Mocnik L, Balakrishna D, Nguyen N, Tucker L, Sarvetnick N (2000) A mechanism for il-10-mediated diabetes in the nonobese diabetic (nod) mouse: Icam-1 deficiency blocks accelerated diabetes. J Immunol 165(12):7330–7337

Barshes NR, Wyllie S, Goss JA (2005) Inflammation-mediated dysfunction and apoptosis in pancreatic islet transplantation: implications for intrahepatic grafts. J Leukoc Biol 77(5):587–597

Bence KK, Delibegovic M, Xue B, Gorgun CZ, Hotamisligil GS, Neel BG, Kahn BB (2006) Neuronal ptp1b regulates body weight, adiposity and leptin action. Nat Med 12(8):917–924. doi:nm1435 [pii] 10.1038/nm1435

Bendtzen K, Mandrup-Poulsen T, Nerup J, Nielsen JH, Dinarello CA, Svenson M (1986) Cytotoxicity of human pi 7 interleukin-1 for pancreatic islets of langerhans. Science 232(4757):1545–1547

Boni-Schnetzler M, Thorne J, Parnaud G, Marselli L, Ehses JA, Kerr-Conte J, Pattou F, Halban PA, Weir GC, Donath MY (2008) Increased interleukin (il)-1beta messenger ribonucleic acid expression in beta-cells of individuals with type 2 diabetes and regulation of il-1beta in human islets by glucose and autostimulation. J Clin Endocrinol Metab 93(10):4065–4074

Boni-Schnetzler M, Boller S, Debray S, Bouzakri K, Meier DT, Prazak R, Kerr-Conte J, Pattou F, Ehses JA, Schuit FC, Donath MY (2009) Free fatty acids induce a proinflammatory response in islets via the abundantly expressed interleukin-1 receptor i. Endocrinology 150(12):5218–5229

Bonny C, Oberson A, Negri S, Sauser C, Schorderet DF (2001) Cell-permeable peptide inhibitors of jnk: novel blockers of beta-cell death. Diabetes 50(1):77–82

Boraschi D, Tagliabue A (2006) The interleukin-1 receptor family. Vitam Horm 74:229–254

Borsello T, Forloni G (2007) Jnk signalling: a possible target to prevent neurodegeneration. Curr Pharm Des 13(18):1875–1886

Bottino R, Fernandez LA, Ricordi C, Lehmann R, Tsan MF, Oliver R, Inverardi L (1998) Transplantation of allogeneic islets of langerhans in the rat liver: effects of macrophage depletion on graft survival and microenvironment activation. Diabetes 47(3):316–323

Butler AE, Jang J, Gurlo T, Carty MD, Soeller WC, Butler PC (2004) Diabetes due to a progressive defect in beta-cell mass in rats transgenic for human islet amyloid polypeptide (hip rat): a new model for type 2 diabetes. Diabetes 53(6):1509–1516

Cai D, Yuan M, Frantz DF, Melendez PA, Hansen L, Lee J, Shoelson SE (2005) Local and systemic insulin resistance resulting from hepatic activation of ikk-beta and nf-kappab. Nat Med 11(2):183–190

Cardozo AK, Ortis F, Storling J, Feng YM, Rasschaert J, Tonnesen M, Van Eylen F, Mandrup-Poulsen T, Herchuelz A, Eizirik DL (2005) Cytokines downregulate the sarcoendoplasmic reticulum pump ca2+ atpase 2b and deplete endoplasmic reticulum ca2+, leading to induction of endoplasmic reticulum stress in pancreatic beta-cells. Diabetes 54(2):452–461

Carpenter L, Cordery D, Biden TJ (2001) Protein kinase cdelta activation by interleukin-1beta stabilizes inducible nitric-oxide synthase mrna in pancreatic beta-cells. J Biol Chem 276(7):5368–5374

Carpenter L, Cordery D, Biden TJ (2002) Inhibition of protein kinase c delta protects rat ins-1 cells against interleukin-1beta and streptozotocin-induced apoptosis. Diabetes 51(2):317–324

Chen J, Fontes G, Saxena G, Poitout V, Shalev A (2009) Lack of txnip protects against mitochondria-mediated apoptosis, but not against fatty acid-induced, er-stress-mediated beta cell death. Diabetes. doi:db09-0949 [pii] 10.2337/db09-0949

Cnop M, Welsh N, Jonas JC, Jorns A, Lenzen S, Eizirik DL (2005) Mechanisms of pancreatic beta-cell death in type 1 and type 2 diabetes: many differences, few similarities. Diabetes 54(Suppl 2):S97–107

Corbett JA, Sweetland MA, Wang JL, Lancaster JR Jr, McDaniel ML (1993) Nitric oxide mediates cytokine-induced inhibition of insulin secretion by human islets of langerhans. Proc Natl Acad Sci USA 90(5):1731–1735

de Luca C, Olefsky JM (2008) Inflammation and insulin resistance. FEBS Lett 582(1):97–105

De Souza CT, Araujo EP, Bordin S, Ashimine R, Zollner RL, Boschero AC, Saad MJ, Velloso LA (2005) Consumption of a fat-rich diet activates a proinflammatory response and induces insulin resistance in the hypothalamus. Endocrinology 146(10):4192–4199. doi:en.2004-1520 [pii] 10.1210/en.2004-1520

Deyerle KL, Sims JE, Dower SK, Bothwell MA (1992) Pattern of il-1 receptor gene expression suggests role in noninflammatory processes. J Immunol 149(5):1657–1665

Dinarello CA (1988) Biology of interleukin 1. FASEB J 2(2):108–115

Dinarello CA (2000) The role of the interleukin-1-receptor antagonist in blocking inflammation mediated by interleukin-1. N Engl J Med 343(10):732–734

Dinarello CA (2009) Immunological and inflammatory functions of the interleukin-1 family. Annu Rev Immunol 27:519–550

Dinarello CA, Wolff SM (1982) Molecular basis of fever in humans. Am J Med 72(5):799–819

Donath MY, Storling J, Maedler K, Mandrup-Poulsen T (2003) Inflammatory mediators and islet beta-cell failure: a link between type 1 and type 2 diabetes. J Mol Med 81(8):455–470

Donath MY, Ehses JA, Maedler K, Schumann DM, Ellingsgaard H, Eppler E, Reinecke M (2005) Mechanisms of {beta}-cell death in type 2 diabetes. Diabetes 54(Suppl 2):S108–113

Donath MY, Weder C, Brunner A, Keller C, Whitmore J, Der K, Scannon PJ, Dinarello CA, Solinger AM (2008) Xoma 052, a potential disease modifying anti-il-1β antibody shows sustained hba1c reductions 3 months after single injection with no increases in safety parameters in subjects with t2dm. Diabetes 58(S1):A30

Ehses JA, Perren A, Eppler E, Ribaux P, Pospisilik JA, Maor-Cahn R, Gueripel X, Ellingsgaard H, Schneider MK, Biollaz G, Fontana A, Reinecke M, Homo-Delarche F, Donath MY (2007) Increased number of islet-associated macrophages in type 2 diabetes. Diabetes 56(9):2356–2370

Ehses JA, Giroix M-H, Coulaud J, Akira S, Homo-Delarche F, Donath MY (2008) Il-1β-myd88 signaling is central to islet chemokine secretion in response to metabolic stress: evidence from a spontaneous model of type 2 diabetes, the gk rat. Diabetologia 50(Suppl 1):S177

Eizirik DL (1988) Interleukin-1 induced impairment in pancreatic islet oxidative metabolism of glucose is potentiated by tumor necrosis factor. Acta Endocrinol (Copenh) 119(3):321–325

Elouil H, Cardozo AK, Eizirik DL, Henquin JC, Jonas JC (2005) High glucose and hydrogen peroxide increase c-myc and haeme-oxygenase 1 mrna levels in rat pancreatic islets without activating nfkappab. Diabetologia 48(3):496–505

Emanuelli B, Glondu M, Filloux C, Peraldi P, Van Obberghen E (2004) The potential role of socs-3 in the interleukin-1{beta}-induced desensitization of insulin signaling in pancreatic beta-cells. Diabetes 53(Suppl 3):S97–S103

Flodstrom M, Welsh N, Eizirik DL (1996) Cytokines activate the nuclear factor kappa b (nf-kappa b) and induce nitric oxide production in human pancreatic islets. FEBS Lett 385(1–2):4–6

Frobose H, Ronn SG, Heding PE, Mendoza H, Cohen P, Mandrup-Poulsen T, Billestrup N (2006) Suppressor of cytokine signaling-3 inhibits interleukin-1 signaling by targeting the traf-6/tak1 complex. Mol Endocrinol 20(7):1587–1596

Gabay C, Dreyer M, Pellegrinelli N, Chicheportiche R, Meier CA (2001) Leptin directly induces the secretion of interleukin 1 receptor antagonist in human monocytes. J Clin Endocrinol Metab 86(2):783–791

Giannoukakis N, Rudert WA, Ghivizzani SC, Gambotto A, Ricordi C, Trucco M, Robbins PD (1999) Adenoviral gene transfer of the interleukin-1 receptor antagonist protein to human islets prevents il-1beta-induced beta-cell impairment and activation of islet cell apoptosis in vitro. Diabetes 48(9):1730–1736

Giannoukakis N, Mi Z, Rudert WA, Gambotto A, Trucco M, Robbins P (2000) Prevention of beta cell dysfunction and apoptosis activation in human islets by adenoviral gene transfer of the insulin-like growth factor i. Gene Ther 7(23):2015–2022

Glas R, Sauter NS, Schulthess FT, Shu L, Oberholzer J, Maedler K (2009) Purinergic p2x(7) receptors regulate secretion of interleukin-1 receptor antagonist and beta cell function and survival. Diabetologia 52(8):1579–1588

Goldgaber D, Harris HW, Hla T, Maciag T, Donnelly RJ, Jacobsen JS, Vitek MP, Gajdusek DC (1989) Interleukin 1 regulates synthesis of amyloid beta-protein precursor mrna in human endothelial cells. Proc Natl Acad Sci USA 86(19):7606–7610

Goshen I, Yirmiya R, Iverfeldt K, Weidenfeld J (2003) The role of endogenous interleukin-1 in stress-induced adrenal activation and adrenalectomy-induced adrenocorticotropic hormone hypersecretion. Endocrinology 144(10):4453–4458. doi:10.1210/en.2003-0338 en.2003-0338 [pii]

Haataja L, Gurlo T, Huang CJ, Butler PC (2008) Islet amyloid in type 2 diabetes, and the toxic oligomer hypothesis. Endocr Rev 29(3):303–316

Hammar E, Parnaud G, Bosco D, Perriraz N, Maedler K, Donath M, Rouiller DG, Halban PA (2004) Extracellular matrix protects pancreatic {beta}-cells against apoptosis: role of short- and long-term signaling pathways. Diabetes 53(8):2034–2041

Heitmeier MR, Arnush M, Scarim AL, Corbett JA (2001) Pancreatic {beta}-cell damage mediated by {beta}-cell production of il-1: a novel mechanism for virus-induced diabetes. J Biol Chem 276:11151–11158

Herder C, Brunner EJ, Rathmann W, Strassburger K, Tabak AG, Schloot NC, Witte DR (2008) Elevated levels of the anti-inflammatory interleukin-1 receptor antagonist (il-1ra) precede the onset of type 2 diabetes (whitehall ii study). Diab Care 32(3):421–423

Holden RJ, Mooney PA (1995) Interleukin-1 beta: a common cause of Alzheimer's disease and diabetes mellitus. Med Hypotheses 45(6):559–571

Homo-Delarche F, Calderari S, Irminger JC, Gangnerau MN, Coulaud J, Rickenbach K, Dolz M, Halban P, Portha B, Serradas P (2006) Islet inflammation and fibrosis in a spontaneous model of type 2 diabetes, the gk rat. Diabetes 55(6):1625–1633

Howard JK, Cave BJ, Oksanen LJ, Tzameli I, Bjorbaek C, Flier JS (2004) Enhanced leptin sensitivity and attenuation of diet-induced obesity in mice with haploinsufficiency of socs3. Nat Med 10(7):734–738. doi:10.1038/nm1072 nm1072 [pii]

Jafarian-Tehrani M, Amrani A, Homo-Delarche F, Marquette C, Dardenne M, Haour F (1995) Localization and characterization of interleukin-1 receptors in the islets of langerhans from control and nonobese diabetic mice. Endocrinology 136(2):609–613

Jorns A, Rath KJ, Bock O, Lenzen S (2006) Beta cell death in hyperglycaemic psammomys obesus is not cytokine-mediated. Diabetologia 49(11):2704–2712

Juge-Aubry CE, Somm E, Giusti V, Pernin A, Chicheportiche R, Verdumo C, Rohner-Jeanrenaud F, Burger D, Dayer JM, Meier CA (2003) Adipose tissue is a major source of interleukin-1 receptor antagonist: upregulation in obesity and inflammation. Diabetes 52(5):1104–1110

Karlsen AE, Ronn SG, Lindberg K, Johannesen J, Galsgaard ED, Pociot F, Nielsen JH, Mandrup-Poulsen T, Nerup J, Billestrup N (2001) Suppressor of cytokine signaling 3 (socs-3) protects beta-cells against interleukin-1beta and interferon-gamma -mediated toxicity. Proc Natl Acad Sci USA 98:12191–12196

Kaufman DB, Platt JL, Rabe FL, Dunn DL, Bach FH, Sutherland DE (1990) Differential roles of mac-1+ cells, and cd4+ and cd8+ t lymphocytes in primary nonfunction and classic rejection of islet allografts. J Exp Med 172(1):291–302

Kaufman DB, Gores PF, Field MJ, Farney AC, Gruber SA, Stephanian E, Sutherland DE (1994) Effect of 15-deoxyspergualin on immediate function and long-term survival of transplanted islets in murine recipients of a marginal islet mass. Diabetes 43(6):778–783

Kim S, Millet I, Kim HS, Kim JY, Han MS, Lee MK, Kim KW, Sherwin RS, Karin M, Lee MS (2007) Nf-kappa b prevents beta cell death and autoimmune diabetes in nod mice. Proc Natl Acad Sci USA 104(6):1913–1918

Koenig W, Khuseyinova N, Baumert J, Thorand B, Loewel H, Chambless L, Meisinger C, Schneider A, Martin S, Kolb H, Herder C (2006) Increased concentrations of c-reactive protein and il-6 but not il-18 are independently associated with incident coronary events in middle-aged men and women: results from the monica/kora augsburg case-cohort study, 1984–2002. Arterioscler Thromb Vasc Biol 26(12):2745–2751

Koerner A, Kratzsch J, Kiess W (2005) Adipocytokines: leptin – the classical, resistin – the controversical, adiponectin – the promising, and more to come. Best Pract Res Clin Endocrinol Metab 19(4):525–546

Lacy PE (1994) The intraislet macrophage and type i diabetes. Mt Sinai J Med 61(2):170–174

Larsen CM, Wadt KA, Juhl LF, Andersen HU, Karlsen AE, Su MS, Seedorf K, Shapiro L, Dinarello CA, Mandrup-Poulsen T (1998) Interleukin-1beta-induced rat pancreatic islet nitric oxide synthesis requires both the p38 and extracellular signal-regulated kinase 1/2 mitogen-activated protein kinases. J Biol Chem 273(24):15294–15300

Larsen CM, Faulenbach M, Vaag A, Volund A, Ehses JA, Seifert B, Mandrup-Poulsen T, Donath MY (2007) Interleukin-1-receptor antagonist in type 2 diabetes mellitus. N Engl J Med 356(15):1517–1526

Liadis N, Salmena L, Kwan E, Tajmir P, Schroer SA, Radziszewska A, Li X, Sheu L, Eweida M, Xu S, Gaisano HY, Hakem R, Woo M (2007) Distinct in vivo roles of caspase-8 in beta-cells in physiological and diabetes models. Diabetes 56(9):2302–2311

Liuwantara D, Elliot M, Smith MW, Yam AO, Walters SN, Marino E, McShea A, Grey ST (2006) Nuclear factor-kappab regulates beta-cell death: a critical role for a20 in beta-cell protection. Diabetes 55(9):2491–2501

Loweth AC, Williams GT, James RF, Scarpello JH, Morgan NG (1998) Human islets of langerhans express fas ligand and undergo apoptosis in response to interleukin-1beta and fas ligation. Diabetes 47(5):727–732

Loweth AC, Watts K, McBain SC, Williams GT, Scarpello JH, Morgan NG (2000) Dissociation between fas expression and induction of apoptosis in human islets of langerhans. Diabetes Obes Metab 2(1):57–60

Lumeng CN, Bodzin JL, Saltiel AR (2007) Obesity induces a phenotypic switch in adipose tissue macrophage polarization. J Clin Invest 117(1):175–184

Luotola K, Paakkonen R, Alanne M, Lanki T, Moilanen L, Surakka I, Pietila A, Kahonen M, Nieminen MS, Kesaniemi YA, Peters A, Jula A, Perola M, Salomaa V (2009) Association of variation in the interleukin-1 gene family with diabetes and glucose homeostasis. J Clin Endocrinol Metab 94(11):4575–4583. doi:jc.2009-0666 [pii] 10.1210/jc.2009-0666

Maedler K, Spinas GA, Lehmann R, Sergeev P, Weber M, Fontana A, Kaiser N, Donath MY (2001) Glucose induces beta-cell apoptosis via upregulation of the fas-receptor in human islets. Diabetes 50:1683–1690

Maedler K, Sergeev P, Ris F, Oberholzer J, Joller-Jemelka HI, Spinas GA, Kaiser N, Halban PA, Donath MY (2002) Glucose-induced beta-cell production of interleukin-1beta contributes to glucotoxicity in human pancreatic islets. J Clin Invest 110:851–860

Maedler K, Sergeev P, Ehses JA, Mathe Z, Bosco D, Berney T, Dayer JM, Reinecke M, Halban PA, Donath MY (2004) Leptin modulates beta cell expression of il-1 receptor antagonist and release of il-1beta in human islets. Proc Natl Acad Sci USA 101(21):8138–8143

Maedler K, Schumann DM, Sauter N, Ellingsgaard H, Bosco D, Baertschiger R, Iwakura Y, Oberholzer J, Wollheim CB, Gauthier BR, Donath MY (2006) Low concentration of interleukin-1{beta} induces flice-inhibitory protein-mediated {beta}-cell proliferation in human pancreatic islets. Diabetes 55(10):2713–2722

Maedler K, Dharmadhikari G, Schumann DM, Størling J (2009) Interleukin-1 beta targeted therapy for type 2 diabetes. Expert Opin Biol Ther Sep 9(9):1177–1188

Mandrup-Poulsen T, Bendtzen K, Nielsen JH, Bendixen G, Nerup J (1985) Cytokines cause functional and structural damage to isolated islets of langerhans. Allergy 40(6):424–429

Mandrup-Poulsen T, Bendtzen K, Nerup J, Dinarello CA, Svenson M, Nielsen JH (1986) Affinity-purified human interleukin i is cytotoxic to isolated islets of langerhans. Diabetologia 29(1):63–67

Mandrup-Poulsen T, Zumsteg U, Reimers J, Pociot F, Morch L, Helqvist S, Dinarello CA, Nerup J (1993) Involvement of interleukin 1 and interleukin 1 antagonist in pancreatic beta-cell destruction in insulin-dependent diabetes mellitus. Cytokine 5(3):185–191

Marculescu R, Endler G, Schillinger M, Iordanova N, Exner M, Hayden E, Huber K, Wagner O, Mannhalter C (2002) Interleukin-1 receptor antagonist genotype is associated with coronary atherosclerosis in patients with type 2 diabetes. Diabetes 51(12):3582–3585

Meier CA, Bobbioni E, Gabay C, Assimacopoulos-Jeannet F, Golay A, Dayer JM (2002) Il-1 receptor antagonist serum levels are increased in human obesity: a possible link to the resistance to leptin? J Clin Endocrinol Metab 87(3):1184–1188

Milanski M, Degasperi G, Coope A, Morari J, Denis R, Cintra DE, Tsukumo DM, Anhe G, Amaral ME, Takahashi HK, Curi R, Oliveira HC, Carvalheira JB, Bordin S, Saad MJ, Velloso LA (2009) Saturated fatty acids produce an inflammatory response predominantly through the activation of tlr4 signaling in hypothalamus: implications for the pathogenesis of obesity. J Neurosci 29(2):359–370. doi:29/2/359 [pii] 10.1523/JNEUROSCI.2760-08.2009

Mine T, Miura K, Okutsu T, Mitsui A, Kitahara Y (2004) Gene expression profile in the pancreatic islets of goto-kakizaki (gk) rats with repeated postprandial hyperglycemia. Diabetes 53(Suppl 2):2475A

Mokhtari D, Myers JW, Welsh N (2008) The mapk kinase kinase-1 is essential for stress-induced pancreatic islet cell death. Endocrinology 149(6):3046–3053

Munzberg H, Flier JS, Bjorbaek C (2004) Region-specific leptin resistance within the hypothalamus of diet-induced obese mice. Endocrinology 145(11):4880–4889. doi:10.1210/en.2004-0726 en.2004-0726 [pii]

Narcisse L, Scemes E, Zhao Y, Lee SC, Brosnan CF (2005) The cytokine il-1beta transiently enhances p2x7 receptor expression and function in human astrocytes. Glia 49(2):245–258

Nicoletti F, Di Marco R, Barcellini W, Magro G, Schorlemmer HU, Kurrle R, Lunetta M, Grasso S, Zaccone P, Meroni P (1994) Protection from experimental autoimmune diabetes in the non-obese diabetic mouse with soluble interleukin-1 receptor. Eur J Immunol 24(8):1843–1847

Ortis F, Cardozo AK, Crispim D, Storling J, Mandrup-Poulsen T, Eizirik DL (2006) Cytokine-induced proapoptotic gene expression in insulin-producing cells is related to rapid, sustained, and nonoscillatory nuclear factor-kappab activation. Mol Endocrinol 20(8):1867–1879

Osborn O, Brownell SE, Sanchez-Alavez M, Salomon D, Gram H, Bartfai T (2008) Treatment with an interleukin 1 beta antibody improves glycemic control in diet-induced obesity. Cytokine 44(1):141–148

Owyang A, Maedler K, Gross L, Yin J, Esposito L, Shu L, Jadhav J, Domsgen E, Bergemann J, Lee S, Kantak S (2010) Xoma 052, an anti-IL-1β monoclonal antibody, improves glucose control and β-cell function in the diet-induced obesity mouse model. Endocrinology 151(6):2515–2527

Oyadomari S, Araki E, Mori M (2002) Endoplasmic reticulum stress-mediated apoptosis in pancreatic beta- cells. Apoptosis 7(4):335–345

Parikh H, Carlsson E, Chutkow WA, Johansson LE, Storgaard H, Poulsen P, Saxena R, Ladd C, Schulze PC, Mazzini MJ, Jensen CB, Krook A, Bjornholm M, Tornqvist H, Zierath JR, Ridderstrale M, Altshuler D, Lee RT, Vaag A, Groop LC, Mootha VK (2007) Txnip regulates peripheral glucose metabolism in humans. PLoS Med 4(5):e158. doi:06-PLME-RA-0918R1 [pii] 10.1371/journal.pmed.0040158

Pavlovic D, Andersen NA, Mandrup-Poulsen T, Eizirik DL (2000) Activation of extracellular signal-regulated kinase (erk)1/2 contributes to cytokine-induced apoptosis in purified rat pancreatic beta-cells. Eur Cytokine Netw 11(2):267–274

Pickersgill LM, Mandrup-Poulsen TR (2009) The anti-interleukin-1 in type 1 diabetes action trial – background and rationale. Diabetes Metab Res Rev 25(4):321–324

Pukel C, Baquerizo H, Rabinovitch A (1988) Destruction of rat islet cell monolayers by cytokines. Synergistic interactions of interferon-gamma, tumor necrosis factor, lymphotoxin, and interleukin 1. Diabetes 37(1):133–136

Rabinovitch A, Sumoski W, Rajotte RV, Warnock GL (1990) Cytotoxic effects of cytokines on human pancreatic islet cells in monolayer culture. J Clin Endocrinol Metab 71(1):152–156

Rasouli N, Kern PA (2008) Adipocytokines and the metabolic complications of obesity. J Clin Endocrinol Metab 93(11 Suppl 1):S64–S73

Richardson SJ, Willcox A, Bone AJ, Foulis AK, Morgan NG (2009) Islet-associated macrophages in type 2 diabetes. Diabetologia 52(8):1686–1688. doi:10.1007/s00125-009-1410-z

Roduit R, Thorens B (1997) Inhibition of glucose-induced insulin secretion by long-term preexposure of pancreatic islets to leptin. FEBS Lett 415(2):179–182

Ronn SG, Billestrup N, Mandrup-Poulsen T (2007) Diabetes and suppressors of cytokine signaling proteins. Diabetes 56(2):541–548

Ronn SG, Borjesson A, Bruun C, Heding PE, Frobose H, Mandrup-Poulsen T, Karlsen AE, Rasschaert J, Sandler S, Billestrup N (2008) Suppressor of cytokine signalling-3 expression inhibits cytokine-mediated destruction of primary mouse and rat pancreatic islets and delays allograft rejection. Diabetologia 51(10):1873–1882

Rothwell NJ, Luheshi GN (2000) Interleukin 1 in the brain: biology, pathology and therapeutic target. Trends Neurosci 23(12):618–625

Rui L, Yuan M, Frantz D, Shoelson S, White MF (2002) Socs-1 and socs-3 block insulin signaling by ubiquitin-mediated degradation of irs1 and irs2. J Biol Chem 277(44):42394–42398

Ruotsalainen E, Salmenniemi U, Vauhkonen I, Pihlajamaki J, Punnonen K, Kainulainen S, Laakso M (2006) Changes in inflammatory cytokines are related to impaired glucose tolerance in offspring of type 2 diabetic subjects. Diab Care 29(12):2714–2720

Saldeen J, Lee JC, Welsh N (2001) Role of p38 mitogen-activated protein kinase (p38 mapk) in cytokine- induced rat islet cell apoptosis. Biochem Pharmacol 61(12):1561–1569

Salmenniemi U, Ruotsalainen E, Pihlajamaki J, Vauhkonen I, Kainulainen S, Punnonen K, Vanninen E, Laakso M (2004) Multiple abnormalities in glucose and energy metabolism and coordinated changes in levels of adiponectin, cytokines, and adhesion molecules in subjects with metabolic syndrome. Circulation 110(25):3842–3848

Sandberg JO, Eizirik DL, Sandler S, Tracey DE, Andersson A (1993) Treatment with an interleukin-1 receptor antagonist protein prolongs mouse islet allograft survival. Diabetes 42(12):1845–1851

Sandberg JO, Andersson A, Eizirik DL, Sandler S (1994) Interleukin-1 receptor antagonist prevents low dose streptozotocin induced diabetes in mice. Biochem Biophys Res Commun 202(1):543–548

Sandberg JO, Eizirik DL, Sandler S (1997) IL-1 receptor antagonist inhibits recurrence of disease after syngeneic pancreatic islet transplantation to spontaneously diabetic non-obese diabetic (nod) mice. Clin Exp Immunol 108(2):314–317

Satoh M, Yasunami Y, Matsuoka N, Nakano M, Itoh T, Nitta T, Anzai K, Ono J, Taniguchi M, Ikeda S (2007) Successful islet transplantation to two recipients from a single donor by targeting proinflammatory cytokines in mice. Transplantation 83(8):1085–1092

Sauter NS, Schulthess FT, Galasso R, Castellani LW, Maedler K (2008) The antiinflammatory cytokine interleukin-1 receptor antagonist protects from high-fat diet-induced hyperglycemia. Endocrinology 149(5):2208–2218

Scarim AL, Arnush M, Hill JR, Marshall CA, Baldwin A, McDaniel ML, Corbett JA (1997) Evidence for the presence of type i il-1 receptors on beta-cells of islets of langerhans. Biochim Biophys Acta 1361(3):313–320

Schenk S, Saberi M, Olefsky JM (2008) Insulin sensitivity: modulation by nutrients and inflammation. J Clin Invest 118(9):2992–3002

Schneider H, Pitossi F, Balschun D, Wagner A, del Rey A, Besedovsky HO (1998) A neuromodulatory role of interleukin-1beta in the hippocampus. Proc Natl Acad Sci USA 95(13):7778–7783

Schott WH, Haskell BD, Tse HM, Milton MJ, Piganelli JD, Choisy-Rossi CM, Reifsnyder PC, Chervonsky AV, Leiter EH (2004) Caspase-1 is not required for type 1 diabetes in the nod mouse. Diabetes 53(1):99–104

Schroder K, Zhou R, Tschopp J (2010) The nlrp3 inflammasome: a sensor for metabolic danger? Science 327(5963):296–300. doi:327/5963/296 [pii] 10.1126/science.1184003

Schumann DM, Maedler K, Franklin I, Konrad D, Storling J, Boni-Schnetzler M, Gjinovci A, Kurrer MO, Gauthier BR, Bosco D, Andres A, Berney T, Greter M, Becher B, Chervonsky AV,

Halban PA, Mandrup-Poulsen T, Wollheim CB, Donath MY (2007) The fas pathway is involved in pancreatic beta cell secretory function. Proc Natl Acad Sci USA 104:2861–2866

Seckinger P, Lowenthal JW, Williamson K, Dayer JM, MacDonald HR (1987a) A urine inhibitor of interleukin 1 activity that blocks ligand binding. J Immunol 139(5):1546–1549

Seckinger P, Williamson K, Balavoine JF, Mach B, Mazzei G, Shaw A, Dayer JM (1987b) A urine inhibitor of interleukin 1 activity affects both interleukin 1 alpha and 1 beta but not tumor necrosis factor alpha. J Immunol 139(5):1541–1545

Seufert J, Kieffer TJ, Leech CA, Holz GG, Moritz W, Ricordi C, Habener JF (1999) Leptin suppression of insulin secretion and gene expression in human pancreatic islets: implications for the development of adipogenic diabetes mellitus. J Clin Endocrinol Metab 84(2):670–676

Sims JE, Dower SK (1994) Interleukin-1 receptors. Eur Cytokine Netw 5(6):539–546

Solinas G, Vilcu C, Neels JG, Bandyopadhyay GK, Luo JL, Naugler W, Grivennikov S, Wynshaw-Boris A, Scadeng M, Olefsky JM, Karin M (2007) Jnk1 in hematopoietically derived cells contributes to diet-induced inflammation and insulin resistance without affecting obesity. Cell Metab 6(5):386–397

Spinas GA, Mandrup-Poulsen T, Molvig J, Baek L, Bendtzen K, Dinarello CA, Nerup J (1986) Low concentrations of interleukin-1 stimulate and high concentrations inhibit insulin release from isolated rat islets of langerhans. Acta Endocrinol (Copenh) 113(4):551–558

Spinas GA, Hansen BS, Linde S, Kastern W, Molvig J, Mandrup-Poulsen T, Dinarello CA, Nielsen JH, Nerup J (1987) interleukin 1 dose-dependently affects the biosynthesis of (pro) insulin in isolated rat islets of langerhans. Diabetologia 30(7):474–480

Spinas GA, Palmer JP, Mandrup-Poulsen T, Andersen H, Nielsen JH, Nerup J (1988) The bimodal effect of interleukin 1 on rat pancreatic beta-cells – stimulation followed by inhibition – depends upon dose, duration of exposure, and ambient glucose concentration. Acta Endocrinol (Copenh) 119(2):307–311

Stassi G, Todaro M, Richiusa P, Giordano M, Mattina A, Sbriglia MS, Lo MA, Buscemi G, Galluzzo A, Giordano C (1995) Expression of apoptosis-inducing cd95 (fas/apo-1) on human beta-cells sorted by flow-cytometry and cultured in vitro. Transplant Proc 27(6):3271–3275

Stassi G, De Maria R, Trucco G, Rudert W, Testi R, Galluzzo A, Giordano C, Trucco M (1997) Nitric oxide primes pancreatic beta cells for fas-mediated destruction in insulin-dependent diabetes mellitus. J Exp Med 186(8):1193–1200

Stoffels K, Gysemans C, Waer M, Laureys J, Bouillon R, Mathieu C (2002) Interleukin-1 receptor antagonist inhibits primary non-function and prolongs graft survival time of xenogeneic islets transplanted in sponaeously diabetic autoimmune nod mice. Diabetologia 45(Suppl 2):424–424

Storling J, Binzer J, Andersson AK, Zullig RA, Tonnesen M, Lehmann R, Spinas GA, Sandler S, Billestrup N, Mandrup-Poulsen T (2005) Nitric oxide contributes to cytokine-induced apoptosis in pancreatic beta cells via potentiation of jnk activity and inhibition of akt. Diabetologia 48(10):2039–2050

Taniguchi CM, Emanuelli B, Kahn CR (2006) Critical nodes in signalling pathways: insights into insulin action. Nat Rev Mol Cell Biol 7(2):85–96

Tellez N, Montolio M, Biarnes M, Castano E, Soler J, Montanya E (2005) Adenoviral over-expression of interleukin-1 receptor antagonist protein increases beta-cell replication in rat pancreatic islets. Gene Ther 12(2):120–128

Tellez N, Montolio M, Estil·les E, Escoriza J, Soler J, Montanya E (2007) Adenoviral overproduction of interleukin-1 receptor antagonist increases beta cell replication and mass in syngeneically transplanted islets, and improves metabolic outcome. Diabetologia 50(3):602–611

Thomas HE, Irawaty W, Darwiche R, Brodnicki TC, Santamaria P, Allison J, Kay TW (2004) Il-1 receptor deficiency slows progression to diabetes in the nod mouse. Diabetes 53(1):113–121

Tilg H, Moschen AR (2006) Adipocytokines: mediators linking adipose tissue, inflammation and immunity. Nat Rev Immunol 6(10):772–783

Touzani O, Boutin H, Chuquet J, Rothwell N (1999) Potential mechanisms of interleukin-1 involvement in cerebral ischaemia. J Neuroimmunol 100(1–2):203–215

Tran PO, Gleason CE, Robertson RP (2002) Inhibition of interleukin-1beta-induced cox-2 and ep3 gene expression by sodium salicylate enhances pancreatic islet beta-cell function. Diabetes 51(6):1772–1778

Vandenabeele P, Fiers W (1991) Is amyloidogenesis during Alzheimer's disease due to an il-1-/ il-6-mediated 'acute phase response' in the brain? Immunol Today 12(7):217–219

Venieratos PD, Drossopoulou GI, Kapodistria KD, Tsilibary EC, Kitsiou PV (2010) High glucose induces suppression of insulin signalling and apoptosis via upregulation of endogenous il-1beta and suppressor of cytokine signalling-1 in mouse pancreatic beta cells. Cell Signal. doi:S0898-6568(10)00009-4 [pii] 10.1016/j.cellsig.2010.01.003

Warner SJ, Auger KR, Libby P (1987) Human interleukin 1 induces interleukin 1 gene expression in human vascular smooth muscle cells. J Exp Med 165(5):1316–1331

Weisberg SP, McCann D, Desai M, Rosenbaum M, Leibel RL, Ferrante AW Jr (2003) Obesity is associated with macrophage accumulation in adipose tissue. J Clin Invest 112(12):1796–1808

Weksler-Zangen S, Raz I, Lenzen S, Jorns A, Ehrenfeld S, Amir G, Oprescu A, Yagil Y, Yagil C, Zangen DH, Kaiser N (2008) Impaired glucose-stimulated insulin secretion is coupled with exocrine pancreatic lesions in the cohen diabetic rat. Diabetes 57(2):279–287

Wellen KE, Hotamisligil GS (2005) Inflammation, stress, and diabetes. J Clin Invest 115(5):1111–1119

Welsh N (1996) Interleukin-1 beta-induced ceramide and diacylglycerol generation may lead to activation of the c-jun nh2-terminal kinase and the transcription factor atf2 in the insulin-producing cell line rinm5f. J Biol Chem 271(14):8307–8312

Welsh N, Cnop M, Kharroubi I, Bugliani M, Lupi R, Marchetti P, Eizirik DL (2005) Is there a role for locally produced interleukin-1 in the deleterious effects of high glucose or the type 2 diabetes milieu to human pancreatic islets? Diabetes 54(11):3238–3244

Wilson HL, Francis SE, Dower SK, Crossman DC (2004) Secretion of intracellular il-1 receptor antagonist (type 1) is dependent on p2x7 receptor activation. J Immunol 173(2):1202–1208

Wolf G, Yirmiya R, Goshen I, Iverfeldt K, Holmlund L, Takeda K, Shavit Y (2003) Impairment of interleukin-1 (il-1) signaling reduces basal pain sensitivity in mice: genetic, pharmacological and developmental aspects. Pain 104(3):471–480. doi:S0304395903000678 [pii]

Xu H, Barnes GT, Yang Q, Tan G, Yang D, Chou CJ, Sole J, Nichols A, Ross JS, Tartaglia LA, Chen H (2003) Chronic inflammation in fat plays a crucial role in the development of obesity-related insulin resistance. J Clin Invest 112(12):1821–1830

Yu X, Park BH, Wang MY, Wang ZV, Unger RH (2008) Making insulin-deficient type 1 diabetic rodents thrive without insulin. Proc Natl Acad Sci USA 105(37):14070–14075

Zeender E, Maedler K, Bosco D, Berney T, Donath MY, Halban PA (2004) Pioglitazone and sodium salicylate protect human {beta}-cells against apoptosis and impaired function induced by glucose and interleukin-1{beta}. J Clin Endocrinol Metab 89(10):5059–5066

Zhou R, Tardivel A, Thorens B, Choi I, Tschopp J (2009) Thioredoxin-interacting protein links oxidative stress to inflammasome activation. Nat Immunol. doi:ni.1831 [pii]10.1038/ni.1831

Zuliani G, Ranzini M, Guerra G, Rossi L, Munari MR, Zurlo A, Volpato S, Atti AR, Ble A, Fellin R (2007) Plasma cytokines profile in older subjects with late onset Alzheimer's disease or vascular dementia. J Psychiatr Res 41(8):686–693

Fructose-1, 6-Bisphosphatase Inhibitors for Reducing Excessive Endogenous Glucose Production in Type 2 Diabetes

Paul D. van Poelje, Scott C. Potter, and Mark D. Erion

Contents

Abstract Fructose-1,6-bisphosphatase (FBPase), a rate-controlling enzyme of gluconeogenesis, has emerged as an important target for the treatment of type 2 diabetes due to the well-recognized role of excessive endogenous glucose production (EGP) in the hyperglycemia characteristic of the disease. Inhibitors of

P.D. van Poelje (✉)
Pfizer Inc., Eastern Point Road, Groton, CT 06340, USA
e-mail: paul.vanpoelje@pfizer.com

S.C. Potter
Lilly AME, 10300 Campus Point Drive, San diego, CA 92121, USA
e-mail: pottersc@lilly.com

M.D. Erion
Merck & Co. Inc., 126 E Lincoln Ave, Rahway, NJ 07065, USA
e-mail: mark_erion@merck.com

M. Schwanstecher (ed.), *Diabetes - Perspectives in Drug Therapy*,
Handbook of Experimental Pharmacology 203,
DOI 10.1007/978-3-642-17214-4_12, © Springer-Verlag Berlin Heidelberg 2011

FBPase are expected to fulfill an unmet medical need because the majority of current antidiabetic medications act primarily on insulin resistance or insulin insufficiency and do not reduce gluconeogenesis effectively or in a direct manner. Despite significant challenges, potent and selective inhibitors of FBPase targeting the allosteric site of the enzyme were identified by means of a structure-guided design strategy that used the natural inhibitor, adenosine monophosphate (AMP), as the starting point. Oral delivery of these anionic FBPase inhibitors was enabled by a novel diamide prodrug class. Treatment of diabetic rodents with CS-917, the best characterized of these prodrugs, resulted in a reduced rate of gluconeogenesis and EGP. Of note, inhibition of gluconeogenesis by CS-917 led to the amelioration of both fasting and postprandial hyperglycemia without weight gain, incidence of hypoglycemia, or major perturbation of lactate or lipid homeostasis. Furthermore, the combination of CS-917 with representatives of the insulin sensitizer or insulin secretagogue drug classes provided enhanced glycemic control. Subsequent clinical evaluations of CS-917 revealed a favorable safety profile as well as clinically meaningful reductions in fasting glucose levels in patients with T2DM. Future trials of MB07803, a second generation FBPase inhibitor with improved pharmacokinetics, will address whether this novel class of antidiabetic agents can provide safe and long-term glycemic control.

Keywords AMP mimetic · Antihyperglycemic agent · Endogenous glucose production · Fructose-1,6-bisphosphatase · Gluconeogenesis · Type 2 diabetes

1 Introduction

Type 2 diabetes (T2DM), a disease that afflicts over 180 million people worldwide, is characterized by insulin insufficiency, insulin resistance, and increased endogenous glucose production (EGP). These three abnormalities cause high plasma glucose levels (hyperglycemia), which in turn cause diabetic complications such as loss of vision, renal impairment, and heart disease, in patients. The majority of current antidiabetic drugs (e.g., insulin sensitizers, sulfonylureas, DPP-IV inhibitors) reduce glucose levels by improving peripheral insulin resistance and/or augmenting insulin secretion. Metformin is the only prescribed drug whose primary mechanism of action is the reduction, albeit indirect, of EGP (Hundal et al. 2000). Because most patients with T2DM fail to achieve recommended treatment goals, there is a need for novel, more effective drugs that act alone or in combination with other antidiabetic agents. Direct inhibitors of EGP represent a drug class that could potentially provide glycemic control across a broad patient population and combine effectively with approved antidiabetic drugs. Hence, the discovery of inhibitors of glycogenolysis and gluconeogenesis has been pursued actively by the pharmaceutical industry.

In this chapter, the physiological rationale for developing inhibitors of the gluconeogenic enzyme fructose-1, 6-bisphosphatase (FBPase) for the treatment of T2DM is described. In addition, the challenges associated with the discovery of inhibitors binding to the allosteric site of FBPase are discussed. Special attention is given to a structure-based design strategy that ultimately led to the discovery of the first potent, selective and orally active FBPase inhibitors. Finally, the preclinical proof-of-concept studies that provided the impetus for the clinical development of the first FBPase inhibitors are summarized as well as the initial clinical profile of these inhibitors.

2 Endogenous Glucose Production in Type 2 Diabetes

The liver is the primary organ responsible for EGP. Glucose is produced by the liver by two pathways: gluconeogenesis (the de novo synthesis of glucose from lactate, alanine and glycerol) and glycogenolysis (the breakdown of glycogen stored in the liver). In healthy individuals, gluconeogenesis accounts for ~ 50% of EGP after an overnight fast and increases progressively to account for over 90% of EGP following 40 h of fasting (Landau et al. 1996; Rothman et al. 1991). The contribution of glycogenolysis to EGP declines reciprocally during fasting periods, reaching a negligible contribution by ~96 h. During the postprandial period, EGP is suppressed by rapid and near complete inhibition of glycogenolysis and slower and more modest inhibition (30–50% within 4 h) of gluconeogenesis (Gastaldelli et al. 2001).

In lean and obese patients with T2DM, the rate of EGP in the fasted state is increased relative to that of healthy individuals (Gastaldelli et al. 2000; Magnusson et al. 1992). Increased EGP during fasting is due solely to increased gluconeogenesis; glycogenolytic rates are either unchanged or slightly reduced in patients with T2DM (Magnusson et al. 1992; Wajngot et al. 2001). The significance of increased gluconeogenesis in T2DM is apparent from the strong correlation between the rate of EGP and fasting hyperglycemia: for each incremental increase in EGP, there is a corresponding increase in the fasting glucose levels (Jeng et al. 1994; Maggs et al. 1998). Insulin resistance, in contrast, correlates poorly with the degree of fasting hyperglycemia (Olefsky 1993). During the postprandial period, glycogenolysis and gluconeogenesis are poorly suppressed in patients with T2DM (Cherrington 1999; Gastaldelli et al. 2001). In contrast to fasting hyperglycemia, factors other than increased glucose production (e.g., impaired glucose disposal) play a quantitatively important role in the etiology of postprandial hyperglycemia.

The main cause of excessive EGP in T2DM is an imbalance in the actions of glucoregulatory hormones at the level of both hepatic and extrahepatic tissues. In healthy individuals, a balance between insulin and glucagon secretion by the pancreas ensures an appropriate rate of glucose production by the liver (Cherrington 1999; Leroith et al. 1996). Insulin inhibits glycogenolysis, stimulates glycogen synthesis, reduces gluconeogenesis, and increases glycolytic metabolism of glucose. These actions, which promote glucose storage and utilization in liver, are opposed by glucagon. In patients with T2DM, a combination of insulin deficiency, relative glucagon excess, and hepatic insulin resistance switches the liver to a sustained glucose output mode. Insulin deficiency and insulin resistance in extrahepatic tissues further promote hepatic glucose production by inducing a catabolic state that increases the availability of substrates. In adipose tissue, uncontrolled lipolysis leads to increased supply of glycerol, a gluconeogenic substrate, and of free fatty acids, which can serve as a source of energy for gluconeogenesis. In muscle tissue, increased protein catabolism results in increased supply of gluconeogenic amino acids such as alanine. Elevated EGP accelerates a self-perpetuating Cori cycle by which glucose produced by the liver is converted to lactate in extrahepatic tissues, which in turn fuels EGP.

In light of the above, therapies that reduce EGP are expected to have considerable therapeutic potential for the treatment of T2DM. The strongest rationale exists for inhibitors of gluconeogenesis rather than inhibitors of glycogenolysis, because gluconeogenesis rates are increased in T2DM and contribute to hyperglycemia in both the fasted and postprandial states. Inhibition of gluconeogenesis, or EGP in general, would not be expected to directly improve other underlying causes of T2DM such as insulin resistance and insulin deficiency. However, with long-term control of EGP by inhibition of gluconeogenesis improved glycemic control could reduce glucotoxicity and consequently increase insulin sensitivity and pancreatic function. Furthermore, inhibitors of gluconeogenesis may complement the activity of current antidiabetic agents acting on the pancreas (e.g., sulfonylureas, DPP-IV inhibitors) or the periphery (insulin sensitizers). Combination therapy with these agents could provide exceptional therapeutic benefits by targeting all three abnormalities of the diabetic phenotype: insulin secretion, insulin resistance, and excessive gluconeogenesis.

3 Enzyme targets in the Gluconeogenic Pathways

Glucose is produced from 3-carbon substrates such as lactate by a series of reactions catalyzed by twelve different enzymes. The majority of the enzymes of gluconeogenesis also catalyze the reverse reactions involved in glycolysis, but there are four unidirectional enzymes, which together with their glycolytic counterparts, form the so-called substrate cycles of the pathway: pyruvate carboxylase (PC) and phosphoenolpyruvate carboxykinase (PEPCK) (lower cycle), FBPase (middle cycle), and glucose 6-phosphatase (G-6-Pase) (upper cycle). The latter three enzymes form the major control points in gluconeogenesis (Fig. 1) and have all been targets of drug discovery efforts.

There are significant mechanistic concerns with inhibitors of PEPCK and G-6-Pase. Genetic knockout of PEPCK, the gluconeogenic enzyme proximal to the early mitochondrial steps of the pathway, revealed multiple potential side effects including increased mitochondrial redox state, inhibition of the tricarboxylic acid cycle, and a reduction in β-oxidation of fats, leading to hepatic steatosis (Burgess et al. 2004). Another limitation of PEPCK as a drug target is that its inhibition does not inhibit EGP from glycerol, a substrate with increased abundance that contributes significantly to glucose production in T2DM. Inhibition of the G-6-Pase step in the

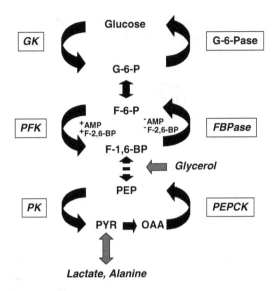

Fig. 1 Substrate cycles in the pathway of gluconeogenesis. Abbreviations: fructose-1,6-bisphosphatase, FBPase; fructose-1,6-bisphosphate, F-1,6-BP; fructose-6-phosphate, F-6-P; glucokinase, GK; glucose-6-phosphatase, G-6-Pase; glucose-6-phosphate, G-6-P, oxaloacetate, OAA; 1-type pyruvate kinase, PK; phosphoenolpyruvate carboxykinase, PEPCK; pyruvate, PYR; 6-phosphofructokinase, PFK. Note that the conversion of pyruvate to oxaloacetate is catalyzed by a mitochondrial enzyme, pyruvate carboxylase. All other enzymes are cytosolic activities. The regulation of PFK and FBPase by the natural effectors adenosine monophosphate (AMP) and fructose-2,6-bisphosphate (F-2,6-BP) is indicated

uppermost substrate cycle is also problematic. G-6-Pase catalyzes the final step common to glucose production by gluconeogenesis as well as glycogenolysis. Simultaneous inhibition of the only two mechanisms of EGP represents a considerable risk for hypoglycemia. None of the PEPCK or G-6-Pase inhibitors identified (Foley et al. 2003; Parker et al. 1998) has progressed beyond initial biological characterization.

FBPase, the gluconeogenic enzyme in the middle substrate cycle (Fig. 1), represents a logical target for pharmacological intervention. As the second-to-last enzyme in gluconeogenesis, FBPase controls the incorporation of all 3-carbon substrates into glucose. Furthermore, the FBPase step is not involved in the breakdown of glycogen and is well removed from the mitochondrial steps of the pathway, theoretically reducing the risk of hypoglycemia and other mechanistic toxicities. An adequate safety margin with respect to hypoglycemia and other theoretical safety concerns such as lacticemia (due to reduced clearance of lactate by gluconeogenesis) and hyperlipidemia (due to shunting of gluconeogenic precursors into lipids) is suggested by the clinical profile of adults with FBPase deficiency (Gitzelmann et al. 1995). FBPase deficiency is a rare autosomal recessive genetic disorder characterized by complete absence of detectable FBPase activity. In adulthood, individuals with FBPase deficiency have near-normal biochemical and clinical parameters, provided they maintain an appropriate diet and avoid prolonged fasting. An additional rationale for targeting FBPase is that its expression and activity are increased in diabetic animal models such as db/db and NZO mice (Kodama et al. 1994; Andrikopoulos et al. 1996). Furthermore, over-expression of human FBPase alone, at least in normal mice, leads to increased gluconeogenesis from glycerol (Lamont et al. 2006), a substrate that enters the pathway just prior to the FBPase step (Fig. 1). These transgenic mice also develop glucose intolerance when fed a high-fat diet (Visinoni et al. 2008), providing additional evidence that increased FBPase activity can contribute to the diabetic phenotype.

4   Structure and function of FBPase

FBPase catalyzes the magnesium-dependent hydrolysis of fructose 1,6-bispho-sphate (F1, 6BP) to fructose 6-phosphate and inorganic phosphate. The enzyme is cytosolic and exists as a tetramer consisting of four identical subunits (36.7 kD), each containing a substrate-binding site, a magnesium-binding site (within the substrate site), and an allosteric regulatory site within 28 Å of the active site (DZugaj and Kochman 1980; El-Maghrabi et al. 1993). Two genes encode FBPase in mammals: a liver FBPase (FBPI), which is expressed primarily in liver and kidney, and a muscle FBPase (FBP2), which is found exclusively in muscle tissue (Skalecki et al. 1995; Tillmann et al. 2002). The two enzymes show ~77% identity at the amino acid level and have almost complete identity in regions of the enzyme involved in the binding of substrate, regulatory molecules, and magnesium. Because muscle tissue lacks glucose-6-phosphatase activity and is consequently nongluco-neogenic, the physiological role of muscle FBPase is unclear. It has been suggested that muscle FBPase may be important for glycogenesis from substrates such as lactate. The liver form of FBPase has a critical role in maintaining blood glucose levels both in liver, the predominant source of de novo synthesized glucose, and the kidney, a relatively minor source of de novo synthesized glucose except during periods of extreme fasting and, reportedly, in the diabetic state (Gerich et al. 2001).

The regulation of liver FBPase has been investigated in detail. Enzyme activity is regulated synergistically by fructose 2,6-bisphosphate (F-2, 6-BP), an inhibitor that binds to the substrate site, and adenosine monophosphate (AMP), an inhibitor that binds to the allosteric site (Gidh-Jain et al. 1994; van Schaftingen and Hers 1981). As deduced from crystallographic studies of the porcine and human liver enzymes, AMP stabilizes an inactive conformation of the enzyme, the T state, while the substrate of FBPase, F1, 6BP, is dephosphorylated only by the active or R state (Ke et al. 1991). The binding of AMP at the allosteric site has been postulated to inhibit the catalytic activity of FBPase by distorting the magnesium-binding site. Since intracellular AMP levels are generally maintained within a narrow range, it is believed that F-2, 6-BP is the major physiological regulator of FBPase. Binding of F-2, 6-BP at the substrate site results in a leftward shift in the inhibition curve for AMP. Intracellular levels of F-2, 6-BP are controlled by a bifunctional enzyme,

6-phosphofructo-2-kinase/fructose-2, 6-bisphosphatase, in such a way that F-2, 6-BP levels are decreased and FBPase activity increased during times of glucose demand (Pilkis 1991). The bifunctional enzyme is under hormonal control by glucagon and insulin. Glucagon increases intracellular cAMP levels which in turn stimulate protein kinase A. Phosphorylation of the bifunctional enzyme by protein kinase A inhibits its kinase activity and stimulates its phosphatase activity, thereby reducing the intracellular levels of F-2, 6-BP. Insulin modulates the effects of glucagon on the activity of the bifunctional enzyme by suppressing the glucagon-induced rise of cAMP. Additional regulation of FBPase is exerted at the genetic level by insulin and glucagon: the expression of the FBPase gene is dependent upon cAMP levels (El-Maghrabi et al. 1991) and, accordingly, the insulin-to-glucagon ratio. The decreased insulin-to-glucagon ratio associated with T2DM likely accounts for the increased expression of FBPase observed in various diabetic animal models.

5   Discovery of inhibitors of FBPase

5.1   Competitive and Uncompetitive Inhibitors

Three classes of inhibitors have been reported: those interacting with the substrate site (competitive inhibitors), a site at the subunit interface (uncompetitive), and the AMP site (noncompetitive). The lack of progress in the discovery of potent and selective competitive inhibitors (Pilkis et al. 1986) is likely a reflection of the highly charged nature of the substrate-binding site and the difficulty of designing suitable carbohydrate phosphate mimetics that can bind with high affinity and compete with the elevated fructose-1, 6-bisphosphate levels that result from the inhibition of FBPase. The discovery of modestly potent uncompetitive inhibitors that bind to a hydrophobic region on the subunit interface of FBPase have also been reported (Choe et al. 2003; Rosini et al. 2006; Wright et al. 2001, 2002). These inhibitors may lack favorable pharmacokinetic properties or have other shortcomings as no data beyond in vitro enzyme inhibition have been described. Competitive and uncompetitive inhibitors will not be discussed further in this chapter.

5.2   Noncompetitive Inhibitors; ZMP

One of the first noncompetitive inhibitors of FBPase identified was 5-Amino-4-imidazolecarboxamide riboside (AICAr) monophosphate (ZMP; Fig. 2), a close analog of AMP (Fig. 2), which has gained widespread use as a tool compound for the study of another antidiabetic target, AMP-activated protein kinase (AMPK). ZMP was discovered serendipitously following observations of glucose lowering in animals and is a relatively weak inhibitor of the rat and human FBPase isoforms with IC_{50} values of 370 and 12 µM, respectively (Erion et al. 2005; Vincent

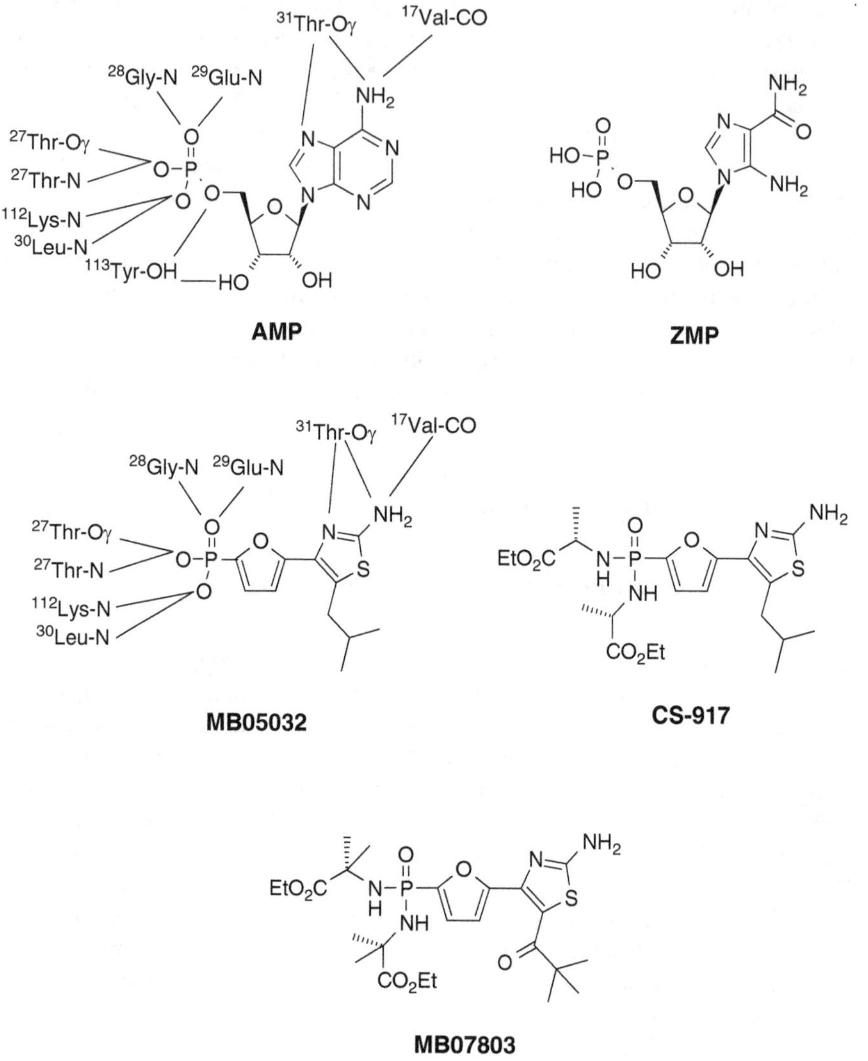

Fig. 2 Structures of inhibitors of FBPase binding to the allosteric site. Key interactions of AMP (a natural inhibitor of the enzyme) and MB05032 (a rationally-designed inhibitor) with amino acid residues in the allosteric-binding pocket are indicated. ZMP is a close structural analog of AMP; CS-917 is an orally bioavailable prodrug of the potent and selective AMP mimetic, MB05032; MB07803 is a second generation compound of the FBPase inhibitor class

et al. 1991). The potent inhibition of gluconeogenesis by AICAr is attributed to its rapid phosphorylation by adenosine kinase in hepatocytes to yield high levels of ZMP. ZMP has poor selectivity for FBPase and, in addition to activating AMPK (EC_{50} 110 μM), modulates the activities of 6-phosphofructo-2-kinase and glycogen phosphorylase (Vincent et al. 1991, 1992; Henin et al. 1996). Although glucose lowering by AICAr in acute and chronic studies in rodents (Pold et al. 2005; Vincent et al. 1996) can be attributed in part to inhibition of FBPase, the poor selectivity of ZMP limits the extent to which the findings validate FBPase as an antidiabetic target or accurately reflect the metabolic side effects associated with FBPase inhibition.

5.3 Noncompetitive Inhibitors; Design of MB05032

High throughput screening of compound libraries as well as structure-based design has been employed to identify noncompetitive inhibitors of FBPase. Although the high-throughput screening efforts led to the identification of inhibitors with a noncompetitive mode of action in vitro (von Geldern et al. 2006; Wright et al. 2003), none of these compounds have been reported to lower blood glucose in vivo. A structure-based drug design strategy using AMP as the starting point, however, yielded the first selective noncompetitive inhibitors of FBPase with potent in vivo activity (Erion et al. 2005). One of the main design hurdles in this structure-based strategy was the achievement of sufficient binding affinity within the largely hydrophilic AMP site in which the majority of interactions are with the phosphate group of AMP (Fig. 2; Reddy and Erion 2005). Use of a phosphate group in the inhibitor scaffold was ruled out due to metabolic instability; phosphates are readily cleaved by phosphatases in vivo. Replacement of the phosphate group of AMP with mimetics such as a carboxylate or phosphonate group resulted in a >1,000-fold loss in potency due to the inability of the mimetics to make the precise and full complement of interactions required to achieve binding affinity within the phosphate-binding pocket (Reddy and Erion 2007). A critical breakthrough in inhibitor design was the realization, based on analysis of the FBPase-AMP complex, that the phosphate-binding site was directly accessible and within 4.24 Å of the C8 position of the purine base of AMP. This allowed the introduction of a phosphonate group into the phosphate-binding site with optimal orientation using a spacer group attached to the C8 position.

Other important aspects in the structure-based design of potent and selective AMP mimetics included the replacement of the ribose ring of AMP with alkyl groups to exploit interactions with a binding surface in a hydrophobic cavity (the side chains of [177]Met, [160]Val, [30]Leu, and [24]Ala) and modification of the pyrimidine portion of the purine ring of AMP (Erion et al. 2007). The latter change was of particular significance since an analysis of binding site interactions of 25 nucleotide-binding enzymes indicated that hydrogen bonds between the proteins and purine base nitrogens N7, [6]NH_2, N1, and N3 were common, whereas only the N7 and the [6]NH_2 group of AMP formed interactions with FBPase (Erion et al. 2007). Accordingly, the removal of N1 and/or N3, in addition to increasing binding affinity by reducing desolvation costs, was expected to improve selectivity for FBPase. Optimization of the phosphonic acid spacer group, the alkyl substituent, and/or the base moiety, aided by analysis of high resolution X-ray structures of

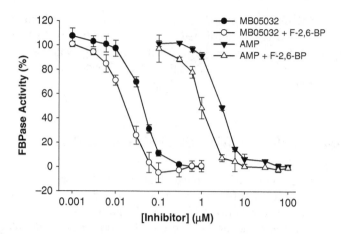

Fig. 3 Inhibition of human liver FBPase by AMP or MB05032 in the absence and presence of fructose-2,6-bisphosphate (F-2,6-BP). AMP, a natural inhibitor, and MB05032, a rationally designed inhibitor, both bind to an allosteric pocket within the enzyme. F-2,6BP is a natural regulator that binds to the substrate site of FBPase. Inhibition of FBPase by AMP and F-2,6-BP is known to be synergistic. The synergism observed between MB05032 and F-2,6-BP confirms the interaction of MB05032 with the allosteric-binding site in a manner analogous to AMP

human FBPase-inhibitor complexes and use of free energy perturbation methodology to predict binding interactions (Erion et al. 2007; Reddy and Erion 2005), resulted in the identification of several lead series, including purine, benzimidazole, and indole AMP mimetics with submicromolar IC_{50} values against human FBPase.

A more extensive structural modification of the lead series identified 2-aminothiazole as an attractive replacement of the base moiety and ultimately MB05032 (Fig. 2), an AMP mimetic that forms multiple favorable interactions with the phosphate and base-binding pockets of the AMP site of human FBPase (six and three hydrogen bonds, respectively) as well as with a hydrophobic cavity within the AMP site (Dang et al. 2007; Erion et al. 2005). MB05032 inhibited human liver FBPase with an IC_{50} of 16 nM, which represents a ~60-fold enhancement in potency relative to AMP. As expected, inhibition was noncompetitive and synergistic with F-2,6-BP (Fig. 3). Importantly and in contrast to AMP, MB05032 did not affect the activity of enzymes such as AMPK, glycogen phosphorylase, or phosphofructokinase at concentrations >1,000-fold higher than the IC_{50} value for human FBPase.

5.4 Discovery of CS-917

The dianionic nature of the phosphonate group of MB05032 at physiological pH, while critical for high affinity to the AMP site of FBPase, was an impediment to cellular penetration and oral absorption (Dang et al 2007; Erion et al. 2005). Oral delivery of MB05032 therefore required a prodrug form to mask the charged phosphonic acid group. Because a variety of traditional prodrug approaches for the

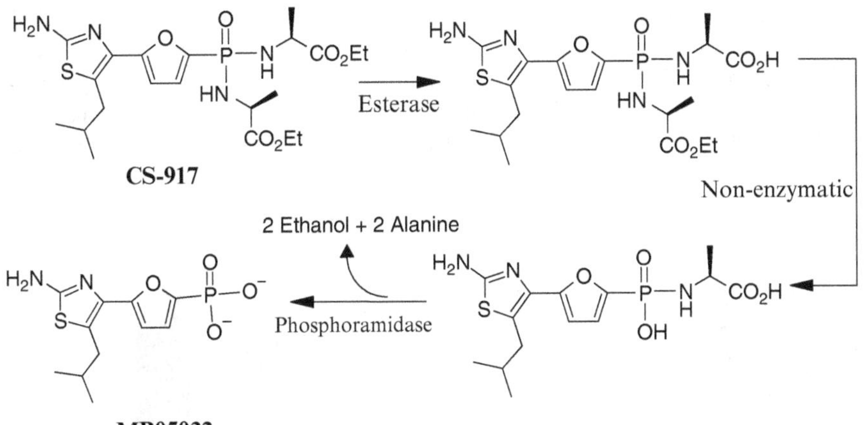

Fig. 4 Enzymatic conversion of prodrug CS-917 to the active FBPase inhibitor MB05032. Both the esterase and phosphoramidase activities are highly expressed in liver, the main site of endogenous glucose production via gluconeogenesis

delivery of phosphonate- or phosphate-containing molecules such as bis-phenyl esters (De Lombaert et al. 1994) or bis-isopropylcarbonate esters (van Gelder et al. 2000) did not improve satisfactorily upon the oral pharmacokinetics of MB05032, emphasis shifted to the design of novel prodrugs. These efforts resulted in the discovery of a new phosphonic diamide prodrug class (Erion et al. 2005; Dang et al. 2007, 2008a) that was chemically stable and achieved the appropriate balance between biological stability (to allow absorption) and susceptibility to enzymatic hydrolysis (to allow in vivo conversion to MB05032). Optimization of the prodrug moiety by synthesis and pharmacokinetic evaluation of prodrugs with various amino acid esters culminated in the identification of CS-917 (Fig. 2), a dialanyl amide of MB05032 (~20% bioavailability) that enabled efficient distribution of MB05032 to the liver after oral administration to rats. CS-917 is cleaved enzymatically to MB05032 by sequential action of hepatic esterase and phosphoramidase activities with generation of the nontoxic byproducts ethanol and alanine (Fig. 4). Of note, CS-917 was converted intracellularly to MB05032 and inhibited glucose production from lactate/pyruvate in human hepatocytes with an EC_{50} of 0.22 µM (Erion et al. 2005).

6 Mechanism of Action, Efficacy, and Safety of CS-917

6.1 Mechanism of Action

Tracer studies in postabsorptive male Zucker diabetic fatty (ZDF) rats confirmed the inhibition of gluconeogenesis following treatment with CS-917. Two different tracer methods were employed; one measured the incorporation of $[^{14}C]$-bicarbonate, and the second, the incorporation of deuterated water into glucose (Erion et al. 2005; van Poelje et al. 2006a). Oral administration of CS-917 resulted in dose-dependent inhibition of [^{14}C]-bicarbonate incorporation into glucose (Fig. 5a), with a 30 mg/kg dose leading to a ~25% reduction in de novo glucose synthesis. In parallel studies, a 30 mg/kg dose also resulted in significant lowering of blood glucose,

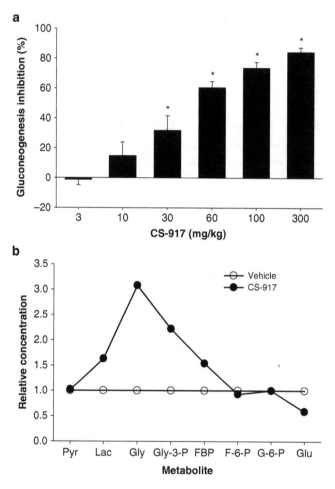

Fig. 5 (a) Inhibition by CS-917 of the incorporation of [^{14}C]-bicarbonate into glucose in post-absorptive, male ZDF rats. Animals were dosed orally with CS-917 2 h prior to intravenous administration of tracer (0.4 µCi/g body weight). Incorporation of label into plasma glucose was assessed at 20 min following exposure to tracer by ion-exchange chromatography. *p <0.05 vs. vehicle (Students t-test). (b) Effect of CS-917 treatment on gluconeogenic precursors, interme-diates, and products in livers of postabsorptive, male ZDF rats. Livers were harvested and extracted with perchloric acid 4 h after an oral dose of CS-917 (300 mg/kg). Liver metabolites were measured by enzyme-coupled spectrophotometric assays. Abbreviations: pyruvate, Pyr; lactate, Lac; glycerol, Gly; glycerol-3-phosphate, Gly-3-P; fructose 1,6-bisphosphate, fructose-6-phosphate, F-6-P; FBP; glucose-6-phosphate, G-6-P; glucose, Glu

indicating that partial inhibition of gluconeogenesis (~25%) is sufficient to elicit the desired pharmacological effect. The deuterated water method, coupled with standard glucose tracer dilution methodology, allowed a quantitative assessment of the relative contribution of gluconeogenesis and glycogenolysis to EGP in treated and untreated animals. Using this methodology, a maximal oral dose of CS-917 (300 mg/kg) was found to result in a ~70% reduction of gluconeogenesis-derived glucose, which was partially countered by a 58% increase in glycogenolysis. The net effect of reduced gluconeogenesis and increased glycogenolysis was an overall 46% reduction of EGP to a rate (~8 μmoles/min/kg) similar to that of nondiabetic rats in a similar nutritional state. The observed reciprocal regulation of gluconeogenesis and glycogenolysis has also been described in humans and is termed "hepatic autoregulation" (Boden et al. 2001; Jenssen et al. 1990). It serves to maintain glucose production in the event that one pathway of glucose production is impaired and thereby prevents undue glucose lowering.

FBPase was demonstrated as the target enzyme of CS-917 in vivo by the measurement of substrates, intermediates, and products of gluconeogenesis in liver, following oral administration of CS-917 to fasted male ZDF rats (Erion et al. 2005). As shown in Fig. 5b, substrates and intermediates prior to the FBPase step were elevated 1.5- to 3.1-fold in CS-917-treated rats relative to vehicle-treated rats, whereas glucose 6-phosphate and glucose were unchanged and 41% decreased, respectively. This pattern of substrates and intermediates is consistent with inhibition of FBPase. Interestingly, a similar pattern was observed in kidneys in treated rats, suggesting that FBPase was inhibited by CS-917 in both gluconeogenic organs (liver and kidney). This conclusion is supported by the observation that MB05032, the active metabolite of CS-917, potently inhibited gluconeogenesis from pyruvate in studies using isolated, perfused kidneys from ZDF rats (van Poelje et al. 2006b).

6.2   Efficacy

6.2.1   Monotherapy

The acute glucose lowering profile of CS-917 in diabetic rats revealed an insulin-independent mode of action with efficacy in both the postabsorptive and postprandial states (Erion et al. 2005; van Poelje et al. 2006a). At a dose of CS-917 that inhibited gluconeogenesis by ~25% (30 mg/kg), rapid glucose lowering (75–100 mg/dL) was achieved with a single administration in the postabsorptive state in young (~9 week-old), hyperinsulinemic as well as aged (>13 week-old), hypoinsulinemic male ZDF rats. The dose response for glucose lowering was relatively steep, with minimal and maximal effects achieved within a ~three-fold dose range (30–100 mg/kg). Consistent with an important contribution of gluconeogenesis to postprandial hyperglycemia, a marked improvement in oral glucose tolerance was observed following single doses of CS-917 (30 mg/kg) in both young and aged male ZDF rats, with the baseline-normalized AUC_{0-3h} of blood glucose reduced to 65% and 35%, respectively.

Evaluations of CS-917 for up to 6 weeks were performed in prediabetic and diabetic, male ZDF rats as well as in female ZDF rats rendered diabetic by feeding a high-fat diet. In studies with prediabetic, male ZDF rats (~6 weeks of age), CS-917 prevented the onset of hyperglycemia in the majority of animals throughout the 6-week treatment period. CS-917 treatment of male ZDF rats with advanced diabetes (10 weeks of age) led to a lowering of blood glucose by ~33 and ~44% relative to controls after 1 and 2 weeks of therapy, respectively. A rapid response to drug treatment was evident in this study from a ~69% reduction in glycosuria after 3 days of treatment. A profound glucose lowering response was also noted during a 2-week evaluation of CS-917 in high-fat-fed female ZDF rats (Table 1). Treatment, in fact, resulted in the normalization of blood glucose in this milder model of T2DM.

6.2.2 Combination Therapy

Although robust and sustained efficacy was observed with CS-917 as a monotherapy, the current practice of treating patients with combinations of antidiabetic drugs prompted the evaluation of CS-917 in combination with glyburide (an insulin secre-tagogue of the sulfonylurea class) and with pioglitazone (an insulin sensitizer of the thiazolidinedione class). In acute studies in male ZDF rats subjected to an oral glucose tolerance test, improvement of glucose tolerance with CS-917 was additive to that achieved with the sulfonylurea glyburide (Erion et al. 2004). CS-917 treatment did not result in increased insulin levels or, when combined with glyburide, affect the insulin response resulting from glyburide treatment. Chronic studies of CS-917 in male ZDF rats in combination with pioglitazone demonstrated significantly improved glycemic control relative to either therapy alone (Fig. 6a) (van Poelje et al. 2006c). CS-917 monotherapy resulted in increased levels of blood lactate, a major substrate of gluco-neogenesis, but there was no evidence of increased lactate levels in the combination group (Fig. 6b). Attenuation of blood lactate levels by pioglitazone co-treatment is likely due to enhanced activity of pyruvate dehydrogenase (PDH), an enzyme that regulates the entry of lactate/pyruvate into the oxidative tricarboxylic acid cycle. Insulin sensitizers are known to increase PDH activity in rats by reducing the expression of a regulatory kinase, PDK (Sugden and Holness 2006). Consistent with this finding, reduced expression of PDK4 was observed in skeletal muscle samples in the pioglitazone and the combination groups in ZDF rats (Fig. 6c). The combination studies suggest that CS-917 may provide additional therapeutic benefits when co-administered with insulin secretagogues or with insulin sensitizers. In addition, insulin sensitizers may have beneficial effects on potential FBPase inhibitor-induced alterations of lactate homeostasis. In support of the latter benefit, rosiglitazone co-treatment has been reported to normalize lactate metabolism in metformin-treated patients (Fonseca et al. 2000).

Table 1 Physiological and metabolic parameters in control and CS-917-treated, high-fat diet-fed, female ZDF rats (nonfasted). CS-917 was administered as a food admixture at concentrations of 0.1% (~100 mg/kg/day) and 0.3% (300 mg/kg/day). *$p < 0.05$ vs. control; ANOVA with Dunnett's post hoc test.

Parameter		Control		CS-917 (0.1%)		CS-917 (0.3%)	
Length of treatment	Days	0	14	0	14	0	14
Blood glucose	mg/dL	289 ± 29	305 ± 34	287 ± 34	166 ± 24	283 ± 27	169 ± 30
Insulin	ng/mL	29.3 ± 4.7	17.2 ± 3.4	28.6 ± 5	14.2 ± 1	26.9 ± 5.3	12.9 ± 1.4
Glucagon	pg/mL	–	122 ± 6	–	114 ± 6	–	108 ± 8
Lactate	mM	1.22 ± 0.18	1.94 ± 0.46	1.12 ± 0.08	1.63 ± 0.51	1.58 ± 0.16	1.55 ± 0.26
NEFA	mM	–	1.2 ± 0.3	–	1.1 ± 0.1	–	1.2 ± 0.1
Triglycerides	mg/dL	1,291 ± 120	1,706 ± 229	1,110 ± 103	1,145 ± 164	1,094 ± 154	1,154 ± 60
Cholesterol	mg/dL	–	133 ± 8	–	120±5	–	158 ± 4
β–hydroxybutyrate	μM	–	133 ± 13	–	132±16	–	117 ± 12
Liver triglycerides	mg/g	–	15.3 ± 2.5	–	13±3	–	12 ± 1
Liver glycogen	μmoles/g	–	276 ± 11	–	244 ± 13	–	222±13*
Alkaline phosphatase	IU/mL	–	196 ± 28	–	110±8*	–	118±14*
Water intake	mL/day	27 ± 5	31 ± 6	29 ± 5	15 ± 2	30 ± 6	16±1
Food intake	g/day	16 ± 1	17 ± 1	15 ± 1	15 ± 1	15 ± 1	13±2
Body weight	g	339 ± 5	388 ± 7	326 ± 4	370 ± 4	323 ± 7	360±9

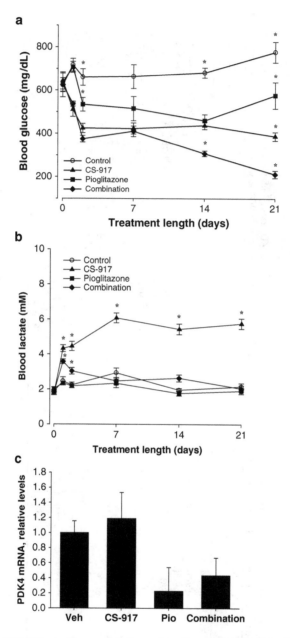

Fig. 6 Effects of CS-917, pioglitazone, and the combination of CS-917 and pioglitazone on blood glucose (**a**) blood lactate (**b**) and skeletal muscle PDK4 mRNA levels (**c**) of male, ZDF rats. Rats were divided into blood glucose-matched groups ($n = 8$/group; blood glucose ~600 mg/dL) at the onset of the study and fed Teklad 5008 chow or an admixture of this chow with CS-917 (0.2%), pioglitazone (0.03%), or the combination of CS-917 and pioglitazone (0.2 and 0.03%, respectively) for 3 weeks. *$p < 0.05$ vs. all groups (ANOVA, Tukey-Kramer)

6.3 Safety

One of the main theoretical concerns with FBPase inhibitors is increased risk of hypoglycemia. It is reassuring that chronic treatment with maximal doses of CS-917 did not elicit hypoglycemia in freely feeding nondiabetic rats or a variety of diabetic rodent models (van Poelje et al. 2006a, c). Of note, inhibition of gluconeogenesis by CS-917 was associated with minimal reduction of hepatic glycogen levels in diabetic animal models (e.g., female ZDF rats; Table 1). This suggests that the direct pathway of glycogen repletion (from glucose) can compensate for reduced glycogen synthesis by the indirect pathway (gluconeogenesis) and can ensure maintenance of adequate glycogen stores to counteract hypoglycemia. In contrast to diabetic rats, significant hypoglycemia was observed with CS-917 treatment of 16-h fasted, Sprague Dawley rats. This fasting period almost fully depletes hepatic glycogen stores in these animals. A greater safety margin for hypoglycemia may be predicted in patients with T2DM since glycogen reserves, although lower than in healthy individuals, are ~50% maintained following a 24-h fast (Magnusson et al. 1992). Relative to other drug classes such as insulin and the sulfonylureas, which inhibit both glucose production and accelerate glucose disposal, one could also speculate that the risk of hypoglycemia should be considerably less with the inhibition of gluconeogenesis alone by means of the FBPase inhibitor approach.

Another potential consequence of FBPase inhibition is lacticemia due to reduced clearance of lactate by gluconeogenesis. Sustained lacticemia, in addition to potentially altering acid-base balance, could result in diversion of lactate into lipogenesis pathways and lead to fatty liver and hypertriglyceridemia. In high-fat-fed, female ZDF rats (Table 1) and in db/db mice, CS-917 treatment did not lead to elevation of lactate or lipids. In male ZDF rats, modest elevation of plasma lactate and lipids was evident only at high doses in animals with advanced disease (van Poelje et al. 2006a). These metabolic ramifications are likely a reflection of the exaggerated rates of gluconeogenesis and lipogenesis unique to this animal model (Lee et al. 2000). Taken together, the studies indicate that excess lactate and lactate-derived substrates are efficiently cleared by alternative pathways (e.g., oxidation) when gluconeogenesis is inhibited. Increased utilization of lactate and lactate-derived substrates is expected following gluconeogenesis inhibition as this reduces the supply of glucose that is normally used as a source of energy.

Lastly, it should be noted that long-term FBPase inhibition in overtly diabetic rodents did not alter food intake or weight gain, or affect clinical chemistry parameters related to kidney or liver function (van Poelje et al. 2006a, c). Overall, long-term FBPase inhibition was found to be safe and well-tolerated in animal models.

Insight into the potential safety of the gluconeogenesis inhibitor approach is also provided by the extensive clinical experience with metformin. Although diverse pharmacological mechanisms have been described for metformin, its primary mechanism of action in humans is believed to be an indirect inhibition or reduction of gluconeogenesis (~33% at the highest dose prescribed, 850 mg TID; Hundal et al. 2000). Metformin therapy effectively lowers glucose levels, generally improves lipid profiles, and does not cause hypoglycemia. However, asymptomatic lacticemia (lactate levels 3-5 mM) is observed in 4% of metformin-treated patients (Cryer et al. 2005). Importantly, lacticemia in these patients is not thought to predispose to lactic acidosis or be clinically significant. Thus, inhibition of gluconeogenesis appears an effective and safe means of reducing glucose levels in patients with T2DM.

7 Clinical Development of FBPase Inhibitors

7.1 CS-917

Ascending single- and multiple-dose studies of CS-917 in overnight-fasted healthy volunteers revealed encouraging tolerability and safety profiles with no incidence of hypoglycemia (Walker et al. 2006a, b). In a subsequent 14-day Phase 2a trial in patients with T2DM, CS-917 treatment was also safe and well tolerated. Furthermore, all doses (50–400 mg), except for the 100 mg dose, resulted in a statistically significant reduction of the AUC_{0-6h} of glucose versus placebo (Triscari et al. 2006; Bruce et al. 2006). The change in fasting plasma glucose levels from baseline on Day 14 in the treated groups ranged from 30–35 mg/dL relative to placebo. Mean plasma lactate levels were slightly elevated at higher doses in the CS-917-treated patients, but remained within normal limits at all doses. Pharmacokinetic analysis in patients indicated rapid absorption of CS-917 and efficient conversion to MB05032, which reached a C_{max} at 2–3 h following drug administration. Exposure for CS-917 and MB05032 was approximately linear for the 50–200 mg dose range. These trials provided a preliminary indication of the therapeutic utility of FBPase inhibitors in patients with T2DM and validate the use of the novel diamide prodrug class for the oral delivery of AMP mimetics to humans.

The clinical development of CS-917 was mired by the outcome of two key trials: a Phase 1 interaction trial with metformin and a Phase 2b monotherapy trial. During the course of the Phase 1 interaction trial, two patients on metformin therapy developed lactic acidosis shortly after the addition of CS-917 therapy. Both cases of lactic acidosis resolved following drug withdrawal but obviously raised concerns of a potential pharmacokinetic and/or pharmacological interaction between metformin and CS-917. CS-917 demonstrated an excellent safety profile in a subsequent 3-month monotherapy trial but did not achieve its clinical endpoint: neither of the two doses evaluated (50 or 100 mg, BID) resulted in a statistically significant reduction in hemoglobin A1c levels (HbA1c; a measure of long term glucose exposure). The extent to which the selection of doses of CS-917 at the lower end of the effective range or the enrollment of patients with significantly lower HbA1c levels than the Phase 2a trials affected the outcome of the Phase 2b trial is unclear. Nevertheless, the disappointing results of this trial, coupled with the unexpected outcome of the metformin interaction trial, led to the discontinuation of the clinical development of CS-917.

7.2   MB07803

Following the completion of the Phase 2a studies of CS-917, a second FBPase inhibitor, MB07803, was entered into clinical development. MB07803 is a second generation compound of a related structural class (Fig. 2) but with improved pharmacokinetic characteristics relative to CS-917. These improvements include higher oral bioavailability, a longer half-life of the active metabolite, and a markedly reduced rate of metabolic transformation to inactive N-acetylated products (Dang et al. 2008b). MB07803 completed successfully a safety and tolerability study in healthy volunteers (Phase 1) as well as an initial proof-of-concept study in patients with T2DM (Phase 2a). In the Phase 2a study, patients received either placebo or an oral dose of MB07803 (10, 50, 100, or 200 mg) once daily for 28 days. At the highest dose of MB07803, statistically and clinically significant lowering of fasting plasma glucose was achieved. Of note, lactate levels remained within normal limits in all patients and no sustained lacticemia (defined as lactate >4.5 mM on two consecutive visits) was observed.

8   Conclusions and Perspectives

Inhibition of excessive EGP is an important strategy for controlling hyperglycemia in patients with T2DM. Because none of the currently marketed drugs directly target the overproduction of glucose by the liver, the discovery of direct and therefore potentially more efficacious inhibitors of glucose production has been of considerable interest to the pharmaceutical industry for many years. FBPase has emerged as a key target for pharmacological intervention due to its pivotal role in controlling gluconeogenesis and the important contribution of gluconeogenesis to fasting and postprandial hyperglycemia in patients with T2DM.

Overall, the efficacy profile in diabetic rodents of CS-917, the best characterized drug in the FBPase inhibitor class, indicates that FBPase inhibitors may be useful for controlling both fasting and postprandial hyperglycemia in the early as well as more advanced stages of T2DM. The insulin-independent mode of action of CS-917 suggests that FBPase inhibitors may provide a more durable treatment than current drugs, the antidiabetic activity of which declines as pancreatic function deteriorates. Efficacy studies of CS-917 in diabetic rodents also indicate that FBPase inhibitors may combine effectively with other agents such as insulin sensitizers and insulin secretagogues. The latter is desirable because patients are increasingly being treated with drug combinations to maintain strict glycemic control.

Evaluation of CS-917 in animal models, healthy human volunteers, and patients with T2DM has provided the initial safety profile of the FBPase inhibitor drug class, particularly with regards to risk of hypoglycemia, lacticemia, and hyperlipidemia. Chronic administration of high doses of CS-917 did not cause hypoglycemia or weight gain in diabetic rodents and was without metabolic ramifications other than profound glucose lowering in the majority of diabetic models evaluated. CS-917 and the second generation FBPase inhibitor MB07803 demonstrated good tolerability and safety profiles in healthy volunteers and in monotherapy trials in patients with T2DM. Preliminary trials of CS-917 and MB07803 also suggest that clinically relevant glucose lowering can be achieved in patients with T2DM. Larger scale and longer-term clinical trials of MB07803 as a monotherapy and in combination with approved agents will provide a definitive assessment of the safety as well as the long-term therapeutic utility of the FBPase inhibitor drug class.

References

Andrikopoulos S, Rosella G, Kaczmarczyk SJ, Zajac JD, Proietto J (1996) Impaired regulation of hepatic fructose-1, 6-biphosphatase in the New Zealand Obese mouse: an acquired defect. Metabolism 45(5):622–626

Boden G, Chen X, Capulong E, Mozzoli M (2001) Effects of free fatty acids on gluconeogenesis and autoregulation of glucose production in type 2 diabetes. Diabetes 50:810–816

Bruce SR, Walker J, Feins K, Tao B, Triscari J (2006) Initial safety, tolerability and glucose lowering of CS-917, a novel fructose 1, 6-bisphosphatase (FBPase) inhibitor, in subjects with type 2 diabetes. Diabetologia 49(Suppl 1):0037

Burgess SC, Hausler N, Merritt M, Jeffrey FM, Storey C, Milde A, Koshy S, Lindner J, Magnuson MA, Malloy CR, Sherry AD (2004) Impaired tricarboxylic acid cycle activity in mouse livers lacking cytosolic phosphoenolpyruvate carboxykinase. J Biol Chem 279:48941–48949

Cherrington AD (1999) Banting Lecture: Control of glucose uptake and release by the liver in vivo. Diabetes 48:1198–1214

Choe JY, Nelson SW, Arienti KL, Axe FU, Collins TL, Jones TK, Kimmich RD, Newman MJ, Norvell K, Ripka WC, Romano SJ, Short KM, Slee DH, Fromm HJ, Honzatko RB (2003) Inhibition of fructose-1, 6-bisphosphatase by a new class of allosteric effectors. J Biol Chem 278:51176–51183

Cryer DR, Nicholas SP, Henry DH, Mills DJ, Stadel BV (2005) Comparative outcomes study of metformin intervention versus conventional approach. Diab Care 28:539–543

Dang Q, Kasibhatla SR, Reddy KR, Jiang T, Reddy MR, Potter SC, Fujitaki JM, van Poelje PD, Huang J, Lipscomb WN (2007) Erion MD (2007) Discovery of potent and specific fructose-1, 6-bisphosphatase inhibitors and a series of orally-bioavailable phosphoramidase-sensitive prodrugs for the treatment of type 2 diabetes. J Am Chem Soc 129(50):15491–15502

Dang Q, Kasibhatla SR, Jiang T, Fan K, Liu Y, Taplin F, Schulz W, Cashion DK, Reddy KR, van Poelje PD, Fujitaki JM, Potter SC, Erion MD (2008a) Discovery of phosphonic diamide prodrugs and their use for the oral delivery of a series of fructose 1, 6-bisphosphatase inhibitors. J Med Chem 51(14):4331–4339

Dang Q, van Poelje PD, Lemus RH, Tian F, Potter SC, Fujitaki JM, Linemeyer D1, Erion ME (2008b) Discovery of a second generation FBPase inhibitor, MB07803, with reduced metabolism and improved oral bioavailability. 235th ACS Meeting, Medi-21

De Lombaert S, Erion MD, Tan J, Blanchard L, el-Chehabi L, Ghai RD, Sakane Y, Berry C, Trapani AJ (1994) N-Phosphonomethyl dipeptides and their phosphonate prodrugs, a new generation of neutral endopeptidase (NEP, EC 3.4.24.11) inhibitors. J Med Chem 37 (4):498–511

Dzugaj A, Kochman M (1980) Purification of human liver fructose-1, 6-bisphosphatase. Biochim Biophys Acta 614(2):407–412

El-Maghrabi MR, Gidh-Jain M, Austin LR, Pilkis SJ (1993) Isolation of a human liver fructose 1, 6-bisphosphatase cDNA and expression of the protein in Escherichia coli. J Biol Chem 268:9466–9472

El-Maghrabi MR, Lange AJ, Kummel L, Pilkis SJ (1991) The rat fructose-1, 6-bisphosphatase gene. Structure and regulation of expression. J Biol Chem 266:2115–2120

Erion MD, Potter SC, van Poelje PD (2004) Fructose 1, 6-bisphosphatase inhibition improves oral glucose tolerance and enhances the antidiabetic action of glyburide in the ZDF rat (Abstract). Diabetologia 47(Suppl 1):0796

Erion MD, van Poelje PD, Dang Q, Kasibhatla SR, Potter SC, Reddy MR, Reddy KR, Jiang T, Lipscomb WN (2005) MB06322 (CS-917): A potent and selective inhibitor of fructose 1, 6-bisphosphatase for controlling gluconeogenesis in type 2 diabetes. Proc Natl Acad Sci 102:7970–7975

Erion MD, Dang Q, Reddy MR, Kasibhatla SR, Huang J, Lipscomb WN, van Poelje PD (2007) Structure-guided design of AMP mimics that inhibit fructose-1, 6-bisphosphatase with high affinity and specificity. J Am Chem Soc 129(50):15480–15490

Foley LH, Wang P, Dunten P, Ramsey G, Gubler ML, Wertheimer SJ (2003) Modified 3-alkyl-1, 8-dibenzylxanthines as GTP-competitive inhibitors of phosphoenolpyruvate carboxykinase. Bioorg Med Chem Lett 13:3607–3610

Fonseca V, Rosenstock J, Patwardhan R, Salzman A (2000) ffect of metformin and rosiglitazone combination therapy in patients with type 2 diabetes mellitus: a randomized controlled trial. JAMA 283(13):1695–1702

Gastaldelli A, Baldi S, Pettiti M, Toschi E, Camastra S, Natali A, Landau BR, Ferrannini E (2000) Influence of obesity and type 2 diabetes on gluconeogenesis and glucose output in humans: a quantitative study. Diabetes 49(8):1367–1373

Gastaldelli A, Toschi E, Pettiti M, Frascerra S, Quiñones-Galvan A, Sironi AM, Natali A, Ferrannini E (2001) Effect of physiological hyperinsulinemia on gluconeogenesis in nondiabetic subjects and in type 2 diabetic patients. Diabetes 50:1807–1812

Gerich JE, Meyer C, Woerle HJ, Stumvoll M (2001) Renal gluconeogenesis: its importance in human glucose homeostasis. Diab Care 24(2):382–391

Gidh-Jain M, Zhang Y, van Poelje PD, Liang JY, Huang S, Kim J, Elliott JT, Erion MD, Pilkis SJ, Raafat el-Maghrabi M et al (1994) The allosteric site of human liver fructose-1, 6-bisphosphatase. Analysis of six AMP site mutants based on the crystal structure. J Biol Chem 269:27732–27738

Gitzelmann R, Steinmann B, Van Den Berghe G (1995) In: Scriver CR, Beaudet AL (eds) The metabolic and molecular basis of inherited disease, vol 1. McGraw-Hill, New York, pp 905–934

Henin N, Vincent MF, Van den Berghe G (1996) Stimulation of rat liver AMP-activated protein kinase by AMP analogues. Biochim Biophys Acta 1290:197–203

Hundal RS, Krssak M, Dufour S, Laurent D, Lebon V, Chandramouli V, Inzucchi SE, Schumann WC, Petersen KF, Landau BR, Shulman GI (2000) Mechanism by which metformin reduces glucose production in type 2 diabetes. Diabetes 49:2063–2069

Jeng CY, Sheu WH, Fuh MM, Chen YD, Reaven GM (1994) Relationship between hepatic glucose production and fasting plasma glucose concentration in patients with NIDDM. Diabetes 43:1440–1444

Jenssen T, Nurjhan N, Consoli A, Gerich JE (1990) Failure of substrate-induced gluconeogenesis to increase overall glucose appearance in normal humans. J Clin Invest 86:489–497

Ke H, Liang JY, Zhang Y, Lipscomb WN (1991) Conformational transition of fructose-1, 6-bisphosphatase: structure comparison between the AMP complex (T form) and the fructose 6-phosphate complex (R form). Biochemistry 30:4412–4420

Kodama H, Fujita M, Yamazaki M, Yamaguchi I (1994) The possible role of age-related increase in the plasma glucagon/insulin ratio in the enhanced hepatic gluconeogenesis and hyperglycemia in genetically diabetic (C57BL/DsJ-db/db) mice. Jpn J Pharmacol 66:281–287

Lamont BJ, Visinoni S, Fam BC, Kebede M, Weinrich B, Papapostolou S, Massinet H, Proietto J, Favaloro J, Andrikopoulos S (2006) Expression of human fructose-1, 6-bisphosphatase in the liver of transgenic mice results in increased glycerol gluconeogenesis. Endocrinology 147:2764–2772

Landau BR, Wahren J, Chandramouli V, Schumann WC, Ekberg K, Kalhan SC (1996) Contributions of gluconeogenesis to glucose production in the fasted state. J Clin Invest 98:378–385

Lee WN et al (2000) Loss of regulation of lipogenesis in the Zucker diabetic (ZDF) rat. Am J Physiol Endocrinol Metab 279:E425–E432

Leroith D, Taylor SI, Olefsky JM (1996) *Diabetes* mellitus, a fundamental and clinical text. Lippincott-Raven Publishers, Philadelphia, PA

Maggs DG, Buchanan TA, Burant CF, Cline G, Gumbiner B, Hsueh WA, Inzucchi S, Kelley D, Nolan J, Olefsky JM, Polonsky KS, Silver D, Valiquett TR, Shulman GI (1998) Metabolic effects of troglitazone monotherapy in type 2 diabetes mellitus: a randomized, double-blind, placebo-controlled trial. Ann Intern Med 128:176–185

Magnusson I, Rothman DL, Katz LD, Shulman RG, Shulman GI (1992) Increased rate of gluconeogenesis in type II diabetes mellitus. A ^{13}C nuclear magnetic resonance study. J Clin Invest 90:1323–1327

Olefsky JM (1993) Insulin resistance and the pathogenesis of non-insulin dependent diabetes mellitus: cellular and molecular mechanisms. In: Ostenson OG et al (eds) New concepts in the pathogenesis of NIDDM. Plenum Press, New York, pp 129–150

Parker JC, van Volkenburg MA, Levy CB, Martin WH, Burk SH, Kwon Y, Giragossian C, Gant TG, Carpino PA, McPherson RK, Vestergaard P, Treadway JL (1998) Plasma glucose levels are reduced in rats and mice treated with an inhibitor of glucose-6-phosphatase. Diabetes 47:1630–1636

Pilkis SJ (1991) Hepatic gluconeogenesis/glycolysis: regulation and structure/function relationships of substrate cycle enzymes. Ann Rev Nutr 11:465–515

Pilkis SJ, McGrane MM, Kountz PD, el-Maghrabi MR, Pilkis J, Maryanoff BE, Reitz AB, Benkovic SJ (1986) The effect of arabinose 1, 5-bisphosphate on rat hepatic 6-phosphofructo-1-kinase and fructose-1, 6-bisphosphatase. Biochem Biophys Res Commun 138:159–166

Pold R, Jensen LS, Jessen N, Buhl ES, Schmitz O, Flyvbjerg A, Fujii N, Goodyear LJ, Gotfredsen CF, Brand CL, Lund S (2005) Long-term AICAR administration and exercise prevents diabetes in ZDF rats. Diabetes 54:928–934

Reddy MR, Erion MD (2005) Computer-aided drug design strategies used in the discovery of fructose 1, 6-bisphosphatase inhibitors. Curr Pharm Des 11:283–294

Reddy MR, Erion MD (2007) Relative binding affinities of fructose-1, 6-bisphosphatase inhibitors calculated using a quantum mechanics-based free energy perturbation method. J Am Chem Soc 129(30):9296–9297

Rosini M, Mancini F, Tarozzi A, Colizzi F, Andrisano V, Bolognesi ML, Hrelia P, Melchiorre C (2006) Design, synthesis, and biological evaluation of substituted 2, 3-dihydro-1H-cyclopenta [b]quinolin-9-ylamine related compounds as fructose-1, 6-bisphosphatase inhibitors. Bioorg Med Chem 14:7846–7853

Rothman DL, Magnusson I, Katz LD, Shulman RG, Shulman GI (1991) Quantitation of hepatic glycogenolysis and gluconeogenesis in fasting humans with ^{13}C-NMR. Science 254:573–576

Skalecki K, Mularczyk W, Dzugaj A (1995) Kinetic properties of D-fructose-1, 6-bisphosphate 1-phosphohydrolase isolated from human muscle. Biochem J 310(Pt 3):1029–1035

Sugden MC, Holness MJ (2006) Mechanisms underlying regulation of the expression and activities of the mammalian pyruvate dehydrogenase kinases. Arch Physiol Biochem 112(3):139–149

Tillmann H, Bernhard D, Eschrich K (2002) Fructose-1, 6-bisphosphatase genes in animals. Gene 291(1–2):57–66

Triscari J, Walker J, Feins K, Tao B, Bruce SR (2006) Multiple ascending doses of CS-917, a novel fructose 1, 6-bisphosphatase (FBPase) inhibitor, in subjects with type 2 diabetes treated for 14 days. Diabetes 55(Suppl 1):444–P

van Gelder J, Deferme S, Annaert P, Naesens L, De Clercq E, Van den Mooter G, Kinget R, Augustijns P (2000) Increased absorption of the antiviral ester prodrug tenofovir disoproxil in rat ileum by inhibiting its intestinal metabolism. Drug Metab Dispos 28:1394–1396 van Poelje PD, Potter SC, Chandramouli VC, Landau BR, Dang Q, Erion MD (2006a) Inhibition of fructose 1, 6-bisphosphatase reduces excessive endogenous glucose production and attenuates hyperglycemia in ZDF rats. Diabetes 55:1747–1754

van Poelje PD, Potter SC, Linemeyer DL, Erion MD (2006b) MB06322 (CS-917) lowers blood glucose in rodents by inhibiting both hepatic and renal gluconeogenesis. Diabetes 55(Suppl 1):575–P

van Poelje PD, Potter SC, Topczewski E, Hou J, Linemeyer DL, Erion MD (2006c) Combination therapy with pioglitazone and a fructose-1, 6-bisphosphatase inhibitor (MB06322) improves glycemic control and lactate homeostasis in male Zucker diabetic fatty (ZDF) rats. Diabetologia 49(Suppl 1):0852

van Schaftingen E, Hers H-G (1981) Inhibition of fructose-1, 6-bisphosphatase by fructose 2, 6-bisphosphate. Proc Natl Acad Sci USA 78:2861–2863

Vincent MF, Bontemps F, Van den Berghe G (1992) Inhibition of glycolysis by 5-amino-4-imidazolecarboxamide riboside in isolated rat hepatocytes. Biochem J 281:267–272

Vincent MF, Erion MD, Gruber HE, Van den Berghe G (1996) Hypoglycaemic effect of AIC Ariboside in mice. Diabetologia 39:1148–1155

Vincent MF, Marangos PJ, Gruber HE, Van den Berghe G (1991) Inhibition by AICA riboside of gluconeogenesis in isolated rat hepatocytes. Diabetes 40:1259–1266

Visinoni S, Fam BC, Blair A, Rantzau C, Lamont BJ, Bouwman R, Watt MJ, Proietto J, Favaloro JM, Andrikopoulos S (2008) Increased glucose production in mice overexpressing human fructose-1, 6-bisphosphatase in the liver. Am J Physiol Endocrinol Metab 295:E1132–E1141

von Geldern TW, Lai C, Gum RJ, Daly M, Sun C, Fry EH, Abad-Zapatero C (2006) Benzoxazole benzenesulfonamides are novel allosteric inhibitors of fructose-1, 6-bisphosphatase with a distinct binding mode. Bioorg Med Chem Lett 16:1811–1815

Wajngot A, Chandramouli V, Schumann WC, Ekberg K, Jones PK, Efendic S, Landau BR (2001) Quantitative contributions of gluconeogenesis to glucose production during fasting in type 2 diabetes mellitus. Metabolism 50:47–52

Walker J, Triscari J, Dmuchowski C, Kaneko T, Bruce SR (2006a) Safety and tolerability of single doses of CS-917, a novel gluconeogenesis inhibitor, in normal male volunteers. Diabetes 55(Suppl 1):2002–PO

Walker J, Triscari J, Dmuchowski C, Kaneko T, Bruce SR (2006b) Safety, tolerability and pharmacodynamics of multiple doses of CS-917 in normal volunteers. Diabetes 55(Suppl 1):2003–PO

Wright SW, Carlo AA, Carty MD, Danley DE, Hageman DL, Karam GA, Levy CB, Mansour MN, Mathiowetz AM, McClure LD, Nestor NB, McPherson RK, Pandit J, Pustilnik LR, Schulte GK, Soeller WC, Treadway JL, Wang IK, Bauer PH (2002) Anilinoquinazoline inhibitors of fructose 1, 6-bisphosphatase bind at a novel allosteric site: synthesis, in vitro characterization, and X-ray crystallography. J Med Chem 45:3865–3877

Wright SW, Carlo AA, Danley DE, Hageman DE, Hageman DL, Karam GA, Mansour MN, McClure LD, Pandit J, Schulte GK, Treadway JL, Wang IK, Bauer PH (2003) 3-(2-carboxyethyl)-4, 6-dichloro-1H-indole-2-carboxylic acid: an allosteric inhibitor of fructose-1, 6-bisphosphatase at the AMP site. Bioorg Med Chem Lett 13:2055–2058

Wright SW, Hageman DL, McClure LD, Carlo AA, Treadway JL, Mathiowetz AM, Withka JM, Bauer PH (2001) Allosteric inhibition of fructose-1, 6-bisphosphatase by anilinoquinazolines. Bioorg Med Chem Lett 11:17–21

AMP-Activated Protein Kinase and Metabolic Control

Benoit Viollet and Fabrizio Andreelli

Contents

Abstract AMP-activated protein kinase (AMPK), a phylogenetically conserved serine/threonine protein kinase, is a major regulator of cellular and whole-body energy homeostasis that coordinates metabolic pathways in order to balance nutrient

---

B. Viollet (✉)
Inserm, U1016, Institut Cochin, Paris, France
and
Cnrs, UMR8104, Paris, France
and
Univ Paris Descartes, Paris, France
e-mail: benoit.viollet@inserm.fr

F. Andreelli
Inserm, U1016, Institut Cochin, Paris, France
and
Cnrs, UMR8104, Paris, France
and
Univ Paris Descartes, Paris, France
and
Department of Diabetology, Pitié-Salpêtrière Hospital (AP-HP), Univ Pierre et Marie Curie-Paris 6, Paris, France

M. Schwanstecher (ed.), *Diabetes - Perspectives in Drug Therapy*,
Handbook of Experimental Pharmacology 203,
DOI 10.1007/978-3-642-17214-4_13, © Springer-Verlag Berlin Heidelberg 2011

supply with energy demand. It is now recognized that pharmacological activation of AMPK improves blood glucose homeostasis, lipid profile, and blood pressure in insulin-resistant rodents. Indeed, AMPK activation mimics the beneficial effects of physical activity or those of calorie restriction by acting on multiple cellular targets. In addition, it is now demonstrated that AMPK is one of the probable (albeit indirect) targets of major antidiabetic drugs including the biguanides (metformin) and thia-zolidinediones, as well as of insulin-sensitizing adipokines (e.g., adiponectin). Taken together, such findings highlight the logic underlying the concept of targeting the AMPK pathway for the treatment of metabolic syndrome and type 2 diabetes.

Keywords AMP-activated protein kinase · Diabetes · Energy balance · Obesity · Therapeutic strategy

1 Introduction

Obesity (defined as a body mass index (BMI) of >30 kg m^{-2}) and the metabolic syndrome are related conditions that can be considered as precursors of type 2 diabetes (T2D) and increase the risk of developing this disease by >20-fold (Willett et al. 1999). Although these conditions clearly have a genetic component, as indicated by the high prevalence in certain ethnic group, the rapid increase in the prevalence of these conditions in populations throughout the world suggests the contribution of environmental factors. A widely accepted explanation for the increasing prevalence of these conditions lays on the frequent consumption of processed foods with high energy and low fiber content and the reduction in physical exercise due to sedentary lifestyle in modern urban environment. Thus, obesity arises due to an imbalance between energy intake and energy expenditure where caloric excess accumulates preferentially as lipids not only in adipose tissue but also in muscle and liver. Disruption of energy balance has led to an increased prevalence of T2D and related comorbidities such as coronary heart disease, heart failure, hypertension, and renal failure (Wing et al. 2001).

T2D has a high prevalence worldwide and its treatment produces considerable costs for the health budgets. Prevention and management of T2D has become a major public health challenge around the world. Diabetes is defined by a fasting plasma glucose higher than 7 mM (Alberti et al. 1998). T2D is characterized by altered lipid and glucose metabolism (fasting or postprandial hyperglycemia and dyslipidemia) as a consequence of combined insulin resistance in skeletal muscle, liver, and adipose tissue and relative defects of insulin secretion by β-cells that may arise due to an imbalance between energy intake and expenditure (Saltiel and Kahn 2001). Insulin is the primary anabolic hormone that stimulates uptake and storage of fuel substrates while inhibiting substrate production in peripheral tissues (Kahn et al. 2006). It lowers blood glucose levels by facilitating glucose uptake, mainly into skeletal muscle and fat tissue, and by inhibiting endogenous glucose produc-tion in the liver. Insulin resistance occurs when a normal dose of insulin is unable to elicit its metabolic responses. Peripheral insulin resistance is associated with lipid partitioning in specific compartments, i.e., muscle and liver, more than with obesity per se (DeFronzo and Tripathy 2009; Unger 1995). In the natural history of T2D, pancreatic β-cells initially compensate for insulin resistance by increasing insulin secretion, but, with time, progressive β-cell failure leads to insulin deficiency, and hyperglycemia ensues (Fonseca 2009).

Lifestyle intervention is now recognized as the first-line strategy for the management of T2D and remains important for optimization of metabolic control. This is supported by observational studies and clinical trials comparing the respective effects of diet, drugs, or exercise in persons at high risk for T2D (Knowler et al. 2002; Pan et al. 1997; Tuomilehto et al. 2001). The Diabetes Prevention Program (DPP) Research Group conducted a large, randomized clinical trial involving adults in the United States who were at high risk for the development of this disease (Knowler et al. 2002). In this study, the lifestyle intervention was particularly effective (and more than an oral hypoglycemic drug) to prevent the onset of diabetes. In clinical practice, when lifestyle modification fails to achieve or sustain adequate glycemic control, insulin or oral antidiabetic agents are typically used to manage the disease (Nathan et al. 2009). Treatment options with oral agents are quite diverse, including metformin, thiazolidinediones (TZDs), α-glucosidase inhibitors, sulfonylureas, DPP-4 inhibitors, and GLP-1 analogs. The currently available classes of oral agents differ in mechanism and duration of action, and the degree to which they lower blood glucose and their side-effect profile (including hypoglycemia, weight gain, edema, fractures, lactic acidosis, and gastrointestinal intolerance). Because it is recognized that T2D is a progressive disease worsening with time, all available drugs can be used alone or in varied associations.

There is a pressing need to develop new therapeutic strategies to prevent and treat T2D. Exciting recent developments have shown that AMP-activated protein kinase (AMPK), a phylogenetically conserved serine/threonine protein kinase, acts as an integrator of regulatory signals monitoring systemic and cellular energy balance, thus providing the emerging concept, as first suggested by Winder and Hardie (1999), that AMPK is an attractive therapeutic target for intervention in many conditions of disordered energy balance including T2D and insulin resistance.

2   Rational for a Pharmacological Management of T2D by Targeting AMPK

Physical activity is an important determinant to prevent and control T2D. Current guidelines recommend practical, regular, and moderate regimens of physical activity. The multiple metabolic adaptations that occur in response to physical activity can improve glycemic control for individuals with T2D or delay the onset of the disease. Indeed, it is now recognized that beneficial effects of physical activity are still maintained in insulin-resistant populations. This suggests that some metabolic actions of exercise (as increase in muscular glucose uptake) are dependent on specific intracellular pathways that bypass signaling altered by insulin resistance. In consequence, any drug inducing favorable changes similar to those of physical exercise on whole-body metabolism are attractive candidates for treatment and prevention of obesity, metabolic syndrome, and T2D. Interestingly, it is now well established that muscle contraction is a prototypical AMPK activator (Hayashi et al. 1998). Thus, it is expected that part of the effect of physical activity in preventing the development of metabolic disorders related to a sedentary lifestyle is due to the activation of AMPK. Indeed, it has been documented that pharmacological AMPK activation may recapitulate some of the exercise-induced short-term adaptations and is likely to mediate beneficial effects of exercise on insulin sensitivity and glucose transport in skeletal muscle (Bergeron et al. 1999; Fisher et al. 2002). In addition, pharmacological AMPK activation resulted in long-term adaptation similar to those induced by endurance exercise training with the induction of genes linked to oxidative metabolism and enhanced running endurance (Narkar et al. 2008).

In the DPP, the incidence of diabetes was reduced by 58% with a low-calorie, low-fat diet, as compared with placebo after 3 years of follow-up (Knowler et al. 2002). The beneficial effect of calorie restriction in reducing T2D incidence was confirmed by other clinical studies (Pan et al. 1997; Tuomilehto et al. 2001). In overweight and obese humans, calorie restriction improves glucose tolerance, lipid profile, and insulin action and reduces mortality associated with T2D (Hammer et al. 2008; Jazet et al. 2008; Larson-Meyer et al. 2006; Weiss et al. 2006). In order to produce a metabolic profile similar to those of calorie restriction in diabetic patients, there is an increased interest in developing pharmacological agents acting as "calorie-restriction" mimetics. Such agents could provide the beneficial metabolic, hormonal, and physiological effects of calorie restriction without altering dietary intake or experiencing any potential adverse consequences of excessive restriction. To this purpose, phytochemicals mimicking the effects of calorie restriction (polyphenols) were recently identified as potent activators for AMPK in vitro and in vivo (Baur et al. 2006; Collins et al. 2007; Zang et al. 2006).

Additionally, it is now recognized that a dysfunction in AMPK signaling pathway might have sustained deleterious effects at the systemic levels and might contribute to the events that lead to the metabolic syndrome. It is interesting to note that there is a strong correlation between low activation state of AMPK with metabolic disorders associated with insulin resistance, obesity, and sedentary activities (Lee et al. 2005a, b; Luo et al. 2005; Martin et al. 2006). Recent studies showed that AMPK is likely to be under both endocrine and autocrine control in rodents. Thus, in addition to exercise and starvation, AMPK is activated by the fat-cell-derived hormones adiponectin and leptin (Minokoshi et al. 2002; Tomas et al. 2002; Yamauchi et al. 2002) and interleukin-6 (IL-6) (Kelly et al. 2004). Conversely, AMPK activity is suppressed in muscle and liver by sustained hyperglycemia, in liver by refeeding after starvation (Assifi et al. 2005), and by increases in the plasma concentration of other adipocyte-derived hormones such as resistin (Banerjee et al. 2004) and tumor necrosis factor-α (TNF-α) (Steinberg et al.

2006). In addition to its role in the periphery, AMPK also regulates energy intake and body weight by mediating opposing effects of anorexigenic and orexigenic signals in the hypothalamus (Andersson et al. 2004; Kim et al. 2004; Kola et al. 2005; Minokoshi et al. 2004). In addition, many therapies that are useful in treating the metabolic syndrome and associated disorders in humans, including TZDs (Fryer et al. 2002; Saha et al. 2004), metformin (Zhou et al. 2001), calorie deprivation, and exercise, have been shown to activate AMPK system. Lastly, the development of transgenic and knockout (KO) mouse models (see below) have made possible to better understand the physiological role of AMPK and confirm that disruption of AMPK pathway in various tissues induces various phenotypes mimicking the metabolic syndrome observed in humans.

By taking together the physiological functions of AMPK and the suspected role of AMPK in metabolic disorders, activation of AMPK pathway appears as a promising tool to prevent and/or to treat metabolic disorders.

3 Structure and Regulation of AMPK

AMPK is a major regulator of cellular and whole-body energy homeostasis that coordinates metabolic pathways in order to balance nutrient supply with energy demand. Activation of AMPK switches off ATP-consuming anabolic pathways and switches on ATP-producing catabolic pathways (Viollet et al. 2003). This would typically occur when AMPK is activated as a result of energy deprivation linked to alterations of the intracellular AMP/ATP ratio (e.g., hypoxia, glucose deprivation and muscle contraction), changes in calcium concentration, as well as the action of various adipocytokines. AMPK is composed of three different subunits α, β, and γ appearing in several isoforms with different action properties (Fig. 1). The α-subunit contains the catalytic site, whereas regulatory β- and γ-subunits are important to maintain the stability of the heterotrimeric complex. The β-subunit contains a central region that allows AMPK complex to bind glycogen. The γ-subunit contains four tandem repeats known as cystathionine β-synthase (CBS) motifs that bind together two molecules of AMP or ATP in a mutually exclusive manner. Binding of AMP (on γ-subunit) activates AMPK via a complex mechanism involving direct allosteric activation and phosphorylation of α-subunit on Thr-172 by upstream kinases as the protein kinase LKB1 (a tumor suppressor whose germline mutations in humans are the cause of Peutz–Jeghers syndrome), the CaMKKβ (calmodulin-dependent protein kinase β), and TAK1 (mammalian transforming growth factor β-activated kinase) (Fig. 2). Although it was originally proposed that AMP binding promoted AMPK phosphorylation by upstream kinases, recent work suggested entire inhibition of dephosphorylation of Thr-172 to be critical (Sanders et al. 2007; Suter et al. 2006) (Fig. 2).

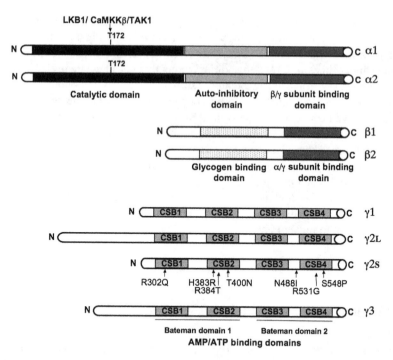

Fig. 1 *Domain organization of the catalytic α- and regulatory β- and γ- subunits of AMPK.* Each AMPK molecule comprises a α-catalytic (α1 and α2) and regulatory β- (β1 and β2) and γ- (γ1, γ2, and γ3) subunits. The catalytic α-subunit is phosphorylated at Thr-172 by upstream kinases (LKB1, CaMKKβ, and TAK1), leading to enzyme activation. The β-subunit contains a glycogen-binding domain. The γ-subunit contains four nucleotide-binding modules (CBS domains) capable of cooperative binding to two molecules of either ATP or AMP. Mutations in the human γ2-subunit gene (PRKAG2) causing cardiac hypertrophy associated with abnormal glycogen accumulation and conduction system disease are shown

4 Beneficial Metabolic Effects of Targeting AMPK Pathway

4.1 Mimicking the Beneficial Effects of Physical Exercise

It has been confirmed in large-scale epidemiological and interventional studies that regular physical activity is of great benefit for the metabolic control of subjects with metabolic syndrome or impaired glucose tolerance or T2D (Knowler et al. 2002; Pan et al. 1997; Tuomilehto et al. 2001). Although appropriate diet and exercise regimes should therefore be the first choice of treatment and prevention of T2D, in some patients such management is not appropriate for other medical reasons, or when compliance is difficult because of social factors or poor motivation. In these cases, drugs that act on the signaling pathways involved in physical activity are attractive candidates for treatment and prevention. It is now clearly demonstrated that AMPK is activated by physical training in an intensity-dependent

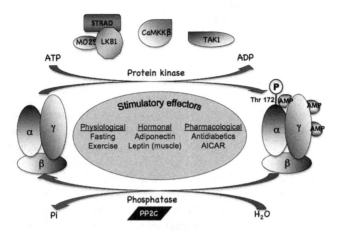

Fig. 2 *Regulation of AMPK by upstream kinases and phosphatases.* The major upstream kinase is a complex between the tumor suppressor kinase LKB1 and two accessory subunits, STRAD and MO25, which appears to be constitutively active. The CaMKKβ could also phosphorylate Thr-172 and activate AMPK following a rise in cytosolic Ca^{2+}. A third potential upstream kinase is TAK1, but its physiological significance is uncertain. Thr-172 phosphorylation is removed by PP2C phosphatase. Physiological, hormonal, and pharmacological stimulatory effectors of AMPK complex are listed

manner both in humans and in rodents (Steinberg and Kemp 2009). AMPK activation during muscle contraction is a physiological adaptation in front of increased energy demand and ATP turnover. It has been demonstrated that AMPK activation may recapitulate some of the exercise-induced adaptations and is likely to mediate not only beneficial effects of exercise on insulin sensitivity and glucose transport in skeletal muscle (Fisher et al. 2002) but also additional metabolic benefits coming from AMPK activation by exercise in liver and in adipose tissue (Park et al. 2002). Conversely, it has also been demonstrated that disruption of muscular AMPK signaling can be a key factor in the pathophysiology of metabolic disorders. Indeed, reduction of muscular AMPK activity exacerbates the development of insulin resistance and glucose intolerance during high-fat feeding, disturbs muscle energy balance during exercise (as indicated by a reduced muscular ATP content during muscle contraction), and abolishes mitochondrial biogenesis (Fujii et al. 2008; Jorgensen et al. 2005; Zong et al. 2002).

As a proof of concept, studies with AMPK activators in animal models of T2D have provided promising results. The first evidence came from in vivo treatment with the pharmacological compound AICAR (5-amino-imidazole-4-carboxamide-1-β-ᴅ-ribofuranoside, metabolized to ZMP, which is an analog of AMP) of various animal models of insulin resistance, causing improvement in most, if not all, of the metabolic disturbances of these animals (Bergeron et al. 2001a; Buhl et al. 2002; Iglesias et al. 2002; Pold et al. 2005; Song et al. 2002). In addition, long-term AICAR administration prevents the development of hyperglycemia in Zucker diabetic fatty (ZDF) rats, improves peripheral insulin sensitivity in skeletal muscle, and delays β-cell dysfunction associated with T2D (Pold et al. 2005). AICAR increases muscle glucose uptake concomitantly with glucose transporter 4 (GLUT4) translocation to the plasma membrane in insulin-resistant animal models and in humans (Merrill et al. 1997; Koistinen et al. 2003; Kurth-Kraczek et al. 1999). Interestingly, AMPK-induced glucose transport occurs through a mechanism distinct from that utilized by the classical insulin signaling pathway because it is not blocked by inhibitors of phosphatidylinositol 3-kinase, and also because the effects of insulin and AMPK activators are additive (Hayashi et al. 1998). This metabolic improvement can be also explained partly by increased expression of specific muscle proteins mimicking some of the effects of exercise training following chronic pharmacological activation of AMPK in vivo. Thus, AICAR or chronic intake of the creatine analog β-guanadinopropionic acid (β-GPA, which competitively inhibits creatine uptake and lowers ATP content) in rodent increases muscle expression of glucose transporter GLUT4 and hexokinase II, an effect partly mediated by the transcriptional coactivator peroxisome proliferator-activated receptor-γ coactivator-1α (PGC-1α) (Holmes et al. 1999; Michael et al. 2001). It has been proposed that the development of skeletal muscle insulin resistance may be partly linked to decreased mitochondrial density (Petersen et al. 2003). Interestingly, chronic activation of AMPK with AICAR or β-GPA increases mitochondrial content and expression of mitochondrial proteins, leading to a mitochondrial biogenesis (Bergeron et al. 2001b; Winder et al. 2000; Zong et al. 2002). All of these data argue for AMPK as a key factor for the metabolic adaptation of skeletal muscle to physical exercise. Supporting this, the effects of chronic activation of AMPK mimicking physical activity on gene expression and mitochondrial biogenesis are abolished in AMPKα2 knockout (KO) and mAMPK-KD (transgenic mice overexpressing a kinase-dead AMPKα2 mutant [K45R mutation] in skeletal muscle) mice (Zong et al. 2002; Holmes et al. 2004; Jorgensen et al. 2005). Increased mitochondrial biogenesis after chronic activation of AMPK is partly explained by increased expression of nuclear respiratory factor-1 and -2 (which are critical regulators of genes encoding electron chain complexes) (Bergeron et al. 2001b). Another critical factor for mitochondrial biogenesis is the inducible coactivator of nuclear receptors, PGC-1α. Regulation of PGC-1α by AMPK is complex. First, it has been demonstrated that AMPK directly phosphorylates and activates PGC-1α (Jager et al. 2007). In addition, activated PGC-1α in turn increased the expression of PGC-1α and of mitochondrial oxidative genes (cytochrome c and uncoupling protein 1). Interestingly, PGC-1α activity and expression are reduced in T2D in humans (Mootha et al. 2003). Thus, AMPK activators could be used in order to reverse this defect. Additionally, activation of AMPK in response to physical exercise has also been observed in extramuscular tissues such as liver and adipose tissue (Park et al. 2002) and might account for additional metabolic benefits. Physical training increases circulating adiponectin and mRNA expression of its receptors in muscle, which may mediate the improvement of insulin resistance and the metabolic syndrome in response to exercise by activation of AMPK.

Lastly, increase in blood supply is critical for physiological adaptation during physical activity. Vasodilatation is a vital mechanism of systemic blood flow regulation that occurs during periods of increased energy demand. Thus, because AMPK plays a central role in the adaptation to metabolic stress, it is tempting to speculate that AMPK could be involved in the regulation of metabolic vasomotion. It is well known that moderate-intensity exercise increases nitric oxide synthase (NOS) activity (Roberts et al. 1999). Interestingly, it has been recently reported that mAMPK-KD mice are unable to increase total NOS activity during moderate-intensity exercise and may cause an impairment in muscle blood flow (Lee-Young et al. 2009). This finding is supported by the close association between AMPK and nNOSμ phosphorylation following moderate-intensity exercise (Chen et al. 2000; Stephens et al. 2002) and reduced expression of nNOSμ in mAMPK-KD mice (Lee-Young et al. 2009). This indicates how changes in tissue metabolism can direct blood flow according to demand. In addition, the lower skeletal muscle capillarization in mAMPK-KD mice might also contribute to the reduced blood flow during exercise (Zwetsloot et al. 2008). Nitric oxide (NO) plays a fundamental role in vascular homeostasis and it has been suggested that impaired NO efflux from contracting mAMPK-KD mice suppressed exercise-induced vascular relaxation (Lee-Young et al. 2009). Furthermore, it has suggested that AMPK activation is in part regulated by endogenous NO in a positive feedback mechanism, such that increased NO activates AMPK, which further augments NOS activity and NO production (Lira et al. 2007; Zhang et al. 2008). Accordingly, the exercise-induced increase in AMPK signaling was ablated in skeletal muscle of eNOS KO mice (Lee-Young et al. 2010). Therefore, AMPK–eNOS interaction might play an important role in the adaptation processes during exercise in order to maintain cellular energy levels by amending vascular function.

4.2 Mimicking the Beneficial Effects of Calorie/Dietary Restriction

Excessive calorie intake increases the risk of developing chronic disease such as obesity, metabolic syndrome, T2D, systemic low-grade inflammation, cardiovascular event, and premature mortality. Conversely, calorie restriction improves glucose tolerance and insulin action and reduces mortality linked to T2D and cardiovascular diseases (Hammer et al. 2008; Jazet et al. 2008; Larson-Meyer et al. 2006; Weiss et al. 2006). Because it is difficult to maintain long-term calorie restriction in modern society, there has been an increased interest in developing pharmacological agents that act as "calorie-restriction" mimetics. Among them, plant-derived polyphenolic compounds, such as resveratrol (which is present in grapes, peanuts, and several other plants) were first recognized as mimicking the effects of calorie restriction in lower eukaryote (Howitz et al. 2003). Additionally, resveratrol administration prevents the deleterious effects of high-calorie intake on insulin resistance and metabolic syndrome components in rodents (Baur et al. 2006; Lagouge et al. 2006; Milne et al. 2007; Sun et al. 2007; Zang et al. 2006).

Resveratrol has been described as a potent activator of the NAD(+)-dependent deacetylases sirtuins including SIRT1, one of the seven mammalian sirtuin genes (Howitz et al. 2003). However, recent findings indicate that resveratrol is not direct SIRT1 activator (Pacholec et al. 2010). Resveratrol, like other polyphenols, also activates AMPK (Baur et al. 2006; Collins et al. 2007; Zang et al. 2006). Acute activation of AMPK by resveratrol appears to be independent of SIRT1 (Dasgupta and Milbrandt 2007), probably through changes in AMP/ATP ratio as resveratrol inhibits the mitochondrial F1 ATPase (Gledhill et al. 2007). Furthermore, resveratrol increased the NAD(+)/NADH ratio in an AMPK-dependent manner, which may explain how it may activate SIRT1 indirectly (Canto et al. 2009; Um et al. 2010). SIRT1 has been suggested to prime the organism in order to reduce the deleterious effects of insulin resistance on energy balance and metabolic homeostasis. Thus, SIRT1 activation increases hepatic insulin sensitivity, decreases whole-body energy requirements (Banks et al. 2008; Sun et al. 2007), promotes adaptation of insulin secretion during insulin resistance development (Bordone et al. 2006; Moynihan et al. 2005), and coordinates lipid mobilization and utilization (Picard et al. 2004). The knowledge of SIRT1 action at the molecular level has been more delineated by using chronic treatments with resveratrol and it has been suggested that SIRT1 promotes LKB1-dependent AMPK stimulation through the direct deacetylation and activation of LKB1 (Hou et al. 2008; Lan et al. 2008). Thus, polyphenols such as resveratrol are now recognized as compounds with great potential to improve and/or delay or prevent metabolic disorders linked to Western lifestyle by activating the complementary metabolic stress sensors SIRT1 and AMPK (Canto et al. 2009). Accordingly, it has been recently established that AMPK acts as the prime initial sensor for fasting-induced adaptations in skeletal muscle and that SIRT1 downstream signaling was blunted in the absence of AMPK (Canto et al. 2010). In addition, recent studies demonstrated that resveratrol failed to increase the metabolic rate, insulin sensitivity, glucose tolerance, mitochondrial biogenesis, and physical endurance in the absence of either AMPKα1 or AMPKα2 (Um et al. 2010).

4.3 Mimicking the Beneficial Effects of Hypoglycemic Agents

4.3.1 AMPK Action in Liver

T2D is the result of an imbalance between glucose production and glucose uptake by peripheral tissues. Elevated hepatic glucose production is a major cause of fasting hyperglycemia in diabetic subjects (Saltiel and Kahn 2001). From various effectors, AMPK signaling is a key factor that controls hepatic glucose production. Indeed, systemic infusion of AICAR in normal and insulin-resistant obese rats leads to the inhibition of hepatic glucose production (Bergeron et al. 2001a). Additionally, short-term hepatic expression of a constitutively active form of the α2-catalytic subunit (AMPKα2-CA) leads to mild hypoglycemia in normal mice (Foretz et al. 2005; Viana et al. 2006) and abolishes hyperglycemia in diabetic *ob/ob* and streptozotocin-induced diabetic mice (Foretz et al. 2005; Viana et al. 2006) by inhibition of gluconeogenesis (Foretz et al. 2005; Lochhead et al. 2000; Viana et al. 2006). This effect is achieved at least to a large extent via the regulation of a transcriptional coactivator, transducer of regulated CREB activity 2 (TORC2) (Koo et al. 2005), which is known to mediate CREB-dependent transcription of PGC1α and its subsequent gluconeogenic targets PEPCK and G6Pase genes. AMPK activation causes TORC2 phosphorylation and sequesters the coactivator in the cytoplasm, thus blunting the expression of the gluconeogenic program (Koo et al. 2005). Control of hepatic glucose production by activated AMPK is also demonstrated in resistin KO mice and in adiponectin-treated rodents (Banerjee et al. 2004; Yamauchi et al. 2002), suggesting that hepatic AMPK is specifically a target of both adipocytokines, the former acting as an AMPK inhibitor and the latter as an activator. This was also demonstrated by lack of systemic adiponectin infusion effect on hepatic glucose production in liver-specific AMPKα2 KO mice (Andreelli et al. 2006).

4.3.2 AMPK Action in Skeletal Muscle

After a meal or during the euglycemic hyperinsulinemic clamp, both situations with high circulating levels of insulin, skeletal muscle is the main site for glucose disposal in the body. This is sustained by the insulin-dependent translocation of glucose transporter GLUT4 from intracellular vesicles to the cell surface, which is impaired in T2D patients. As described above, it has been clearly demonstrated that muscular AMPK activation, either by exercise or by AICAR, stimulates muscle glucose uptake. Interestingly, even if AMPK and insulin act through phosphorylation of downstream target of Akt (Akt substrate of 160 kDa, AS160) (Dreyes et al. 2008), AMPK-dependent and insulin-dependent GLUT4 translocation are distinct pathways (Treebak et al. 2007). Additionally, both exercise-induced muscular AMPK activation and AS160 phosphorylation are reduced in obese nondiabetic and obese T2D subjects (Musi et al. 2001) but maintained in lean T2D patients (Bruce et al. 2005), suggesting that dysregulation of muscular AMPK is more dependent on obesity than on hyperglycemia. Discovery of muscular AMPK activators in order to mimic regular physical activity metabolic effects is an important challenge. It was first demonstrated that some adipokines stimulate glucose transport in skeletal muscle in an AMPK-dependent manner. Indeed, leptin is known to stimulate glucose uptake in peripheral tissue (Kamohara et al. 1997; Minokoshi et al. 1999) by stimulating AMPKα2 phosphorylation and activation in skeletal muscle (Minokoshi et al. 2002). Adiponectin, another adipokine, has also been shown to increase glucose transport in both lean and obese skeletal muscle, although the effect was less significant in obese skeletal muscle (Bruce et al. 2005). It has also recently been recognized that IL-6 [also called "myokine" (Febbraio and Pedersen 2005)] is released acutely from the skeletal muscle during prolonged exercise, activates AMPK, and improves peripheral glucose uptake and insulin sensitivity at the whole-body level (Glund et al. 2007). In contrast, chronic exposure of IL-6 (as observed in obesity) promotes insulin resistance both in vitro and in vivo (Nieto-Vazquez et al. 2008). The dual effect of IL-6 on insulin sensitivity probably explains some conflicting results recently discussed in detail elsewhere (Nieto-Vazquez et al. 2008). Importantly, it has also been suggested that AICAR, in addition to activating AMPK, suppresses chronic IL-6 release by an AMPK-independent mechanism in insulin-resistant models (Glund et al. 2009). This strongly suggests that AMPK activators can act at a multitissular level in order to restore metabolic interorgan cooperation.

Interestingly, available hypoglycemic drugs as metformin and TZDs have been reported to activate AMPK (Fryer et al. 2002; Zhou et al. 2001). Even if it was postulated that blood glucose-lowering effects of metformin are mediated by AMPK activation from studies of mice that are deficient in the upstream AMPK kinase, LKB1, (Shaw et al. 2005) recent studies have shown that LKB1 phosphory-lates and activates at least 12 AMPK-related kinases in the liver. These data raised the question whether the glucose-lowering function of LKB1 is mediated by AMPK-related kinases rather than AMPK itself.

Because circulating levels of adiponectin are decreased in individuals with obesity and insulin resistance, adiponectin replacement in humans may be a promising approach. It has been demonstrated that full-length adiponectin activates AMPK in the liver, while globular adiponectin did so in both muscle and the liver (Yamauchi et al. 2002). Blocking AMPK activation by the use of a dominant-negative mutant inhibited the action of full-length adiponectin on glucose hepatic production (Yamauchi et al. 2002). In addition, lack of action of adiponectin on hepatic glucose production when AMPKα2 catalytic subunit is missing strongly supports the concept that adiponectin effect is strictly dependent on AMPK (Andreelli et al. 2006). While awaiting adiponectin analog development, alternative ways to restore adiponectin effects have been suggested recently. Improved metabolic disorders following TZD administration are in part mediated through adiponectin-dependent activation of AMPK since activation of AMPK by rosigli-tazone treatment is diminished in adiponectin KO mice (Nawrocki et al. 2006). TZDs can markedly enhance the expression and secretion of adiponectin in vitro and in vivo through the activation of its promoter and also antagonize the suppressive effect of TNF-α on the production of adiponectin (Maeda et al. 2001). Interestingly, in human adipose tissue, AICAR has been shown to increase the expression of adiponectin (Lihn et al. 2004; Sell et al. 2006), while no change in serum adiponectin concentration or adipocyte adiponectin content was found in T2D patients treated with metformin (Phillips et al. 2003).

4.3.3 AMPK Action in β-Cells

β-Cell failure is a strong determinant in the pathogenesis of T2D. This defect inexorably aggravates with time as demonstrated in prospective clinical studies (U.K. Prospective Diabetes Study Group 1995). According to the glucolipotoxicity hypothesis, (Prentki et al. 2002) chronic high glucose dramatically influences β-cell metabolism. Indeed, it has been observed in high glucose condition that an increase of cytosolic fatty acyl-CoA partitioning toward potentially toxic cellular products (e.g., diacylglycerol, ceramide, and lipid peroxides), lead to impaired insulin secretory response to glucose and ultimately apoptosis (Donath et al. 2005). Indeed, decrease in β-cell mass is likely to play a role in the pathogenesis of human T2D (Butler et al. 2003) as it does in rodent models of the disease (Kaiser et al. 2003; Rhodes 2005).

Pathways regulating β-cell turnover are also implicated in β-cell insulin secretory function. In consequence, decrease in β-cell mass is not dissociable from an intrinsic secretory defect. Because AMPK is important for the balance of intracellular energy homeostasis, it was interesting to analyze to what extent AMPK regulates β-cell function/survival. AICAR, dose dependently, improves β-cell function probably by reducing apoptosis induced by prolonged hyperglycemia (Nyblom et al. 2008). In addition to AICAR, β-cell AMPK activation (by metformin, TZDs or adenovirus-mediated overexpression of AMPKα1-CA) favors fatty acid β-oxidation and prevents glucolipotoxicity-induced insulin secretory dysfunction in β-cells (Eto et al. 2002; El-Assaad et al. 2003; Higa et al. 1999; Lupi et al. 2002). In contrast, the role of AMPK in the control of β-cell death survival remains controversial (Kefas et al. 2003a, b; Kim et al. 2007; Riboulet-Chavey et al. 2008; Richards et al. 2005).

Beyond a potential role of AMPK for long-term regulation of β-cell function and survival, AMPK may also acutely regulate insulin secretion. Thus, AMPK activity is rapidly decreased when glucose levels are increased over the physiological range, suggesting that AMPK could be one of the regulator of insulin secretion through its capacity to sense intracellular energy (da Silva Xavier et al. 2003; Leclerc et al. 2004). Interestingly, activation of AMPK by AICAR, berberine, metformin, and TZDs or by overexpression of AMPKα1-CA markedly reduced glucose-stimulated insulin secretion in β-cell lines and in rodent and human islets (Eto et al. 2002; Leclerc et al. 2004; Wang et al. 2007; Zhou et al. 2008). Similarly, activation of AMPK selectively in β-cells in AMPKα1-CA transgenic mice decreased glucose-stimulated insulin secretion (Sun et al. 2010). This could be considered as a deleterious effect of AMPK activation. But it is hypothesized that pharmacological activation of AMPK and its subsequent decrease in insulin secretion could be appropriate in insulin-resistant conditions characterized by high insulin levels. Indeed, it has been suggested that reduction of the pathological hyperinsulinemia is potentially a mechanism to protect β-cell mass. Consistent with this assumption, systemic AICAR infusion in prediabetic Zucker fatty rats prevented the development of hyperglycemia and preserved β-cell mass (Pold et al. 2005).

Taken together, these data suggest that AMPK is an emergent factor that could protect by different ways β-cell function and β-cell mass from the deleterious effects of glucolipotoxicity.

4.4 Mimicking the Beneficial Effects of Hypolipidemic Agents

Dyslipidemia of both insulin resistance and T2D is a recognized risk factor for cardiovascular disease. Diabetic dyslipidemia is a cluster of potentially atherogenic lipid and lipoprotein abnormalities that are metabolically interrelated. Activated AMPK inhibits cholesterol and fatty acid synthesis. Thus, AMPK suppresses expression of lipogenesis-associated genes such as fatty acid synthase, pyruvate kinase and acetyl-CoA carboxylase (ACC) (Foretz et al. 1998, 2005; Leclerc et al. 1998, 2001; Woods et al. 2000), and 3-hydroxy-3-methylglutaryl-coenzyme A reductase (HMG-CoA reductase). HMG-CoA reductase activity is inhibited by phosphorylation of Ser-872 by AMPK (Clarke and Hardie 1990). Adiponectin activates AMPK and inhibits cholesterol synthesis in vivo, suggesting that AMPK is a key regulator of cholesterol pathways (Ouchi et al. 2001). Inhibition of ACC by AMPK leads to a drop in malonyl-CoA content and a subsequent decrease in fatty acid synthesis and increase in fatty acid oxidation, thus reducing excessive storage of triglycerides. Consistently, overexpression of AMPKα2-CA in the liver or treatment with AICAR, metformin, or A769662 (a small-molecule AMPK activator) in lean and obese rodents decreases plasma triglyceride levels, concomitantly with an increase in plasma β-hydroxybutyrate levels, suggesting elevated hepatic lipid oxidation (Bergeron et al. 2001a; Cool et al. 2006; Foretz et al. 2005; Zhou et al. 2001). Conversely, liver-specific AMPKα2 deletion leads to increased plasma triglyceride levels and enhanced hepatic lipogenesis (Andreelli et al. 2006). These data emphasize the critical role for AMPK in the control of hepatic lipid deposition via decreased lipogenesis and increased lipid oxidation, thus improving lipid profile in T2D.

It is well documented that changes in adipose tissue mass are frequently associated with alterations in insulin sensitivity (Eckel et al. 2005; Katsuki et al. 2003). AMPK evidenced recently as a regulator of fat mass. Indeed, activation of AMPK in white adipocytes is concomitant with a decreased expression of genes coding lipogenic enzyme (Orci et al. 2004) and leads to a decreased lipogenic flux and a decreased triglyceride synthesis (Daval et al. 2005; Sullivan et al. 1994). In white adipocytes, AMPK activation using AICAR or overexpression of AMPK-CA has been shown to inhibit β-adrenergic-induced lipolysis (Corton et al. 1995; Sullivan et al. 1994). Hormone-sensitive lipase (HSL), one of the key proteins responsible for the lipolytic activity, is activated by PKA phosphorylation at serines 563, 659, and 660 (Anthonsen et al. 1998). AMPK reduces this activation through phosphorylation at Ser-565 (Garton et al. 1989; Garton and Yeaman 1990). This effect has been demonstrated both in white adipocytes and in skeletal muscle in both resting and contracting conditions (Muoio et al. 1999; Smith et al. 2005; Watt et al. 2006). Thus, inhibition of HSL by AMPK represents a mechanism to limit this recycling and ensure that the rate at which fatty acids are released by lipolysis does not exceed the rate at which they could be disposed of by export or by internal oxidation.

Beyond its hypolipidemic properties, AMPK system can also be a regulator of ectopic lipid metabolism. Depot of lipids in tissue is a hallmark defect in metabolic syndrome in humans. According to this lipotoxicity hypothesis, insulin resistance develops when excess lipids are deposited in insulin-sensitive cell types. The balance between lipid oxidation and lipid storage in cells is mainly regulated by malonyl-CoA, generated by ACC. Malonyl-CoA is known to inhibit transport of fatty acids into mitochondria via allosteric regulation of carnitine palmitoyltransferase-1, thereby preventing them from being metabolized. Activated AMPK inhibits malonyl-CoA synthesis and shifts the balance toward mitochondrial fatty acid oxidation and away from fat storage. Several studies have shown that activation of AMPK with AICAR, α-lipoic acid, leptin, adiponectin, and IL-6 enhances muscle fatty acid β-oxidation (Carey et al. 2006; Lee et al. 2005b; Merrill et al. 1997; Minokoshi et al. 2002; Yamauchi et al. 2002). Chronic leptin treatment increases skeletal muscle fatty acid oxidation in an AMPK-dependent manner by increasing AMP/ATP ratio in oxidative muscle fibers and by increasing AMPKα2 nuclear translocation and PPARα transcription (Suzuki et al. 2007). Studies in transgenic animals support these observations since expression of the activating AMPKγ3 R225Q mutation in muscle increased fatty acid oxidation and protected against excessive triglyceride accumulation and insulin resistance in skeletal muscle (Barnes et al. 2004). Interestingly, recent data have shown that resistin lowers AMPK signaling in muscle cells and that this reduction is associated with suppressed fatty acid oxidation (Palanivel and Sweeney 2005).

Nonalcoholic fatty liver disease is a serious consequence of obesity, increasing the risk of liver cancer or cirrhosis. The origin of this disease is unknown and probably multifactorial. Nevertheless, because insulin resistance is recognized as an associate and/or promoting mediator of the disease, management of insulin resistance becomes an important challenge. For this specific point and because AMPK is a key factor in lipid partitioning (balance between synthesis and oxidation), management of nonalcoholic fatty liver disease by activators of AMPK represents a new therapeutic strategy. Adiponectin treatment restores insulin sensitivity and decreases hepatic steatosis of obese mice (Xu et al. 2003). This effect is linked to an activation of AMPK in the liver that decreases fatty acid biosynthesis and increases mitochondrial fatty acid oxidation (Yamauchi et al. 2001). Reduction of liver steatosis when AMPK is activated has also been confirmed by a decrease in liver triglyceride content in lean and obese rodents during AICAR infusion (Bergeron et al. 2001a; Cool et al. 2006) and after treatment with small-molecule AMPK activators (Cool et al. 2006). The synthesis of triglycerides is regulated by the supply of both glycerol-3-phosphate (from carbohydrate metabolism) and fatty acyl-coenzyme A. The first step of triglyceride synthesis is catalyzed by glycerol-3-phosphate acyl-transferase (GPAT). AICAR or exercise-induced AMPK activation reduces hepatic GPAT activity and triglyceride esterification (Muoio et al. 1999; Park et al. 2002). Fasting that increases hepatic AMPK inhibits GPAT activity (Witters et al. 1994). In the same way, AMPK activation by resveratrol protects against lipid accumulation in the liver of diabetic mice (Zang et al. 2006) in association with increased mitochondrial number (Baur et al. 2006) and SIRT1-dependent deacetylation of peroxisome proliferator-activated receptor coactivator (PGC)-1α, a master regulator of mitochondrial biogenesis (Baur et al. 2006; Rodgers and Puigserver 2007). The efficacy of metformin as a treatment for fatty liver disease has been confirmed in obese (ob/ob) mice, which develop hyperinsulinemia, insulin resistance, and fatty livers (Lin et al. 2000).

The discovery of new strategies of management of hepatic steatosis in humans is of considerable interest. AMPK activation could be one of them as suggested by recent clinical studies in T2D patients. Indeed, it has been demonstrated that AICAR infusion results in significant decline in circulating plasma nonesterified fatty acid (NEFA) levels, suggesting stimulation of hepatic fatty acid oxidation and/ or a reduction in whole-body lipolytic rate (Boon et al. 2008). Management of hepatic steatosis by targeting AMPK is also suggested by recent successes in treating this disorder with diet, exercise, and TZDs all known as AMPK activators (Carey et al. 2002; Neuschwander-Tetri and Caldwell 2003). Other studies are needed to analyze the beneficial effect of AMPK activation for the management of fatty liver diseases in humans.

4.5 Mimicking the Beneficial Effects of an Antiobesity Drug

Weight reduction is best achieved by behavioral change to reduce energy intake and by increasing physical activity to enhance energy expenditure. Therefore, the AMPK system may be an important pharmacological target to reduce fatty acid storage in adipocytes and to treat obesity. By inducing fatty acid oxidation within the adipocyte, activation of AMPK would reduce fat cell size and also prevent fatty acids from being exported to peripheral tissues and cause deleterious effects. Direct evidence linking AMPK activation to diminished adiposity was first obtained by chronic administration of AICAR to lean and obese rats, an effect attributable, at least in part, to an increase in energy expenditure (Buhl et al. 2002; Winder et al. 2000). Furthermore, the antiobesity hormone leptin increases fatty acid oxidation in skeletal muscle by activating AMPK (this process involves an increase in the AMP/ATP ratio) (Minokoshi et al. 2002) and depletes body fat stores by activating AMPK activity and by increasing uncoupling mitochondrial protein (UCP)-1 and UCP-2 expression (Orci et al. 2004). β3-Adrenoceptor (β3-AR) agonists were also found to have remarkable antiobesity and antidiabetic effects in rodents and these compounds were found to stimulate AMPK in fat cells (Moule and Denton 1998). In addition, overexpression of UCP-1 in adipocytes leads to an increase in the AMP/ATP ratio and activation of AMPK, inactivation of ACC, and a decreased lipogenesis (Matejkova et al. 2004). Additionally, a strong mitochondrial biogenesis in response to increased UCP-1 expression in adipocytes has been demonstrated (Orci et al. 2004; Rossmeisl et al. 2002), features that could enhance the fatty acid oxidation capacity of adipocytes in response to AMPK activation. During chronic AICAR treatment, activated AMPK increases UCP-3 expression in muscle independently of changes in mito-chondrial biogenesis (Stoppani et al. 2002; Zhou et al. 2000). This effect can also explain changes in energy expenditure during AMPK activation.

5   Benefits of Targeting AMPK Pathway for Metabolic Complications

5.1   AMPK and Ischemic Heart

T2D is recognized as an important risk factor for cardiovascular diseases and mortality. In ischemic heart, balance between glucose and lipids is altered. In this situation, activation of AMPK is considered as a metabolic adaptation to rescue energy supply. Indeed, AMPK stimulates glycolysis and sustains energy supply during ischemic stress. Convincing evidence suggests that the more the AMPK is activated in ischemic myocardial tissue, the more the size of infarcted tissue is reduced. Because the size of myocardial infarcted tissue is one of the variables that determine the risk of sudden death and the risk of cardiac insufficiency in humans, reduction of the volume of ischemic tissue is an important therapeutic challenge. Thus, promotion of glucose oxidation or inhibition of fatty acid oxidation in ischemic/reperfused hearts could be a promising novel therapeutic approach during myocardial ischemia. Such a mechanism has been demonstrated during the phenomenon called ischemic preconditioning. This phenomenon (consisting in repeated brief episodes of myocardial ischemia) (Murry et al. 1986) induces endogenous protective mechanisms in the heart that becomes more resistant to subsequent ischemic episodes. The molecular mechanism of this protective effect is based on AMPK activation in a PKC-dependent manner and promotion of glucose utilization in myocardial cells (Nishino et al. 2004). Attractively, adiponectin protects the heart from ischemia by activating AMPK and increasing the energy supply to heart cells (Shibata et al. 2005). For example, high blood levels of adiponectin are associated with a lower risk of heart attack, and vice versa (Pischon et al. 2004). Additionally, adiponectin levels rapidly decline after the onset of acute myocardial infarction. Similarly, in mice, deletion of adiponectin induces increased heart damage after reperfusion that was associated with diminished AMPK signaling in the myocardium (Shibata et al. 2005). In addition, it has also been reported that adiponectin attenuated cardiac hypertrophy through activation of AMPK signaling pathway (Liao et al. 2005; Shibata et al. 2004a). These findings clearly show that adiponectin has a cardioprotective role in vivo during ischemia through AMPK-dependent mechanisms.

Since AMPK regulates the balance between glucose and fatty acid metabolism at the cellular level, the metabolic response of the heart to global ischemia was studied in AMPKα2$-/-$ mice. These hearts displayed a more rapid onset of ischemic contracture, which was associated with a decrease in ATP content, in lactate production, in glycogen content, and in the phosphorylation state of ACC (Zarrinpashneh et al. 2006). Importance of metabolic adaptation via AMPK activation during ischemia was also documented in another transgenic mouse model overexpressing a dominant-negative form of AMPKα2 in the heart (Russell et al. 2004). These studies indicate that the α2-isoform of AMPK is required for the metabolic response of the heart to ischemia, suggesting that AMPK is cardioprotective. Thus, AMPK activators could be of particular interest for the management of myocardial ischemia. Nevertheless, inappropriate activation of AMPK can have deleterious consequences in the heart. Indeed, in humans, a variety of mutations in the γ2-subunit (Fig. 1) have been shown to produce a glycogen storage cardiomyopathy characterized by ventricular preexcitation, conduction defects, and cardiac hypertrophy (Dyck and Lopaschuk 2006). This argues for a restrictive use of AMPK activators during the acute phase of heart ischemia and not for a chronic activation of cardiac AMPK. Thus, the balance between benefits and deleterious cardiac effects of AMPK activation has to be studied in detail.

5.2 AMPK and Endothelial Dysfunction

Endothelial cell dysfunction, as manifested by impaired vascular relaxation or an increase in circulating vascular cell adhesion molecules, is present in patients with T2D, and it is thought to be one component of the inflammatory process that initiates atherogenesis (Van Gaal et al. 2006). Based on studies using genetically modified mice, the production of NO via eNOS is crucial in the regulation of vascular tone (Lau et al. 2000; Maxwell et al. 1998). The activity of eNOS is largely determined by posttranslational modifications such as multisite phosphorylation and protein interactions. Interestingly, AMPK enhances eNOS activity by direct phosphorylation of Ser-1177 (Chen et al. 1999, 2000) and Ser-633 (Chen et al. 2009) and by promoting its association with heat shock protein 90 (Davis et al. 2006), leading to endothelial NO production. In this respect, metformin has been proposed to improve endothelium function in diabetes by favoring phosphorylation of eNOS by AMPK activation (Davis et al. 2006). Metformin was also shown to relax endothelium-denuded rat aortic rings precontracted with phenylephrine, showing that AMPK can induce vasorelaxation in an endothelium- and NOS-independent manner (Majithiya and Balaraman 2006). Accordingly, AMPK activation in response to hypoxia or metabolic challenge can induce vasorelaxation of big vessels (Evans et al. 2005; Rubin et al. 2005), thereby favoring blood flow. Interestingly, AMPK-dependent adiponectin vascular effects have been demonstrated for angiogenic repair in an ischemic hind limb model (Shibata et al. 2004b). Similarly, α-lipoic acid improves vascular dysfunction by normalizing triglyceride and lipid peroxide levels and NO synthesis in endothelial cells from obese rat by activating AMPK (Lee et al. 2005a). Attractively, adiponectin exhibits potent anti-atherosclerotic effects and suppresses endothelial cell proliferation via AMPK activation (Kubota et al. 2002; Yamauchi et al. 2003).

Beyond the vascular effects of AMPK activation, it has been recently demonstrated that AMPK can regulate blood pressure. Thus, long-term administration of AICAR reduces systolic blood pressure in an insulin-resistant animal model (Buhl et al. 2002). In this process, a potential role for AMPK could be the regulation of ion channels or sodium cotransporters including ENaC and the Na–K–2Cl cotransporter (Carattino et al. 2005; Fraser et al. 2003). These data provide additional support to the hypothesis that AMPK activation might be a potential future pharmacological strategy for treating the cardiovascular risk factors linked to the metabolic syndrome.

6 Conclusion

Lifestyle modifications are recognized as an important preventive and therapeutic intervention for impaired glucose tolerance, insulin resistance, and T2D patients. AMPK activators are potential new therapeutic agents for the treatment of T2D by mimicking the beneficial effects of physical activity and of calorie restriction. Accordingly, AMPK-activating agents could also be used as regulators of hyperglycemia, obesity, lipid disorders, lipotoxicity, and cardiovascular risk by targeting specific cellular pathways (Fig. 3). Resveratrol, metformin, TZDs, adiponectin, and leptin are now considered as AMPK activators. However, many other effects of AMPK activation should be carefully evaluated and many questions are not resolved: Are new AMPK activators tissue specific? What are the consequences of a long-term pharmacological AMPK activation? Additional studies are required to address these critical points.

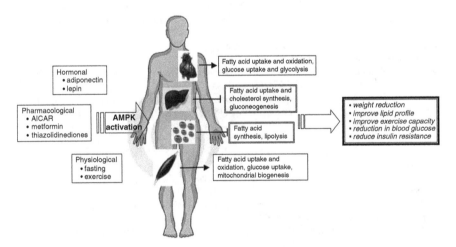

Fig. 3 *AMPK, a potential therapeutic target in metabolic disease.* AMPK pathway has become the focus of a great deal of attention as a novel therapeutic target in metabolic disease because it has been demonstrated that physiological and pharmacological activation of AMPK results in remodeling different metabolic pathways. AMPK has several important metabolic effects, mimicking the beneficial effects of exercise, including modulation of lipid metabolism, enhanced muscle glucose uptake, increased mitochondrial biogenesis, improvement in insulin sensitivity, and reduction in blood glucose. Activation of AMPK by pharmacological agents presents a unique challenge to prevent and treat the metabolic abnormalities associated with the metabolic syndrome

Acknowledgments This work was supported by the European Commission integrated project (LSHM-CT-2004-005272/exgenesis), Agence Nationale de la Recherche (ANR-06-PHYSIO-026), Association Française contre les Myopathies (AFM), Association pour l'Etude des Diabètes et des Maladies Métaboliques (ALFEDIAM), and Institut Benjamin Delessert.

References

Alberti KG, Zimmet PZ (1998) New diagnostic criteria and classification of diabetes–again? Diabet Med 15:535–536

Andersson U, Filipsson K et al (2004) AMP-activated protein kinase plays a role in the control of food intake. J Biol Chem 279:12005–12008

Andreelli F, Foretz M et al (2006) Liver adenosine monophosphate-activated kinase-alpha2 catalytic subunit is a key target for the control of hepatic glucose production by adiponectin and leptin but not insulin. Endocrinology 147:2432–2441

Anthonsen MW, Ronnstrand L et al (1998) Identification of novel phosphorylation sites in hormone-sensitive lipase that are phosphorylated in response to isoproterenol and govern activation properties in vitro. J Biol Chem 273:215–221

Assifi MM, Suchankova G et al (2005) AMP-activated protein kinase and coordination of hepatic fatty acid metabolism of starved/carbohydrate-refed rats. Am J Physiol Endocrinol Metab 289: E794–E800

Banerjee RR, Rangwala SM et al (2004) Regulation of fasted blood glucose by resistin. Science 303:1195–1198

Banks AS, Kon N et al (2008) SirT1 gain of function increases energy efficiency and prevents diabetes in mice. Cell Metab 8:333–341

Barnes BR, Marklund S et al (2004) The 5'-AMP-activated protein kinase gamma3 isoform has a key role in carbohydrate and lipid metabolism in glycolytic skeletal muscle. J Biol Chem 279:38441–38447

Baur JA, Pearson KJ et al (2006) Resveratrol improves health and survival of mice on a high-calorie diet. Nature 444:337–342

Bergeron R, Russell RR 3rd et al (1999) Effect of AMPK activation on muscle glucose metabolism in conscious rats. Am J Physiol 276:E938–E944

Bergeron R, Previs SF et al (2001a) Effect of 5-aminoimidazole-4-carboxamide-1-beta-D-ribofur-anoside infusion on in vivo glucose and lipid metabolism in lean and obese Zucker rats. Diabetes 50:1076–1082

Bergeron R, Ren JM et al (2001b) Chronic activation of AMP kinase results in NRF-1 activation and mitochondrial biogenesis. Am J Physiol Endocrinol Metab 281:E1340–E1346

Boon H, Bosselaar M et al (2008) Intravenous AICAR administration reduces hepatic glucose output and inhibits whole body lipolysis in type 2 diabetic patients. Diabetologia 51: 1893–1900

Bordone L, Motta MC et al (2006) Sirt1 regulates insulin secretion by repressing UCP2 in pancreatic beta cells. PLoS Biol 4:e31

Bruce CR, Mertz VA et al (2005) The stimulatory effect of globular adiponectin on insulin-stimulated glucose uptake and fatty acid oxidation is impaired in skeletal muscle from obese subjects. Diabetes 54:3154–3160

Buhl ES, Jessen N et al (2002) Long-term AICAR administration reduces metabolic disturbances and lowers blood pressure in rats displaying features of the insulin resistance syndrome. Diabetes 51:2199–2206

Butler AE, Janson J et al (2003) Beta-cell deficit and increased beta-cell apoptosis in humans with type 2 diabetes. Diabetes 52:102–110

Canto C, Gerhart-Hines Z et al (2009) AMPK regulates energy expenditure by modulating NAD+ metabolism and SIRT1 activity. Nature 458:1056–1060

Canto C, Jiang LQ et al (2010) Interdependence of AMPK and SIRT1 for metabolic adaptation to fasting and exercise in skeletal muscle. Cell Metab 11:213–219

Carattino MD, Edinger RS et al (2005) Epithelial sodium channel inhibition by AMP-activated protein kinase in oocytes and polarized renal epithelial cells. J Biol Chem 280:17608–17616

Carey DG, Cowin GJ et al (2002) Effect of rosiglitazone on insulin sensitivity and body composition in type 2 diabetic patients [corrected]. Obes Res 10:1008–1015

Carey AL, Steinberg GR et al (2006) Interleukin-6 increases insulin-stimulated glucose disposal in humans and glucose uptake and fatty acid oxidation in vitro via AMP-activated protein kinase. Diabetes 55:2688–2697

Chen ZP, Mitchelhill KI et al (1999) AMP-activated protein kinase phosphorylation of endothelial NO synthase. FEBS Lett 443:285–289

Chen ZP, McConell GK et al (2000) AMPK signaling in contracting human skeletal muscle: acetyl-CoA carboxylase and NO synthase phosphorylation. Am J Physiol Endocrinol Metab 279:E1202–E1206

Chen Z, Peng IC et al (2009) AMP-activated protein kinase functionally phosphorylates endothelial nitric oxide synthase Ser633. Circ Res 104:496–505

Clarke PR, Hardie DG (1990) Regulation of HMG-CoA reductase: identification of the site phosphorylated by the AMP-activated protein kinase in vitro and in intact rat liver. EMBO J 9:2439–2446

Collins QF, Liu HY et al (2007) Epigallocatechin-3-gallate (EGCG), a green tea polyphenol, suppresses hepatic gluconeogenesis through 5′-AMP-activated protein kinase. J Biol Chem 282:30143–30149

Cool B, Zinker B et al (2006) Identification and characterization of a small molecule AMPK activator that treats key components of type 2 diabetes and the metabolic syndrome. Cell Metab 3:403–416

Corton JM, Gillespie JG et al (1995) 5-aminoimidazole-4-carboxamide ribonucleoside. A specific method for activating AMP-activated protein kinase in intact cells? Eur J Biochem 229:558–565

da Silva Xavier G, Leclerc I et al (2003) Role for AMP-activated protein kinase in glucose-stimulated insulin secretion and preproinsulin gene expression. Biochem J 371:761–774

Dasgupta B, Milbrandt J (2007) Resveratrol stimulates AMP kinase activity in neurons. Proc Natl Acad Sci USA 104:7217–7222

Daval M, Diot-Dupuy F et al (2005) Anti-lipolytic action of AMP-activated protein kinase in rodent adipocytes. J Biol Chem 280:25250–25257

Davis BJ, Xie Z et al (2006) Activation of the AMP-activated kinase by antidiabetes drug metformin stimulates nitric oxide synthesis in vivo by promoting the association of heat shock protein 90 and endothelial nitric oxide synthase. Diabetes 55:496–505

DeFronzo RA, Tripathy D (2009) Skeletal muscle insulin resistance is the primary defect in type 2 diabetes. Diabetes Care 32(Suppl 2):S157–S163

Donath MY, Ehses JA et al (2005) Mechanisms of beta-cell death in type 2 diabetes. Diabetes 54(Suppl 2):S108–S113

Dreyer HC, Drummond MJ et al (2008) Resistance exercise increases human skeletal muscle AS160/TBC1D4 phosphorylation in association with enhanced leg glucose uptake during postexercise recovery. J Appl Physiol 105:1967–1974

Dyck JR, Lopaschuk GD (2006) AMPK alterations in cardiac physiology and pathology: enemy or ally? J Physiol 574:95–112

Eckel RH, Grundy SM et al (2005) The metabolic syndrome. Lancet 365:1415–1428

El-Assaad W, Buteau J et al (2003) Saturated fatty acids synergize with elevated glucose to cause pancreatic beta-cell death. Endocrinology 144:4154–4163

Eto K, Yamashita T et al (2002) Genetic manipulations of fatty acid metabolism in beta-cells are associated with dysregulated insulin secretion. Diabetes 51(Suppl 3):S414–S420

Evans AM, Mustard KJ et al (2005) Does AMP-activated protein kinase couple inhibition of mitochondrial oxidative phosphorylation by hypoxia to calcium signaling in O2-sensing cells? J Biol Chem 280:41504–41511

Febbraio MA, Pedersen BK (2005) Contraction-induced myokine production and release: is skeletal muscle an endocrine organ? Exerc Sport Sci Rev 33:114–119

Fisher JS, Gao J et al (2002) Activation of AMP kinase enhances sensitivity of muscle glucose transport to insulin. Am J Physiol Endocrinol Metab 282:E18–E23

Fonseca VA (2009) Defining and characterizing the progression of type 2 diabetes. Diabetes Care 32(Suppl 2):S151–S156

Foretz M, Carling D et al (1998) AMP-activated protein kinase inhibits the glucose-activated expression of fatty acid synthase gene in rat hepatocytes. J Biol Chem 273:14767–14771

Foretz M, Ancellin N et al (2005) Short-term overexpression of a constitutively active form of AMP-activated protein kinase in the liver leads to mild hypoglycemia and fatty liver. Diabetes 54:1331–1339

Fraser SA, Mount PF et al (2003) Inhibition of the Na-K-2Cl cotransporter by novel interaction with the metabolic sensor AMP-activated protein kinase. J Am Soc Nephrol 14:545A

Fryer LG, Parbu-Patel A et al (2002) The anti-diabetic drugs rosiglitazone and metformin stimulate AMP-activated protein kinase through distinct signaling pathways. J Biol Chem 277:25226–25232

Fujii N, Ho RC et al (2008) Ablation of AMP-activated protein kinase alpha2 activity exacerbates insulin resistance induced by high-fat feeding of mice. Diabetes 57:2958–2966

Garton AJ, Yeaman SJ (1990) Identification and role of the basal phosphorylation site on hormone-sensitive lipase. Eur J Biochem 191:245–250

Garton AJ, Campbell DG et al (1989) Phosphorylation of bovine hormone-sensitive lipase by the AMP-activated protein kinase. A possible antilipolytic mechanism. Eur J Biochem 179:249–254

Gledhill JR, Montgomery MG et al (2007) Mechanism of inhibition of bovine F1-ATPase by resveratrol and related polyphenols. Proc Natl Acad Sci USA 104:13632–13637

Glund S, Deshmukh A et al (2007) Interleukin-6 directly increases glucose metabolism in resting human skeletal muscle. Diabetes 56:1630–1637

Glund S, Treebak JT et al (2009) Role of adenosine 5'-monophosphate-activated protein kinase in interleukin-6 release from isolated mouse skeletal muscle. Endocrinology 150:600–606

Hammer S, Snel M et al (2008) Prolonged caloric restriction in obese patients with type 2 diabetes mellitus decreases myocardial triglyceride content and improves myocardial function. J Am Coll Cardiol 52:1006–1012

Hayashi T, Hirshman MF et al (1998) Evidence for 5' AMP-activated protein kinase mediation of the effect of muscle contraction on glucose transport. Diabetes 47:1369–1373

Higa M, Zhou YT et al (1999) Troglitazone prevents mitochondrial alterations, beta cell destruction, and diabetes in obese prediabetic rats. Proc Natl Acad Sci USA 96:11513–11518

Holmes BF, Kurth-Kraczek EJ et al (1999) Chronic activation of 5'-AMP-activated protein kinase increases GLUT-4, hexokinase, and glycogen in muscle. J Appl Physiol 87:1990–1995

Holmes BF, Lang DB et al (2004) AMP kinase is not required for the GLUT4 response to exercise and denervation in skeletal muscle. Am J Physiol Endocrinol Metab 287:E739–E743

Hou X, Xu S et al (2008) SIRT1 regulates hepatocyte lipid metabolism through activating AMP-activated protein kinase. J Biol Chem 283:20015–20026

Howitz KT, Bitterman KJ et al (2003) Small molecule activators of sirtuins extend *Saccharomyces cerevisiae* lifespan. Nature 425:191–196

Iglesias MA, Ye JM et al (2002) AICAR administration causes an apparent enhancement of muscle and liver insulin action in insulin-resistant high-fat-fed rats. Diabetes 51:2886–2894

Jager S, Handschin C et al (2007) AMP-activated protein kinase (AMPK) action in skeletal muscle via direct phosphorylation of PGC-1alpha. Proc Natl Acad Sci USA 104:12017–12022

Jazet IM, Schaart G et al (2008) Loss of 50% of excess weight using a very low energy diet improves insulin-stimulated glucose disposal and skeletal muscle insulin signalling in obese insulin-treated type 2 diabetic patients. Diabetologia 51:309–319

Jorgensen SB, Wojtaszewski JF et al (2005) Effects of alpha-AMPK knockout on exercise-induced gene activation in mouse skeletal muscle. FASEB J 19:1146–1148

Kahn SE, Hull RL et al (2006) Mechanisms linking obesity to insulin resistance and type 2 diabetes. Nature 444:840–846

Kaiser N, Leibowitz G et al (2003) Glucotoxicity and beta-cell failure in type 2 diabetes mellitus. J Pediatr Endocrinol Metab 16:5–22

Kamohara S, Burcelin R et al (1997) Acute stimulation of glucose metabolism in mice by leptin treatment. Nature 389:374–377

Katsuki A, Sumida Y et al (2003) Increased visceral fat and serum levels of triglyceride are associated with insulin resistance in Japanese metabolically obese, normal weight subjects with normal glucose tolerance. Diabetes Care 26:2341–2344

Kefas BA, Cai Y et al (2003a) AMP-activated protein kinase can induce apoptosis of insulin-producing MIN6 cells through stimulation of c-Jun-N-terminal kinase. J Mol Endocrinol 30:151–161

Kefas BA, Heimberg H et al (2003b) AICA-riboside induces apoptosis of pancreatic beta cells through stimulation of AMP-activated protein kinase. Diabetologia 46:250–254

Kelly M, Keller C et al (2004) AMPK activity is diminished in tissues of IL-6 knockout mice: the effect of exercise. Biochem Biophys Res Commun 320:449–454

Kim MS, Park JY et al (2004) Anti-obesity effects of alpha-lipoic acid mediated by suppression of hypothalamic AMP-activated protein kinase. Nat Med 10:727–733

Kim WH, Lee JW et al (2007) AICAR potentiates ROS production induced by chronic high glucose: roles of AMPK in pancreatic beta-cell apoptosis. Cell Signal 19:791–805

Knowler WC, Barrett-Connor E et al (2002) Reduction in the incidence of type 2 diabetes with lifestyle intervention or metformin. N Engl J Med 346:393–403

Koistinen HA, Galuska D et al (2003) 5-amino-imidazole carboxamide riboside increases glucose transport and cell-surface GLUT4 content in skeletal muscle from subjects with type 2 diabetes. Diabetes 52:1066–1072

Kola B, Hubina E et al (2005) Cannabinoids and ghrelin have both central and peripheral metabolic and cardiac effects via AMP-activated protein kinase. J Biol Chem 280:25196–25201

Koo SH, Flechner L et al (2005) The CREB coactivator TORC2 is a key regulator of fasting glucose metabolism. Nature 437:1109–1111

Kubota N, Terauchi Y et al (2002) Disruption of adiponectin causes insulin resistance and neointimal formation. J Biol Chem 277:25863–25866

Kurth-Kraczek EJ, Hirshman MF et al (1999) 5′ AMP-activated protein kinase activation causes GLUT4 translocation in skeletal muscle. Diabetes 48:1667–1671

Lagouge M, Argmann C et al (2006) Resveratrol improves mitochondrial function and protects against metabolic disease by activating SIRT1 and PGC-1alpha. Cell 127:1109–1122

Lan F, Cacicedo JM et al (2008) SIRT1 modulation of the acetylation status, cytosolic localization, and activity of LKB1. Possible role in AMP-activated protein kinase activation. J Biol Chem 283:27628–27635

Larson-Meyer DE, Heilbronn LK et al (2006) Effect of calorie restriction with or without exercise on insulin sensitivity, beta-cell function, fat cell size, and ectopic lipid in overweight subjects. Diabetes Care 29:1337–1344

Lau KS, Grange RW et al (2000) nNOS and eNOS modulate cGMP formation and vascular response in contracting fast-twitch skeletal muscle. Physiol Genomics 2:21–27

Leclerc I, Kahn A et al (1998) The 5′-AMP-activated protein kinase inhibits the transcriptional stimulation by glucose in liver cells, acting through the glucose response complex. FEBS Lett 431:180–184

Leclerc I, Lenzner C et al (2001) Hepatocyte nuclear factor-4alpha involved in type 1 maturity-onset diabetes of the young is a novel target of AMP-activated protein kinase. Diabetes 50:1515–1521

Leclerc I, Woltersdorf WW et al (2004) Metformin, but not leptin, regulates AMP-activated protein kinase in pancreatic islets: impact on glucose-stimulated insulin secretion. Am J Physiol Endocrinol Metab 286:E1023–E1031

Lee WJ, Lee IK et al (2005a) Alpha-lipoic acid prevents endothelial dysfunction in obese rats via activation of AMP-activated protein kinase. Arterioscler Thromb Vasc Biol 25:2488–2494

Lee WJ, Song KH et al (2005b) Alpha-lipoic acid increases insulin sensitivity by activating AMPK in skeletal muscle. Biochem Biophys Res Commun 332:885–891

Lee-Young RS, Griffee SR et al (2009) Skeletal muscle AMP-activated protein kinase is essential for the metabolic response to exercise in vivo. J Biol Chem 284:23925–23934

Lee-Young RS, Ayala JE et al (2010) Endothelial nitric oxide synthase is central to skeletal muscle metabolic regulation and enzymatic signaling during exercise in vivo. Am J Physiol Regul Integr Comp Physiol 298:R1399–R1408

Liao Y, Takashima S et al (2005) Exacerbation of heart failure in adiponectin-deficient mice due to impaired regulation of AMPK and glucose metabolism. Cardiovasc Res 67:705–713

Lihn AS, Jessen N et al (2004) AICAR stimulates adiponectin and inhibits cytokines in adipose tissue. Biochem Biophys Res Commun 316:853–858

Lin HZ, Yang SQ et al (2000) Metformin reverses fatty liver disease in obese, leptin-deficient mice. Nat Med 6:998–1003

Lira VA, Soltow QA et al (2007) Nitric oxide increases GLUT4 expression and regulates AMPK signaling in skeletal muscle. Am J Physiol Endocrinol Metab 293:E1062–E1068

Lochhead PA, Salt IP et al (2000) 5-aminoimidazole-4-carboxamide riboside mimics the effects of insulin on the expression of the 2 key gluconeogenic genes PEPCK and glucose-6-phosphatase. Diabetes 49:896–903

Luo Z, Saha AK et al (2005) AMPK, the metabolic syndrome and cancer. Trends Pharmacol Sci 26:69–76

Lupi R, Del Guerra S et al (2002) Lipotoxicity in human pancreatic islets and the protective effect of metformin. Diabetes 51(Suppl 1):S134–S137

Maeda N, Takahashi M et al (2001) PPARgamma ligands increase expression and plasma concentrations of adiponectin, an adipose-derived protein. Diabetes 50:2094–2099

Majithiya JB, Balaraman R (2006) Metformin reduces blood pressure and restores endothelial function in aorta of streptozotocin-induced diabetic rats. Life Sci 78:2615–2624

Martin TL, Alquier T et al (2006) Diet-induced obesity alters AMP kinase activity in hypothalamus and skeletal muscle. J Biol Chem 281:18933–18941

Matejkova O, Mustard KJ et al (2004) Possible involvement of AMP-activated protein kinase in obesity resistance induced by respiratory uncoupling in white fat. FEBS Lett 569:245–248

Maxwell AJ, Schauble E et al (1998) Limb blood flow during exercise is dependent on nitric oxide. Circulation 98:369–374

Merrill GF, Kurth EJ et al (1997) AICA riboside increases AMP-activated protein kinase, fatty acid oxidation, and glucose uptake in rat muscle. Am J Physiol 273:E1107–E1112

Michael LF, Wu Z et al (2001) Restoration of insulin-sensitive glucose transporter (GLUT4) gene expression in muscle cells by the transcriptional coactivator PGC-1. Proc Natl Acad Sci USA 98:3820–3825

Milne JC, Lambert PD et al (2007) Small molecule activators of SIRT1 as therapeutics for the treatment of type 2 diabetes. Nature 450:712–716

Minokoshi Y, Haque MS et al (1999) Microinjection of leptin into the ventromedial hypothalamus increases glucose uptake in peripheral tissues in rats. Diabetes 48:287–291

Minokoshi Y, Kim YB et al (2002) Leptin stimulates fatty-acid oxidation by activating AMP-activated protein kinase. Nature 415:339–343

Minokoshi Y, Alquier T et al (2004) AMP-kinase regulates food intake by responding to hormonal and nutrient signals in the hypothalamus. Nature 428:569–574

Mootha VK, Lindgren CM et al (2003) PGC-1alpha-responsive genes involved in oxidative phosphorylation are coordinately downregulated in human diabetes. Nat Genet 34:267–273

Moule SK, Denton RM (1998) The activation of p38 MAPK by the beta-adrenergic agonist isoproterenol in rat epididymal fat cells. FEBS Lett 439:287–290

Moynihan KA, Grimm AA et al (2005) Increased dosage of mammalian Sir2 in pancreatic beta cells enhances glucose-stimulated insulin secretion in mice. Cell Metab 2:105–117

Muoio DM, Seefeld K et al (1999) AMP-activated kinase reciprocally regulates triacylglycerol synthesis and fatty acid oxidation in liver and muscle: evidence that sn-glycerol-3-phosphate acyltransferase is a novel target. Biochem J 338(Pt 3):783–791

Murry CE, Jennings RB et al (1986) Preconditioning with ischemia: a delay of lethal cell injury in ischemic myocardium. Circulation 74:1124–1136

Musi N, Fujii N et al (2001) AMP-activated protein kinase (AMPK) is activated in muscle of subjects with type 2 diabetes during exercise. Diabetes 50:921–927

Narkar VA, Downes M et al (2008) AMPK and PPARdelta agonists are exercise mimetics. Cell 134:405–415

Nathan DM, Buse JB et al (2009) Medical management of hyperglycaemia in type 2 diabetes mellitus: a consensus algorithm for the initiation and adjustment of therapy: a consensus statement from the American Diabetes Association and the European Association for the Study of Diabetes. Diabetologia 52:17–30

Nawrocki AR, Rajala MW et al (2006) Mice lacking adiponectin show decreased hepatic insulin sensitivity and reduced responsiveness to peroxisome proliferator-activated receptor gamma agonists. J Biol Chem 281:2654–2660

Neuschwander-Tetri BA, Caldwell SH (2003) Nonalcoholic steatohepatitis: summary of an AASLD Single Topic Conference. Hepatology 37:1202–1219

Nieto-Vazquez I, Fernandez-Veledo S et al (2008) Dual role of interleukin-6 in regulating insulin sensitivity in murine skeletal muscle. Diabetes 57:3211–3221

Nishino Y, Miura T et al (2004) Ischemic preconditioning activates AMPK in a PKC-dependent manner and induces GLUT4 up-regulation in the late phase of cardioprotection. Cardiovasc Res 61:610–619

Nyblom HK, Sargsyan E et al (2008) AMP-activated protein kinase agonist dose dependently improves function and reduces apoptosis in glucotoxic beta-cells without changing triglyceride levels. J Mol Endocrinol 41:187–194

Orci L, Cook WS et al (2004) Rapid transformation of white adipocytes into fat-oxidizing machines. Proc Natl Acad Sci USA 101:2058–2063

Ouchi N, Kihara S et al (2001) Adipocyte-derived plasma protein, adiponectin, suppresses lipid accumulation and class A scavenger receptor expression in human monocyte-derived macrophages. Circulation 103:1057–1063

Pacholec M, Bleasdale JE et al (2010) SRT1720, SRT2183, SRT1460, and resveratrol are not direct activators of SIRT1. J Biol Chem 285:8340–8351

Palanivel R, Sweeney G (2005) Regulation of fatty acid uptake and metabolism in L6 skeletal muscle cells by resistin. FEBS Lett 579:5049–5054

Pan XR, Li GW et al (1997) Effects of diet and exercise in preventing NIDDM in people with impaired glucose tolerance. The Da Qing IGT and Diabetes Study. Diabetes Care 20:537–544

Park H, Kaushik VK et al (2002) Coordinate regulation of malonyl-CoA decarboxylase, sn-glycerol-3-phosphate acyltransferase, and acetyl-CoA carboxylase by AMP-activated protein kinase in rat tissues in response to exercise. J Biol Chem 277:32571–32577

Petersen KF, Befroy D et al (2003) Mitochondrial dysfunction in the elderly: possible role in insulin resistance. Science 300:1140–1142

Phillips SA, Ciaraldi TP et al (2003) Modulation of circulating and adipose tissue adiponectin levels by antidiabetic therapy. Diabetes 52:667–674

Picard F, Kurtev M et al (2004) Sirt1 promotes fat mobilization in white adipocytes by repressing PPAR-gamma. Nature 429:771–776

Pischon T, Girman CJ et al (2004) Plasma adiponectin levels and risk of myocardial infarction in men. JAMA 291:1730–1737

Pold R, Jensen LS et al (2005) Long-term AICAR administration and exercise prevents diabetes in ZDF rats. Diabetes 54:928–934

Prentki M, Joly E et al (2002) Malonyl-CoA signaling, lipid partitioning, and glucolipotoxicity: role in beta-cell adaptation and failure in the etiology of diabetes. Diabetes 51(Suppl 3):S405–S413

Rhodes CJ (2005) Type 2 diabetes-a matter of beta-cell life and death? Science 307:380–384

Riboulet-Chavey A, Diraison F et al (2008) Inhibition of AMP-activated protein kinase protects pancreatic beta-cells from cytokine-mediated apoptosis and CD8+ T-cell-induced cytotoxicity. Diabetes 57:415–423

Richards SK, Parton LE et al (2005) Over-expression of AMP-activated protein kinase impairs pancreatic {beta}-cell function in vivo. J Endocrinol 187:225–235

Roberts CK, Barnard RJ et al (1999) Acute exercise increases nitric oxide synthase activity in skeletal muscle. Am J Physiol 277:E390–E394

Rodgers JT, Puigserver P (2007) Fasting-dependent glucose and lipid metabolic response through hepatic sirtuin 1. Proc Natl Acad Sci USA 104:12861–12866

Rossmeisl M, Barbatelli G et al (2002) Expression of the uncoupling protein 1 from the aP2 gene promoter stimulates mitochondrial biogenesis in unilocular adipocytes in vivo. Eur J Biochem 269:19–28

Rubin LJ, Magliola L et al (2005) Metabolic activation of AMP kinase in vascular smooth muscle. J Appl Physiol 98:296–306

Russell RR 3rd, Li J et al (2004) AMP-activated protein kinase mediates ischemic glucose uptake and prevents postischemic cardiac dysfunction, apoptosis, and injury. J Clin Invest 114:495–503

Saha AK, Avilucea PR et al (2004) Pioglitazone treatment activates AMP-activated protein kinase in rat liver and adipose tissue in vivo. Biochem Biophys Res Commun 314:580–585

Saltiel AR, Kahn CR (2001) Insulin signalling and the regulation of glucose and lipid metabolism. Nature 414:799–806

Sanders MJ, Grondin PO et al (2007) Investigating the mechanism for AMP activation of the AMP-activated protein kinase cascade. Biochem J 403:139–148

Sell H, Dietze-Schroeder D et al (2006) Cytokine secretion by human adipocytes is differentially regulated by adiponectin, AICAR, and troglitazone. Biochem Biophys Res Commun 343:700–706

Shaw RJ, Lamia KA et al (2005) The kinase LKB1 mediates glucose homeostasis in liver and therapeutic effects of metformin. Science 310:1642–1646

Shibata R, Ouchi N et al (2004a) Adiponectin-mediated modulation of hypertrophic signals in the heart. Nat Med 10:1384–1389

Shibata R, Ouchi N et al (2004b) Adiponectin stimulates angiogenesis in response to tissue ischemia through stimulation of amp-activated protein kinase signaling. J Biol Chem 279:28670–28674

Shibata R, Sato K et al (2005) Adiponectin protects against myocardial ischemia-reperfusion injury through AMPK- and COX-2-dependent mechanisms. Nat Med 11:1096–1103

Smith AC, Bruce CR et al (2005) AMP kinase activation with AICAR further increases fatty acid oxidation and blunts triacylglycerol hydrolysis in contracting rat soleus muscle. J Physiol 565:547–553

Song XM, Fiedler M et al (2002) 5-Aminoimidazole-4-carboxamide ribonucleoside treatment improves glucose homeostasis in insulin-resistant diabetic (ob/ob) mice. Diabetologia 45:56–65

Steinberg GR, Kemp BE (2009) AMPK in health and disease. Physiol Rev 89:1025–1078

Steinberg GR, Michell BJ et al (2006) Tumor necrosis factor alpha-induced skeletal muscle insulin resistance involves suppression of AMP-kinase signaling. Cell Metab 4:465–474

Stephens TJ, Chen ZP et al (2002) Progressive increase in human skeletal muscle AMPKalpha2 activity and ACC phosphorylation during exercise. Am J Physiol Endocrinol Metab 282:E688–E694

Stoppani J, Hildebrandt AL et al (2002) AMP-activated protein kinase activates transcription of the UCP3 and HKII genes in rat skeletal muscle. Am J Physiol Endocrinol Metab 283:E1239–E1248

Sullivan JE, Brocklehurst KJ et al (1994) Inhibition of lipolysis and lipogenesis in isolated rat adipocytes with AICAR, a cell-permeable activator of AMP-activated protein kinase. FEBS Lett 353:33–36

Sun C, Zhang F et al (2007) SIRT1 improves insulin sensitivity under insulin-resistant conditions by repressing PTP1B. Cell Metab 6:307–319

Sun G, Tarasov AI et al (2010) Ablation of AMP-activated protein kinase alpha1 and alpha2 from mouse pancreatic beta cells and RIP2.Cre neurons suppresses insulin release in vivo. Diabetologia 53(5):924–936

Suter M, Riek U et al (2006) Dissecting the role of 5'-AMP for allosteric stimulation, activation, and deactivation of AMP-activated protein kinase. J Biol Chem 281:32207–32216

Suzuki A, Okamoto S et al (2007) Leptin stimulates fatty acid oxidation and peroxisome proliferator-activated receptor alpha gene expression in mouse C2C12 myoblasts by changing the subcellular localization of the alpha2 form of AMP-activated protein kinase. Mol Cell Biol 27:4317–4327

Tomas E, Tsao TS et al (2002) Enhanced muscle fat oxidation and glucose transport by ACRP30 globular domain: acetyl-CoA carboxylase inhibition and AMP-activated protein kinase activation. Proc Natl Acad Sci USA 99:16309–16313

Treebak JT, Birk JB et al (2007) AS160 phosphorylation is associated with activation of alpha2-beta2gamma1- but not alpha2beta2gamma3-AMPK trimeric complex in skeletal muscle during exercise in humans. Am J Physiol Endocrinol Metab 292:E715–E722

Tuomilehto J, Lindstrom J et al (2001) Prevention of type 2 diabetes mellitus by changes in lifestyle among subjects with impaired glucose tolerance. N Engl J Med 344:1343–1350

U.K. Prospective Diabetes Study Group (1995) U.K. prospective diabetes study 16: Overview of 6 years' therapy of type II diabetes: a progressive disease. Diabetes 44:1249–1258

Um JH, Park SJ et al (2010) AMP-activated protein kinase-deficient mice are resistant to the metabolic effects of resveratrol. Diabetes 59:554–563

Unger RH (1995) Lipotoxicity in the pathogenesis of obesity-dependent NIDDM. Genetic and clinical implications. Diabetes 44:863–870

Van Gaal LF, Mertens IL et al (2006) Mechanisms linking obesity with cardiovascular disease. Nature 444:875–880

Viana AY, Sakoda H et al (2006) Role of hepatic AMPK activation in glucose metabolism and dexamethasone-induced regulation of AMPK expression. Diabetes Res Clin Pract 73:135–142

Viollet B, Andreelli F et al (2003) Physiological role of AMP-activated protein kinase (AMPK): insights from knockout mouse models. Biochem Soc Trans 31:216–219

Wang X, Zhou L et al (2007) Troglitazone acutely activates AMP-activated protein kinase and inhibits insulin secretion from beta cells. Life Sci 81:160–165

Watt MJ, Holmes AG et al (2006) Regulation of HSL serine phosphorylation in skeletal muscle and adipose tissue. Am J Physiol Endocrinol Metab 290:E500–E508

Weiss EP, Racette SB et al (2006) Improvements in glucose tolerance and insulin action induced by increasing energy expenditure or decreasing energy intake: a randomized controlled trial. Am J Clin Nutr 84:1033–1042

Willett WC, Dietz WH et al (1999) Guidelines for healthy weight. N Engl J Med 341:427–434

Winder WW, Hardie DG (1999) AMP-activated protein kinase, a metabolic master switch: possible roles in type 2 diabetes. Am J Physiol 277:E1–E10

Winder WW, Holmes BF et al (2000) Activation of AMP-activated protein kinase increases mitochondrial enzymes in skeletal muscle. J Appl Physiol 88:2219–2226

Wing RR, Goldstein MG et al (2001) Behavioral science research in diabetes: lifestyle changes related to obesity, eating behavior, and physical activity. Diabetes Care 24:117–123

Witters LA, Gao G et al (1994) Hepatic 5'-AMP-activated protein kinase: zonal distribution and relationship to acetyl-CoA carboxylase activity in varying nutritional states. Arch Biochem Biophys 308:413–419

Woods A, Azzout-Marniche D et al (2000) Characterization of the role of AMP-activated protein kinase in the regulation of glucose-activated gene expression using constitutively active and dominant negative forms of the kinase. Mol Cell Biol 20:6704–6711

Xu A, Wang Y et al (2003) The fat-derived hormone adiponectin alleviates alcoholic and nonalcoholic fatty liver diseases in mice. J Clin Invest 112:91–100

Yamauchi T, Kamon J et al (2001) The fat-derived hormone adiponectin reverses insulin resistance associated with both lipoatrophy and obesity. Nat Med 7:941–946

Yamauchi T, Kamon J et al (2002) Adiponectin stimulates glucose utilization and fatty-acid oxidation by activating AMP-activated protein kinase. Nat Med 8:1288–1295

Yamauchi T, Kamon J et al (2003) Globular adiponectin protected ob/ob mice from diabetes and ApoE-deficient mice from atherosclerosis. J Biol Chem 278:2461–2468

Zang M, Xu S et al (2006) Polyphenols stimulate AMP-activated protein kinase, lower lipids, and inhibit accelerated atherosclerosis in diabetic LDL receptor-deficient mice. Diabetes 55: 2180–2191

Zarrinpashneh E, Carjaval K et al (2006) Role of the alpha2 isoform of AMP-activated protein kinase in the metabolic response of the heart to no-flow ischemia. Am J Physiol Heart Circ Physiol 291(6):H2875–H2883

Zhang J, Xie Z et al (2008) Identification of nitric oxide as an endogenous activator of the AMP-activated protein kinase in vascular endothelial cells. J Biol Chem 283:27452–27461

Zhou M, Lin BZ et al (2000) UCP-3 expression in skeletal muscle: effects of exercise, hypoxia, and AMP-activated protein kinase. Am J Physiol Endocrinol Metab 279:E622–E629

Zhou G, Myers R et al (2001) Role of AMP-activated protein kinase in mechanism of metformin action. J Clin Invest 108:1167–1174

Zhou L, Wang X et al (2008) Berberine acutely inhibits insulin secretion from beta-cells through 3′, 5′-cyclic adenosine 5′-monophosphate signaling pathway. Endocrinology 149:4510–4518

Zong H, Ren JM et al (2002) AMP kinase is required for mitochondrial biogenesis in skeletal muscle in response to chronic energy deprivation. Proc Natl Acad Sci USA 99:15983–15987

Zwetsloot KA, Westerkamp LM et al (2008) AMPK regulates basal skeletal muscle capillarization and VEGF expression, but is not necessary for the angiogenic response to exercise. J Physiol 586:6021–6035

Mitochondria as Potential Targets in Antidiabetic Therapy

Paula I. Moreira and Catarina R. Oliveira

Contents

Abstract A growing body of evidence suggests that mitochondrial abnormalities are involved in diabetes and associated complications. This chapter gives an overview about the effects of diabetes in mitochondrial function of several tissues including the pancreas, skeletal and cardiac muscle, liver, and brain. The realization that mitochondria are at the intersection of cells' life and death has made them a promising target for drug discovery and therapeutic interventions. Here, we also discuss literature that examined the potential protective effect of insulin, insulin-sensitizing drugs, and mitochondrial-targeted antioxidants.

Keywords Cardiac muscle · Diabetes · Insulin · Insulin-sensitizing drugs · Liver · Mitochondria · Mitochondria-targeted antioxidants · Nervous tissue · Pancreas · Skeletal muscle

P.I. Moreira (✉) and C.R. Oliveira (✉)
Faculty of Medicine and Center for Neuroscience and Cell Biology, University of Coimbra, 3004-517, Coimbra, Portugal
e-mail: pismoreira@gmail.com, catarina.n.oliveira@gmail.com

M. Schwanstecher (ed.), *Diabetes - Perspectives in Drug Therapy*,
Handbook of Experimental Pharmacology 203,
DOI 10.1007/978-3-642-17214-4_14, © Springer-Verlag Berlin Heidelberg 2011

1   Introduction

Diabetes mellitus is a complex metabolic disorder characterized by hyperglycemia and is a major cause of morbidity and mortality affecting nearly 8% of the population in the world. Type 1 diabetes mellitus (T1DM) affects 5–10% of diabetic patients and results from the specific destruction of pancreatic β cells by the immune system culminating in hypoinsulinemic and hyperglycemic states. Type 2 diabetes mellitus (T2DM) is the most common form of diabetes and affects 80–95% of diabetic patients. Worldwide, approximately 200 million people currently have T2DM, a prevalence that has been predicted to increase to 366 million by 2030 (Kalofoutis et al. 2007). T2DM is complex in etiology and is characterized by a relative insulin deficiency, reduced insulin action and insulin resistance of glucose transport, especially in skeletal muscle and adipose tissue. T2DM is a polygenic disease that results from a complex interplay between genetic predisposition and environmental factors such as diet, degree of physical activity and age (Ross et al. 2004). Diabetic metabolic changes result in macrovascular complications leading to accelerated atherosclerosis and coronary heart or peripheral arterial disease, or both, and in microvascular complications leading to retinopathy, nephropathy, and neuropathy (Hudson et al. 2005).

Several clinical prospective trials designed for investigating whether diabetes complications are related to glycemia regulation clearly established that hyperglycemia causes vascular and tissue damage in patients with diabetes (The Diabetes Control and Complications Trial Research Group 1993; UK Prospective Diabetes Study (UKPDS) Group 1998), which can be worsened by genetic factors of individual susceptibility and by associated pathologic factors, such as hypertension or hyperhomocysteinemia. It is generally admitted that repeated acute changes in blood glucose and cellular glucidic metabolism, as well as cumulative long-term alterations of cellular and extracellular constituents, represent the mechanisms that mediate the damaging effects of hyperglycemia. A strict glycemic control may reduce in part the incidence of microvascular complications but is less effective in preventing the progression of macrovascular diseases that are largely associated with hypertension (Fowler et al. 2008). The damaging effects of hyperglycemia affect mainly certain cell types that are unable to maintain their intracellular glucose concentration in hyperglycemic conditions (endothelial cells in the vascular system, mesangial cells in the kidney, neurons and neuroglia in the nervous system, and pancreatic β-cells) interfering with cell constituents namely mitochondria.

Both T1DM and T2DM, as well as other types of diabetes, are associated with similar long-term complications that, at least in part, appear to result from pathogenic processes at the mitochondrial level. However, mitochondrial abnormalities associated with diabetes are tissue-specific. Bugger et al. (2009) compared the effect of diabetes in several tissues obtained from the type 1 diabetic akita mice. The authors observed that the fatty acid oxidation (FAO) proteins were less abundant in liver mitochondria, whereas FAO protein content was induced in mitochondria from all other tissues. Kidney mitochondria showed coordinate induction of tricarboxylic acid (TCA) cycle enzymes, whereas TCA cycle proteins were repressed in cardiac mitochondria. In addition, the levels of oxidative phosphorylation system (OXPHOS) subunits were coordinately increased in liver mitochondria, whereas mitochondria of other tissues were unaffected. Mitochondrial respiration, ATP synthesis, and morphology were unaffected in liver and kidney mitochondria. In contrast, state 3 respiration, ATP synthesis, and mitochondrial cristae density were decreased in cardiac mitochondria and accompanied by coordinate repression of OXPHOS and peroxisome proliferator-activated receptor (PPAR)-γ coactivator (PGC)-1α transcripts (Bugger et al. 2009). We also observed that diabetes promoted a significant decrease in kidney and brain mitochondrial coenzyme Q9 (CoQ9) content, while an increase of CoQ9 was observed in heart mitochondria (Moreira et al. 2006). Furthermore, diabetes induced a significant increase in hydrogen peroxide (H_2O_2) production in kidney mitochondria, this effect being accompanied by a significant increase in glutathione peroxidase (GPx) and reductase (GR) activities. In addition, brain mitochondria presented a lower ATP content and ability to accumulate Ca^{2+}. In contrast, heart and kidney mitochondria presented a slight higher capacity to accumulate this cation (Moreira et al. 2006). These studies clearly show that diabetes impacts mitochondria differently from different tissues.

Here we address the close association between mitochondrial alterations and *diabetes*, putting emphasis on T2DM, in different tissues. The potential protective role of insulin, insulin-sensitizing drugs, and mitochondrial target antioxidants will be also debated.

2 Mitochondrial Abnormalities in Diabetes

Mitochondria are increasingly recognized as subcellular organelles that are essential for generating the energy that fuels normal cellular function while, at the same time, they monitor cellular health in order to make a rapid decision (if necessary) to initiate programmed cell death (Fig. 1). As such, the mitochondria sit at a strategic position in the hierarchy of cellular organelles to continue the healthy life of the cell or to terminate it. Mitochondria are involved in the generation of cellular ATP via OXPHOS. However, OXPHOS is a major source of endogenous toxic free radicals, including H_2O_2, hydroxyl (HO•), and superoxide (O_2^-•) radicals that are products of normal cellular respiration. With inhibition of electron transport chain (ETC), electrons accumulate in complex I and coenzyme Q, where they can be donated directly to molecular oxygen (O_2) to give O_2^-• that can be detoxified by the mitochondrial manganese superoxide dismutase (MnSOD) to give H_2O_2 that, in turn, can be converted to H_2O by glutathione peroxidase (GPx). However, O_2^-• in the presence of nitric oxide (NO•), formed during the conversion of arginine to citrulline by nitric oxide synthase (NOS), can lead to peroxynitrite ($ONOO^-$). H_2O_2 in the presence of reduced transition metals can be converted to toxic HO• via Fenton and/or Haber Weiss reactions. Several efficient enzymatic processes are

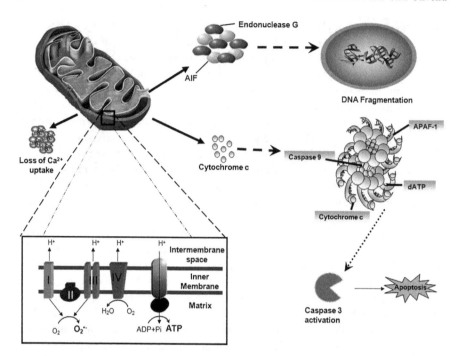

Fig. 1 *Mitochondrial (dys)function.* Besides the fundamental role of mitochondria in the generation of energy (ATP), these organelles are also the main producers of oxygen free radicals. If the
defense mechanisms are debilitated, these reactive species initiate a cascade of deleterious events
within the cell. Mitochondrial abnormalities associated with enhanced oxidative stress have long
been recognized to play a major role in degenerative disorders like diabetes. Indeed, under adverse
conditions, mitochondria suffer profound alterations that lead to a reduced generation of ATP and
an enhanced production of reactive oxygen species. Mitochondria also lose the Ca^{2+} buffering
capacity, which can initiate a cascade of deleterious events within the cell. Impaired mitochondria
also release several proapoptotic factors upon induction of apoptosis. These factors are either
directly triggering apoptosis by associating with cytosolic factors to form the apoptosome. Finally,
some mitochondrial, proapoptotic proteins translocate into the nucleus to induce DNA fragmentation. Altogether these mitochondrial alterations contribute to cell degeneration and death. *AIF*
apoptosis inducing factor; *APAF-1* protease-activating factor 1; *dATP* 2'-deoxyadenosine 5'-
triphosphate; *ROS* reactive oxygen species

continuously operational to quench the reactive species including SOD, GPx,
superoxide reductase, catalase, peroxiredoxin, and thioredoxin/thioredoxin reductase. Inevitably, if the amount of free radical species produced overwhelms the cell
capacity to neutralize them, oxidative stress occurs, followed by mitochondrial
dysfunction and cell damage. Reactive species generated by mitochondria have
several cellular targets including mitochondrial components themselves. The lack of
histones in mitochondrial DNA (mtDNA) and diminished capacity for DNA repair
render mitochondria an easy target to oxidative stress events (Moreira et al. 2010).

Mitochondria also serve as high capacity Ca^{2+} sinks, which allows them to stay
in tune with changes in cytosolic Ca^{2+} loads and aid in maintaining cellular Ca^{2+}

homeostasis that is required for normal cellular function. Conversely, excessive Ca^{2+} uptake into mitochondria has been shown to increase ROS production, inhibit ATP synthesis, induce mitochondrial permeability transition pore (PTP), and release small proteins that trigger the initiation of apoptosis, such as cytochrome c and apoptosis-inducing factor (AIF), from the mitochondrial intermembrane space into the cytoplasm. Released cytochrome c binds apoptotic protease activating factor 1 (Apaf-1) and activates the caspase cascade (Fig. 1) (Moreira et al. 2010). Such alterations in mitochondrial function have been proposed to be involved in the pathogenesis of several disorders including diabetes.

The next subsections are devoted to discuss diabetes-associated mitochondrial abnormalities in pancreas, skeletal and cardiac muscle, liver and nervous tissue.

2.1 Pancreas

The role of deficient β-cell mass in T2DM has been debated for decades (Ahrén 2005). At the moment, the strongest data available suggest that there is, indeed, a loss of β-cell mass in T2DM due to increased apoptosis. While this is most likely a fact, it cannot by itself account for the perturbation of insulin secretion in the disease. β-Cell loss in diabetic obese and lean individuals at most reaches $\approx$60% and 40%, respectively, compared with obese and lean nondiabetic individuals (Butler et al. 2003). Reasonably, if the observed loss of β-cell mass plays a role in the pathogenesis of T2DM, it does so in combination with a functional defect.

It is widely agreed that insulin secretion is largely controlled by metabolism of fuels, foremost glucose, in the pancreatic β-cell (Muoio and Newgard 2008). Other factors, such as circulating hormones, paracrine and autocrine mechanisms and neuronal control, all combine to modulate insulin secretion elicited by *β-cell* metabolism (Ahrén 2000). In this context, mitochondria play a key role. Both rapid (first phase) and more prolonged (second phase) insulin secretion (O'Connor et al. 1980) are dependent on glucose metabolism and mitochondrial oxidative capacity; glucose oxidation increases the ATP/ADP ratio, inhibiting plasma membrane ATP-sensitive K^+ channels and allowing voltage-gated Ca^{2+} channels to open. Increased cytoplasmic Ca^{2+} then triggers exocytosis of plasma-membrane docked insulin granules (first phase). Subsequent recruitment of granules to the plasma membrane (second phase) appears to depend on mitochondrial metabolites produced by anaplerosis (Hasan et al. 2008). Mitochondrial metabolism is also required for the transient, controlled production of ROS, which is required for the mitochondrial signaling pathways that trigger granule exocytosis (Evans et al. 2003; Leloup et al. 2009).

Mitochondria also play a critical role in the control of β-cell mass. Data suggest that increased apoptosis underlies the loss of β cells observed in islets from patients with T2DM (Butler et al. 2003; Trifunovic and Larsson 2008). Indeed, the apoptotic pathways converge in mitochondria, where caspase 3 activation and cytochrome c

release seem to precipitate these events. Furthermore, aging potentiates mitochondrial abnormalities that occur in T2DM (Lee et al. 2007; Trifunovic and Larsson 2008).

A study performed on islets obtained from 14 cadaveric donors with T2DM showed that the isolated islets are smaller and contain a reduced proportion of β-cells (Deng et al. 2004). Insulin secretion in response to glucose is impaired with respect to threshold, rate, and total amount released. In contrast, the maximal response, as elicited by KCl, is unchanged. These results suggest that diabetic islets exhibit a specific metabolic impairment, since the secretory dysfunction is restricted to glucose. Anello et al. (2005) examined insulin secretion and mitochondrial function in islets from seven T2DM donors. The authors confirmed that glucose-stimulated, but not arginine-stimulated, insulin secretion is impaired in those individuals supporting the idea that in T2DM the β-cells are characterized by metabolic impairments. While ATP levels in the basal state are elevated in diabetic islets, they fail to trigger insulin secretion. Moreover, glucose is less effective to hyperpolarize the mitochondrial membrane in the diabetic islets. The authors explain this finding with an increase in the mitochondrial uncoupling protein (UCP) 2 levels. This protein is thought to uncouple the respiratory chain from ATP production by allowing protons to flow back into the mitochondrial matrix without giving rise to ATP. Interestingly, these functional alterations are paralleled by changes in mitochondrial structure. Also expression of complexes I and IV is increased in the islets from patients with T2DM (Anello et al. 2005).

There are also several studies performed with diabetic animals showing that mitochondrial alterations are intimately associated with β-cells dysfunction. Using an antibody specific for toxic oligomers and cryo-immunogold labeling in human islet amyloid polypeptide (IAPP) transgenic mice, human insulinoma and pancreas from humans with and without T2DM, Gurlo et al. (2010) tried to establish the abundance and sites of formation of IAPP toxic oligomers. The authors concluded that IAPP toxic oligomers are formed intracellularly within the secretory pathway in T2DM. Most striking, IAPP toxic oligomers appeared to disrupt membranes of the secretory pathway and then, when adjacent to mitochondria, disrupted mitochondrial membranes (Gurlo et al. 2010). Toxic oligomer-induced secretory pathway and mitochondrial membrane disruption is a novel mechanism to account for cellular dysfunction and apoptosis in T2DM. Lu et al. (2010) investigated the changes in islet mitochondrial function and morphology during progression from insulin resistance (3 weeks old), immediately before hyperglycemia (5 weeks old), and after diabetes onset (10 weeks old) in transgenic MKR mice, a model of T2DM, compared with controls. At 3 weeks, MKR mice were hyperinsulinemic but normoglycemic and β-cells showed negligible mitochondrial or morphological changes. At 5 weeks, MKR islets displayed abrogated hyperpolarization of mitochondrial membrane potential ($\Delta\Psi m$), reduced mitochondrial Ca^{2+} uptake, slightly enlarged mitochondria, and reduced glucose-stimulated insulin secretion. By 10 weeks, MKR mice were hyperglycemic and hyperinsulinemic and β-cells contained swollen mitochondria with disordered cristae. β-Cells displayed impaired stimulus–secretion coupling including reduced hyperpolarization of $\Delta\Psi m$, impaired Ca^{2+}-signaling, and reduced glucose-stimulated ATP/ADP and insulin release. Furthermore, decreased complex IV-dependent O_2 consumption and signs of oxidative stress were observed in diabetic islets. Protein profiling of diabetic islets revealed that 36 mitochondrial proteins were differentially expressed, including inner membrane proteins of the ETC (Lu et al. 2010). This study provides novel evidence for a critical role of defective mitochondrial oxidative phosphorylation and morphology in the pathology of insulin resistance-induced β-cell failure.

2.2 Skeletal Muscle

Mitochondria are particularly important for skeletal muscle function, given the high oxidative demands imposed on this tissue by intermittent contraction. Mitochondria play a critical role in ensuring adequate levels of ATP needed for contraction by the muscle sarcomere. Skeletal muscle is the largest insulin-sensitive organ in humans, accounting for more than 80% of insulin-stimulated glucose disposal. Thus, insulin resistance in this tissue has a major impact on whole-body glucose homeostasis. In skeletal muscle, disruption of mitochondrial biology is evident in some insulin-resistant subjects, years before they develop diabetes (Befroy et al. 2007; Patti et al. 2003; Petersen et al. 2004). Furthermore, the impairment of skeletal muscle mitochondria is recognized as a hallmark of established T2DM (Lowell and Shulman 2005). However, whether perturbations in these organelles are central to the pathophysiology of insulin resistance in the skeletal muscle is robustly debated (Boushel et al. 2007; Holloszy 2009; Kraegen et al. 2008). Moreover, in light of the lack of macrovascular benefit of exclusively targeting glucose control in T2DM (Gerstein et al. 2008; Patel et al. 2008), and the strong association of elevated insulin levels with T2DM complications, it is crucial to better understand the pathophysiology of insulin resistance.

A panoply of animal and human studies show the occurrence of skeletal mitochondrial abnormalities related to diabetes and its complications. Due to space limitations and abundance of studies performed in human skeletal muscle, here we only discuss human data. To study whether perturbations in skeletal muscle mitochondrial biology are evident in subjects at risk for diabetes, investigators have studied offspring of subjects with T2DM. Lean offspring of diabetic patients have been shown to have a significant reduction in skeletal muscle ATP synthesis in response to insulin stimulation as measured by magnetic resonance spectroscopy saturation transfer. These measurements reflected impaired baseline activity of mitochondrial OXPHOS (Petersen et al. 2004, 2005). Insulin-resistant offspring of diabetic parents also showed a reduction in skeletal muscle density and content (Morino et al. 2005). The lean offspring of diabetic parents in these studies may represent a narrowly selected cohort of individuals that do not necessarily reflect the early pathophysiology of T2DM in this heterozygous disease process. In addition, the measurement of basal oxidative phosphorylation flux did not reflect maximal oxidative capacity of skeletal muscle, and whether baseline perturbations are sufficient to initiate the development of skeletal muscle insulin resistance has been questioned (Holloszy 2009; Kemp 2008).

Interestingly, overweight subjects with maintained insulin sensitivity exhibit increased lipid oxidation and the maintenance of normal myocellular lipid content (Perseghin et al. 2002). Conversely, nondiabetic extremely obese subjects often do exhibit insulin resistance (Thyfault et al. 2004) that is linked, in part, to the accumulation of fatty acid esters in the skeletal muscle (Cooney et al. 2002) and the capacity of the skeletal muscle to oxidize fatty acid substrates is significantly blunted (Kim et al. 2000; Thyfault et al. 2004).

The study of skeletal muscle from diabetic individuals shows the coordinate down-regulation of genes encoding OXPHOS enzymes (Mootha et al. 2003), lower levels of the β-subunit of ATP synthase protein (Højlund et al. 2003), decreased mitochondrial respiration (Mogensen et al. 2007), and evidence of reduced bioenergetic capacity, as illustrated by slower postexercise recovery of skeletal muscle high-energy phosphate stores compared with nondiabetic controls (Scheuermann-Freestone and Clarke 2003). These later data have been confirmed, comparing overweight diabetic with nondiabetic subjects where in vivo skeletal muscle phosphocreatine recovery half-life after exercise is blunted in the diabetic subjects (Phielix et al. 2008; Schrauwen-Hinderling et al. 2007). Other studies show that subjects with insulin resistance or T2DM have diminished type I oxidative muscle fiber content (Lillioja et al. 1987) and an increased skeletal muscle glycolytic to oxidative phosphorylation enzyme ratio (Simoneau and Kelley 1997). It was also shown that the total activity of NADH-oxidase in biopsy obtained from lean individuals is significantly higher than corresponding activity for obese or T2DM individuals. The specific activity of NADH-oxidase and NADH-oxidase/citrate synthase and NADH-oxidase/β-hydroxyacyl-CoA dehydrogenase ratios are reduced by two- to threefolds in both T2DM and obesity (Ritov et al. 2009). The analysis of direct respirometry to measure mitochondrial O_2 consumption in the skeletal muscle aligns with this concept in that diabetic individuals have diminished complexes I and II substrate-driven oxidative capacity in gastrocnemius muscle compared with controls when normalized to muscle mass; however, when normalized to mitochondrial genomic content, these respirometry differences were abolished (Boushel et al. 2007). These results suggest that the function of mitochondria does not differ in diabetes but that the overall muscle content of mitochondria is diminished. Recently, Rabøl et al. (2010) compared mitochondrial respiration and markers of mitochondrial content in the skeletal muscle of arm and leg in patients with T2DM and obese control subjects. The authors observed that in the arm, mitochondrial respiration and citrate synthase activity did not differ between groups, but mitochondrial respiration per milligram of muscle was significantly higher in the leg muscle of the control subjects compared to T2DM. Fiber type compositions in arm and leg muscles were not different between the T2DM and control group, and maximum rate of O_2 consumption did not differ between the groups (Rabøl et al. 2010). These results demonstrate that reduced mitochondrial function in T2DM is only present in the leg musculature suggesting that mitochondrial dysfunction is not a primary defect affecting all skeletal muscle but could be related to a decreased response to locomotor muscle use in T2DM. Despite some disparate results, the majority of the studies demonstrate that diabetes, as well as other insulin-resistant conditions, are intimately associated with skeletal muscle mitochondria impairment.

2.3   Cardiac Muscle

The heart is voracious in its appetite for energy, to such extent that it generates and consumes a mass of ATP daily that surpasses cardiac mass itself by approximately 5- to 10-fold (Opie 2004). This demand for energy reflects the continuous contractile functioning of the heart to sustain systemic circulation and nutrient supply. This high energy flux translates into the cardiomyocyte having a mitochondrial volume between 23% and 32% of myocellular volume (Sack 2009).

The literature shows that diabetes directly affects the function of cardiac mitochondria. Boudina et al. (2007) examined the function of heart mitochondria in saponin-permeabilized heart muscle fibers isolated from insulin-resistant, diabetic, leptin receptor-deficient *db/db* mice compared with lean controls. These investigators reported decreased respiration on complex I substrates and palmitoyl-carnitine, associated with proportionately reduced ATP production and therefore no change in ADP/O index. These investigators also reported a decreased content of the F1 α-subunit of ATP synthase and an increase in fatty acid-induced proton conductance based on proton-leak kinetics (Boudina et al. 2007). These results indicate a reduced cardiac muscle function in *db/db* mice. However, other studies also performed in *db/db* mice show that these rodents exhibited a diminished cardiac glucose oxidation and increased mitochondrial FAO and myocardial O_2 consumption (Tabbi-Anneni et al. 2008). These perturbations resulted in decreased mitochondrial efficiency (uncoupling) and a reduced capacity to respond to increased cardiac workload (Boudina et al. 2005). It was recently shown that the transcription factor, nuclear factor kappa-B (NF-kB)-induced oxidative stress contributes to mitochondrial and cardiac dysfunction in *db/db* mice (Mariappan et al. 2010).

Oliveira et al. (2003) reported that streptozotocin (STZ)-induced diabetes, a model of T1DM, facilitates the PTP in cardiac mitochondria, resulting in decreased mitochondrial Ca^{2+} accumulation. The authors also observed that cardiac mitochondria from diabetic rats had depressed O_2 consumption during the phosphorylative state. The authors suggested that the reduced mitochondrial Ca^{2+} uptake observed in heart mitochondria from diabetic rats is related to an enhanced susceptibility to PTP rather than to damage to the Ca^{2+} uptake machinery. It was also shown that type 1 diabetic cardiomyopathy in the Akita mouse model is characterized by lipotoxicity and diastolic dysfunction with preserved systolic function (Basu et al. 2009).

Clinical data are being accumulated that diabetes does impair cardiac function, independent of other risk factors (Fox et al. 2007; Galderisi et al. 1991; Ishihara et al. 2001). It has been shown using proton magnetic resonance spectroscopy (^{1}H-MRS) that subjects with diabetes or glucose intolerance have increased cardiac steatosis (McGavock et al. 2007), which may indirectly reflect diminished mitochondrial metabolic capacity. The measurement of high-energy phosphates in diabetic and nondiabetic subjects without clinical coronary artery disease and normal echocardiographic studies indirectly suggest that diabetic subjects have diminished cardiac energetics (Scheuermann-Freestone and Clarke 2003). Hopefully, future studies in human cardiac tissue will clarify the interaction between diabetes and mitochondrial function.

2.4 Liver

The liver plays a central, unique role in carbohydrate, protein, and fat metabolism. It is critical for maintaining glucose homeostasis (1) during fuel availability, via storage of glucose as glycogen or conversion to lipid export and storage in adipose tissue and (2) in the fasting state, via catabolism of glycogen, synthesis of glucose from noncarbohydrate sources such as amino acids (gluconeogenesis) and ketogenesis. In turn, these responses are regulated by the key hormones insulin and glucagon, which modulate signaling pathways and gene expression, leading to inhibition or stimulation of glucose production, respectively.

Abnormalities of liver mitochondria could potentially affect multiple cellular functions within hepatocytes, both directly such as reduced ATP generation, alterations in oxidative stress, reduced capacity for FAO, and indirectly via effects on energy-requiring processes, including gluconeogenesis, synthesis of urea, bile acids, cholesterol and proteins, and detoxification (Patti and Corvera 2010).

Recently, Gnoni et al. (2010) reported that liver mitochondria from STZ-induced diabetic rats presented reduced expression and activity of citrate carrier. It was also shown that ZDF rats, an obese model of T2DM, have a defect in the mitochondrial metabolism (Satapati et al. 2008).

Although human liver studies have been limited due to lack of tissue biopsy samples from otherwise healthy individuals, two groups have examined hepatic gene expression related to mitochondrial function in both obesity and T2DM (Misu et al. 2007; Pihlajamäki et al. 2009; Takamura et al. 2008). In the first study (Misu et al. 2007), severe obesity (mean BMI 52 kg/m^2) was associated with reduced expression of 7 of the 25 genes encoding proteins of OXPHOS and the expression of these genes was inversely correlated with hepatic lipid accumulation and paralleled by reduced expression of PGC-1α and genes known to be regulated by thyroid hormone. Similar patterns were observed in obese subjects with established T2DM. Interestingly, reduced expression of OXPHOS genes was also observed in mice fed a high-fat diet and normalized by acute therapy with thyroid hormone T3 suggesting that functional hepatic thyroid hormone resistance could contribute to reduced expression of mitochondrial oxidative genes in this context (Pihlajamäki et al.

2009). In contrast, studies in Japanese individuals with established diabetes and modest obesity (BMI 27 kg/m^2) reported a modest increase in the expression of multiple genes of OXPHOS complexes, in parallel with BMI and insulin resistance (Takamura et al. 2008). The up-regulation of these OXPHOS genes was also positively associated with expression of several genes associated with mitochondrial biogenesis, ROS generation, and antioxidant defenses. Thus, increased ROS related to increased FAO and/or hyperglycemia might contribute to the up-regulation of OXPHOS gene expression in coexisting obesity and T2DM. Studies of individuals with nonalcoholic steatohepatitis (NASH) provide additional opportunities to identify potential interactions between hepatic lipid accumulation, insulin resistance, and mitochondrial function in humans. Indeed, enzymatic activities of complexes I–IV were reduced in liver extracts from patients with NASH and were inversely correlated with BMI and insulin resistance (Greenfield et al. 2008; Pérez-Carreras et al. 2003). Moreover, NASH is characterized by prominent abnormalities in mitochondrial ultrastructure, with increased size, loss of cristae, and paracrystalline inclusion bodies similar to those observed in some mitochondrial myopathies (Sanyal et al. 2001). Although these data cannot address whether such changes are indeed pathogenic, it is interesting that reduced OXPHOS activity in this setting is accompanied by increased tissue long-chain acylcarnitines and reduced short-chain acylcarnitines, despite normal carnitine palmitoyltransferase-1 (CPT-1) activity and β-oxidation genes (Kohjima et al. 2007; Misu et al. 2007). Similarly, circulating β-hydroxybutyrate levels are increased in NASH (Sanyal et al. 2001). Together, these data suggest excessive, but incomplete, FAO is potentially limited by reduced availability to NAD^+ and FAD. In summary, available data indicate that hepatic lipid accumulation and insulin resistance are intimately linked with mitochondrial dysfunction.

2.5 Nervous Tissue

Neuronal cells have a high demand for energy to maintain ion gradients across the plasma membrane that is critical for the generation of action potentials. This intense energy requirement is continuous; even brief periods of oxygen or glucose deprivation result in neuronal death. Mitochondria are essential for neuronal function because the limited glycolytic capacity of these cells makes them highly dependent on OXPHOS for their energetic needs.

Previous studies from our laboratory show that brain mitochondria isolated from STZ diabetic rats possess a lower content of CoQ9 indicating a deficit in antioxidant defenses in diabetic animals and, consequently, an increased probability for the occurrence of oxidative stress (Moreira et al. 2005b). Schmeichel et al. (2003) suggested that oxidative stress leads to oxidative injury of dorsal root ganglion (DRG) neurons, mitochondria being a specific target. Leinninger et al. (2006) reported that mitochondria in DRG neurons undergo hyperglycemia-mediated injury through the proapoptotic proteins Bim and Bax and the fission protein dynamin-regulated protein 1 (DRP1). It was also shown that the fission-mediated fragmentation of mitochondria is associated with enhanced production of mito-chondrial ROS and cell injury in hyperglycemic conditions (Yu et al. 2008). The role of mitochondrial fission in the pathogenesis of diabetic neuropathy was recently addressed (Edwards et al. 2010). The authors observed a greater mitochon-drial biogenesis in DRG neurons from diabetic compared with nondiabetic mice. An essential step in mitochondrial biogenesis is mitochondrial fission, regulated by the mitochondrial fission protein DRP1. Evaluation of diabetic neurons in vivo indicated small, fragmented mitochondria suggesting an increased fission. In vitro studies revealed that short-term hyperglycemic exposure increased levels of DRP1 protein. The knockdown of the DRP1 gene decreased the susceptibility to hyper-glycemic damage (Edwards et al. 2010).

Increasing data support the idea that mitochondrial function declines with aging and in age-related diseases. We have previously reported an age-related impairment of the respiratory chain and an uncoupling of OXPHOS in brain mitochondria isolated from GK rats, a model of T2DM (Moreira et al. 2003). Furthermore, we also showed that aging exacerbates the decrease in the energetic levels promoted by diabetes (Moreira et al. 2003). The maintenance of OXPHOS capacity is extremely important in the brain since about 90% of the ATP required for the normal functioning of neurons is provided by mitochondria. Because CNS depends so heavily on ATP production, the inhibition of OXPHOS will affect this system before any other system. For example, CNS requires a large amount of ATP for the transmission of impulses along the neural pathway, thus mitochondrial function impairment may contribute to the loss of neuronal metabolic control and, conse-quently, to neurodegeneration.

Brain mitochondria of GK rats also presented an age-related susceptibility to Ca^{2+}, indicating that aging predisposes the diabetic rats' mitochondria to the opening of PTP. The PTP opening might be also associated with osmotic swelling of mitochondria leading to structural changes of these organelles. Indeed, in peripheral nerves of diabetic humans, the existence of mitochondrial ballooning and disruption of internal cristae is observed, although this is localized to Schwann cells and is rarely observed in axons (Kalichman et al. 1998). Similar structural abnormalities in mitochondria have been described in Schwann cells of galactose-fed rats (Kalichman et al. 1998) and DRG neurons of long-term STZ diabetic rats (Sasaki et al. 1997). One hypothesis is that high glucose concentrations potentiate OXPHOS that may result in damaging amounts of ROS that lead to changes in mitochondrial structure and function (Nishikawa et al. 2000).

We also observed that diabetes-related mitochondrial dysfunction is exacerbated by the presence of Aβ, a neurotoxic peptide that is intimately involved with Alzheimer's disease (AD) pathophysiology (Moreira et al. 2003, 2005a, b). In recent years, it has been demonstrated that diabetes is a risk factor for several neurodegenerative disorders including AD, Parkinson's (PD) and Huntington's (HD) diseases, and amyotrophic lateral sclerosis (ALS) (Cardoso et al. 2009a; Moreira et al. 2007a, 2009).

3 Mitochondria as Potential Targets

Although mitochondria are one major source of ROS, they are also specific targets of oxidative damage. The accumulation of oxidative damage could contribute to mitochondrial dysfunction and cell death in a range of degenerative diseases, including diabetes. The knowledge that mitochondrial dysfunction has a preponderant role in several diseases opened a window for new therapeutic strategies aimed to preserve/ameliorate mitochondrial function.

3.1 Insulin and Insulin-Sensitizing Drugs

Insulin/insulin growth factor 1 (IGF-1) signaling pathway is involved in the balance of the physiological processes that control aging, development, growth, reproduction, metabolism, and resistance to oxidative stress (Cardoso et al. 2009a, b). In contrast, the inhibition of this signaling pathway reduces cell survival by promoting oxidative stress, mitochondrial dysfunction, and prodeath signaling cascade activation (Cardoso et al. 2009b).

It was previously shown that aged rats present a decrease in the mitochondrial $\Delta\Psi m$ and ATPase activity and an increase in mitochondrial oxidative damage (Puche et al. 2008). However, aged rats treated with IGF-1 presented an improved mitochondrial function associated with an increased ATP production and reduced free radical generation, oxidative damage, and apoptosis (Puche et al. 2008). It was also shown that IGF-1 protects from hyperglycemia-induced oxidative stress and neuronal injury by regulating $\Delta\Psi m$, possibly by the involvement of UCP3 (Gustafsson et al. 2004). Exposure to high glucose dose- and time-dependently induced apoptotic changes (DNA fragmentation, altered $\Delta\Psi m$, and cytochrome c release) in human umbilical vein endothelial cells (HUVECs) (Li et al. 2009b). Addition of IGF-1 blocked the high glucose effects in a manner dependent on the expression of IGF-1 receptor (IGF-1R) since silencing of this receptor with small interference RNA could diminish the antiapoptosis effects of IGF-1 (Li et al. 2009b). Other studies demonstrated that stimulation of different cell types with IGF-1 or insulin leads to Akt translocation to mitochondria and GSK-3β phosphorylation (Bijur and Jope 2003), supporting a direct action of IGF-1 and/or insulin in mitochondria. Boirie et al. (2001) reported that insulin selectively stimulates mitochondrial protein synthesis in skeletal muscle and activates mitochondrial enzyme activity. Kuo et al. (2009) reported that insulin replacement not only prevents activation of the cardiac mitochondrial-dependent apoptotic and calcineurin-nuclear factor activation transcription 3 (NFAT3) hypertrophic pathways, promoted by STZ-induced diabetes, but it also enhances the cardiac insulin/IGF-1R-phosphatidylinositol 3′-kinase (PI3K)-Akt survival pathway. Previous studies from our laboratory show that insulin treatment attenuates diabetes-induced mitochondrial alterations by improving the oxidative phosphorylation efficiency and protecting against the increase in oxidative stress (Moreira et al. 2005a, 2006). In the reperfused brain, insulin regulates cytochrome c release through PI3K/Akt activation, promoting the binding between Bax and Bcl-xl, and preventing Bax translocation to the mitochondria (Sanderson et al. 2008). A recent study reported that mitochondrial respiratory chain dysfunction in DRG of STZ-induced diabetic rats is improved by insulin (Roy Chowdhury et al. 2010).

Insulin-sensitizing thiazolidinediones (TZDs) are generally considered to work as agonists for PPARγ. However, Bolten et al. (2007) reported that diabetic mice treated for 1 week with TZDs presented an increased expression of an array of mitochondrial proteins and PGC1α, the master regulator of mitochondrial biogenesis. Thus, the pharmacology of the insulin-sensitizing TZDs may involve acute actions on the mitochondria that are independent of direct activation of the nuclear receptor PPARγ. These findings suggest a potential alternative route to the discovery of novel insulin-sensitizing drugs. A previous study demonstrated that pioglitazone and ciglitazone, two TZDs, attenuated hyperglycemia-induced ROS production and increased the expression of nuclear respiratory factor 1 (NRF-1), mitochondrial transcription factor A (TFAM), and MnSOD mRNA in HUVECs (Fujisawa et al. 2009). Moreover, pioglitazone also increased mtDNA and mitochondrial density (Fujisawa et al. 2009). These results suggest that TZDs normalize hyperglycemia-induced mitochondrial ROS production by induction of MnSOD and promotion of mitochondrial biogenesis by activating PGC-1α. It was also shown that pioglitazone induced mitochondrial biogenesis in subcutaneous adipose tissue (Bogacka et al. 2005) as well as in a neuron-like cell line associated with a reduced mitochondrial oxidative stress (Ghosh et al. 2007). A recent randomized, double-blind, parallel study shows that pioglitazone increased plasma adiponectin levels, stimulated muscle AMPK signaling and increased the expression of genes involved in adiponectin signaling, mitochondrial function, and FAO in skeletal muscle (Coletta et al. 2009). Fuenzalida et al. (2007) reported that neuronal cells treated with rosiglitazone up-regulated Bcl-2, thereby stabilizing mitochondrial potential and protecting against apoptosis. Similar results were obtained by Wu et al. (2009) who demonstrated that rosiglitazone protected cells against oxygen–glucose deprivation (OGD)-induced cytotoxicity and apoptosis by suppressing H_2O_2 production, maintaining $\Delta\Psi m$, attenuating cytochrome c release and inhibiting activation of caspases 3 and 9. Moreover, OGD caused a significant suppression of Bcl-2 and Bcl-xl proteins levels that were restored by pretreatment with rosiglitazone (Wu et al. 2009).

It has also been shown that PPARγ activation protected rat hippocampal neurons against Aβ toxicity (Combs et al. 2000; Inestrosa et al. 2005), induced up-regulation of Bcl-2 pathway, protected mitochondrial function, and prevented neuronal degeneration induced by Aβ exposure and oxidative stress (Fuenzalida et al. 2007). Indeed, rosiglitazone beneficial effects in memory and cognition seem to be mediated by the improvement of mitochondrial function, since it leads to an increase in mitochondria number and metabolic efficiency (Kummer and Heneka 2008). Furthermore, brain mitochondrial biogenesis induced by rosiglitazone (Strum et al. 2007) could be possibly due to PGC-1α since this PPARγ coactivator regulates mitochondrial function and metabolism (Handschin and Spiegelman

2006). Furthermore, it was reported that rosiglitazone protects human neuroblastoma cells against acetaldehyde, an inhibitor of mitochondrial function (Jung et al. 2006). This protection was mediated by the induction of antioxidant enzymes and increased expression of Bcl-2 and Bax (Jung et al. 2006). Recently, the same authors demonstrated that rosiglitazone protects SH-SY5Y cells against 1-methyl-4-phenylpyridinium (MPP^+)-induced cytotoxicity, an experimental model of PD, by preventing mitochondrial dysfunction and oxidative stress (Jung et al. 2007). These results suggest that PPARγ agonists provide neuroprotection by regulating mitochondrial antioxidant enzymes expression and maintaining the balance between pro- and antiapoptotic gene expression. Furthermore, PPARγ agonists are known to regulate the expression of UCP2 (Hunter et al. 2008), mitochondrial proteins that attenuate mitochondrial ROS production and limit ROS-induced cellular damage. Recently, Quintanilla et al. (2008) reported that cells expressing mutant huntingtin (htt), an experimental model of HD, presented significant defects in the PPARγ signaling pathway in comparison with cells expressing wild-type htt protein. In addition, the authors observed that the pretreatment with rosiglitazone prevented the loss of $\Delta\Psi$, mitochondrial Ca^{2+} deregulation and oxidative stress (Quintanilla et al. 2008). Evidence shows that PGC-1α is a strong suppressor of ROS production and induces the expression of ROS scavenging enzymes (St-Pierre et al. 2006). Moreover, it has been reported that mutant htt can affect mitochondrial function through the inhibition of PGC-1α expression (Cui et al. 2006; Weydt et al. 2006).

In summary, evidence shows that insulin and insulin-sensitizing agents can be useful in the treatment of diabetes and other degenerative diseases, mitochondria being an important target in these pathological conditions.

3.2   Metabolic Antioxidants

Metabolic antioxidants are involved in cellular energy production and act as cofactors of several metabolic enzymes (Fig. 2).

Lipoic acid (LA) is a coenzyme for mitochondrial pyruvate and α-ketoglutarate dehydrogenase. Furthermore, it is a powerful antioxidant and can recycle other antioxidants, such as vitamins C and E and glutathione. Recently, Balkis Budin et al. (2009) reported that LA prevents the alteration of vascular morphology in diabetic rats probably through the improvement of glycemic status, dyslipidemia, and antioxidant enzymatic activities. It was also shown that LA effectively attenuates mitochondria-dependent cardiac apoptosis and exerts a protective role against the development of diabetic cardiomyopathy (Li et al. 2009a). The ability of LA to suppress mitochondrial oxidative damage is concomitant with an enhancement of MnSOD activity and an increase in the glutathione (GSH) content of myocardial mitochondria (Li et al. 2009a). It was also shown that LA and acetyl-L-carnitine (ALCAR; a compound that acts as an intracellular carrier of acetyl groups across the inner mitochondrial membrane) complementarily promote mitochondrial biogenesis in murine 3T3-L1 adipocytes (Shen et al. 2008b). Furthermore, we have

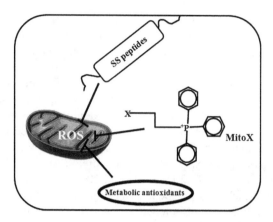

Fig. 2 *Mitochondrial-directed therapies.* Metabolic antioxidants are involved in cellular energy-production and act as cofactors of several metabolic enzymes. In addition, they have also potent antioxidant actions avoiding damage of lipids, proteins, and DNA. Mitochondria-targeted anti-oxidants and SS peptides are selectively accumulated into mitochondria, a major source of reactive oxygen species (ROS) protecting mitochondrial and cellular components against oxidative damage

recently shown that LA and/or *N*-acetyl cysteine (NAC; an antioxidant and gluta-thione precursor) decrease mitochondrial-related oxidative stress in AD patient fibroblasts (Moreira et al. 2007b). Suh and collaborators (2004) reported that old rats injected with LA presented an improvement in glutathione redox status of both cerebral and myocardial tissues when compared with control rats. In turn, Aliev et al. (2009) reported that LA and ALCAR supplementation significantly reduced the number of severely damaged mitochondria and increased the number of intact mitochondria in the hippocampus of aged rats.

The treatment of diabetic GK rats with a combination of LA, ALCAR, nicotin-amide, and biotin effectively improved glucose tolerance, decreased the basal insulin secretion and the levels of circulating free fatty acids (FFA), and prevented the reduction of mitochondrial biogenesis in skeletal muscle (Shen et al. 2008a). This treatment also significantly increased mRNA levels of genes involved in lipid metabolism, including PPARα, PPARδ, and CPT-1 and the activity of mitochon-drial complexes I and II in skeletal muscle. All of these effects of mitochondrial nutrients were comparable to that of pioglitazone (Shen et al. 2008a).

CoQ (also known as ubiquinone) is a respiratory chain component that accepts electrons from complexes I or II, to form the reduced product ubiquinol, which donates electrons to mitochondrial complex III. The in vivo ubiquinone pool exists largely in a reduced ubiquinol form, acting as an antioxidant and a mobile electron transfer. Ubiquinol has been reported to function as an antioxidant by donating a hydrogen atom from one of its hydroxyl groups to a lipid peroxyl radical, thereby decreasing lipid peroxidation within the mitochondrial inner membrane (Ernster et al. 1992). Recently, Sena et al. (2008) reported that CoQ10 and vitamin E decrease glycated hemoglobin HbA1c and pancreatic lipid peroxidation. In addition, it has been reported that CoQ10 prevents high glucose-induced oxidative stress in HUVECs (Tsuneki et al. 2007). It has also been shown that CoQ10 treatment significantly improved deranged carbohydrate and lipid metabolism of experimental chemically induced diabetes in rats (Modi et al. 2006). Previous studies from our laboratory also show that CoQ10 treatment attenuated the decrease in oxidative phosphorylation efficiency and avoided the increase in H_2O_2 production induced by the neurotoxic peptide Aβ40 (Moreira et al. 2005b). Altogether these results suggest that the delivery of antioxidants, that protect mitochondria and avoid oxidative stress-related events, could be a promising therapeutic approach for the treatment of diabetes and associated complications.

3.3   Mitochondria-Targeted Antioxidants and SS Peptides

A major limitation in using antioxidant therapies to treat diabetes and other degenerative diseases has been the inability to enhance the antioxidant levels within mitochondria. However, in the last years, considerable progress has been made in developing mitochondria-targeted antioxidants (i.e., antioxidants that are selectively accumulated into mitochondria) (Fig. 2). Several mitochondria-targeted antioxidants have been developed by conjugating the lipophilic triphenylphosphonium (TPP^+) cation to an antioxidant moiety, such as CoQ (MitoQ) and vitamin E (MitoVitE) (Fig. 2) (Murphy and Smith 2000). This approach makes use of the potential gradient across the mitochondrial inner membrane. As a result of the proton gradient, a negative potential from 150 to 180 mV is generated across the mitochondrial inner membrane. Lipophilic cations may therefore accumulate 100- to 1,000-fold in mitochondria. However, to function as therapies, mitochondria-targeted antioxidants must be delivered to mitochondria within cells in patients, preferably following oral administration. As TPP^+ cations pass easily through phospholipid bilayers, they should be able to pass from the gut to the bloodstream and from there to most tissues. Indeed, previous studies show that TPP^+-derived compounds are orally bioavailable to mice; this was shown by feeding mice tritiated methylTPP (MitoQ and MitoVitE) in their drinking water, which led to uptake into the plasma and from there into the heart, brain, liver, kidney, and muscle (Smith et al. 2003). The methylTPP was shown to be cleared from all organs at a similar rate by a first-order process with a half-life of about 1.5 days (Smith et al. 2003). These findings are consistent with the distribution of orally administered alkyl TPP compounds to all organs because of their facile permeation through biological membranes.

Data from the literature show that MitoVitE is taken up by mitochondria ≈80-fold more than vitamin E (Smith et al. 1999). The authors observed that MitoVitE is far more effective to protect mitochondria against oxidative stress than vitamin E itself. Furthermore, it has been shown that MitoVitE is 800-fold more potent than idebenone in protecting against GSH depletion in cultured fibroblasts from patients with Friedreich ataxia (a disease characterized by slowly progressive neurodegeneration and cardiomyopathy) and is 350-fold more potent than trolox (Jauslin et al. 2003). Recently, it has been shown that MitoVitE mitigates ethanol-induced accumulation of intracellular oxidants and counteracts suppression of glutathione peroxidase/glutathione reductase functions, protein expression of γ-glutamylcysteine synthetase, and total cellular glutathione levels in cerebellar granule cells (Siler-Marsiglio et al. 2005).

MitoQ is a promising therapeutic antioxidant that has been successfully targeted to mitochondria. MitoQ excessively accumulates in the mitochondria and reduces toxic insults from free radicals in the mitochondria. This effect ultimately leads to the protection of cells from age- and/or disease-related mitochondrial insults. Recently, the effects of MitoQ on mitochondria in several in vitro cell models were tested (Bedogni et al. 2003; Dhanasekaran et al. 2004; Hwang et al. 2001; Jauslin et al. 2003). In cultured fibroblasts from Friedreich ataxia patients, MitoQ prevented cell death known to be caused by endogenous oxidative stress (Jauslin et al. 2003). Low concentrations of MitoQ selectively inhibited serum deprivation-induced apoptosis in PC12 cells (Bedogni et al. 2003). MitoQ reduces ROS formation and preserves mitochondrial function after glutathione depletion, even in the cells lacking mtDNA (Lu et al. 2008). These studies suggest that MitoQ may reduce free radicals, decrease oxidative damage, and maintain mitochondrial function. Since oxidative damage is intimately involved with the pathophysiology of AD, there is strong interest in determining whether mitochondria-targeted antioxidants decrease oxidative damage in the neurons of AD patients (Reddy 2006). In phase I trials, MitoQ showed good pharmacokinetic behavior with oral dosing at 80 mg (1 mg/kg), resulting in a plasma $C_{max}=33.15$ ng/ml and $T_{max}=1$ h. This formulation is now in phase II clinical trial for PD and Friedreich ataxia (Antipodean Pharmaceuticals Inc., San Francisco, CA, USA).

There is a novel class of small peptide antioxidants that target mitochondria in a membrane potential-independent manner; i.e., these peptides do not use the negative potential generated across the mitochondrial inner membrane to accumulate into the mitochondria (Fig. 2). The structural motif of these SS peptides centers on alternating aromatic residues and basic amino acids (Szeto 2006). These peptides are easy to synthesize, readily soluble in water and resistant to peptidase degradation. Despite carrying 3+ net charge at physiological pH, these peptides have been shown to readily penetrate cell membranes of a variety of cell types (Zhao et al. 2003). Their cellular uptake appears to be concentration-dependent, nonsaturable, and not requiring energy (Zhao et al. 2003). SS-31 has a remarkable potency that can be explained by its extensive cellular uptake and selective partitioning into mitochondria. Intracellular concentrations of [^{3}H]SS-31 were sixfold higher than extracellular concentrations. Studies using isolated mitochondria revealed that [^{3}H] SS-31 was concentrated 5,000-fold in the mitochondrial pellet. By concentrating in the inner mitochondrial membrane, SS-31 became localized to the site of ROS production, and protected against mitochondrial oxidative damage and against further ROS production (Zhao et al. 2004). The SS peptides that possess a tyrosine residue, such as SS-31, can dose dependently scavenge H_2O_2, hydroxyl radical, and peroxynitrite (Szeto 2008; Zhao et al. 2004). By scavenging hydroxyl radical, SS-31 also inhibits lipid peroxidation (Zhao et al. 2004). SS-20, which does not contain a tyrosine residue, lacks scavenging activity but can still reduce H_2O_2 production (Szeto 2008) suggesting that SS-20 may reduce mitochondrial ROS production via another mechanism.

SS-31 protects neuronal cells against tert-butyl-hydroperoxide-induced mitochondrial depolarization and apoptotic cell death by reducing intracellular ROS, decreasing markers of apoptotic cell death, and caspase activity (Zhao et al. 2005). It decreases mitochondrial ROS production and inhibits PTP induction and mitochondrial depolarization in isolated mitochondria (Zhao et al. 2004). It has been shown that daily injections of SS-31 to G93A SOD1 mutants, an animal model of ALS, before onset of symptoms lead to a significant increase in survival and improvement of motor performance (Petri et al. 2006). Recently, Yang et al. (2009) examined the ability of SS-31 and SS-20, to protect against 1-methyl-4-phenyl-1,2,3,6-tetrahydropyride (MPTP) neurotoxicity in mice, an animal model of PD. SS-31 produced dose-dependent complete protection against loss of dopamine and its metabolites in striatum, as well as loss of tyrosine hydroxylase immunoreactive neurons in substantia nigra. SS-20 also demonstrated significant neuroprotective effects on dopaminergic neurons of MPTP-treated mice. Both SS-31 and SS-20 were very potent in preventing MPP^+-induced cell death in cultured dopamine cells (Yang et al. 2009). Studies with isolated mitochondria showed that both SS-31 and SS-20 prevented MPP^+-induced inhibition of O_2 consumption and ATP production and mitochondrial swelling (Yang et al. 2009). Recently, Thomas et al. (2007) reported that SS-31 readily penetrates intact mouse islets, preserves mitochondrial polarization, reduces islet cell apoptosis, and increases islet cell yield. These findings provide strong evidence that SS peptides, which target both mitochondrial dysfunction and oxidative damage, are a promising approach for the treatment of degenerative disorders including diabetes.

4 Conclusions

Mitochondria play an important role in controlling the life and death of a cell. Consequently, mitochondrial dysfunction leads to a range of degenerative disorders. Therefore, the development of approaches to avoid or decrease mitochondrial dysfunction may have therapeutic potential. Insulin, insulin-sensitizing drugs, metabolic antioxidants and mitochondria-directed antioxidants, and SS peptides proved to be effective in preclinical and small clinical studies. However, larger clinical trials with larger numbers of participants will provide more definitive information on the therapeutic efficacy of these compounds. In the future, mitochondria-directed antioxidants will be expected to open new avenues for the manipulation of mitochondrial function allowing protection against degenerative diseases including diabetes.

References

Ahrén B (2000) Autonomic regulation of islet hormone secretion–implications for health and disease. Diabetologia 43:393–410

Ahrén B (2005) Type 2 diabetes, insulin secretion and beta-cell mass. Curr Mol Med 5:275–286

Aliev G, Liu J, Shenk JC et al (2009) Neuronal mitochondrial amelioration by feeding acetyl-L-carnitine and lipoic acid to aged rats. J Cell Mol Med 13:320–333

Anello M, Lupi R, Spampinato D et al (2005) Functional and morphological alterations of mitochondria in pancreatic beta cells from type 2 diabetic patients. Diabetologia 48:282–289

Balkis Budin S, Othman F, Louis SR et al (2009) Effect of alpha lipoic acid on oxidative stress and vascular wall of diabetic rats. Rom J Morphol Embryol 50:23–30

Basu R, Oudit GY, Wang X et al (2009) Type 1 diabetic cardiomyopathy in the Akita (Ins2WT/C96Y) mouse model is characterized by lipotoxicity and diastolic dysfunction with preserved systolic function. Am J Physiol Heart Circ Physiol 297:H2096–H2108

Bedogni B, Pani G, Colavitti R et al (2003) Redox regulation of cAMPresponsive element-binding protein and induction of manganous superoxide dismutase in nerve growth factor-dependent cell survival. J Biol Chem 278:16510–16519

Befroy DE, Petersen KF, Dufour S et al (2007) Impaired mitochondrial substrate oxidation in muscle of insulin-resistant offspring of type 2 diabetic patients. Diabetes 56:1376–1381

Bijur GN, Jope RS (2003) Rapid accumulation of Akt in mitochondria following phosphatidylinositol 3-kinase activation. J Neurochem 87:1427–1435

Bogacka I, Xie H, Bray GA, Smith SR (2005) Pioglitazone induces mitochondrial biogenesis in human subcutaneous adipose tissue in vivo. Diabetes 54:1392–1399

Boirie Y, Short KR, Ahlman B et al (2001) Tissue-specific regulation of mitochondrial and cytoplasmic protein synthesis rates by insulin. Diabetes 50:2652–2658

Bolten CW, Blanner PM, McDonald WG et al (2007) Insulin sensitizing pharmacology of thiazolidinediones correlates with mitochondrial gene expression rather than activation of PPARgamma. Gene Regul Syst Bio 1:73–82

Boudina S, Sena S, O'Neill BT et al (2005) Reduced mitochondrial oxidative capacity and increased mitochondrial uncoupling impair myocardial energetics in obesity. Circulation 112:2686–2695

Boudina S, Sena S, Theobald H et al (2007) Mitochondrial energetics in the heart in obesity-related diabetes: direct evidence for increased uncoupled respiration and activation of uncoupling proteins. Diabetes 56:2457–2466

Boushel R, Gnaiger E, Schjerling P et al (2007) Patients with type 2 diabetes have normal mitochondrial function in skeletal muscle. Diabetologia 50:790–796

Bugger H, Chen D, Riehle C et al (2009) Tissue-specific remodeling of the mitochondrial proteome in type 1 diabetic akita mice. Diabetes 58:1986–1997

Butler AE, Janson J, Bonner-Weir S et al (2003) Beta-cell deficit and increased beta-cell apoptosis in humans with type 2 diabetes. Diabetes 52:102–110

Cardoso S, Correia S, Santos RX et al (2009a) Insulin is a two-edged knife on the brain. J Alzheimers Dis 18:483–507

Cardoso S, Santos R, Correia S et al (2009b) Insulin and insulin-sensitizing drugs in neurodegeneration: mitochondria as therapeutic targets. Pharmaceuticals 2:250–286

Coletta DK, Sriwijitkamol A, Wajcberg E et al (2009) Pioglitazone stimulates AMP-activated protein kinase signalling and increases the expression of genes involved in adiponectin signalling, mitochondrial function and fat oxidation in human skeletal muscle in vivo: a randomised trial. Diabetologia 52:723–732

Combs CK, Johnson DE, Karlo JC et al (2000) Inflammatory mechanisms in Alzheimer's disease: inhibition of beta-amyloid-stimulated proinflammatory responses and neurotoxicity by PPARgamma agonists. J Neurosci 20:558–567

Cooney GJ, Thompson AL, Furler SM et al (2002) Muscle long-chain acyl CoA esters and insulin resistance. Ann NY Acad Sci 967:196–207

Cui L, Jeong H, Borovecki F et al (2006) Transcriptional repression of PGC-1alpha by mutant huntingtin leads to mitochondrial dysfunction and neurodegeneration. Cell 127:59–69

Deng S, Vatamaniuk M, Huang X et al (2004) Structural and functional abnormalities in the islets isolated from type 2 diabetic subjects. Diabetes 53:624–632

Dhanasekaran A, Kotamraju S, Kalivendi SV et al (2004) Supplementation of endothelial cells with mitochondria-targeted antioxidants inhibit peroxide-induced mitochondrial iron uptake, oxidative damage, and apoptosis. J Biol Chem 279:37575–37587

Edwards JL, Quattrini A, Lentz SI et al (2010) Diabetes regulates mitochondrial biogenesis and fission in mouse neurons. Diabetologia 53:160–169

Ernster L, Forsmark P, Nordenbrand K (1992) The mode of action of lipid-soluble antioxidants in biological membranes: relationship between the effects of ubiquinol and vitamin E as inhibitors of lipid peroxidation in submitochondrial particles. Biofactors 3:241–248

Evans JL, Goldfine ID, Maddux BA, Grodsky GM (2003) Are oxidative stress-activated signaling pathways mediators of insulin resistance and beta-cell dysfunction? Diabetes 52:1–8

Fowler SP, Williams K, Resendez RG et al (2008) Fueling the obesity epidemic? Artificially sweetened beverage use and long-term weight gain. Obesity (Silver Spring) 16:1894–1900

Fox CS, Coady S, Sorlie PD et al (2007) Increasing cardiovascular disease burden due to diabetes mellitus: the Framingham Heart Study. Circulation 115:1544–1550

Fuenzalida K, Quintanilla R, Ramos P et al (2007) Peroxisome proliferator-activated receptor γ up-regulates the Bcl-2 anti-apoptotic protein in neurons and induces mitochondrial stabilization and protection against oxidative stress and apoptosis. J Biol Chem 282:37006–37015

Fujisawa K, Nishikawa T, Kukidome D et al (2009) TZDs reduce mitochondrial ROS production and enhance mitochondrial biogenesis. Biochem Biophys Res Commun 379:43–48

Galderisi M, Anderson KM, Wilson PW, Levy D (1991) Echocardiographic evidence for the existence of a distinct diabetic cardiomyopathy (the Framingham Heart Study). Am J Cardiol 68:85–89

Gerstein HC, Miller ME, Byington RP et al (2008) Effects of intensive glucose lowering in type 2 diabetes. N Engl J Med 358:2545–2559

Ghosh S, Patel N, Rahn D et al (2007) The thiazolidinedione pioglitazone alters mitochondrial function in human neuron-like cells. Mol Pharmacol 71:1695–1702

Gnoni GV, Giudetti AM, Mercuri E et al (2010) Reduced activity and expression of mitochondrial citrate carrier in streptozotocin-induced diabetic rats. Endocrinology 151:1551–1559

Greenfield V, Cheung O, Sanyal AJ (2008) Recent advances in nonalcholic fatty liver disease. Curr Opin Gastroenterol 24:320–327

Gurlo T, Ryazantsev S, Huang CJ et al (2010) Evidence for proteotoxicity in beta cells in type 2 diabetes: toxic islet amyloid polypeptide oligomers form intracellularly in the secretory pathway. Am J Pathol 176:861–869

Gustafsson H, Söderdahl T, Jönsson G et al (2004) Insulin-like growth factor type 1 prevents hyperglycemia-induced uncoupling protein 3 down-regulation and oxidative stress. J Neurosci Res 77:285–291

Handschin C, Spiegelman BM (2006) Peroxisome proliferator-activated receptor gamma coactivator 1 coactivators, energy homeostasis, and metabolism. Endocr Rev 27:728–735

Hasan NM, Longacre MJ, Stoker SW et al (2008) Impaired anaplerosis and insulin secretion in insulinoma cells caused by small interfering RNA-mediated suppression of pyruvate carboxylase. J Biol Chem 283:28048–28059

Højlund K, Wrzesinski K, Larsen PM et al (2003) Proteome analysis reveals phosphorylation of ATP synthase beta -subunit in human skeletal muscle and proteins with potential roles in type 2 diabetes. J Biol Chem 278:10436–10442

Holloszy JO (2009) Skeletal muscle "mitochondrial deficiency" does not mediate insulin resistance. Am J Clin Nutr 89:463S–466S

Hudson BI, Wendt T, Bucciarelli LG et al (2005) Diabetic vascular disease: it's all the RAGE. Antioxid Redox Signal 7:1588–1600

Hunter RL, Choi DY, Ross SA, Bing G (2008) Protective properties afforded by pioglitazone against intrastriatal LPS in Sprague-Dawley rats. Neurosci Lett 432:198–201

Hwang PM, Bunz F, Yu J et al (2001) Ferredoxin reductase affects p53-dependent, 5-fluorouracil-induced apoptosis in colorectal cancer cells. Nat Med 7:1111–1117

Inestrosa NC, Godoy JA, Quintanilla RA et al (2005) Peroxisome proliferator-activated receptor gamma is expressed in hippocampal neurons and its activation prevents beta-amyloid neuro-degeneration: role of Wnt signaling. Exp Cell Res 304:91–104

Ishihara M, Inoue I, Kawagoe T et al (2001) Diabetes mellitus prevents ischemic preconditioning in patients with a first acute anterior wall myocardial infarction. J Am Coll Cardiol 38:1007–1011

Jauslin ML, Meier T, Smith RA, Murphy MP (2003) Mitochondria targeted antioxidants protect Friedreich Ataxia fibroblasts from endogenous oxidative stress more effectively than untargeted antioxidants. FASEB J 17:1972–1974

Jung TW, Lee JY, Shim WS et al (2006) Rosiglitazone protects human neuroblastoma SH-SY5Y cells against acetaldehyde-induced cytotoxicity. Biochem Biophys Res Commun 340:221–227

Jung TW, Lee JY, Shim WS et al (2007) Rosiglitazone protects human neuroblastoma SH-SY5Y cells against MPP+ induced cytotoxicity via inhibition of mitochondrial dysfunction and ROS production. J Neurol Sci 253:53–60

Kalichman MW, Powell HC, Mizisin AP (1998) Reactive, degenerative, and proliferative Schwann cell responses in experimental galactose and human diabetic neuropathy. Acta Neuropathol 95:47–56

Kalofoutis C, Piperi C, Kalofoutis A et al (2007) Type II diabetes mellitus and cardiovascular risk factors: current therapeutic approaches. Exp Clin Cardiol 12:17–28

Kemp GJ (2008) The interpretation of abnormal 31P magnetic resonance saturation transfer measurements of Pi/ATP exchange in insulin-resistant skeletal muscle. Am J Physiol Endocrinol Metab 294:E640–622

Kim JY, Hickner RC, Cortright RL et al (2000) Lipid oxidation is reduced in obese human skeletal muscle. Am J Physiol Endocrinol Metab 279:E1039–E1044

Kohjima M, Enjoji M, Higuchi N et al (2007) Re-evaluation of fatty acid metabolism-related gene expression in nonalcoholic fatty liver disease. Int J Mol Med 20:351–358

Kraegen EW, Cooney GJ, Turner N (2008) Muscle insulin resistance: a case of fat overconsumption, not mitochondrial dysfunction. Proc Natl Acad Sci USA 105:7627–7628

Kummer MP, Heneka MT (2008) PPARs in Alzheimer's disease. PPAR Res 2008:403896

Kuo WW, Chung LC, Liu CT et al (2009) Effects of insulin replacement on cardiac apoptotic and survival pathways in streptozotocin-induced diabetic rats. Cell Biochem Funct 27:479–487

Lee S, Jeong SY, Lim WC et al (2007) Mitochondrial fission and fusion mediators, hFis1 and OPA1, modulate cellular senescence. J Biol Chem 282:22977–22983

Leinninger GM, Backus C, Sastry AM et al (2006) Mitochondria in DRG neurons undergo hyperglycemic mediated injury through Bim, Bax and the fission protein Drp1. Neurobiol Dis 23:11–22

Leloup C, Tourrel-Cuzin C, Magnan C et al (2009) Mitochondrial reactive oxygen species are obligatory signals for glucose-induced insulin secretion. Diabetes 58:673–681

Li CJ, Zhang QM, Li MZ et al (2009a) Attenuation of myocardial apoptosis by alpha-lipoic acid through suppression of mitochondrial oxidative stress to reduce diabetic cardiomyopathy. Chin Med J 122:2580–2586

Li Y, Wu H, Khardori R et al (2009b) Insulin-like growth factor-1 receptor activation prevents high glucose-induced mitochondrial dysfunction, cytochrome-c release and apoptosis. Biochem Biophys Res Commun 384:259–264

Lillioja S, Young AA, Culter CL et al (1987) Skeletal muscle capillary density and fiber type are possible determinants of in vivo insulin resistance in man. J Clin Invest 80:415–424

Lowell BB, Shulman GI (2005) Mitochondrial dysfunction and type 2 diabetes. Science 307:384–387

Lu C, Zhang D, Whiteman M, Armstrong JS (2008) Is antioxidant potential of the mitochondrial targeted ubiquinone derivative MitoQ conserved in cells lacking mtDNA? Antioxid Redox Signal 10:651–660

Lu H, Koshkin V, Allister EM et al (2010) Molecular and metabolic evidence for mitochondrial defects associated with beta-cell dysfunction in a mouse model of type 2 diabetes. Diabetes 59:448–459

Mariappan N, Elks CM, Sriramula S et al (2010) NF-kappaB-induced oxidative stress contributes to mitochondrial and cardiac dysfunction in type II diabetes. Cardiovasc Res 85:473–483

McGavock JM, Lingvay I, Zib I et al (2007) Cardiac steatosis in diabetes mellitus: a 1H-magnetic resonance spectroscopy study. Circulation 116:1170–1175

Misu H, Takamura T, Matsuzawa N et al (2007) Genes involved in oxidative phosphorylation are coordinately upregulated with fasting hyperglycaemia in livers of patients with type 2 diabetes. Diabetologia 50:268–277

Modi K, Santani DD, Goyal RK, Bhatt PA (2006) Effect of coenzyme Q10 on catalase activity and other antioxidant parameters in streptozotocin-induced diabetic rats. Biol Trace Elem Res 109:25–34

Mogensen M, Sahlin K, Fernström M et al (2007) Mitochondrial respiration is decreased in skeletal muscle of patients with type 2 diabetes. Diabetes 56:1592–1599

Mootha VK, Lindgren CM, Eriksson KF et al (2003) PGC-1alpha-responsive genes involved in oxidative phosphorylation are coordinately downregulated in human diabetes. Nat Genet 34:267–273

Moreira PI, Santos MS, Moreno AM (2003) Increased vulnerability of brain mitochondria in diabetic (Goto-Kakizaki) rats with aging and amyloid-beta exposure. Diabetes 52:1449–1456

Moreira PI, Santos MS, Sena C et al (2005a) Insulin protects against amyloid beta-peptide toxicity in brain mitochondria of diabetic rats. Neurobiol Dis 18:628–637

Moreira PI, Santos MS, Sena C et al (2005b) CoQ10 therapy attenuates amyloid beta-peptide toxicity in brain mitochondria isolated from aged diabetic rats. Exp Neurol 196:112–119

Moreira PI, Rolo AP, Sena C et al (2006) Insulin attenuates diabetes-related mitochondrial alterations: a comparative study. Med Chem 2:299–308

Moreira PI, Santos MS, Seiça R, Oliveira CR (2007a) Brain mitochondrial dysfunction as a link between Alzheimer's disease and diabetes. J Neurol Sci 257:206–214

Moreira PI, Harris PLR, Zhu X et al (2007b) Lipoic acid and N-acetyl cysteine decrease mitochondrial-related oxidative stress in Alzheimer disease patient fibroblasts. J Alzheimer Dis 12:195–206

Moreira PI, Cardoso SM, Pereira CM et al (2009) Mitochondria as a therapeutic target in Alzheimer's disease and diabetes. CNS Neurol Disord Drug Targets 8:492–511

Moreira PI, Carvalho C, Zhu X et al (2010) Mitochondrial dysfunction is a trigger of Alzheimer's disease pathophysiology. Biochim Biophys Acta 1802:2–10

Morino K, Petersen KF, Dufour S et al (2005) Reduced mitochondrial density and increased IRS-1 serine phosphorylation in muscle of insulin-resistant offspring of type 2 diabetic parents. J Clin Invest 115:3587–3593

Muoio DM, Newgard CB (2008) Mechanisms of disease: molecular and metabolic mechanisms of insulin resistance and beta-cell failure in type 2 diabetes. Nat Rev Mol Cell Biol 9:193–205

Murphy MP, Smith RA (2000) Drug delivery to mitochondria: the key to mitochondrial medicine. Adv Drug Deliv Rev 41:235–250

Nishikawa T, Edelstein D, Du XL et al (2000) Normalizing mitochondrial superoxide production blocks three pathways of hyperglycaemic damage. Nature 404:787–790

O'Connor MD, Landahl H, Grodsky GM (1980) Comparison of storage- and signal-limited models of pancreatic insulin secretion. Am J Physiol 238:R378–R389

Oliveira PJ, Seiça R, Coxito PM et al (2003) Enhanced permeability transition explains the reduced calcium uptake in cardiac mitochondria from streptozotocin-induced diabetic rats. FEBS Lett 554:511–514

P.I. Moreira and C.R. Oliveira

Opie LH (2004) Fuels: aerobic and anaerobic metabolism. In: Opie LH (ed) The heart, physiology, from cell to circulation, 4th edn. Lippincott–Raven, Philadelphia

Patel A, MacMahon S, Chalmers J et al (2008) Intensive blood glucose control and vascular outcomes in patients with type 2 diabetes. N Engl J Med 358:2560–2572

Patti ME, Corvera S (2010) The role of mitochondria in the pathogenesis of type 2 diabetes. Endocrinol Rev 31(3):364–395

Patti ME, Butte AJ, Crunkhorn S et al (2003) Coordinated reduction of genes of oxidative metabolism in humans with insulin resistance and diabetes: potential role of PGC1 and NRF1. Proc Natl Acad Sci USA 100:8466–8471

Pérez-Carreras M, Del Hoyo P, Martín MA et al (2003) Defective hepatic mitochondrial respiratory chain in patients with nonalcoholic steatohepatitis. Hepatology 38:999–1007

Perseghin G, Scifo P, Danna M et al (2002) Normal insulin sensitivity and IMCL content in overweight humans are associated with higher fasting lipid oxidation. Am J Physiol Endocrinol Metab 283:E556–E564

Petersen KF, Dufour S, Befroy D et al (2004) Impaired mitochondrial activity in the insulin-resistant offspring of patients with type 2 diabetes. N Engl J Med 350:664–671

Petersen KF, Dufour S, Shulman GI (2005) Decreased insulin-stimulated ATP synthesis and phosphate transport in muscle of insulin-resistant offspring of type 2 diabetic parents. PLoS Med 2:e233

Petri S, Kiaei M, Damiano M et al (2006) Cell-permeable peptide antioxidants as a novel therapeutic approach in a mouse model of amyotrophic lateral sclerosis. J Neurochem 98:1141–1148

Phielix E, Schrauwen-Hinderling VB, Mensink M et al (2008) Lower intrinsic ADP-stimulated mitochondrial respiration underlies in vivo mitochondrial dysfunction in muscle of male type 2 diabetic patients. Diabetes 57:2943–2949

Pihlajamäki J, Boes T, Kim EY et al (2009) Thyroid hormone-related regulation of gene expression in human fatty liver. J Clin Endocrinol Metab 94:3521–3529

Puche JE, García-Fernández M, Muntané J et al (2008) Low doses of insulin-like growth factor-I induce mitochondrial protection in aging rats. Endocrinology 149:2620–2627

Quintanilla RA, Jin YN, Fuenzalida K et al (2008) Rosiglitazone treatment prevents mitochondrial dysfunction in mutant huntingtin-expressing cells: possible role of peroxisome proliferator-activated receptor-gamma (PPARgamma) in the pathogenesis of Huntington disease. J Biol Chem 283:25628–25637

Rabøl R, Larsen S, Højbjerg PM et al (2010) Regional anatomic differences in skeletal muscle mitochondrial respiration in type 2 diabetes and obesity. J Clin Endocrinol Metab 95:857–863

Reddy PH (2006) Mitochondrial oxidative damage in aging and Alzheimer's disease: implications for mitochondrially targeted antioxidant therapeutics. J Biomed Biotechnol 3:31372

Ritov VB, Menshikova EV, Azuma K et al (2009) Deficiency of electron transport chain in human skeletal muscle mitochondria in type 2 diabetes mellitus and obesity. Am J Physiol Endocrinol Metab 298:E49–E58

Ross SA, Gulve EA, Wang M (2004) Chemistry and biochemistry of type 2 diabetes. Chem Rev 104:1255–1282

Roy Chowdhury SK, Zherebitskaya E, Smith DR et al (2010) Mitochondrial respiratory chain dysfunction in dorsal root ganglia of streptozotocin-induced diabetic rats and its correction by insulin treatment. Diabetes 59(4):1082–1091

Sack MN (2009) Type 2 diabetes, mitochondrial biology and the heart. J Mol Cell Cardiol 46:842–849

Sanderson TH, Kumar R, Sullivan JM, Krause GS (2008) Insulin blocks cytochrome c release in the reperfused brain through PI3-K signaling and by promoting Bax/Bcl-XL binding. J Neurochem 106:1248–1258

Sanyal AJ, Campbell-Sargent C, Mirshahi F et al (2001) Nonalcoholic steatohepatitis: association of insulin resistance and mitochondrial abnormalities. Gastroenterology 120:1183–1192

Sasaki H, Schmelzer JD, Zollman PJ, Low PA (1997) Neuropathology and blood flow of nerve, spinal roots and dorsal root ganglia in longstanding diabetic rats. Acta Neuropathol 93:118–128

Satapati S, He T, Inagaki T et al (2008) Partial resistance to peroxisome proliferator-activated receptor-alpha agonists in ZDF rats is associated with defective hepatic mitochondrial metabolism. Diabetes 57:2012–2021

Scheuermann-Freestone M, Clarke K (2003) Abnormal cardiac high-energy phosphate metabolism in a patient with type 2 diabetes mellitus. J Cardiometab Syndr 1:366–368

Schmeichel AM, Schmelzer JD, Low PA (2003) Oxidative injury and apoptosis of dorsal root ganglion neurons in chronic experimental diabetic neuropathy. Diabetes 52:165–171

Schrauwen-Hinderling VB, Kooi ME, Hesselink MK et al (2007) Impaired in vivo mitochondrial function but similar intramyocellular lipid content in patients with type 2 diabetes mellitus and BMI-matched control subjects. Diabetologia 50:113–120

Sena CM, Nunes E, Gomes A et al (2008) Supplementation of coenzyme Q10 and alpha-tocopherol lowers glycated hemoglobin level and lipid peroxidation in pancreas of diabetic rats. Nutr Res 28:113–121

Shen W, Hao J, Tian C et al (2008a) A combination of nutriments improves mitochondrial biogenesis and function in skeletal muscle of type 2 diabetic Goto-Kakizaki rats. PLoS One 3:e2328

Shen W, Liu K, Tian C et al (2008b) R-alpha-lipoic acid and acetyl-L-carnitine complementarily promote mitochondrial biogenesis in murine 3T3-L1 adipocytes. Diabetologia 51:165–174

Siler-Marsiglio KI, Pan Q, Paiva M et al (2005) Mitochondrially targeted vitamin E and vitamin E mitigate ethanol-mediated effects on cerebellar granule cell antioxidant defense systems. Brain Res 1052:202–211

Simoneau JA, Kelley DE (1997) Altered glycolytic and oxidative capacities of skeletal muscle contribute to insulin resistance in NIDDM. J Appl Physiol 83:166–171

Smith RA, Porteous CM, Gane AM, Murphy MP (1999) Selective targeting of an antioxidant to mitochondria. Eur J Biochem 263:709–716

Smith RA, Porteous CM, Gane AM, Murphy MP (2003) Delivery of bioactive molecules to mitochondria in vivo. Proc Natl Acad Sci USA 100:5407–5412

St-Pierre J, Drori S, Uldry M et al (2006) Suppression of reactive oxygen species and neurodegeneration by the PGC-1 transcriptional coactivators. Cell 127:397–408

Strum JC, Shehee R, Virley D, Richardson J et al (2007) Rosiglitazone induces mitochondrial biogenesis in mouse brain. J Alzheimers Dis 11:45–51

Suh JH, Wang H, Liu RM, Liu J, Hagen TM (2004) (R)-alpha-lipoic acid reverses the age-related loss in GSH redox status in post-mitotic tissues: evidence for increased cysteine requirement for GSH synthesis. Arch Biochem Biophys 423:126–135

Szeto HH (2006) Cell-permeable, mitochondrial-targeted, peptide antioxidants. AAPS J 8: E277–E283

Szeto HH (2008) Mitochondria-targeted cytoprotective peptides for ischemia-reperfusion injury. Antioxid Redox Signal 10:601–619

Tabbi-Anneni I, Buchanan J, Cooksey RC, Abel ED (2008) Captopril normalizes insulin signaling and insulin-regulated substrate metabolism in obese (ob/ob) mouse hearts. Endocrinology 149:4043–4050

Takamura T, Misu H, Matsuzawa-Nagata N et al (2008) Obesity upregulates genes involved in oxidative phosphorylation in livers of diabetic patients. Obesity (Silver Spring) 16:2601–2609

The Diabetes Control and Complications Trial Research Group (1993) The effect of intensive treatment of diabetes on the development and progression of long-term complications in insulin-dependent diabetes mellitus. N Engl J Med 329:977–986

Thomas DA, Stauffer C, Zhao K et al (2007) Mitochondrial targeting with antioxidant peptide SS-31 prevents mitochondrial depolarization, reduces islet cell apoptosis, increases islet cell yield, and improves post-transplantation function. J Am Soc Nephrol 18:213–222

Thyfault JP, Kraus RM, Hickner RC et al (2004) Impaired plasma fatty acid oxidation in extremely obese women. Am J Physiol Endocrinol Metab 287:E1076–E1081

Trifunovic A, Larsson NG (2008) Mitochondrial dysfunction as a cause of ageing. J Intern Med 263:167–178

Tsuneki H, Sekizaki N, Suzuki T et al (2007) Coenzyme Q10 prevents high glucose-induced oxidative stress in human umbilical vein endothelial cells. Eur J Pharmacol 566:1–10

UK Prospective Diabetes Study (UKPDS) Group (1998) Intensive blood-glucose control with sulphonylureas or insulin compared with conventional treatment and risk of complications in patients with type 2 diabetes (UKPDS 33). Lancet 352:837–853

Weydt P, Pineda VV, Torrence AE et al (2006) Thermoregulatory and metabolic defects in Huntington's disease transgenic mice implicate PGC-1alpha in Huntington's disease neurodegeneration. Cell Metab 4:349–362

Wu JS, Lin TN, Wu KK (2009) Rosiglitazone and PPAR-gamma overexpression protect mitochondrial membrane potential and prevent apoptosis by upregulating anti-apoptotic Bcl-2 family proteins. J Cell Physiol 220:58–71

Yang L, Zhao K, Calingasan NY et al (2009) Mitochondria targeted peptides protect against 1-methyl-4-phenyl-1, 2, 3, 6-tetrahydropyridine neurotoxicity. Antioxid Redox Signal 11:2095–2104

Yu T, Sheu SS, Robotham JL, Yoon Y (2008) Mitochondrial fission mediates high glucose-induced cell death through elevated production of reactive oxygen species. Cardiovasc Res 79:341–351

Zhao K, Luo G, Zhao GM et al (2003) Transcellular transport of a highly polar 3+ net charge opioid tetrapeptide. J Pharmacol Exp Ther 304:425–432

Zhao K, Zhao GM, Wu D et al (2004) Cell-permeable peptide antioxidants targeted to inner mitochondrial membrane inhibit mitochondrial swelling, oxidative cell death, and reperfusion injury. J Biol Chem 279:34682–34690

Zhao K, Luo G, Giannelli S, Szeto HH (2005) Mitochondria-targeted peptide prevents mitochondrial depolarization and apoptosis induced by tert-butyl hydroperoxide in neuronal cell lines. Biochem Pharmacol 70:1796–1806

Research and Development of Glucokinase Activators for Diabetes Therapy: Theoretical and Practical Aspects

Franz M. Matschinsky, Bogumil Zelent, Nicolai M. Doliba, Klaus H. Kaestner, Jane M. Vanderkooi, Joseph Grimsby, Steven J. Berthel, and Ramakanth Sarabu

Contents

Dedication and Support Acknowledgement: Dedicated to Elke Matschinsky, "sine qua none"; Supported by NIH grants NIDDK 22122 and 19525.

F.M. Matschinsky (✉) and N.M. Doliba
Department of Biochemistry and Biophysics, University of Pennsylvania, Institute for Diabetes, Obesity and Metabolism, 415 Curie Blvd, 605 CRB, Philadelphia, PA 19104, USA
e-mail: matsch@mail.med.upenn.edu; nicolai@mail.med.upenn.edu

B. Zelent
Department of Biochemistry and Biophysics, University of Pennsylvania, 422 Curie Boulevard, 910 Stellar Chance Building, Philadelphia, PA 19104-6059, USA
e-mail: zelentb@mail.med.upenn.edu

K.H. Kaestner
Department of Genetics, University of Pennsylvania, Institute for Diabetes, Obesity and Metabolism, 415 Curie Blvd, 752B CRB, Philadelphia, PA 19104-6145, USA
e-mail: kaestner@mail.med.upenn.edu

J.M. Vanderkooi
Department of Biochemistry and Biophysics, University of Pennsylvania, 422 Curie Boulevard, 913A Stellar Chance Building, Philadelphia, PA 19104-6059, USA
e-mail: vanderko@mail.med.upenn.edu

J. Grimsby
Department of Metabolic Diseases, Roche Research Center, 340 Kingsland Street, Nutley, NJ 07110-1199, USA
e-mail: joseph.grimsby@roche.com

S.J. Berthel and R. Sarabu
Department of Discovery Chemistry, Roche Research Center, 340 Kingsland Street, Nutley, NJ 07110-1199, USA
e-mail: steven.berthel@roche.com; ramakanth.sarabu@roche.com

M. Schwanstecher (ed.), *Diabetes - Perspectives in Drug Therapy*,
Handbook of Experimental Pharmacology 203,
DOI 10.1007/978-3-642-17214-4_15, © Springer-Verlag Berlin Heidelberg 2011

Abstract Glucokinase (GK; EC 2.7.1.1.) phosphorylates and regulates glucose metabolism in insulin-producing pancreatic beta-cells, hepatocytes, and certain cells of the endocrine and nervous systems allowing it to play a central role in glucose homeostasis. Most importantly, it serves as glucose sensor in pancreatic beta-cells mediating glucose-stimulated insulin biosynthesis and release and it governs the capacity of the liver to convert glucose to glycogen. Activating and inactivating mutations of the glucokinase gene cause autosomal dominant hyper-insulinemic hypoglycemia and hypoinsulinemic hyperglycemia in humans, respectively, illustrating the preeminent role of glucokinase in the regulation of blood glucose and also identifying the enzyme as a potential target for developing anti-diabetic drugs. Small molecules called glucokinase activators (GKAs) which bind to an allosteric activator site of the enzyme have indeed been discovered and hold great promise as new antidiabetic agents. GKAs increase the enzyme's affinity for glucose and also its maximal catalytic rate. Consequently, they stimulate insulin biosynthesis and secretion, enhance hepatic glucose uptake, and augment glucose metabolism and related processes in other glucokinase-expressing cells. Manifestations of these effects, most prominently a lowering of blood glucose, are observed in normal laboratory animals and man but also in animal models of diabetes and patients with type 2 diabetes mellitus (T2DM). These compelling concepts and results sustain a strong R&D effort by many pharmaceutical companies to generate GKAs with characteristics allowing for a novel drug treatment of T2DM.

Keywords GK Activators (GKAs) · Glucokinase (GK) · Glucose Homeostasis · Hyperglycemia · Type 2 Diabetes Mellitus (T2DM)

1   Introduction to the problem

R&D of new pharmacological agents for the treatment of chronic diseases including type 2 diabetes mellitus (T2DM), essential vascular hypertension, or obesity must be based on at least four fundamental considerations: (1) the pathologies and epidemiology of the disease, (2) biochemical and physiological corollaries of the pathologies, (3) biochemical genetics and pathophysiologies of the disease, (4) standard medical care and pharmacotherapies. The present assessment of the potential of the newly discovered glucokinase activators (GKAs) in the treatment of patients with T2DM is made on the basis of such a general approach.

The current status of fundamental knowledge about T2DM can perhaps be summarized by stating that the precise biochemical genetic basis of the disease remains largely unknown even though many "diabetes genes" have been identified and essential pathophysiological processes have been elucidated (Doria et al. 2008; LeRoith et al. 2004; Lyssenko et al. 2008, 2009; McCarthy and Froguel 2002; Meigs et al. 2008; Pearson 2009; Pilgaard et al. 2009; Ridderstråle and Groop 2009; Sparsø et al. 2009; Weedon et al. 2006). The relative contributions of "diabetes genes" to the disease are uncertain and do probably explain only a mere fraction of cases. It is also not clear whether the known diabetes genes lead to the discovery of targets that are suitable for drug development and, if druggable, have a medically significant impact factor. Furthermore, currently available drugs for the treatment of the disease are insufficient to control the disease consistently and persistently. This situation requires a strategy that follows the principle used successfully in the treatment of vascular hypertension, i.e., to develop a very broad spectrum of antidiabetic compounds with different mechanism of action (MOA) affecting different signaling pathways involved in fuel homeostasis. Physicians treating hypertension have a wide choice of drugs with distinct MAO which they can prescribe as monotherapy or in increasingly complex combination therapies usually with success, even though there is rarely a clear understanding of the primary cause or causes of this disease.

Currently available antidiabetic drugs fall into the following categories (Nolte and Karam 2007): (1) insulin and insulino-mimetics, (2) insulin sensitizers including metformin, (3) sulfonylurea and related compounds, (4) GLP1 and related compounds including DPPIV inhibitors, and (5) agents that retard digestion and absorption of carbohydrates from the intestine. The search for novel antidiabetic agents is currently pursued with great intensity because these available drug therapies are far from satisfactory (Desouza and Fonseca 2009).

2   Finding New Drug Targets from Exploring T2DM, Hyperinsulinism, and Glucose Homeostasis Generally

T2DM is a genetic disease which afflicts the middle or older age population but in recent years it is being diagnosed not infrequently in teens. Its incidence is dramatically increased by overeating and physical inactivity usually associated with obesity (LeRoith et al. 2004; Reaven 1988). Numerous diabetes genes have been discovered but their relative contribution to the disease has not been fully determined (Doria et al. 2008; Lyssenko et al. 2008, 2009; McCarthy and Froguel 2002;

Meigs et al. 2008; Pearson 2009; Pilgaard et al. 2009; Ridderstråle and Groop 2009; Sparsø et al. 2009; Weedon et al. 2006). It is anticipated that many more diabetes genes will be found. Thus, despite the impressive advances in this field, T2DM remains the "Geneticists Nightmare." Progression of the disease is usually slow, sometimes measured in decades, from impaired fasting blood glucose (IFG) to impaired glucose tolerance (IGT) to overt T2DM defined by a fasting blood glucose higher than 7 mM and a 2-h glucose value larger than 11 mM in a glucose tolerance test. Long-term complications include micro- and macro-vascular disease afflicting primarily the eyes, the kidney, the heart, and the peripheral nervous system, may be debilitating but are difficult to treat (Calcutt et al. 2009). T2DM could be largely prevented by drastic restriction of caloric intake and regular physical activity. The underlying pathophysiology is briefly sketched as follows: Overnutrition and limited physical activity interfere with the action of insulin in liver, muscle, and adipose tissue (LeRoith et al. 2004; Reaven 1988). This increased peripheral resistance to insulin places a greater burden on the genetically compromised pancreatic beta-cells which not only fail to adapt sufficiently to the demand of overcoming insulin resistance by mass expansion of the islet organ and functionally manifest in enhanced insulin secretion but show, with time, impaired cell function and decreased pancreatic islet mass (Kahn 2003; Kahn et al. 2009). A striking illustration of the remarkable normal capacity of the endocrine pancreas for adaptation to increased demand is uncomplicated obesity in either sex or pregnancy in the healthy women who may experience many cycles of expansion and involution of her pancreatic beta-cell tissue without ever suffering negative effects. An equally striking example of the beta-cell failure to adapt is "diabetes in pregnancy" which indicates the presence of diabetes genes causing transient or even persistent disease. Hyperglycemia persisting for years causes diabetic complications in the majority of cases by little understood but hotly debated biochemical mechanisms. This description points to a genetically based failure of the endocrine pancreas to adapt to increased peripheral resistance (whatever its cause) but provides no compelling clue for prominent molecular defects of the disease that could suggest specific and promising targets for correction by new pharmacological agents. One solution of the problem is to develop a broad spectrum of antidiabetic drugs with distinctly different MOA that might enhance beta-cell function and augment insulin action by targeting physiological mechanisms (Desouza and Fonseca 2009; Nolte and Karam 2007). The development of GKAs is one striking example of this approach.

Since a failure of the beta-cell to adapt to insulin resistance seems to be a root cause of T2DM it is reasonable to develop drugs that have a potential remedy for this particular deficiency. Very specific clues about promising drug targets in beta-cells can be found in "experiments of nature" resulting in monogenic hyperinsulinemic hypoglycemia in children (PHHI or HI, a more widely used abbreviation) (Dunne et al. 2004; González-Barroso et al. 2008; Kapoor et al. 2009; Palladino et al. 2008). These syndromes have been thoroughly explored in the last two decades and mutations of at least six genes have been discovered that are causing these hypoglycemia syndromes, including the SUR-1 receptor of the K-ATP channel (Dunne et al. 2004; Palladino et al. 2008), the pancreatic beta-cell glucose sensor GK

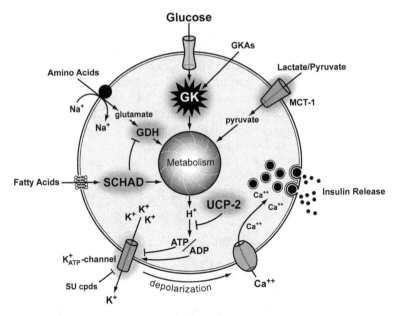

Fig. 1 Biochemical genetic studies of HI syndromes uncovered unique clues for drug targets in beta-cell signaling pathways that are physiologically involved in stimulus secretion coupling. Regulated fuel stimulation of insulin biosynthesis and secretion is central to glucose homeostasis. The figure shows six proteins (or protein complexes) and the associated signaling pathways that have high regulatory impact physiologically. These proteins can be the cause of enhanced insulin secretion when altered by activating or inactivating mutations. Two of these, i.e., GK and K_{ATP}-channel, have been successfully drug targeted by GKAs and sulfonyl urea compounds, respectively. Note that the discovery of the latter agents preceded by decades the identification of the corresponding receptor whereas in the case of GKAs the drug was tailor made as guided by known physiological chemistry. Some of the others are potential targets

(Dunne et al. 2004; Palladino et al. 2008), the enzyme glutamate-dehydrogenase (GDH) (Dunne et al. 2004; Palladino et al. 2008), the monocarboxylic transporter (MCT) (Dunne et al. 2004; Palladino et al. 2008), the mitochondrial uncoupling protein UCP-2 (González-Barroso et al. 2008), and short chain hydroxy-acyl-CoA dehydrogenase (SCHAD) (Kapoor et al. 2009) as illustrated in Fig. 1. The blood glucose lowering sulfonylurea compounds, discovered about 50 years ago and still widely used, target SUR-1 as one of these genes. GK, another member of this group of genes, is now considered an outstanding target for developing GKAs as antidiabetic agents and is the topic of the present chapter. It is not unreasonable to explore the possibility of targeting GDH, MCT, SCHAD, and UCP-2 or components of the associated signaling chains. Hypersecretion of insulin in these forms of HI is triggered by molecular processes that cause either direct activation, de-inhibition, or inhibition of proteins in these signaling pathways which could provide opportunities for targeting novel pharmaceuticals. It is noteworthy that there is currently no evidence suggesting that these HI syndromes are associated with an increased incidence of T2DM that might result from overstimulation of the pancreatic beta-cells nor is there evidence that the relative hyperinsulinism of HI leads to insulin resistance. Similar to the monogenic HI syndromes described above, monogenic MODY forms (maturity onset diabetes of the young) are caused by genetic defects of beta-cell function and promise to provide useful guidance in the search of targets for developing antidiabetic pharmaceuticals because of their relatively high control strength in regulating blood glucose compared to the genes associated with T2DM (http://www.ncbi.nlm.nih.gov/entrez/dispomim.cgi?id=606391).

Other clues for potential beta-cell drug targets come from studies of the normal physiological chemistry of insulin secretion (Fig. 2) (Ahrén 2009; Gromada 2006; Henquin 2000; Matschinsky et al. 2006; Newgard and Matschinsky 2001; Wess et al. 2007). Activation of the GLP-1 receptor is now considered a promising approach and has resulted in the development of synthetic exendin-4 and DPP-IV

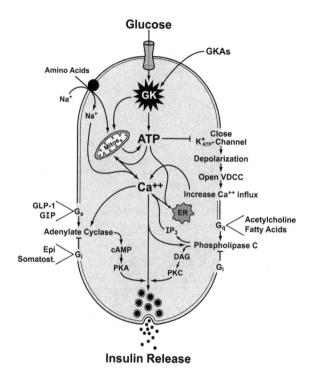

Fig. 2 Physiology and pharmacology of stimulus secretion coupling in pancreatic beta-cells. The figure depicts the interconnections between fuel stimulation of insulin secretion and its modification by the neuroendocrine system and illustrates the absolute glucose and GK requirement for neuron-endocrine stimulation of beta-cell functions. Note: Fatty acids are considered here as membrane receptor agonists while their potential fuel function is ignored because of the low rate of oxidation compared to that of glucose and amino acids. The following abbreviations are used: *DAG* diacylglycerol; *Epi* epinephrine; *ER* endoplasmic reticulum; *G i, q and s* G-proteins i, q and s; *GLP-1* glucagon like peptide 1; *GIP* gastric inhibitory peptide; *IP3* inositol-triphosphate; *PKA* proteinkinase A; *PKC* proteinkinase C; *Somatost.* somatostatin; *VDCC* voltage dependent calcium channels

inhibitors (Ahrén 2009; Desouza and Fonseca 2009). These drugs increase cAMP in the beta-cell and sensitize it to glucose and perhaps amino acids. G-protein coupled receptors that lead to activation of phospholipase C and coupling to the PKC pathway include the fatty acid receptor GPR-40 and the muscarinic acetylcholine receptor M3 (Desouza and Fonseca 2009; Gromada 2006; Wess et al. 2007). The activation of either one of these results in glucose-dependent insulin release. Pharmacological activators of GPR-40 have been discovered and have been shown to stimulate insulin release (Desouza and Fonseca 2009; Gromada 2006). Using M3 as an antidiabetic drug target is difficult because of the broad involvement of this receptor in cardiovascular and gastrointestinal physiology. Steps downstream of M3, for example, regulators of G-protein signaling (RGPS) might be more useful in this regard (Wess et al. 2007). The absolute glucose dependency of all these receptor-mediated stimulations of insulin release is remarkable and stresses again the overarching role of the GK glucose sensor for beta-cell function and thus the high potential of GKAs for diabetes therapy.

3   Slow Evolution of the Idea That Glucokinase Might Serve as *Glucose Sensor* and as Drug Receptor

It took nearly three decades from the time GK was discovered in liver (1963) and then in the pancreatic islets (1968) to the time the enzyme was recognized as a drug target for developing oral antidiabetic medications (Table 1) (Cuesta-Munoz et al. 2001; Doliba et al. 2001; Grimsby et al. 2001, 2003; Matschinsky and Ellerman 1968; Sharma et al. 1964; Sols et al. 1964; Walker and Rao 1964). The evolution of GK as a drug receptor occurred against great odds. A central role of GK in glucose homeostasis was not universally accepted because it had vocal detractors (Malaisse and Sener 1985) and it was even more difficult to envision how GK could be activated equally well in pancreatic beta-cells and hepatocytes such that its pharmacological actions could exert a maximal glucose lowering action without causing serious side effects as, for example, hyperlipidemia (Desai et al. 2001; O'Doherty et al. 1999). It was also uncertain at that time how GK expression and stability in the pancreatic beta-cell might be influenced by the severity of T2DM raising the specter of a target molecule that might fail or decline as the disease progresses. This in mind and cognizant of the enormous risk in selecting any new drug target for R&D the decision in the early 1990s to focus on GK was indeed courageous. However, the kinetics and the physiological chemistry of GK were well understood and it is fortunate that the knowledge about the genetics, biochemistry, and biophysics of the enzyme advanced greatly while the drug development program was underway reinforcing the undertaking (Matschinsky 2009; Matschinsky et al. 2006).

GK (ATP:D-glucose 6-phosphotransferase, also known as hexokinase IV or D) is one of four hexokinase isoenzymes (HK I–IV or HK A–D) which catalyzes the

Table 1 Milestones in glucokinase research

Year	Discovery	References
1963/1964	Glucokinase discovered in liver	Sharma et al. (1964), Sols et al. (1964), Walker and Rao (1964)
1968	Glucokinase identified in mouse pancreatic islets	Matschinsky (1990)
1975/1976	Sigmoidal glucose dependency of glucokinase discovered	Cárdenas (1995), Cornish-Bowden and Cárdenas (2004), Neet and Ainslie (1980)
1977/1980	Mnemonic and slow transition models of Gk kinetics	Cárdenas (1995), Cornish-Bowden and Cárdenas (2004), Neet and Ainslie (1980)
1984/1986	The glucokinase glucose sensor paradigm published	Meglasson and Matschinsky (1984, 1986)
1986	Differential expression control of hepatic and islet GK	Bedoya et al. (1986b),
1986	Detection of GK in human islets	Bedoya et al. (1986a)
1989	GKRP discovered	Detheux et al. (1991), Vandercammen and Van Schaftingen (1990, 1991)
1989/1991	Rat and human liver cDNA cloned	Cárdenas 1995, Iynedjian (2004), Postic et al. (2001)
1993	GK linkage of MODY reported	Froguel et al. (1992), Hattersley et al. (1992)
1998	GK linked PHHI described	Glaser et al. (1998)
2001	GK linked PNDM described	Njølstad et al. (2001)
2001/2003	First reports on discovery of GKAs	Cuesta-Munoz et al. (2001), Doliba et al. (2001), Grimsby et al. (2001, 2003)
2003/2004	Crystal structures of GK reported	Grimsby et al. (2003), Kamata et al. (2004)
2008/2010	First reports of GKA use in human diabetics	Bonadonna et al. (2010), Zhi et al. (2008)

Historical milestones in GK research and evolution of the concept that GK might be a drug receptor candidate. The time line covers a period of 45 years from the discovery of hepatic GK in 1963/1964 to the first report about the successful use of GKAs in human diabetic subjects in 2008/2009

phosphorylation of D-glucose by MgATP according to the following equation (Cárdenas 1995; Cárdenas et al. 1998; Wilson 2004):

$$R - CH_2OH + MgATP^{2-} \longrightarrow R - H_2 - O - PO_3^{2-} + MgADP^- + H^+.$$

The biochemical and kinetic characteristics of the enzyme are well established (Cárdenas 1995; Matschinsky 1996; Matschinsky et al. 2006; Wilson 2004; Xu et al. 1995). It has a molecular weight of about 50 kD and exists as a monomer. It is not specific for D-glucose, reacting also with D-mannose and D-fructose. The enzyme has cooperative kinetics with its substrates glucose and mannose as manifest by a Hill coefficient (nH, a measure of cooperativity) of 1.7. The kinetic constants and the tissue distribution clearly distinguish GK from the other hexokinases. The $S_{0.5}$ (the concentration supporting half maximal catalytic rate) for D-glucose is about 8.0 mM, at least 100-fold lower than that of the other hexokinases. In contrast to hexokinases, GK is not controlled by product inhibition. GK is, however, inhibited by the hepatic GK regulatory protein (GKRP), a nuclear protein that binds the enzyme and facilitates its translocation from the cytosol to the nucleus (Detheux et al. 1991; Vandercammen and Van Schaftingen 1990, 1991; Veiga-da-Cunha and Van Schaftingen 2002). This process is enhanced by fructose-6-P and blocked or reversed by glucose and fructose-1-P. GK has an allosteric activator site which mediates an increased catalytic rate (kcat), decreases of the glucose $S_{0.5}$ and the nH but raises the ATP Km, the concentration of ATP supporting halfmaximal rate (Dunten et al. 2004; Efanov et al. 2005; Grimsby et al. 2004; Grimsby et al. 2003; Kamata et al. 2004; Matschinsky et al. 2006). This activator site is located about 20 Å away from the substrate binding site. An endogenous activator has been postulated but has not been identified. GKAs activate GK by binding to this activator site. The allosteric activator site is not accessible to most GKAs unless permissive levels of glucose are present (Matschinsky 2009). Note, however, that some GKAs bind to GK with low affinity in the absence of glucose (Antoine et al. 2009). The enzyme forms complexes with other proteins and cellular organelles, for example, it is bound and activated by 6PF2K/F2,6P2ase (6-phosphofructo-2-kinase/fructose-2,6-biphosphatase) (Baltrusch et al. 2004) and also associates reversibly with BAD (a proapoptotic protein factor (Danial et al. 2008)) in a multiprotein complex that binds to mitochondria and enhances respiration. Crystal structures of the free and ligand bound form of GK have been published (Dunten et al. 2004; Efanov et al. 2005; Grimsby et al. 2003, 2004; Kamata et al. 2004). In the absence of substrates or GKAs, the enzyme exists in a so-called wide open or super open conformation which changes to a closed conformation when glucose and GKA are bound in a 1:1:1 ternary complex (Fig. 3). Crystal structures in presence of glucose alone have not been reported. The conformational change induced by glucose (and other sugars) can be observed conveniently by monitoring tryptophan fluorescence (Molnes et al. 2008; Zelent et al. 2008). The quantum yield increases twofold upon saturation with sugar (Fig. 4). GK contains three tryptophans (W99, W167, and W257) which are located at strategic sites of the enzyme. Glucose binding greatly increases the fluorescence of W99 and W167 but not that of W257. Based on enzyme kinetics, crystal structures, molecular dynamics, and tryptophan measurements, a model of GK which attempted to explain the unique cooperative or sigmoidal glucose dependency curve of the enzyme slowly evolved (Cárdenas 1995; Cornish-Bowden and Cárdenas 2004; Heredia et al. 2006; Kim et al. 2007; Lin and Neet 1990; Neet and Ainslie 1980; Sarabu and Grimsby 2005). Depending on the absence or presence of glucose, GK exists in one or two conformations with low and high affinity for glucose, respectively, although recent biophysical studies suggest that in the absence of glucose, GK might already exist in an equilibrium of multiple conformations with different ligand affinities and that glucose shifts the equilibrium from the more open to the more closed forms (Antoine et al. 2009). Whatever the molecular details, glucose induces a reversible, concentration dependent slow transition from the low affinity to the high affinity forms of GK. The ensuing catalytic cycles, initiated by binding of the second substrate MgATP, are much faster than the return of the activated free enzyme to the inactivated low affinity form resulting in sigmoidal kinetics. This process was termed by A. Cornish-Bowden the "mnemonic mechanism" (Cornish-Bowden and Cárdenas 2004) and by K. Neet the "ligand induced

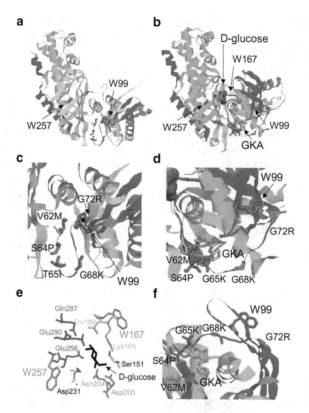

Fig. 3 Essential structural features of the GK molecule explaining its glucose sensor and drug receptor capabilities. (**a, b**) Wide open (**a**) and closed (**b**) conformations of GK. Note that in (**a**) parts of the molecule are not definable in the crystal structure. Tryptophans W99, W167, and W257 are presented because measurements of their fluorescence provide deep insights into the structure and function of GK. The locations of glucose and GKA are noted in (**b**) because both were required for successful crystallization of the closed structure. Panels (**c**) and (**d**) are cutouts to illustrate details of the configurations of the allosteric activator binding site in the open (**c**) and in the closed (**d**) conformations. Several missense mutations that activate the enzyme are used to delineate the allosteric activator site. The loop connecting beta-1 with beta-2 and the alpha helices 5 and 13 which together with connecting loop I encapsulate the GKA binding site are highlighted. Panels (**e**) and (**f**) give details about the glucose binding site and show how GK activation results in a dramatic translocation of W99 explaining much of the fluorescence enhancement caused by glucose binding

slow transition mechanism" (1980) of cooperative kinetics, because the enzyme does apparently remember its activated status between catalytic cycles or shows very slow transitions between states of activation, respectively. Sigmoidicity of glucose dependency (i.e., an nH of 1.7) is critical for GK function both in glucose sensor cells and hepatocytes because the enzyme gains its highest response range for catalytic activity between 3 and 7 mM glucose rather than 0 and 3 mM for a hyperbolic enzyme with a comparably low glucose affinity, a physiologically highly relevant fact (Matschinsky 1996, 2009; Matschinsky et al. 2006). GK is encoded by

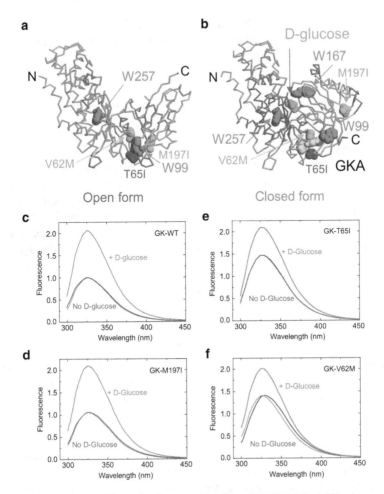

Fig. 4 Structural impact of the activation caused by HI and MODY linked GK missense mutations as monitored by tryptophan fluorescence. T65I and M197I cause HI and both are located within the activator site in the open conformation, however, M197I undergoes a large lateral move in the course of the glucose-induced conformational change. V62M, although activating, causes MODY because it is thermally unstable. V62M and T65I have a high quantum yield in the basal state without glucose and do not respond well to GKAs and GKRP suggesting a more compact conformation in the basal state. They do, however, fully close when saturated with glucose. In the case of V62M, glucose shifts the F-max from a relatively red position to the blue such that it reaches the F-peak wavelength of the wild type enzyme as indicated by the normalized spectra in the presence (*red*) and absence (*blue*) of glucose. This suggests that the structure of V62M is locally loosened consistent with its moderate thermal instability. M197I activates GK slightly, sufficiently though to cause HI. Spectrally, it is indistinguishable from the wild type and responds normally to GKAs and GKRP. The M197I case illustrates how even minor activation of one allele affects glucose homeostasis markedly

one gene but transcription is regulated in a tissue-specific manner by two promoters, the upstream or neuroendocrine promoter operating in glucose sensor cells as defined below and the downstream or hepatic promoter operating exclusively in the liver (Bedoya et al. 1986b; Iynedjian 1993, 2004, 2008; Jetton et al. 1994; Postic et al. 2001). The neuroendocrine promoter regulates constitutive expression at a relatively low basal level and the hepatic promoter regulates insulin-driven expression resulting in GK levels that might be an order of magnitude higher than in glucose sensor cells.

These features of the GK molecule have the hallmarks of a classical drug receptor and justify the description of GK as glucose sensor in neurons of the hypothalamus, in endocrine cells of the pancreas and gut, or as metabolic regulator of hepatic glucose metabolism, in all instances serving as GKA drug receptor. It is of historical interest that in the late 1960s Dr. Philipp Randle was distinguishing between two alternative models for glucose sensing in the pancreatic beta-cells involving a substrate or a regulator site, exemplified by a glucose-metabolizing enzyme or an unspecified allosteric site for glucose, respectively, that would stimulate metabolism and insulin release (Randle et al. 1968). We are now able to combine these two models into one concept, that of GK functioning intracellularly as a glucose sensor and as a receptor for nonessential allosteric activator drugs (and perhaps endogenous activators) which stimulate glucose metabolism in GK-expressing cells thereby affecting glucose homeostasis profoundly.

4 Biological Systems Analysis of GK (GK in Pancreatic Islet Beta-Cells, Liver, and Neuroendocrine Cells Is Central to Understanding Glucose Homeostasis and GKA Action)

At this point, it is essential to digress and expand on the central role of GK in glucose homeostasis in order to permit a comprehensive discussion of the potential that GKAs might have in the treatment of patients with T2DM. It is obvious that the process of glucose phosphorylation and its regulation by insulin is essential to glucose homeostasis. Cellular glucose phosphorylation and glucose transport are tightly coupled and show large and characteristic differences from tissue to tissue as illustrated by three groupings (Cárdenas 1995; Joost et al. 2002; Mueckler 1994; Tal et al. 1992; Thorens et al. 1990; Wilson 2004): (1) brain and red blood cells, (2) muscle (both heart and skeletal) and adipose tissue, and (3) liver and insulin producing pancreatic beta-cells. In brain and red blood cells, glucose transport (by Glut1 or 3) is insulin independent and not limiting and phosphorylation is accomplished primarily by hexokinase 1. In heart, skeletal muscle, and adipocytes, glucose transport (by Glut4) is limiting and insulin regulated, and phosphorylation is catalyzed primarily by hexokinase 2. In hepatocytes and pancreatic beta-cells, glucose transport (primarily by Glut2) has a capacity that is much higher than that of the rate-limiting GK, the predominant hexokinase in these tissues. GKAs have thus the potential of stimulating glucose phosphorylation in liver and pancreatic islets thereby influencing processes downstream of GK but have in general no direct effect on most brain cells, red and white blood cells, heart, skeletal, and smooth muscle or fat cells. The roles of GK in pancreatic beta-cells and in liver are fundamentally different but complementary and thus the pharmacological effects of GKAs in these organs are also different.

In the beta-cells, GK plays the role as glucose sensor and thus controls all their glucose-dependent functions which includes (Matschinsky 1996; Matschinsky and Ellerman 1968; Matschinsky et al. 2006; Newgard and Matschinsky 2001) (Fig. 2) (1) glucose-stimulated insulin release and biosynthesis; (2) glucose-dependent fuel stimulation of insulin release by fatty acids and amino acids; (3) glucose-dependent receptor-mediated stimulation of insulin release by acetylcholine, GLP1, GIP, and by fatty acids via the GPR40 receptor; (4) beta-cell survival and replication. It seems that in man and common laboratory rodents, glucose is not replaceable by other fuels at physiological concentrations. This profound impact of GK and glucose metabolism on beta-cell function is explained by the unique enzymatic, biophysical, and molecular biological characteristics of the beta-cell isoform of GK. GK expression in beta-cells is controlled by the upstream (or neuroendocrine) promoter of the GK gene and is constitutive but enhanced by glucose such that its basal activity as observed at about 4-mM glucose can be increased about fivefold by saturation levels of the sugar (Bedoya et al. 1986b; Liang et al. 1990, 1992; Matschinsky 1996; Zelent et al. 2005). This effect is due in part to substrate stabilization of the protein. A possible role of insulin in the process of glucose induction of islet GK remains controversial because it is difficult or nearly impossible to study GK induction in beta-cells by glucose without interference with extracellular insulin. Results with the beta-cell insulin receptor knock-out (βIRKO) mice which are practically normal as young animals and develop a diabetic phenotype only as they age suggest, however, that insulin signaling is not absolutely required to maintain beta-cell expression of the GK glucose sensor (Gleason et al. 2007; Kulkarni et al. 1999; Leibiger et al. 2001; Okada et al. 2007). In beta-cells, GK is found in the cytosol but also associated with insulin granules (Arden et al. 2004; Miwa et al. 2004) and with mitochondria (Arden et al. 2006; Danial et al. 2008) but not in association with GKRP in the nucleus, in striking contrast to hepatocytes (see below). The association with mitochondria depends on the presence of the proapoptotic factor BAD (Danial et al. 2008). It appears that BAD, which also binds to the antiapoptotic factors BCL2 and BCL-x which neutralize the proapoptotic mediators BAK and BAX, serves two roles, one in metabolism thereby enhancing GSIR, and another in cell survival enhancing apoptosis and cell turnover. It seems that the equilibrium between these two roles is determined by GK and other partners of the complex facilitating the association of BAD with mitochondria and it is also positively regulated by glucose levels. GK is probably complexed by the bifunctional enzyme fructose-6-P-2-kinase/fructose-2,6-P$_2$-2-phosphatase which serves de facto as a GK activator (Baltrusch et al. 2004). However, the glucose metabolism of the beta-cells is primarily determined by the basic kinetic characteristics of GK as discussed above. These constants and the glucose controlled level of the enzyme protein determine the glucose dependency of beta-cell glycolysis and glucose oxidation.

The ATP Km is about 0.15 mM at 5-mM glucose and is slightly increased as the glucose rises and assures that the enzyme is always fully saturated with the second substrate. The rate of glucose oxidation of isolated islets is about 1/3 of the glycolytic rate indicating that other factors besides GK control the rate of ATP production during stimulation with high glucose (Sweet et al. 1996). Still, GK has a control strength in the regulation of glucose metabolism of pancreatic islet tissue (both of glycolysis and oxidation) approaching unity implying that even small changes of the effective enzyme activity (kcat/glucose $S_{0.5}$), by as little as 10–20%, have a marked effect on glucose metabolism (Matschinsky 1990; Meglasson and Matschinsky 1984, 1986). The coupling between glucose metabolism and stimulation of insulin secretion is absolute and is mediated by coupling factors (Henquin 2000; Matschinsky 1996; Matschinsky et al. 2006; Newgard and Matschinsky 2001; Prentki and Matschinsky 1987). It is still hotly debated which metabolites and cofactors qualify as coupling factors in this process. Unanimity exists about the critical roles of ATP^{4-} and $MgADP^{-}$, but there is argument whether to include NAD(P)H, 5'-AMP, acyl-CoA, and diacyl-glycerol in this category. In any case, the ATP/ADP ratio determines the membrane potential by regulating the K_{ATP} channel such that an increase of the ratio in the course of increased glucose metabolism inhibits K-efflux and depolarizes the beta-cell, opening voltage-sensitive Ca^{++} channels at a glucose threshold of about 5.0 mM and triggering insulin release. The associated elevation of cytosolic Ca^{++} activates adenylate cyclase and enhances the PKA signaling pathway whereas DAG (generated in the course of glucose metabolism and by Ca^{++} induced P-lipase C activation) enhances the PKC pathway. These protein kinases greatly increase the effectiveness of the Ca^{++} trigger. Prolonged exposure to high physiological or abnormally high glucose levels of 5–10 mM or up to 25 mM induces GK and augments GSIR (Liang et al. 1990, 1992; Zelent et al. 2005). This induction of GK enzymatic activity in beta-cells can occur in the absence of changing mRNA levels for GK in evident contrast to the situation in the liver. This difference is strikingly illustrated by the finding that mannoheptulose, a competitive GK inhibitor that cannot be phosphorylated, induces beta-cell GK many fold (see discussion below and Fig. 16).

The liver contains 99.9% of the body's GK complement (Matschinsky et al. 2006). The enzyme sustains the high capacity process in the liver of clearing glucose from the portal blood postprandially as the first step of glycogen biosynthesis via the direct pathway and also enhancing glycolysis to sustain the indirect pathway of glycogen synthesis (Fig. 5). Regulation of GK in the liver is long term and short term (Agius 1998, 2008; Agius and Peak 1993; Agius et al. 1995; Detheux et al. 1991; Iynedjian 1993, 2004; Vandercammen and Van Schaftingen 1990, 1991; Veiga-da-Cunha and Van Schaftingen 2002). Expression of GK is primarily controlled by insulin such that GK falls markedly in diabetes and is induced in hyperinsulinemic states in striking contrast to the situation in the beta-cells. Insulin induction is driven by the downstream hepatic promoter of the GK gene such that transcription and translation are increased manifold in less than 1 h and turnover of mRNA and GK protein are very high. The underlying mechanisms of acute regulation of GK are complex and depend on a subtle interplay of many factors.

The effect of elevated glucose on the GK/GKRP complex in the nuclear compartment of the hepatocytes is central to stimulation of glycolysis and glycogen synthesis (Detheux et al. 1991; Vandercammen and Van Schaftingen 1990, 1991; Veiga-da-Cunha and Van Schaftingen 2002). GKRP is a nuclear protein that binds GK. This complex formation is enhanced by fructose-6-P and counteracted by glucose and fructose-1-P, the two most important regulators of this process. GKRP binds preferentially to GK and inhibits the enzyme when glucose is low and the super open conformation of GK predominates, resulting in a sequestration of GK in the nucleus under fasting conditions. Postprandially elevated glucose causes a dissociation of the GK/GKRP complex and release of the active enzyme to

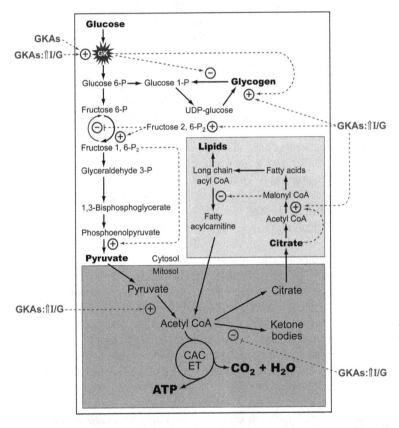

Fig. 5 Ramifications of altered GK activity for hepatic intermediary metabolism. The effects of GK activation by insulin-induced expression or pharmacological stimulation by GKAs on energy, glycogen and lipid metabolism are sketched. End products of main pathways are highlighted. Essential hormonal (i.e., the elevated insulin/glucagon (I/G) ratio) and metabolite signal molecules (i.e., G6P and F2,6P2) are indicated. Gluconeogenesis is ignored except for the reference to the F1,6P2 phosphatase step

the cytosol. Alimentary fructose potentiates the effect of glucose even at rather low concentrations of less than 1.0 mM because its product fructose-1-P generated by fructokinase dissociates the GK/GKRP complex very efficiently. Glucose-6-P, the product of the GK reaction, is an allosteric activator of glycogen synthase and glucose renders phosphorylase-a a better substrate for inactivation by protein phosphatase (Agius 1998, 2008; Agius and Peak 1993; Agius et al. 1995). Not surprisingly, glucose-6-P also enhances glycogen synthesis by providing substrate in the form of UDP-glucose for the synthase. It needs to be postulated that the glucose-6-phosphatase system is inhibited under these conditions to avoid futile cycling by mechanisms not well understood. Gluconeogenesis is also curbed as a result of the elevated insulin/glucagon ratio.

While GK containing cells of the endocrine pancreas and the liver clearly constitute the central axis of the glucose homeostatic system in man and common laboratory animals, the presence and possible role of GK in cells of several other fuel sensing tissues needs careful attention in the present context (Matschinsky et al. 2006). GK has been discovered in the glucagons-producing alpha-cells of the pancreas (Bedoya et al. 1987; Reimann et al. 2008), nuclei of the hypothalamus (Levin et al. 2004, 2008), the tractus solitarius and the raphe nuclei of the brain stem (Matschinsky et al. 2006), and in the gonadotropes and thyrotropes of the pituitary gland (Hille et al. 1995; Matschinsky 2008; Sorenson et al. 2007; Zelent et al. 2006). There is convincing evidence that the enzyme may exist in the GLP1-secreting L-cells of the intestine (Reimann et al. 2004, 2006, 2008) but its biological significance has been disputed, even though not compellingly (Murphy et al. 2008). Much indirect evidence indicates that glucose sensor cells of the hepatoportal vasculature may contain GK participating in the generation of the "portal signal" that influences glucose homeostasis (Cherrington 1999; Donovan et al. 1994; Thorens 2004). The discussion of the role that GK may play in these tissues requires a clear definition of the terms "glucose sensing", "glucose sensor", or "glucoreceptor" (Matschinsky 2009). GK per se serves as the "glucose sensor" or "glucoreceptor" molecule and GK-containing cells are considered "glucose sensor cells" or "glucoreceptor cells" involved in "glucose sensing" if glucose stimulation results in the metabolism-dependent generation of a clearly defined neural, endocrine, or paracrine signal usually associated with altered electrical activity of the cell. Using this narrow definition, we include endocrine cells of the pancreas, gonadotropes and thyrotropes, intestinal L-cells, portal glucose sensor cells, the carotid body (perhaps), and neurons in the brainstem and hypothalamus. But we exclude hepatic parenchymal cells which serve as fuel depot storing glucose in the form of glycogen but seem to lack a GK-mediated signaling function as defined here. Another distinction between these two types of GK-containing cells seems to be the nature of GK expression control, constitutive and independent of elevated insulin levels in all GK-containing glucose sensor cells to be contrasted with the hepatocytes, where GK expression is entirely insulin controlled. It is important to appreciate that at least 99.9% of the total GK content of the body is located in the liver. This illustrates the involvement of the enzyme in the insulin dependent, high capacity process of hepatic glucose clearance from the blood postprandially in contrast to its regulatory role in the largely insulin independent neuroendocrine system, which regulates in an indirect manner glucose usage and storage in large organs and systems, i.e., liver, skeletal muscle, heart, and adipose tissue. The molecular details and physiological significance of glucose sensing in cells other than the pancreatic beta-cells remain largely undetermined except perhaps the role of GK containing hypothalamic neurons (Levin et al. 2004, 2008) which seem to govern counterregulatory processes that curb hypoglycemia. However, in the context of the present discussion assessing the therapeutic potential and the possible dangers and side effects of GKAs these insufficiently explored GK cells cannot be ignored.

5 Glucokinase Disease and the Status of GK in Type I and II Diabetes

Genetic studies of GK-linked diabetes and hypoglycemia syndromes (often referred to as "Glucokinase Disease") and the results of studies exploring the status of pancreatic and hepatic GK in T1DM and T2DM have demonstrated that GK has the hallmarks of a promising antidiabetic drug target. The physiological chemical exploration of glucokinase resulted in the prediction that even minor alterations in its enzymatic activity would lead to diabetes mellitus or hypoglycemia owing to its central role in fuel stimulation and its obligatory priming role in neuroendocrine control of insulin secretion. This prediction was verified by the discoveries of GK gene linked PHHI, MODY2, and PNDM as defined above resulting in the wide range of defects in glucose homeostasis, from hypoglycemia that might be severe enough to cause seizures (PHHI) to lethal forms of diabetes mellitus due to inhibitory mutations affecting both alleles of the gene (PNDM) (Christesen et al. 2002; Cuesta-Muñoz et al. 2004; Danial et al. 2008; Edghill and Hattersley 2008; Froguel et al. 1992; Glaser et al. 1998; Gloyn et al. 2003; Hattersley et al. 1992; Njølstad et al. 2001, 2003). The biochemical genetics of activating mutants are particularly relevant for this discussion of GKAs. Thirteen distinct activating GK mutants have been discovered so far, all with autosomal dominant inheritance affecting only one allele demonstrating again the high control strength of the enzyme on glucose homeostasis and the potential that activating the enzyme pharmacologically might have for diabetes therapy (Dunne et al. 2004; Matschinsky 2009; Palladino et al. 2008). It is remarkable that homozygocity of activating mutants has not been seen so far and that all known heterozygous cases are severe enough to require treatment with diazoxide, partial pancreatectomy, or both. It is speculated that subclinical forms of GK-linked hypoglycemia exist which have escaped detection for obvious reasons. The activating mutants are located in the allosteric activator site (with one exception, i.e., M197I (Sayed et al. 2009)) and many of them were found to have reduced responsiveness or are totally refractory to GKAs and to GKRP.

It is obvious that the presence of functionally active GK in the pancreatic beta-cells and the hepatocytes of diabetics is a prerequisite for GKA-based antidiabetic therapy. Information on this critical point is, however, limited. Considering beta-cells in human diabetics, it is reasonable to extrapolate from countless clinical studies and from limited results of glucose-stimulated insulin release perifusion studies in islets isolated from diabetic donors that GK is indeed functional, even though activity might be partially reduced (Deng et al. 2004). This conclusion is warranted because total lack of the enzyme would result in absolute glucose refractoriness and insulin-dependent diabetes with ketosis. In two studies of islets from T2DM donors, GK mRNA was found to be reduced (Del Guerra et al. 2005; Li et al. 2009). Studies in several animal models of T2DM also suggest that beta-cell GK remains functional and that there is no clear relationship between the status of GK in beta-cells and the severity of the disease (Liang et al. 1994). In this context it is of interest that islet tissue microdissected from pancreas of alloxan or streptozotocin diabetic rats showed GK activity that is comparable to that of control tissue suggesting the presence of the enzyme in alpha and delta cells of the islets apparently expressed independently of insulin (Bedoya et al. 1987). Considering the GK status in hepatic parenchymal cells, it is well documented that the enzyme is lost in untreated animal models of T1DM and by extrapolation also in humans with the disease when not treated with insulin but that the enzyme is readily induced by insulin. The assessment of the hepatic GK status in T2DM in man is based on two publications. The available data indicate that hepatic GK in individuals with IFG, IGT, or mild diabetes mellitus is normal or actually increased (Wilms et al. 1970) whereas it is drastically reduced in liver tissue of morbidly obese with diabetes mellitus (Caro et al. 1995). These latter results resemble the finding in diabetic fatty rats (Torres et al. 2009). Altogether this information suggests that GK is operative in the beta-cells and the hepatocytes of individuals with IFG, with glucose intolerance and mild forms of diabetes and that GK-activating drugs could interact with the desired target and influence glucose metabolism accordingly.

6 Discovery of GKAs by High-Throughput Screening

The plan to develop antidiabetic drugs that might increase GK activity and thereby stimulate insulin release and/or enhance hepatic glucose usage was conceived in 1990 by a team of scientists associated with Hoffmann–La Roche Inc. At this time the decision was based entirely on the results of fundamental research in enzymology, molecular biology, and physiological chemistry using laboratory animals but the possibility that mutations of GK could cause diabetes mellitus had already been recognized. While the R&D of this program was underway linkage of MODY (maturity onset diabetes of the young) to the GK gene (Hattersley et al. 1992) and cases of PHHI (persistent hyperinsulinemic hypoglycemia in infancy) also linked to GK were discovered (Dunne et al. 2004; Palladino et al. 2008) greatly streng thening the case of GK as a drug target for antidiabetic therapy. The human biochemical genetic findings were soon confirmed and extended by generating the corresponding mouse knockout and transgenic models (Magnuson and Kim 2004). It was also recognized that activating GK mutants causing PHHI clustered in the same area of the enzyme that bound GKAs and was remote from the glucose/MgATP binding site suggesting the existence of a hitherto unrecognized allosteric activator site of this enzyme (Christesen et al. 2002; Cuesta-Muñoz et al. 2004; Glaser et al. 1998; Gloyn et al. 2003). These parallel developments reinforced each other in the conclusion that GK activation had a unique potential as a means of increasing insulin secretion from the beta-cell and stimulating hepatic glucose phosphorylation.

Two tactics of screening had been used in the search for drugs that might activate GK, both based on the same rationale that this activation might be best achieved by blocking physiological or pathological inhibition of the enzyme. Acyl-CoA (Qian-Cutrone et al. 1999) and GKRP (GK regulatory protein found in liver) (Grimsby et al. 2003) were employed as inhibitors but their choice greatly influenced the outcome. With the screening test employing acyl-CoA as inhibitors, investigators at Bristol–Myers Squibb discovered lipid binding molecules in *Streptomyces* and *Nocardia* strains which reactivated GK but this initial lead was not further pursued apparently because acyl-CoA does not seem to play a physiological or pathological role in the regulation of GK and perhaps also for practical reasons. With the GKRP-based screening method, investigators at Hoffmann–La Roche discovered small molecules that counteracted GKRP inhibition of the enzyme but also activated GK directly. These compounds were termed GK activators or GKAs. Optimization of the lead compound resulted in GKAs with the following characteristics: GKAs usually increase the kcat of GK by 50–100%, but in some instances reduce kcat; they increase the affinity for glucose as indicated by a decrease of the $S_{0.5}$ in many cases by factors of 5–10; they lower the affinity for the second substrate MgATP (i.e., increase the ATPKm) when the measurement is performed at physiological glucose levels of 5 mM but have no effect on this parameter when the measurement is done with saturating levels of glucose; many of them do lower the nH of the enzyme's glucose dependency curve; and they counteract the inhibition of GK by GKRP. One wonders why in the first screening paradigm using GK inhibited by acyl-CoA, potential GKAs were overlooked because they should be detectable independent of the chemical nature of a reversible inhibitor. In the following section, the structures of various GKA classes and their actions will be discussed in detail.

7 What Are GKAs Chemically?

In the mid-1990s, Roche embarked on a high-throughput screen (HTS) of 120,000 compounds that led to the identification of a single, direct activator of GK (Fig. 6). The hit compound was considered "drug-like" based on its low molecular weight

Fig. 6 An acyl urea
derivative GKA hit molecule
of high-throughput screening
and the optimized lead
RO0281675

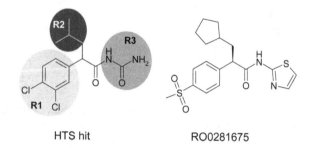

HTS hit   RO0281675

(303.19), balance of polar and hydrophobic moieties (clogP – 3.86), and potency. Structure–activity relationships were established through the synthesis of analogs with different substitution patterns on the aryl (R1), alkyl (R2), and acyl urea (R3) regions of the molecule. These efforts culminated in the identification of RO0281675 (Fig. 6) as a potent GKA lead molecule, with excellent pharmacokinetics and efficacy profiles in several preclinical animal models of type 2 diabetes (Grimsby et al. 2003).

Extensive interest in GK activation as a potential mechanism to treat T2D was triggered after the publishing of several patent applications from Roche beginning in 2000. To date, over 100 patent applications (Fig. 7) have been published from several leading pharmaceutical research organizations (Sarabu et al. 2008). In general, most GKAs adhere to a common pharmacophore model with related structural motifs as shown in Fig. 8a. The model consists of a center, which can be a carbon or an aromatic ring, and three attachments (R1–R3) to it. Typically, two of these attachments (R1 and R2) are hydrophobic groups, with at least one of them being an aromatic ring. The third attachment (R3) is predominately a 2-aminoheterocycle or a N-acyl urea moiety. This group provides a key recognition element between activators and the protein through a hydrogen-bond donor (amide NH) and acceptor (urea carbonyl or imine).

GKA patents can be organized into the following structural types: I. "carbon" centered (saturated, olefin, and cyclopropyl subtypes), II. aromatic ring centered (standard and atypical hydrogen bond donor/acceptor), III. amino acid based, and IV. "nitrogen" centered as shown in Fig. 8b. To date, the majority of GKA patents are related to the carbon centered and aromatic ring centered analogs (Fig. 9, red and blue, respectively).

Despite this wide variety of structural types observed in the patent literature, the R3 amide/heterocycle side chain is well conserved. More recently, deviations from this motif have appeared in the aromatic centered compounds (Fig. 9, aromatic atypical). In these molecules, the hydrogen bond donor acceptor pharmacophore, usually satisfied by an 2-amino heterocycle, is replaced by a biheteroaryl moiety. Banyu, for example, has disclosed a series of 2-pyridyl-benzimidazoles lacking the amide sidechain (Fig. 10). In this case, the benzimidazole NH and the pyridiyl "N" are the hydrogen-bond donor acceptor recognition element (Banyu Pharamceutical Co. Ltd 2005).

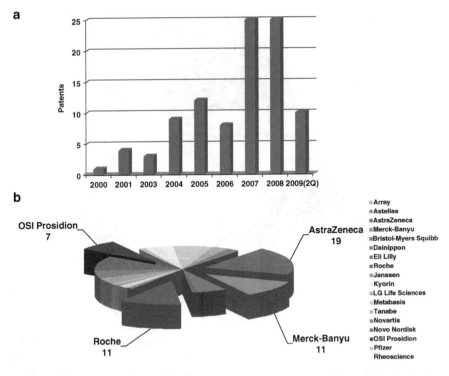

Fig. 7 Number of GK activator patents published per year since 2000 and distribution by pharmaceutical company (data compiled from Prous Integrity 8/09)

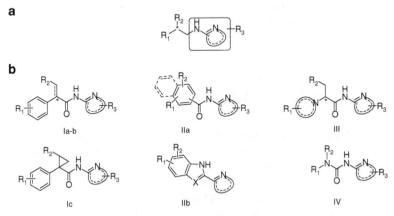

* = a chiral sp³ carbon or an olefinic sp² carbon, or an aromatic moiety;

R1 = aromatic or aliphatic or 'N' containing ring;

R2 = aliphatic or aromatic ring

R3 = heterocyclic N-acyl amine or amine with a potential for H-bond donor-acceptor interaction

Fig. 8 GK activator pharmacophore model (**a**) and common GK activator structural motifs (**b**)

Fig. 9 Distribution of GKA patents by structural class

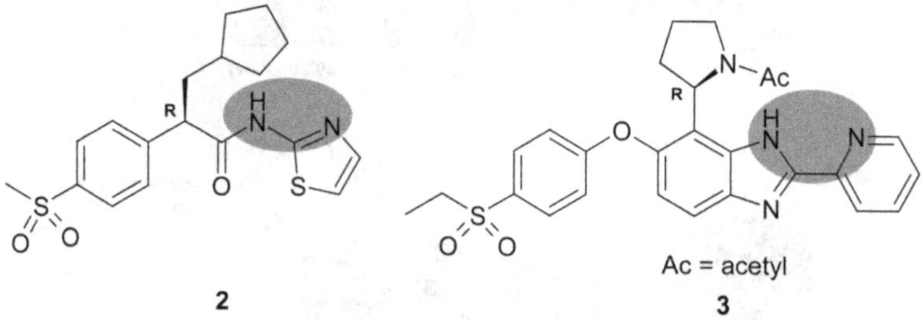

Fig. 10 Donor acceptor pair (*gray* highlight) in prototypical GKA RO0281675 (**2**) and Banyu compound (**3**) (Banyu Pharamceutical Co. Ltd 2005)

Crystallographic studies on GKAs from different series have confirmed the molecular recognition interactions required for binding to the allosteric activator site of the protein. In addition to the R3 H-bond donor–acceptor interaction with the Arg[63] backbone carbonyl and amide NH, hydrophobic interactions of R1 and R2 are important. The R1 aryl group makes hydrophobic, and possibly pi–pi interactions, with the Tyr[214] and Tyr[215] residues while the R2 hydrophobic group interacts with the Met[235] side chain, two important elements (Fig. 11). In aromatic centered analogs that do not posses two distinct R1 and R2 hydrophobes, the core itself may act as one of these elements.

8 How Do GKAs Activate GK at the Molecular Level?

Not surprisingly GKAs mimic all effects that have been discovered in the biochemical genetic studies of recombinant human activating GK with mutations causing PHHI (Fig. 12) (Banyu Pharamceutical Co. Ltd 2005; Brocklehurst et al. 2004; Coghlan and Leighton 2008; Dunten et al. 2004; Efanov et al. 2005; Futamura et al. 2006; Fyfe et al. 2007; Grimsby et al. 2003, 2004; Guertin and Grimsby 2006; Kamata et al. 2004; Leighton et al. 2007; Nakamura et al. 2007; Sagen et al. 2006; Sarabu and Grimsby 2005; Sarabu et al. 2008; Sorhede Winzell et al. 2007). The predominant effect is a lowering of the glucose $S_{0.5}$ to as low as 10% of the control. Many activators increase the kcat as much as twofold but in some instances the kcat may be somewhat decreased by the drug. GKAs lower the Hill coefficient to varying

Fig. 11 Interactions
of composite GKA
pharmacophore with
allosteric activator binding
site. R3 donor acceptor
hydrogen bonds are depicted
in *dashed* lines, R1 and R2
hydrophobes highlighted in
gray.

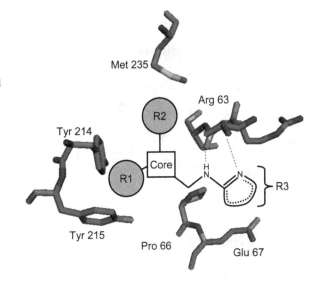

degrees in some cases to a value approaching unity. It should be realized that
small effects on the nH may be overlooked in the kinetic analysis. Alterations of the
Km for the second substrate MgATP are seen only when the analysis is performed
at physiological glucose levels such that the normal Km of about 0.15 mM might
be increased two to threefold to values that are observed with saturating concentrations of glucose. GKAs enhance the dissociation of the GK/GKRP complex in
the presence of physiological levels of glucose (Futamura et al. 2006) and stabilize
the enzyme in thermolability tests performed at physiological glucose levels. It is
important to realize that most GKAs studied so far do not bind to the enzyme in the
absence of glucose. A striking exception is a strongly activating Merck–Banyu
compound which was found to bind effectively in the absence of glucose, a finding
greatly facilitated by the fluorescence characteristics of the molecule (Antoine et al.
2009). GKA actions have also been studied with a number of spontaneous activating and inactivating mutants of GK causing glucokinase disease. These explorations resulted in the striking finding that decreased responsiveness to GKAs is
usually associated with refractoriness to GKRP inhibition. Two outstanding examples
are V62M (Gloyn et al. 2003) and G72R (Sagen et al. 2006), both mutants that are
kinetically activating but paradoxically, cause hyperglycemia (Fig. 4). Since these
two mutant molecules also show increased thermal instability in solution-based
assays as well as in biological assays it was speculated that decreased responsiveness to a putative endogenous activator and/or structural and catalytic instability
may be part of the explanation of the hyperglycemic phenotype. The lack of
response to GKRP and to GKAs in these two cases is explained by the hypothesis
that these mutations, both located in a critical connecting loop between the large
and the small lobe and part of the allosteric site, induce the more compact conformation of the activated state precluding GKRP binding and obviating activation by

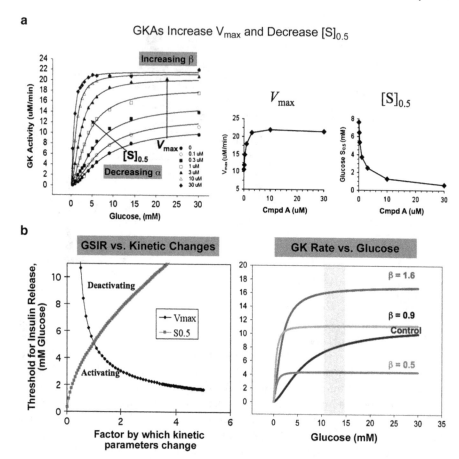

Fig. 12 GKAs increase Vmax and decrease the glucose $S_{0.5}$. Panel (**a**) shows the effect of rising concentrations of a GKA (piragliatin) on the kinetics of recombinant pancreatic islet GK leading to a twofold increase of the Vmax and a nearly tenfold decrease of the glucose $S_{0.5}$. In Panel (**b**) (*left*) the impact of a GKA on the threshold of glucose-stimulated insulin secretion (GSIR) is illustrated. Decreasing the glucose $S_{0.5}$ by 50% or increasing the Vmax by a factor of 2 lowers the threshold from normally 5 mM to 3 mM glucose. Also in Panel (**b**) (*right*) the impact of three different GKAs with different beta – but similar alpha values is shown on relative enzymatic activity (KG rate), beta referring to changes of kcat and alpha to changes of $S_{0.5}$

GKAs. The majority of known activating GK mutants show a reduction or total lack of responsiveness to GKRP and GKAs.

The allosteric activator site and the contact amino acids for GKAs in this site are defined in several published crystal structures (Dunten et al. 2004; Efanov et al. 2005; Grimsby et al. 2003, 2004; Kamata et al. 2004). The site is formed by a loop connecting beta 1 and beta 2 (including V62 to G72) and by the alpha helix 5 and the C-terminal alpha helix 13 (Fig. 3). Depending on the chemical nature of the drug it involves the following amino acids: V62, R63, E210, I211, Y214, Y215, M235, V452, V455, and A456. Mutations of many of these and neighboring amino acids cause activation of this enzyme and were identified in patients with PHHI. Connecting loop I, approximately between V62 and G72, forms a hairpin loop and is occluded in the super open form impeding or preventing GKA binding but opens and accepts the drug when the substrate site is occupied by glucose. Helices alpha-5 containing amino acids 210–215 and alpha-13 containing amino acids 445–465 contribute to the drug binding site. In the wide open conformation, these two helices are positioned parallel to each other but following glucose binding alpha-13 swings into a position which is perpendicular to alpha-5 and is threaded behind connecting loop I.

Biophysical studies, including differential scanning calorimetry (DSC), scintillation proximity assays (SPA), and tryptophan fluorescence, have shown that GKA binding to the enzyme is greatly enhanced by D-glucose (Fig. 4) (Dunten et al. 2004; Heredia et al. 2006; Ralph et al. 2008; Zelent et al. 2008). In fact some GKAs do not bind at all in the absence of a sugar ligand. This conclusion is also supported by the finding that GKAs alone do not stabilize the GK protein in thermolability tests but are able to enhance the stabilizing effect of glucose. These observations are explained by crystallographic studies. Three teams have crystallized GK as a ternary complex containing GKA and D-glucose (1:1:1) and have delineated the allosteric activator site, identifying as many as nine contact amino acids depending on the chemistry of the drug as detailed above (Dunten et al. 2004; Efanov et al. 2005; Grimsby et al. 2003, 2004; Kamata et al. 2004). This drug binding site was not accessible in GK crystals prepared without glucose, at least not with the compounds then available. However, in the presence of near saturating levels of a particular activator, which on its own does not alter fluorescence, D-mannoheptulose initiates a slow and relatively large ligand induced tryptophan fluorescence increase (Fig. 13). The transition process is temperature dependent and allows an assessment of the activation energy of the slow ligand induced transition from a more open to a more closed configuration. This ligand induced "slow transition" (measured in terms of min) is also observed with the physiological substrate D-glucose but the phenomenon is optimally studied with the inhibitory sugar. These results are interpreted as manifestation of the transition from a "super open" to the "closed" GK conformation, suggesting that the sugar binding site and the allosteric drug binding site interact cooperatively to bring about this slow transition from one structure to the other.

GKAs greatly potentiate the competitive inhibition of GKRP action on glucokinase by glucose, thereby releasing the enzyme from the nuclear compartment of the hepatocyte. This GKA interference with GKRP inhibition of glucokinase finds its corollary in the large impact of activating mutations on the apparent Ki of GKRP which may increase maximally more than 100-fold depending on the nature of the mutation. Quantitative comparisons of published results of in vitro studies of different compounds are complicated by the fact that their K_d and EC_{50} values are inversely related to the glucose levels and may vary as much as 10–100 fold, respectively, in comparing low and high glucose saturation (Ralph et al. 2008). In order to overcome this difficulty, the relative GK activity index (%-AI) could be used to compare the efficacy of different GKAs. The AI (which is defined as ($kcat$/glucose $S_{0.5}^{nH}$) $\times$ (2.5/2.5+ATPKm)) is an approximate expression of the enzyme's catalytic efficiency. Its use is suggested because the %-AI was instrumental

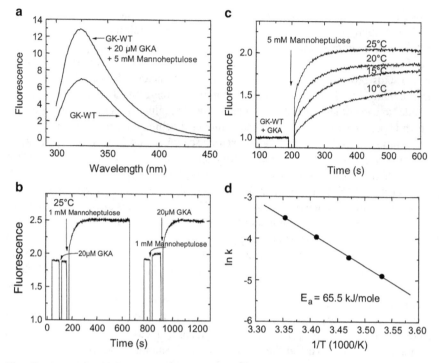

Fig. 13 Tryptophan fluorescence (TF) of normal GK in the presence of mannoheptulose and a GKA. In all instances about 1 microM recombinant GK was used dissolved in 5 mM phosphate buffer (pH 7.3) with 100 mM KCl and 1 mM dithiothreitol. In panel A the TF of GK in the basal state and in the presence of 5 mM mannoheptulose and 20 microM of GKA is shown. TF is 1.92 times basal and the F-peak wavelength is not changed after addition of the sugar. Panel (**b**) shows the slow transition kinetics of TF induced by 1 mM mannoheptulose plus 20 microM GKA Note only the combination of the two ligands initiates the slow transition and also note that the K_d for mannoheptulose in the absence of GKA is 20 mM but is 1.25 mM in the presence of the drug. Panel C shows the temperature dependency of the slow transition and panel D presents an Arrhenius plot of the results given in panel C, allowing the calculation of 65.5 kJ/mole of activation energy for the ligand-induced conformational change

in attempts to characterize the impact of activating and inhibiting GK mutants on the glucose threshold for insulin secretion, allowing predictions to be made about the severity of the respective phenotypes of the mutant carriers (Gloyn et al. 2004). Mathematical modeling experiments illustrate the enormous range of GK stimulation by GKAs which have different effects on kcat, glucose $S_{0.5}$, and the Hill coefficient (Matschinsky 2009).

9 Effects of GKAs at the Cellular and Organ Levels

All known biological effects of GKAs are entirely predictable from the central role of GK in glucose homeostasis and from the effects of the drug at the molecular level as discussed above (Bonadonna et al. 2008; Brocklehurst et al. 2004; Coghlan and

Leighton 2008; Coope et al. 2006; Efanov et al. 2005; Futamura et al. 2006; Fyfe et al. 2007; Grimsby et al. 2003, 2004; Guertin and Grimsby 2006; Johnson et al. 2007; Leighton et al. 2007; Migoya et al. 2009; Nakamura et al. 2007; Ohyama et al. 2009; Sarabu and Grimsby 2005; Sarabu et al. 2008; Sorhede Winzell et al. 2007; Walker and Rao 1964; Zhi et al. 2008). This is demonstrated by the action of GKAs on extracorporeal pancreatic beta-cell or liver preparations. GKAs shift the concentration dependency curve of glucose-stimulated insulin release to the left (see example in Fig. 14). Many of them also augment maximal secretory activity moderately. They enhance glycolysis and glucose oxidation and concomitantly increase glucose-induced stimulation of respiration, effects again characterized by marked left shifts of the glucose concentration dependency curves and moderately increased maximal rates. As a result, GKAs increase the *phosphate potential* of pancreatic beta-cells as manifested by increased levels of ATP and P-creatine but decreased levels of ADP and inorganic phosphate (Grimsby et al. 2004). The drugs have no influence on the metabolism of nonglucose fuel stimulants of beta-cells (e.g., amino acids). GKAs increase the concentration of free cytosolic calcium concentrations but only when glucose is present (Fig. 14). GKAs enhance the induction of GK expression by glucose in isolated cultured pancreatic rat islets of Langerhans and thus cause a sustained sensitization of the beta-cells to glucose even after the drug is cleared (Fig. 15). GKAs enhance the induction of GK by glucose and the nonmetabolizable mannoheptulose, a heptose, most likely an expression of protein compaction and stabilization by the ligand in the substrate site (Table 2, Fig. 16). It is noteworthy, however, that induction of glucokinase by mannoheptulose in the absence of glucose (with or without GKA present) is not sufficient to maintain the beta-cell capacity for glucose-stimulated insulin secretion. It seems then that some basal glucose metabolism is required to preserve normal stimulus secretion coupling. GK activation by GKAs in the presence of glucose protects against hydrogen peroxide-induced cell death in model pancreatic beta-cells thus showing a marked antiapoptotic effect (Futamura et al. 2009). GKAs enhance glucose-induced insulin release in isolated cultured human islets from normal organ donors and also from those with type 2 diabetes (Johnson et al. 2007) and unpublished results by the authors). The drugs augment glycogen synthesis in isolated hepatocyte preparations explained by two distinct actions, i.e., the direct activation of cytosolic GK and the enhanced release of inactive GK sequestered in the nuclear compartment (see above) (Brocklehurst et al. 2004; Efanov et al. 2005). It is important to note here that the predictable influence of GKAs on all other GK-expressing cells and tissues that comprise the complex glucose sensing network described above remains to be explored (Matschinsky et al. 2006).

10   Effects of GKAs on Glucose Homeostasis of Normal and Diabetic Laboratory Animals and Humans

Whole body studies with various GKAs in normal laboratory animals and humans and, more importantly, in animal models of type 2 diabetes and humans with this disease had positive outcomes as predicted (Bonadonna et al. 2008;

Fig. 14 Effect of a GKA
(piragliatin) on glucose-
induced respiration, insulin
release, and intracellular
calcium of isolated cultured
mouse islets. Islets were
cultured in RPMI containing
10 mM glucose for 3–4 days
and were thereafter studied.
Panel A shows the insulin
release patterns with glucose
stimulation using stepwise
increasing levels, both in the
absence and presence of
piragliatin. Panel B shows the
effect of 3-microM piragliatin
on islet respiration following
a stepwise increase of glucose
from zero to 3, 6, 12 and
24 mM glucose followed by
treatment with 5 microM of
the uncoupler of OxPhos
FCCP and 1 mM Na-azide. O_2
consumption was determined
with a method based on
phosphorescence quenching
of metalloporphyrins by
oxygen. Panel C shows
corresponding changes of
intracellular Ca^{++} due to
stepwise increases of glucose
from zero to 1, 3, and 9 mM.
The Fura-2 method was
employed

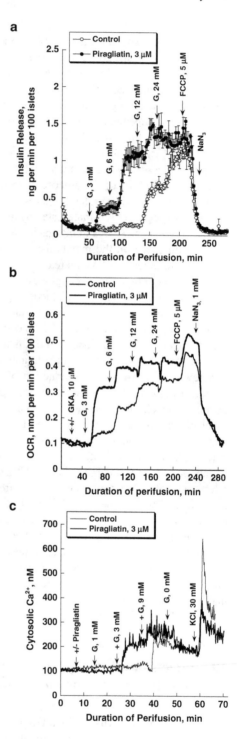

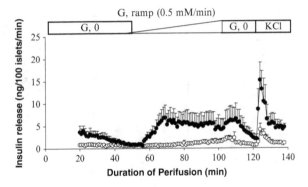

Fig. 15 Enhanced glucose induction of GK by piragliatin in cultured rat islets augments glucose-stimulated insulin release. The marked functional impact of piragliatin on GK induction by 6 mM glucose is shown. Isolated islets were first cultured for 3–4 days with 6 mM glucose in the presence and absence of 3 microM piragliatin and were then tested in a perifusion experiment stimulating them with a glucose ramp in the absence of the activator from 50 to 100 min of the perifusion. *Open circles:* Islets cultured with vehicle (0.1% DMSO) in 6 mM glucose; *Solid circles:* Islets cultured with GK activator (piragliatin at 3 microM) in 6 mM glucose. Note the left shift of the dose–response curve and the marked increase of maximal release

Table 2 Pancreatic islet glucokinase induction by the GK activator piragliatin

Culture conditions	Control (pmol/ug protein/hour)	GKA, 3 μM	Control (glucose $S_{0.5}$, mM)	GKA, 3 μM
1 mM Glucose	82.4 ± 21.9	123 ± 6.37	6.72 ± 3.20	7.90 ± 0.67
3 mM Glucose	120 ± 11.2	340 ± 65.8	8.32 ± 0.62	6.90 ± 0.66
6 mM Glucose	156 ± 8.90	367 ± 109	7.54 ± 0.86	6.13 ± 0.26
9 mM Glucose	214 ± 12.6	452 ± 113	7.04 ± 0.62	6.28 ± 0.54
12 mM Glucose	391 ± 46.0	501 ± 87.0	7.65 ± 0.03	7.22 ± 0.12

Piragliatin enhances glucose induction of rat pancreatic islets GK. Isolated islets were cultured for 3–4 days in RPMI at different glucose concentrations in the presence and absence of 3-microM piragliatin. After removal of the culture medium islets were stored at −80°C. They were then homogenized and the homogenate was briefly centrifuged to remove particulate material. GK was then measured with a method previously described which provides Vmax, glucose $S_{0.5}$, and nH data while also testing the response to GKAs (Zelent et al. 2006). At the assay step the GKA present in the culture medium is diluted at least 1,000 fold such that the drug does not interfere with the analysis. Note that increases of the Vmax are not paralleled by an alteration of the glucose $S_{0.5}$ which cluster around the expected value of about 8 mM. The nH values were also unchanged and the enzyme was activatable by GKAs to the same degree under all conditions (not shown).

Camacho et al. 2009; Coope et al. 2006; Fyfe et al. 2007; Grimsby et al. 2003, 2004; Johnson et al. 2007; Leighton et al. 2007; Migoya et al. 2009; Nakamura et al. 2007; Ohyama et al. 2009; Sorhede Winzell et al. 2007; Walker and Rao 1964; Zhi et al. 2008). In acute studies, GKAs lowered blood sugar in a dose-dependent manner under normal conditions. The drug effect was usually attributed to a dual action, enhancement of insulin release, and facilitation of glucose clearance by the liver. High dosages often resulted in hypoglycemia especially in normal animals (Fig. 17) and human subjects. In the great majority of studies with type 2 diabetic

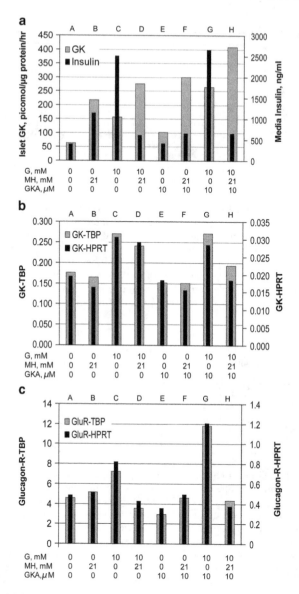

Fig. 16 GK induction in cultured rat pancreatic islets stimulated with glucose and mannoheptu-lose, with and without the GKA piragliatin present. The basic DMEM culture medium contained 10% glucose free fetal calf serum and was supplemented with 7 mM each of glutamine and leucine. Panel A shows results of GK activity measurements of islet homogenates and the insulin levels as measured in the media after 3–4 days of islet culture. Panel B presents the results of quantitative GK mRNA measurements and panel C provides data on quantitative glucagon receptor mRNA measurements for comparison. The results represent means of five separate, highly reproducible measurements of five cultured rat islet isolates. The glucagon receptor data are given as example of protein induction that seems to parallel insulin release and perhaps glucose metabolism in contrast to GK induction which is not. Note that the insulin levels of the culture

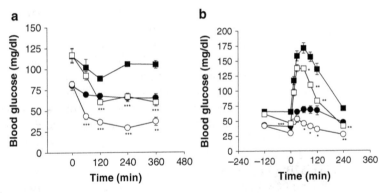

Fig. 17 Acute effects of orally administered GKA, RO0281675, on basal blood glucose levels and glucose tolerance in normal and diabetic rats. (**a**) Glucose lowering effects in 7-week old normal male Wistar rats (Charles River Laboratories) (*circles*) and diabetic Goto-Kakizaki (GK) rats (Charles River Laboratories) (*squares*) orally administered vehicle ($n = $ 5–6/time point) (*filled symbols*) or 50 mg/kg RO0281675 ($n = $ 5–6/time point) (*open symbols*). (**b**) Oral glucose tolerance test in fasted 8-week old male Wistar rats (*circles*) and GK rats (*squares*) orally administered vehicle ($n = $ 4–5/time point) (*filled symbols*) or 50 mg/kg RO0281675 ($n = $ 4–5/time point) (*open symbols*) 120 min prior to glucose administration (2 g/kg). Experimental details previously described (Grimsby et al. 2003). All results are reported as the mean ± SEM. A Student's t-test was used to test for statistical significance (*, $P < 0.05$; **, $P < 0.01$; and ***, $P < 0.005$)

animals and human subjects, GKAs lowered the elevated blood glucose dose dependently, usually, such that normoglycemia was achieved at the highest dosages. Treatment with GKAs was associated with an increased insulinogenic index interpreted as a manifestation of a dual action of the drug causing enhanced insulin secretion which impinges on all insulin sensitive tissues and direct activation of hepatic GK resulting in augmented hepatic glucose uptake and glycogen synthesis. It is noteworthy that the absolute insulin concentrations did not always rise (but may actually fall!) when compared to the untreated diabetic condition, an indication of the dual pancreatic and hepatic action of the activators with reduced insulin resistance. It should be remembered that in GK-linked HI, absolute insulin levels are not necessarily elevated even though they are certainly high relative to the blood glucose concentrations measured in these cases. GKAs may thus, paradoxically, reduce hyperstimulation of the diabetic beta-cells by reestablishing a normal set point of glucose homeostasis (Palladino et al. 2008). Only a small selection of specific examples for in vivo experiments with GKAs is given here and should suffice to provide sufficient proof for the high validity of the proposed MOA and

←

Fig. 16 (continued) medium are unphysiologically high because islets synthesize and release the hormone very actively for a period of days apparently without feedback inhibition of secretion. Abbreviations: *G*, glucose; *GK-TPB*, GK mRNA rel. to TATA binding protein mRNA; *GK-HPRT*, GK mRNA rel. to hypoyxanthine-phosphoribosyl-transferase mRNA; *Glucagon-R-TBP*, glucagon receptor mRNA rel. to TATA binding protein mRNA; *Glucagon-R-HPRT*, glucagon-receptor mRNA rel to hypoxanthine-phosphopribosyl-transferase mRNA; *MH* mannoheptulose

principle of concept of GKA-based diabetes therapy. The first peer reviewed report in 2003 demonstrated the therapeutic potential of allosteric stimulation of GK by GKAs on hand of extensive in vivo studies with normal and diabetic rodents (Grimsby et al. 2003). In normal animals, GKAs lowered the blood sugar by their dual action of stimulating insulin release and augmenting hepatic glucose uptake while decreasing glucose production. In several models of T2DM, GKAs normalized the elevated blood glucose. Numerous studies using GKAs of a wide variety of chemical structures confirmed, complemented, and expanded on these preclinical observations. Of special significance is the ameliorating effect of chronic GKA treatment on the development of hyperglycemia in the course of diet-induced obesity (DIO) in C57Bl/6J mice (Grimsby et al. 2004). A recent report in abstract form showed that a GKA was able to lower blood glucose in rats desensitized to sulfonylurea agents (Ohyama et al. 2009). And finally, it was disclosed at the 2008 meeting of the EASD in Rome that treatment with the experimental GKA *Piragliatin* (RO4389620) lowers blood glucose dose dependently in normal volunteers and patients with T2DM (Bonadonna et al. 2008; Walker and Rao 1964; Zhi et al. 2008). Piragliatin administered orally to healthy male subjects at a dose of 5, 10, and 25 mg was well tolerated causing a dose-dependent reduction of fasting plasma glucose. Furthermore, a single dose of piragliatin lowered both fasting and post challenge glucose in a group of patients with T2DM by improving beta-cell function, hepatic glucose production, and peripheral glucose utilization (Zhi et al. 2008). And finally, multiple doses of piragliatin resulted in a rapid dose-dependent blood glucose reduction over a period of 24 h in individuals with T2DM (Bonadonna et al. 2008). In all these studies, the drug was safe and well tolerated, even though mild-moderate hypoglycemia was observed at the highest doses tested. At the 2009 ADA meeting in New Orleans, investigators from Merck reported that a novel GKA lowered the blood glucose in normal volunteers (Migoya et al. 2009).

11 Critical Assessment of GKA's Potential for Diabetes Therapy

In July 2009, at the time this review was completed, R&D of GKAs as potential new antidiabetic drugs had reached a critical milestone in its progress. New chemical entities with GKA characteristics counting in the hundreds had been discovered and patented. The MOA of GKAs at the molecular, cellular, and organismic level had been explored in much detail and the results of these investigations had proven that the biological and medical concepts on which the R&D of GKAs was based are valid (Coghlan and Leighton 2008; Guertin and Grimsby 2006; Matschinsky 2009; Sarabu and Grimsby 2005; Sarabu et al. 2008). In reports at international meetings in 2008 and 2009 (Bonadonna et al. 2008; Migoya et al. 2009; Walker and Rao 1964; Zhi et al. 2008), strong evidence was presented that GKAs of different chemical structures lower blood glucose markedly in healthy humans and do even normalize hyperglycemia in patients with T2DM in week long trials without showing medically significant side effects except moderate hypoglycemia at high drug dosages. The indications were that an impressive number of leading pharmaceutical houses had embarked on the next phase of developing reliable oral antidiabetic medicines based on the experience with GKAs known at the time.

The present environment remains highly challenging for launching antidiabetic drugs with a novel MOA because of the pressure to develop medicines which not only lower blood glucose markedly and persistently and show advantages over existing therapies but also improve outcome of the cardiovascular diseases which are associated with T2DM (Desouza and Fonseca 2009). GKAs are well positioned to meet this challenge owing to their unique MOA. In view of the ideas and experimental data discussed in this article, it is very reasonable to expect that GKAs have great potential for monotherapy but are also uniquely suited for combination regimens with metformin, GLP-1 analogues, DPP-IV inhibitors, insulin and insulin sensitizers, and perhaps even with sulfonylurea compounds when treatment with these agents is failing. This positive assessment draws its strength from the central role of GK in glucose homeostasis in health and disease and from the uniformly beneficial preclinical studies and early clinical studies with GKAs of very different chemical structures. Of particular importance are the observations which demonstrate that GKAs do not elevate blood lipids, contrary to the predictions and fears that had been expressed in the early stages of developing GK-based therapies. This outcome of lipid neutrality of GKAs is consistent with their ability to prevent the development of diabetes in DIO mice and in the absence of hepatic glycogen and triglyceride accumulation (unpublished). Any lasting impact on cardiovascular health of diabetics will depend on the *quality of blood sugar control* that can be achieved with GKAs alone or in combination regimens in comparison to other drug treatments.

It seems reasonable to predict that several GKAs will enter late-stage clinical trials in the next 2–3 years. It is also safe to predict that proof of MOA and of medical concept will be fully confirmed in these anticipated clinical studies in patients with T2DM and also T1DM. However, it is obvious that the outcome of such trials has to be awaited before final judgment can be made about a significant durable benefit that GKAs may provide for diabetes management compared to established and other experimental therapies. Attention will have to paid to the following questions: (1) Can the blood sugar of diabetics be normalized with GKAs alone or in combinations with approved or currently experimental drugs in a predictable and persistent manner while avoiding medically significant hypoglycemic events? (2) Do GKAs predispose to weight gain? (3) What is the long-term impact on blood lipids in patient with and without statin medication? (4) Do GKAs change the frequency or severity of cardiovascular events? In addition to addressing these major issues, investigators should consider the possibility that GKAs modify the function of other tissues expressing the enzyme besides pancreas and liver including enteroendocrine cells that secrete GLP-1 and GIP, nuclei of the CNS in the hypothalamus and other areas in all probability involved in counterregulation and finally in the gonadotropes of the pituitary.

While answers to the clinically highly relevant questions will be sought in the anticipated clinical trials it is very likely that GKAs will also prove to be of great value in basic biophysical and biochemical studies of the recombinant enzyme and of the central role of GK in glucose homeostasis using normal and diabetic animals or isolated pancreatic islets, enteroendocrine, pituitary, and liver tissue preparations. It can be expected that crystal structures of binary GK complexes with GKAs that bind without glucose and with glucose in the absence of GKAs will become available. Much progress can be expected from the application of DSC and isothermal titration calorimetry (ITC) as well as the use of tryptophan fluorescence based on rapid mixing and equilibrium binding studies. Of foremost importance will be explorations of the incompletely understood cooperative kinetics of GK and of the detailed molecular basis of GK/GKRP and GK/BAD interactions. GKAs promise to serve as powerful tools to assess the biological significance of GK, if any, in neurons, enteroendocrine cells, hepatoportal glucose sensor cells, and gonadotropes.

A burning question that deserves great attention is whether GKAs are beneficial in the preservation of the genetically compromised pancreatic beta-cells in T2DM (Bell and Polonsky 2001; Kahn 2003; Kahn et al. 2009) (Fig. 18). It has been demonstrated in mouse islets that GK interacts with the proapoptotic mediator BAD forming

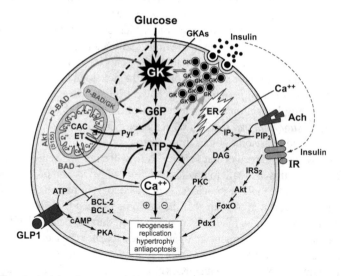

Fig. 18 Hypothetical involvement of GK and therefore GKAs in the preservation and growth promotion of pancreatic beta-cells. The interaction between GK and protein factors governing apoptosis (BCL-2, BCL-xL, BAD, BAK, BAX (the latter two proapoptotic factors situated most downstream in the pathway but not shown)) is highlighted. The P-BAD/GK complex and its association with mitochondria is shown. The glucose and thus GK dependency of GLP-1, insulin, and acetylcholine initiated signaling processes is sketched in. It is hypothesized that activation of these pathways and their enhancement by GKAs protects beta-cells from proapoptotic diabetogenic factors and enhances beta-cell replication or neogenesis. It is further speculated that persistent deviations of free intracellular calcium from a basal set point may favor apoptosis but that transient, perhaps oscillatory changes of calcium may be beneficial (Danial 2007; Orrenius et al. 2003; Scorrano et al. 2003; Whyte et al. 1993). Selected abbreviations: *Akt or PKB* proteinkinase B; *Ach* acetylcholine; *BAD, P-BAD (phospo-BAD), BCL-2 and BCL-x* mediators of apoptosis and antiapoptosis; *CAC* citric acid cycle; *DAG* diacylglycerol; *ER* endoplasmic reticulum; *ET* electrontransport and oxidative phosphorylation; *FoxO* mediator of insulin and insulin like growth factor signaling; *GLP-1* glucagon like peptide 1; *IRS2* insulin receptor substrate 2; *IP3* trisphosphoinositol; *Pdx-1 or IPF-1* insulin promoter factor 1; *PIP2* phosphotidyl inositol biphosphate; *PKA* proteinkinase A; *PKC* proteinkinase C; *Pyr* pyruvate

complexes that are attached to beta-cell mitochondria in effect neutralizing this mediator of cell death (Danial et al. 2008). The GK/BAD complex seems to facilitate the enhancement of glucose stimulation of beta-cell respiration and insulin secretion. It needs to be assessed whether and if so how GKAs influence this apparent inter-compartmental shuttle of GK and BAD. The reported protective effect of GKAs against hydrogen peroxide-induced cell death could be interpreted as indicating that proapoptotic mediators are neutralized by activated GK (Futamura et al. 2009). In this context, it should be recalled that agents which are now being explored as to their potential of preserving beta-cell mass and function in T2DM, for instance GLP-1 and DPP-IV inhibitors (Desouza and Fonseca 2009), are totally dependent on glucose and thus normal GK activity in beta-cells. GKAs may therefore be synergistic in this regard. This is probably also true for the action of other beta-cell growth factors as for instance insulin, IGF, and perhaps the activation of the muscarinic signaling pathway. There is indeed experimental evidence that GK is a critical component together with IRS2 and PKA of pathways that control the adaptive hyperplasia response of the beta-cell to increased peripheral resistance, for example, induced by DIO (Takamoto et al. 2008; Terauchi et al. 2007).

The present status report and brief future perspective remarks concerning GK and GKAs in glucose homeostasis illustrate in an extraordinary manner how significant basic biological and medical progress can be achieved through the close and iterative interaction of basic biochemistry, biochemical genetics, medicinal chemistry, and pharmacology resulting in the discovery and characterization of a promising new class of antidiabetic agents. The challenge is now to transform these chemicals of great promise into useful medicines.

12 Addendum at the time of Revision (Fall 2010)

This postscript lists and briefly evaluates selected pertinent publications that have appeared since this manuscript was first prepared (Bebernitz et al. 2009; Bonadonna et al. 2010). We identified only one peer reviewed extensive report on a mechanistic study of GKA action in patients with type 2 diabetes mellitus (T2DM) (Bonadonna et al. 2010). The GKA piragliatin (RO4389620) had an acute glucose lowering effect in patients with T2DM mediated by increased insulin secretion, decreased endogenous glucose output, and raised glucose use. Furthermore, Array Biopharma reported in considerable detail on a one day single dose ascending dose study in patients with T2DM showing that blood glucose was lowered dose dependently with normalization of blood glucose being achieved at the highest dose and that this effect was associated with increased insulin release (News release of 8/10/09). Among the human biochemical genetic reports three deserve listing here as partic-ularly relevant. KK Osbak et al. have published an update on GK mutations in which a total of 620 mutations in the gck gene in a total of 1,441 families are described and the implications for the clinical course and management of GK linked disorders are discussed. The report also lists 13 ethyl-nitroso-urea (ENU)

GK mutants in the mouse. These diabetes models are potentially useful for exploring in depth the role of GK in glucose homeostasis and as drug target (Osbak et al. 2009). In a remarkable letter to the editor of the *N. Engl. J. Med.* it was reported that the presence of a GK activating mutation (V91L) causing hyperinsulinemic hypoglycemia was associated with large islets and beta-cell proliferation and the hypothesis was advanced that chronically enhanced glucose metabolism is a critical factor regulating pancreatic beta-cell replication (Kassem et al. 2010). In a study designed to assess cardiovascular risk factors it was found that polymorphism of the GK-activating GKRP (hepatic GK regulatory protein) was associated with elevated serum levels of free fatty acids and triglycerides but not with an elevated cardiovascular risk leading to the conclusion that long-term activation of GK (e.g., by GKAs) may not contribute to cardiovascular risk, a reasonable concern when using this drug (Kozian et al. 2010). Among recently published preclinical studies, several pharmaceutical chemistry reports deserve attention (Haynes et al. 2010; Sidduri et al. 2010; Zhang et al. 2009). Of particular interest is a new trend to develop liver-specific GKAs designed to reduce the danger of hypoglycemia (Bebernitz et al. 2009). In a practically highly relevant pharmacological study, investigators from Merck/Tsukuba have tested whether a GKA lowers blood glucose in sulfonylurea-desensitized rats. By demonstrating that GKAs remain fully effective after sulfonylurea failure, this detailed report illustrates clearly the unique features and perhaps advantages of GKAs as compared to these widely used oral antidiabetic agents (Ohyama et al. 2010). As final highlight, we reference two publications which address the possibility that GKAs may stimulate beta-cell proliferation strengthening our views graphically expressed in Fig. 18. (Nakamura et al. 2009; Salpeter et al. 2010); also see Kassem et al. 2010). From this selection, one can conclude with some confidence that the academic and pharmaceutical communities maintain a keen interest in the central role of GK in glucose homeostasis in large part based on the expectation that GKAs have a high potential for the treatment of T2DM and perhaps also T1DM.

References

Agius L (1998) The physiological role of glucokinase binding and translocation in hepatocytes. Adv Enzyme Regul 38:303–331

Agius L (2008) Glucokinase and molecular aspects of liver glycogen metabolism. Biochem J 414:1–18

Agius L, Peak M (1993) Intracellular binding of glucokinase in hepatocytes and translocation by glucose, fructose and insulin. Biochem J 296:785–796

Agius L, Peak M, Van Schaftingen E (1995) The regulatory protein of glucokinase binds to the hepatocyte matrix, but, unlike glucokinase, does not translocate during substrate stimulation. Biochem J 309:711–713

Ahrén B (2009) Islet G protein-coupled receptors as potential targets for treatment of type 2 diabetes. Nat Rev Drug Discov 8(5):369–385, Epub 2009 Apr 14. Review. Erratum in: Nat Rev Drug Discov. 8(6):516

Antoine M, Boutin JA, Ferry G (2009) Binding kinetics of glucose and allosteric activators to human glucokinase reveal multiple conformational states. Biochemistry 48(23):5466–5482

Arden C, Harbottle A, Baltrusch S, Tiedge M, Agius L (2004) Glucokinase is an integral component of the insulin granules in glucose-responsive insulin secretory cells and does not translocate during glucose stimulation. Diabetes 53:2346–2352

Arden C, Baltrusch S, Agius L (2006) Glucokinase regulatory protein is associated with mitochondria in hepatocytes. FEBS Lett 580:2065–2070

Baltrusch S, Wu C, Okar DA, Tiedge M, Lange AJ (2004) Interaction of GK with the bifunctional enzyme 6-phosphofructo-2-kinase/fructose-2,6-bisphosphatase (6PF2K/F26P2ase). In: Matschinsky FM, Magnuson MA (eds) Glucokinase and glycemic disease: from basics to novel therapeutics. Front diabetes, vol 16. Karger, Basel, pp 262–274

Banyu Pharamceutical Co. Ltd., Novel 2-heteroaryl-substituted benzimidazole derivative. WO 2005063738 A1, 2005

Bedoya FJ, Wilson JM, Gosh AK, Finegold D, Matchinsky FM (1986a) The glucokinase glucose sensor in human pancreatic islet tissue. Diabetes 35:61–67

Bedoya FJ, Matschinsky FM, Shimizu T, O'Neil JJ, Appel MC (1986b) Differential regulation of glucokinase activity in pancreatic islets and liver of the rat. J Biol Chem 261:10760–10764

Bedoya FJ, Oberholtzer JC, Matschinsky FM (1987) Glucokinase in B-cell-depleted islets of Langerhans. J Histochem Cytochem 35:1089–1093

Bell GI, Polonsky KS (2001) Diabetes mellitus and genetically programmed defects in β-cell function. Nature 414:788–791

Bonadonna RC, Kapitza C, Heinse T, Avogaro A, Boldrin M, Grimsby J, Mulligan ME, Arbet-Engles C, Balena R (2008) Glucokinase activator RO4389620 improves beta cell function and plasma glucose indexes in patients with type 2 diabetes. Diabetologia 51(Suppl 1):S371 (Abstract # 927)

Brocklehurst KJ, Payne VA, Davies RA, Carroll D, Vertigan HL, Wightman HJ, Aiston S, Waddell ID, Leighton B, Coghlan MP, Agius L (2004) Stimulation of hepatocyte glucose metabolism by novel small molecule glucokinase activators. Diabetes 53:535–541

Calcutt NA, Cooper ME, Kern TS, Schmidt AM (2009) Therapies for hyperglycaemia-induced diabetic complications: from animal models to clinical trials. Nat Rev Drug Discov 8(5): 417–429

Camacho RC, Qureshi SA, Yang X, Eiki J-I, Zhang BB (2009) Stimulation of insulin secretion and enhancement of insulin action, in vivo, by a small molecule glucokinase activator. 69th ADA Scientific Sessions. Diabetes 58 (Suppl 1), Abstract #1501-P, A388

Cárdenas ML (1995) "Glucokinase": its regulation and role in liver metabolism (Molecular biology intelligence unit). R G Landes Company, Austin, TX

Cárdenas ML, Cornish-Bowden A, Ureta T (1998) Evolution and regulatory role of the hexokinases. Biochim Biophys Acta 1401:242–264

Caro JF, Triester S, Patel VK, Tapscott EB, Frazier NL, Dohm GL (1995) Liver glucokinase: decreased activity in patients with type II diabetes. Horm Metab Res 27(1):19–22

Cherrington AD (1999) Banting lecture 1997. Control of glucose uptake and release by the liver in vivo. Diabetes 48:1198–1214

Christesen H, Jacobsen B, Odili S, Buettger C, Cuesta-Munoz A, Hansen T, Brusgaard K, Massa O, Magnuson MA, Shiota C, Matschinsky FM, Barbetti F (2002) The second activating glucokinase mutation (A456V): implications for glucose homeostasis and diabetes therapy. Diabetes 51:1240–1246

Coghlan M, Leighton B (2008) Glucokinase activators in diabetes management. Expert Opin Investig Drugs 17(2):145–167

Coope GJ, Atkinson AM, Allott C, McKerrecher D, Johnstone C, Pike KG, Holme PC, Vertigan H, Gill D, Coghlan MP, Leighton B (2006) Predictive blood glucose lowering efficacy by Glucokinase activators in high fat fed female Zucker rats. Br J Pharmacol 149:328–335

Cornish-Bowden A, Cárdenas ML (2004) Glucokinase: a monomeric enzyme with positive cooperativity. In: Matschinsky FM, Magnuson MA (eds) Glucokinase and glycemic disease: from basics to novel therapeutics. Front diabetes, vol 16. Karger, Basel, pp 125–134

Cuesta-Munoz AL, Boettger CW, Davis E, Shiota C, Magnuson MA, Grippo JF, Grimsby J, Matschinsky FM (2001) Novel pharmacological glucokinase activators partly or fully reverse the catalytic defects of inactivating glucokinase missense mutants that cause MODY-2 (Abstract 436-P; 61st ADA Meeting Philadelphia). Diabetes 50(Suppl 2):A109

Cuesta-Muñoz AL, Huopio H, Otonkoski T, Gomez-Zumaquero JM, Näntö-Salonen K, Rahier J, López-Enriquez S, García-Gimeno MA, Sanz P, Soriguer FC, Laakso M (2004) Severe persistent hyperinsulinemic hypoglycemia due to a de novo glucokinase mutation. Diabetes 53:2164–2168

Danial NN, Walensky LD, Zhang C-Y, Choi CS, Fisher JK, Molina AJA, Datta SR, Pitter KL, Bird GH, Wikstrom JD, Deeney JT, Robertson K, Morash J, Kulkarni A, Neschen S, Kim S, Greenberg ME, Corkey BE, Shirihai OS, Shulman GI, Lowell BB, Korsmeyer SJ (2008) Dual role of proapoptotic BAD in insulin secretion and beta cell survival. Nat Med 14:144–153

Danial NN (2007) BCL-2 family proteins: critical checkpoints of apoptotic cell death. Clin Cancer Res 13(24):7254–7263

Del Guerra S, Lupi R, Marselli L, Masini M, Bugliani M, Sbrana S, Torri S, Pollera M, Boggi U, Mosca F (2005) Functional and molecular defects of pancreatic islets in human type 2 diabetes. Diabetes 54(3):727–735

Deng S, Vatamaniuk M, Huang X, Doliba N, Lian M-M, Frank A, Velidedeoglu E, Desai NM, Koeberlein B, Wolf B, Barker CF, Naji A, Matschinsky FM, Markmann JF (2004) Structural and functional abnormalities in the islets isolated from type 2 diabetic subjects. Diabetes 53:624–632

Desai UJ, Slosberg ED, Boettcher BR, Caplan SL, Fanelli B, Stephan Z, Gunther VJ, Kaleko M, Connelly S (2001) Phenotypic correction of diabetic mice by adenovirus-mediated glucokinase expression. Diabetes 50(10):2287–2295

Desouza C, Fonseca V (2009) Therapeutic targets to reduce cardiovascular disease in type 2 diabetes. Nat Rev Drug Discov 8(5):361–367

Detheux M, Vandercammen A, Van Schaftingen E (1991) Effectors of the regulatory protein acting on liver glucokinase: a kinetic investigation. Eur J Biochem 200:553–561

Doliba N, Vatamaniuk M, Najafi H, Buettger C, Collins H, Sarabu R, Grippo JF, Grimsby J, Matschinsky FM (2001) Novel pharmacological glucokinase activators enhance glucose metabolism, respiration and insulin release in isolated pancreatic islets demonstrating a unique therapeutic potential (Abstract 1495-P; 61st ADA Meeting Philadelphia). Diabetes 50(Suppl 2):A359

Donovan CM, Hamilton-Wessler M, Halter JB, Bergman RN (1994) Primacy of liver glucosensors in the sympathetic response to progressive hypoglycemia. Proc Natl Acad Sci USA 91:2863–2867

Doria A, Patti ME, Kahn CR (2008) The emerging genetic architecture of type 2 diabetes. Cell Metab 8(3):186–200

Dunne MJ, Cosgrove KE, Shepherd RM, Aynsley-Green A, Lindley KJ (2004) Hyperinsulinism in infancy: from basic science to clinical disease. Physiol Rev 84(1):239–275

Dunten P, Swain A, Kammlott U, Crowther R, Lukacs CM, Levin W, Reik L, Grimsby J, Corbett WL, Magnuson MA, Matschinsky FM, Grippo JF (2004) Crystal structure of human liver glucokinase bound to a small molecule allosteric activator. In: Matschinsky FM, Magnuson MA (eds) Glucokinase and glycemic disease: from basics to novel therapeutics. Front diabetes, vol 16. Karger, Basel, pp 145–154

Edghill EL, Hattersley AT (2008) Genetic disorders of the pancreatic beta cell and diabetes (Permanent neonatal diabetes and maturity-onset diabetes of the young). In: Seino S, Bell GI (eds) Pancreatic beta cell in health and disease. Springer, Berlin, pp 399–430

Efanov AM, Barrett DG, Brenner MB, Briggs SL, Delaunois A, Durbin JD, Giese U, Guo H, Radloff M, Gil GS, Sewing S, Wang Y, Weichert A, Zaliani A, Gromada J (2005) A novel glucokinase activator modulates pancreatic islet and hepatocyte function. Endocrinology 146:3696–3701

Froguel P, Vaxillaire M, Sun F, Velho G, Zouali H, Butel MO, Lesage S, Vionnet N, Clement K, Fougerousse F et al (1992) Close linkage of glucokinase locus on chromosome 7p to early-onset non-insulin-dependent diabetes mellitus. Nature 356:162–164

Futamura M, Hosaka H, Kadotani A, Shimazaki H, Sasaki K, Ohyama S, Nishimura T, Eiki J-I, Nagata Y (2006) An allosteric activator of glucokinase impairs the interaction of glucokinase and glucokinase regulatory protein and regulates glucose metabolism. J Biol Chem 281: 37668–37674

Futamura M, Maruki H, Shimazaki H, Hosaka H, Kubota J, Nakamura T, Yamashita R, Iino T, Nishimura T, Nagata Y, Zhang BB, Eiki J-I (2009) Protective effect of glucokinase activation against hydrogen peroxide-induced cell death in a model of pancreatic β-Cells. 69th ADA Scientific Sessions, Diabetes 58(Suppl 1), Abstract #1569-P, A405

Fyfe MCT, White JR, Taylor A, Chatfield R, Wargent E, Printz RL, Sulpice T, McCormack JG, Procter MJ, Reynet C, Widdowson PS, Wong-Kai-In P (2007) Glucokinase activator PSN-GK1 displays enhanced antihyperglycaemic and insulinotropic actions. Diabetologia 50: 1277–1287

Glaser B, Kesavan P, Heyman M, Davis E, Cuesta A, Buchs A, Stanley CA, Thornton PS, Permutt MA, Matschinsky FM, Herold KC (1998) Familial hyperinsulinism caused by an activating glucokinase mutation. N Engl J Med 338(4):226–230

Gleason CE, Gross DN, Birnbaum MJ (2007) When the usual insulin is just not enough. Proc Natl Acad Sci USA 104(21):8681–8682

Gloyn AL, Noordam K, Willemsen MAAP, Ellard S, Lam WWK, Campbell IW, Midgley P, Shiota C, Buettger C, Magnuson MA, Matschinsky FM, Hattersley AT (2003) Insights into the biochemical and genetic basis of glucokinase activation from naturally occurring hypoglycemia mutations. Diabetes 52:2433–2440

Gloyn AL, Odili S, Buettger C, Njølstad PR, Shiota C, Magnuson MA, Matschinsky FM (2004) Glucokinase and the regulation of blood sugar. A mathematical model predicts the threshold for glucose stimulated insulin release for GCK gene mutations that cause hyper- and hypoglycemia. In: Matschinsky FM, Magnuson MA (eds) Glucokinase and glycemic disease: from basics to novel therapeutics. Front diabetes, vol 16. Karger, Basel, pp 92–109

González-Barroso MM, Giurgea I, Bouillaud F, Anedda A, Bellanné-Chantelot C, Hubert L, de Keyzer Y, de Lonlay P, Picquier D (2008) Mutations in UCP2 in congenital hyperinsulinism reveal a role for regulation of insulin secretion. PLoS ONE 3(12):e3850 (1–8)

Grimsby J, Sarabu R, Bizzarro FT, Coffey JW, Chu C-A, Corbett WL, Dvorozniak MT, Guertin KR, Haynes N-E, Hilliard DW, Kester RF, Matschinsky FM, Mahaney PE, Marcus LM, Qi L, Spence CL, Tengi JP, Grippo JF (2001) Allosteric activation of islet and hepatic glucokinase: a potential new approach to diabetes therapy (Abstract 460-P; 61st ADA Meeting Philadelphia). Diabetes 50(Suppl 2):A115

Grimsby J, Matschinsky FM, Grippo JF (2004) Discovery and actions of glucokinase activators. In: Matschinsky FM, Magnuson MA (eds) Glucokinase and glycemic disease: from basics to novel therapeutics. Front diabetes, vol 16. Karger, Basel, pp 360–378

Grimsby J, Sarabu R, Corbett WL, Haynes NE, Bizzarro FT, Coffey JW, Guertin KR, Hilliard DW, Kester RF, Mahaney PE, Marcus L, Qi L, Spence CL, Tengi J, Magnuson MA, Chu CA, Dvorozniak MT, Matschinsky FM, Grippo JF (2003) Allosteric activators of glucokinase: potential role in diabetes therapy. Science 301:370–373

Gromada J (2006) The free fatty acid receptor GPR40 generates excitement in pancreatic beta-cells. Endocrinology 147(2):672–673

Guertin KR, Grimsby J (2006) Small molecule glucokinase activators as glucose lowering agents: a new paradigm for diabetes therapy. Curr Med Chem 13:1839–1843

Hattersley AT, Turner RC, Permutt MA, Patel P, Tanizawa Y, Chiu KC, O'Rahilly S, Watkins PJ, Wainscoat JS (1992) Linkage of type 2 diabetes to the glucokinase gene. Lancet 339: 1307–1310

Henquin JC (2000) Triggering and amplifying pathways of regulation of insulin secretion by glucose. Diabetes 49:1751–1760

Heredia VV, Thomson J, Nettleton D, Sun S (2006) Glucose-induced conformational changes in glucokinase mediate allosteric regulation: transient kinetic analysis. Biochemistry 45: 7553–7562

Hille B, Tse A, Tse FW, Bosma MM (1995) Signaling mechanisms during the response of pituitary gonadotropes to GnRH. Recent Prog Horm Res 50:75–95

Iynedjian PB (1993) Mammalian glucokinase and its gene. Biochem J 293:1–13

Iynedjian PB (2004) Molecular biology of glucokinase regulation. In: Matschinsky FM, Magnuson MA (eds) Glucokinase and glycemic disease: from basics to novel therapeutics. Front diabetes, vol 16. Karger, Basel, pp 155–168

Iynedjian PB (2008) Molecular physiology of mammalian glucokinase. Cell Mol Life Sci 66:27–42. doi:10.1007/s00018-008-8322-9

Jetton TL, Liang Y, Pettepher CC, Zimmerman EC, Cox FG, Horwath K, Matschinsky FM, Magnuson MA (1994) Analysis of upstream glucokinase promoter activity in transgenic mice and identification of glucokinase in rare neuroendocrine cells in the brain and gut. J Biol Chem 269:3641–3654

Johnson D, Shepherd RM, Gill D, Gorman T, Smith DM, Dunne MJ (2007) Glucose-dependent modulation of insulin secretion and intracellular calcium ions by GKA50, a glucokinase activator. Diabetes 56:1694–1702

Joost HG, Bell GI, Best JD, Birnbaum MJ, Charron MJ, Chen YT, Doege H, James DE, Lodish HF, Moley KH (2002) Nomenclature of the GLUT/SLC2A family of sugar/polyol transport facilitators. Am J Physiol Endocrinol Metab 282:E974–E976

Kahn SE (2003) The relative contributions of insulin resistance and beta-cell dysfunction to the pathophysiology of type 2 diabetes. Diabetologia 46:3–19

Kahn SE, Zraika S, Utzschneider KM, Hull RL (2009) The beta cell lesion in type 2 diabetes: there has to be a primary functional abnormality. Diabetologia 52(6):1003–1012

Kamata K, Mitsuya M, Nishimura T, Eiki J, Nagata Y (2004) Structural basis for allosteric regulation of the monomeric allosteric enzyme human glucokinase. Structure 12:429–438

Kapoor RR, James C, Flanagan SE, Ellard S, Eaton S, Hussain K (2009) 3-Hydroxyacyl-coenzyme a dehydrogenase deficiency and hyperinsulinemic hypoglycemia: characterization of a novel mutation and severe dietary protein sensitivity. J Clin Endocrinol Metab 94(7):2221–2225

Kim YB, Kalinowski SS, Marcinkeviciene J (2007) A pre-steady state analysis of ligand binding to human glucokinase: evidence for a preexisting equilibrium. Biochemistry 46:1423–1431

Kulkarni RN, Bruning JC, Winnay JN, Postic C, Magnuson MA, Kahn CR (1999) Tissue-specific knockout of the insulin receptor in pancreatic beta cells creates an insulin secretory defect similar to that in type 2 diabetes. Cell 96:329–339

Leibiger B, Leibiger IB, Moede T, Kemper S, Kulkarni RN, Kahn CR, de Vargas LM, Berggren PO (2001) Selective insulin signaling through A and B insulin receptors regulates transcription of insulin and glucokinase genes in pancreatic beta cells. Mol Cell 7:559–570

Leighton B, Atkinson A, Coope GJ, Coghlan MP (2007) Improved glycemic control after sub-acute administration of a glucokinase activator to male Zucker (fa/fa) rats. Diabetes 56: 0377–OR

LeRoith D, Taylor SI, Olefsky JM (eds) (2004) Diabetes mellitus: a fundamental and clinical text, 3rd edn. Lippincott Williams and Wilkins, Philadelphia, PA

Levin BE, Routh VH, Kang L, Sanders NM, Dunn-Meynell AA (2004) Neuronal glucosensing: what do we know after 50 years? Diabetes 53:2521–2528

Levin BE, Becker TC, Eiki J-I, Zhang BB, Dunn-Meynell AA (2008) Ventromedial hypothalamic glucokinase is an important mediator of the counterregulatory response to insulin-induced hypoglycemia. Diabetes 57:1371–1379

Li C, Liu C, Nissim I, Chen P, Doliba N, Nissim I, Daikhin Y, Stokes D, Yudkoff M, Stanley CA, Matschinsky FM, Naji A (2009) Islets of human type 2 diabetics have decreased gaba shunt and dysregulated glucose metabolism. 69th ADA scientific sessions. Diabetes 58(Suppl 1), Abstract #648-P, A424

Liang Y, Najafi H, Matschinsky FM (1990) Glucose regulates glucokinase activity in cultured islets from rat pancreas. J Biol Chem 265:16863–16866

Liang Y, Najafi H, Smith RM, Zimmerman EC, Magnuson MA, Tal M, Matschinsky FM (1992) Concordant glucose induction of glucokinase, glucose usage and glucose stimulated insulin release in pancreatic islets maintained in organ culture. Diabetes 41:792–806

Liang Y, Bonner-Weir S, Wu YJ, Berdanier CD, Berner DK, Efrat S, Matschinsky FM (1994) In situ glucose uptake and glucokinase activity of pancreatic islets in diabetic and obese rodents. J Clin Invest 93:2473–2481

Lin SX, Neet KE (1990) Demonstration of a slow conformational change in liver glucokinase by fluorescence spectroscopy. J Biol Chem 265:9670–9675

Lyssenko V, Jonsson A, Almgren P, Pulizzi N, Isomaa B, Tuomi T, Berglund G, Altshuler D, Nilsson P, Groop L (2008) Clinical risk factors, DNA variants, and the development of type 2 diabetes. N Engl J Med 359(21):2220–2232

Lyssenko V, Nagorny CL, Erdos MR, Wierup N, Jonsson A, Spégel P, Bugliani M, Saxena R, Fex M, Pulizzi N, Isomaa B, Tuomi T, Nilsson P, Kuusisto J, Tuomilehto J, Boehnke M, Altshuler D, Sundler F, Eriksson JG, Jackson AU, Laakso M, Marchetti P, Watanabe RM, Mulder H, Groop L (2009) Common variant in MTNR1B associated with increased risk of type 2 diabetes and impaired early insulin secretion. Nat Genet 41(1):82–88

Magnuson MA, Kim K-A (2004) Mouse models of altered glucokinase gene expression. In: Matschinsky FM, Magnuson MA (eds) Glucokinase and glycemic disease: from basics to novel therapeutics. front diabetes, vol 16. Karger, Basel, pp 289–300

Malaisse WJ, Sener A (1985) Glucokinase is not the pancreatic beta-cell glucoreceptor. Diabetologia 28:520–527

Matschinsky FM (1990) Glucokinase as glucose sensor and metabolic signal generator in pancreatic β-cells and hepatocytes. Diabetes 39:647–652

Matschinsky FM (1996) BANTING LECTURE 1995: a lesson in metabolic regulation inspired by the glucokinase glucose sensor paradigm. Diabetes 45:223–241

Matschinsky FM (2008) Glucokinase in glucose homeostasis, diabetes mellitus, hypoglycemia, and as drug receptor. In: Susumu S, Bell GI (eds) Pancreatic beta cell in health and disease. Springer, Berlin, pp 451–463

Matschinsky FM (2009) Assessing the potential of glucokinase activators in diabetes therapy. Nat Rev Drug Discov 8(5):399–416

Matschinsky FM, Ellerman JE (1968) Metabolism of glucose in the islets of Langerhans. J Biol Chem 243:2730–2736

Matschinsky FM, Magnuson MA, Zelent D, Jetton TL, Doliba N, Han Y, Taub R, Grimsby J (2006) The network of glucokinase-expressing cells in glucose homeostasis and the potential of glucokinase activators for diabetes therapy. Diabetes 55(1):1–12

McCarthy MI, Froguel P (2002) Genetic approaches to the molecular understanding of type 2 diabetes. Am J Physiol Endocrinol Metab 283:E217–E225

Meglasson MD, Matschinsky FM (1984) New perspectives on pancreatic islet glucokinase. Am J Physiol Endocrinol Metab 246:E1–E13

Meglasson MD, Matschinsky FM (1986) Pancreatic islet glucose metabolism and regulation of insulin secretion. Diabetes Metab Rev 2:163–214

Meigs JB, Shrader P, Sullivan LM, McAteer JB, Fox CS, Dupuis J, Manning AK, Florez JC, Wilson PW, D'Agostino RB Sr, Cupples LA (2008) Genotype score in addition to common risk factors for prediction of type 2 diabetes. N Engl J Med 359(21):2208–2219

Migoya EM, Miller J, Maganti L, Gottesdeiner K, Wagner JA (2009) The Glucokinase (GK) Activator MK-0599 Lowers plasma glucose concentrations in healthy non-diabetic subjects. 69th ADA scientific sessions. Diabetes 58(Suppl 1), Abstract #116-OR, A31

Miwa I, Toyoda Y, Yoshie S (2004) Glucokinase in β-cell insulin-secretory granules. In: Matschinsky FM, Magnuson MA (eds) Glucokinase and glycemic disease: from basics to novel therapeutics. front diabetes, vol 16. Karger, Basel, pp 350–359

Molnes J, Bjørkhaug L, Søvik O, Njølstad PR, Flatmark T (2008) Catalytic activation of human glucokinase by substrate binding: residue contacts involved in the binding of D-glucose to the super-open form and conformational transitions. FEBS J 275(10):2467–2481

Mueckler M (1994) Facilitative glucose transporters. Eur J Biochem 219:713–725

Murphy R, Tura A, Clark PM, Holst JJ, Mari A, Hattersley AT (2008) Glucokinase, the pancreatic glucose sensor, is not the gut glucose sensor. Diabetologia 52:154–159

Nakamura A, Takamoto I, Kubota N, et al (2007) Impact of small molecule glucokinase activator on glucose metabolism in response to high fat diet in mice with ß-cell specific haploinsufficiency of glucokinase gene. Diabetes 56:529-P

Neet KE, Ainslie GR (1980) Hysteretic enzymes. Methods Enzymol 64:192–226

Newgard CB, Matschinsky FM (2001) Substrate control of insulin release: the endocrine system. In: Handbook of physiology, vol II, The endocrine pancreas and regulation of metabolism. Oxford University Press, Oxford, pp 125–151

Njølstad PR, Sovic O, Cuesta-Munoz A, Bjorkhaug L, Massa O, Barbetti F, Undlien DE, Shiota C, Magnuson MA, Molven A, Matschinsky FM, Bell GI (2001) Neonatal diabetes mellitus due to complete glucokinase deficiency. N Engl J Med 344(21):1588–1592

Njølstad PR, Sagen JV, Bjørkhaug L, Odili S, Shehadeh N, Bakry D, Sarici SU, Molnes J, Molven A, Søvik O, Matschinsky FM (2003) Permanent neonatal diabetes mellitus due to glucokinase deficiency – an inborn error of the glucose-insulin signaling pathway. Diabetes 52:2854–2860

Nolte MS, Karam JH (2007) Pancreatic hormones and antidiabetic drugs. In: Katzung BG (ed) Basic and clinical pharmacology, 10th edn. The McGraw-Hill Companies, Boston, MA, pp 683–705

O'Doherty RM, Lehman DL, Telemaque-Potts S, Newgard CB (1999) Metabolic impact of glucokinase overexpression in liver: lowering of blood glucose in fed rats is accompanied by hyperlipidemia. Diabetes 48:2022–2027

Ohyama S, Takano H, Satow A, Fukuroda T, Iino T, Nishimura T, Zhang BB, Eiki J-I (2009) a small molecule glucokinase activator lowers blood glucose in rats desensitized to sulfonylurea agents 69th ADA scientific sessions. Diabetes 58(Suppl 1), Abstract #1537-P, A397

Okada T, Liew CW, Hu J, Hinault C, Michael MD, Krutzfeldt J, Yin C, Holzenberger M, Stoffel M, Kulkarni RN (2007) Insulin receptors in beta-cells are critical for islet compensatory growth response to insulin resistance. Proc Natl Acad Sci USA 104:8977–8982

Online Mendelian Inheritance in Man (OMIM) #606391 Maturity-Onset Diabetes of the Young; MODY; http://www.ncbi.nlm.nih.gov/entrez/dispomim.cgi?id=606391

Orrenius S, Zhivotovsky B, Nicotera P (2003) Regulation of cell death: the calcium-apoptosis link. Nat Rev Mol Cell Biol 4:552–564

Palladino AA, Bennett MJ, Stanley CA (2008) Hyperinsulinism in infancy and childhood: when an insulin level is not always enough. Clin Chem 54(2):256–263

Pearson ER (2009) Translating TCF7L2: from gene to function. Diabetologia 52(7):1227–1230

Pilgaard K, Jensen CB, Schou JH, Lyssenko V, Wegner L, Brøns C, Vilsbøll T, Hansen T, Madsbad S, Holst JJ, Vølund A, Poulsen P, Groop L, Pedersen O, Vaag AA (2009) The T allele of rs7903146 TCF7L2 is associated with impaired insulinotropic action of incretin hormones, reduced 24 h profiles of plasma insulin and glucagon, and increased hepatic glucose production in young healthy men. Diabetologia 52(7):1298–1307

Postic C, Shiota M, Magnuson MA (2001) Cell-specific roles of glucokinase in glucose homeostasis. Recent Prog Horm Res 56:195–218

Prentki M, Matschinsky FM (1987) Ca2+, cAMP and phosphoinositide derived messengers in the coupling mechanisms of insulin secretion. Physiol Rev 67:1185–1248

Qian-Cutrone J, Ueki T, Huang S, Mookhtiar KA, Ezekiel R, Kalinowski SS, Brown KS, Golik J, Lowe S, Pirnik DM, Hugill R, Veitch JA, Klohr SE, Whitney JL, Manly SP (1999) Glucolipsin A and B, two new glucokinase activators produced by Streptomyces purpurogenisclEroticus and Nocardia vaccinii. J Antibiot (Tokyo) 52(3):245–255

Ralph EC, Thomson J, Almaden J, Sun S (2008) Glucose modulation of glucokinase activation by small molecules. Biochemistry 47(17):5028–5036

Randle PJ, Ashcroft SJH, Gill R (1968) Carbohydrate metabolism and release of hormones. In: Dickens F, Randle PJ, Whelan WJ (eds) Carbohydrate metabolism and its disorders, vol 1. Academic, London, pp 427–447

Reaven GM (1988) Role of insulin resistance in human disease. Diabetes 37:1595–1607

Reimann F, Williams L, da Silva XG, Rutter GA, Gribble FM (2004) Glutamine potently stimulates Glucagon-Like Peptide-1 secretion from GLUTag cells. Diabetologia 47:1592–1601

Reimann F, Ward PS, Gribble FM (2006) Signalling mechanisms underlying the release of glucagon-like peptide-1. Diabetes 55(Suppl 2):S78–S85

Reimann F, Habib AM, Tolhurst G, Parker HE, Rogers GJ, Gribble FM (2008) Glucose sensing in L cells: a primary cell study. Cell Metab 8(6):532–539

Ridderstråle M, Groop L (2009) Genetic dissection of type 2 diabetes. Mol Cell Endocrinol 297(1–2):10–17

Sagen JV, Odili S, Bjorkhaug L, Zelent D, Buettger C, Kwagh J, Stanley C, Dahl-Jorgensen K, de Beaufort C, Bell GI, Han Y, Grimsby J, Taub R, Molven A, Sovik O, Njolstad PR, Matschinsky FM (2006) From clinicogenetic studies of maturity-onset diabetes of the young to unraveling complex mechanisms of glucokinase regulation. Diabetes 55(6):1713–1722

Sarabu R, Grimsby J (2005) Targeting glucokinase activation for the treatment of type 2 diabetes – a status review. Curr Opin Drug Discov Devel 8(5):631–637

Sarabu R, Berthel SJ, Kester RF, Tilley JW (2008) Glucokinase activators as new type 2 diabetes therapeutic agents. Expert Opin Ther Pat 18(7):759–768

Sayed S, Langdon DR, Odili S, Chen P, Buettger C, Schiffman AB, Suchi M, Taub R, Grimsby J, Matschinsky FM, Stanley CA (2009) Extremes of clinical and enzymatic phenotypes in children with hyperinsulinism caused by glucokinase activating mutations. Diabetes 58:1419–1427

Scorrano L, Oakes SA, Opferman JT, Cheng EH, Sorcinelli MD, Pozzan T, Korsmeyer SJ (2003) Regulation of endoplasmic reticulum Ca2+ dynamics by BAX, BAK is a control point for apoptosis. Science 300:135–139

Sharma C, Manjeshwar R, Weinhouse S (1964) Hormonal and dietary regulation of hepatic glucokinase. Adv Enzyme Regul 2:189–200

Sols A, Salas M, Vinuela E (1964) Induced biosynthesis of liver glucokinase. Adv Enzyme Regul 2:177–188

Sorenson RL, Stout LE, Brelje TC, Jetton T, Matschinsky FM (2007) Immunohistochemical evidence for the presence of glucokinase in the gonadotropes and thyrotropes of the anterior pituitary gland of rat and monkey. J Histochem Cytochem 55(6):555–566

Sorhede Winzell M, Coghlan MP, et al (2007) Glucokinase activation reduces glycemia and improves glucose tolerance in mice with high-fat diet-induced insulin resistance. Diabetes 56:1482–P

Sparsø T, Grarup N, Andreasen C, Albrechtsen A, Holmkvist J, Andersen G, Jørgensen T, Borch-Johnsen K, Sandbaek A, Lauritzen T, Madsbad S, Hansen T, Pedersen O (2009) Combined analysis of 19 common validated type 2 diabetes susceptibility gene variants shows moderate discriminative value and no evidence of gene-gene interaction. Diabetologia 52(7):1308–1314

Sweet IR, Najafi H, Li G, Berner DK, Matschinsky FM (1996) Effect of a glucokinase inhibitor on energy production and insulin release in pancreatic islets. Am J Physiol 271:E606–E625

Takamoto I, Terauchi Y, Kubota N, Ohsugi M, Ueki K, Kadowaki T (2008) Crucial role of insulin receptor substrate-2 in compensatory beta-cell hyperplasia in response to high fat diet-induced insulin resistance. Diabetes Obes Metab 10(Suppl 4):147–156

Tal M, Liang Y, Najafi H, Lodish HF, Matschinsky FM (1992) Expression and function of GLUT-1 and GLUT-2 glucose transporter isoforms in cells of cultured rat pancreatic islets. J Biol Chem 267:17241–17247

Terauchi Y, Takamoto I, Kubota N, Matsui J, Suzuki R, Komeda K, Hara A, Toyoda Y, Miwa I, Aizawa S (2007) Glucokinase and IRS-2 are required for compensatory beta cell hyperplasia in response to high-fat diet-induced insulin resistance. J Clin Invest 117:246–257

Thorens B, Charron MJ, Lodish HF (1990) Molecular physiology of glucose transporters. Diab Care 13(3):209–218

Thorens B (2004) The hepatoportal glucose sensor. In: Matschinsky FM, Magnuson MA (eds) Glucokinase and glycemic disease: from basics to novel therapeutics. Front diabetes, vol 16. Karger, Basel, pp 327–338

Torres TP, Cathin KL, Chau R, Fujimoto Y, Sasaki N, Printz RL, Newgard CB, Shiota M (2009) Restoration of hepatic glucokinase expression corrects hepatic glucose flux and normalizes plasma glucose in zucker diabetic fatty rats. Diabetes 58:78–86

Vandercammen A, Van Schaftingen E (1990) The mechanism by which rat liver glucokinase is inhibited by the regulatory protein. Eur J Biochem 191:483–489

Vandercammen A, Van Schaftingen E (1991) Competitive inhibition of liver glucokinase by its regulatory protein. Eur J Biochem 200:545–551

Veiga-da-Cunha M, Van Schaftingen E (2002) Identification of fructose 6-phosphate- and fructose 1-phosphate-binding residues in the regulatory protein of glucokinase. J Biol Chem 277: 8466–8473

Walker DG, Rao S (1964) The role of glucokinase in the phosphorylation of glucose by rat liver. Biochem J 90:360–368

Weedon MN, McCarthy MI, Hitman G, Walker M, Groves CJ, Zeggini E, Rayner NW, Shields B, Owen KR, Hattersley AT, Frayling TM (2006) Combining information from common type 2 diabetes risk polymorphisms improves disease prediction. PLoS Med 3(10):e374

Wess J, Eglen RM, Gautam D (2007) Muscarinic acetylcholine receptors: mutant mice provide new insights for drug development. Nat Rev Drug Discov 6:721–733

Whyte MKB, Hardwick SJ, Meagher LC, Savill JS, Haslett C (1993) Transient elevation of cytosolic free calcium retard subsequent apoptosis in neutrophils in vitro. J Clin Invest 92:446–455

Wilms B, Ben-Ami P, Söling HD (1970) Hepatic enzyme activities of glycolysis and gluconeogenesis in diabetes of man and laboratory animals. Horm Metab Res 2:135–141

Wilson JE (2004) The hexokinase gene family. In: Matschinsky FM, Magnuson MA (eds) Glucokinase and glycemic disease: from basics to novel therapeutics. front diabetes, vol 16. Karger, Basel, pp 18–30

Xu LZ, Weber IT, Harrison RW, Gidh-Jain M, Pilkis SJ (1995) Sugar specificity of human pancreatic b-cell glucokinase. Biochemistry 34:6083–6092

Zelent D, Najafi H, Odili S, Buettger C, Weik-Collins H, Li C, Doliba N, Grimsby J, Matschinsky FM (2005) Glucokinase and glucose homeostasis: proven concepts and new ideas. Biochem Soc Trans 33:306–310

Zelent D, Golson ML, Koeberlein B, Quintens R, van Lommel L, Buettger C, Weik-Collins H, Taub R, Grimsby J, Schuit F, Kaestner KH, Matschinsky FM (2006) A glucose sensor role for glucokinase in anterior pituitary cells. Diabetes 55(7):1923–1929

Zelent B, Odili S, Buettger C, Shiota C, Grimsby J, Taub R, Magnuson MA, Vanderkooi JM, Matschinsky FM (2008) Sugar binding to recombinant wild-type and mutant glucokinase monitored by kinetic measurement and tryptophan fluorescence. Biochem J 413(2):269–280

Zhai S, Mulligan ME, Grimsby J, Arbet-Engels C, Boldrin M, Balena R, Zhi J (2008) Phase I assessment of a novel glucose activator RO4389620 in healthy male volunteers. Diabetologia S1(Suppl 1):S372 (Abstract #928)

Zhang J, Li C, Chen K et al (2006) Conformational transition pathway in the allosteric process of human glucokinase. Proc Natl Acad Sci USA 103:13368–13373

Zhi J, Zhai S, Mulligan ME, Grimsby J, Arbet-Engels C, Boldrin M, Balena R (2008) A novel glucokinase activator RO4389620 improved fasting and postprandial plasma glucose in type 2 diabetic patients. Diabetologia 51(Suppl 1):S23 (Abstract # 42)

References for Addendum

Bebernitz GR, Beaulieu V, Dale BA, Deacon R, Duttaroy A, Gao J, Grondine MS, Gupta RC, Kakmak M, Kavana M, Kirman LC, Liang J, Maniara WM, Munshi S, Nadkarni SS, Schuster HF, Stams T, St Denny I, Taslimi PM, Vash B, Caplan SL (2009) Investigation of functionally liver selective glucokinase activators for the treatment of type 2 diabetes. J Med Chem 52(19):6142–6152

Bonadonna RC, Heise T, Arbet-Engels C, Kapitza C, Avogaro A, Grimsby J, Zhi J, Grippo JF, Balena R (2010) Piragliatin (RO4389620), a novel glucokinase activator, lowers plasma glucose both in the postabsorptive state and after a glucose challenge in patients with type

2 diabetes mellitus: a mechanistic study. J Clin Endocrinol Metab. Aug 25, as doi: 10.1210/ji.2010-1041

Haynes NE, Corbett WL, Bizzarro FT, Guertin KR, Hilliard DW, Holland GW, Kester RF, Mahaney PE, Qi L, Spence CL, Tengi J, Dvorozniak MT, Railkar A, Matschinsky FM, Grippo JF, Grimsby J, Sarabu R (2010) Discovery, structure-activity relationships, pharmacokinetics, and efficacy of glucokinase activator (2R)-3-cyclopentyl-2-(4-methanesulfonylphenyl)-N-thiazol-2-yl-propionamide (RO0281675). J Med Chem 53(9):3618–3625

Kassem S, Heyman M, Glaser B, Bhandari S, Motaghedi R, Maclaren NK, García-Gimeno MA, Sanz P, Rahier J, Rodríguez-Bada P, Cobo-Vuilleumier N, Cuesta-Muñoz AL (2010) Large islets, beta-cell proliferation, and a glucokinase mutation. N Engl J Med 362(14):1348–1350

Kozian DH, Barthel A, Cousin E, Brunnhöfer R, Anderka O, März W, Böhm B, Winkelmann B, Bornstein SR, Schmoll D (2010) Glucokinase-activating GCKR polymorphisms increase plasma levels of triglycerides and free fatty acids, but do not elevate cardiovascular risk in the Ludwigshafen Risk and Cardiovascular Health Study. Horm Metab Res 42(7):502–506

Nakamura A, Terauchi Y, Ohyama S, Kubota J, Shimazaki H, Nambu T, Takamoto I, Kubota N, Eiki J, Yoshioka N, Kadowaki T, Koike T (2009) Impact of small-molecule glucokinase activator on glucose metabolism and beta-cell mass. Endocrinology 150(3):1147–1154

News release of 8/10/09; thangeto@arraybiopharma.com

Ohyama S, Takano H, Iino T, Nishimura T, Zhou YP, Langdon RB, Zhang BB, Eiki J (2010) A small-molecule glucokinase activator lowers blood glucose in the sulfonylurea-desensitized rat. Eur J Pharmacol 640(1–3):250–256

Osbak KK, Colclough K, Saint-Martin C, Beer NL, Bellanné-Chantelot C, Ellard S, Gloyn AL (2009) Update on mutations in glucokinase (GCK), which cause maturity-onset diabetes of the young, permanent neonatal diabetes, and hyperinsulinemic hypoglycemia. Hum Mutat 30(11):1512–1526, Review

Salpeter SJ, Klein AM, Huangfu D, Grimsby J, Dor Y (2010) Glucose and aging control the quiescence period that follows pancreatic beta cell replication. Development 137(19):3205–3213

Sidduri A, Grimsby JS, Corbett WL, Sarabu R, Grippo JF, Lou J, Kester RF, Dvorozniak M, Marcus L, Spence C, Racha JK, Moore DJ (2010) 2, 3-Disubstituted acrylamides as potent glucokinase activators. Bioorg Med Chem Lett 20(19):5673–5676

Zhang L, Li H, Zhu Q, Liu J, Chen L, Leng Y, Jiang H, Liu H (2009) Benzamide derivatives as dual-action hypoglycemic agents that inhibit glycogen phosphorylase and activate glucokinase. Bioorg Med Chem 17(20):7301–7312

Index

A

Acarbose, 106
ACCORD trial, 67–68
Acetyl-CoA carboxylase, 222
Acetyl-CoA:diacylglycerol acyltransferase (DGAT), 225
Adenosine monophosphate kinase (AMPK), 13–14, 17, 189
Adipocytes, 82
Adipokine, 37
Adiponectin, 211, 220, 221, 230, 306, 310, 313, 314, 317, 319–321
Adipose tissue, 2–8, 17, 18
β-Adrenoceptor agonists, 206
Affinity, 107–109, 111–113
AFM studies, 111
2-AG. *See* 2-Arachidonoylglycerol
Aglucone binding, 111–114
   sites, 113
Agouti-related peptide (AGRP), 12
AGRP. *See* Agouti-related peptide
AICAR. *See* 5-Aminoimidazole–4-carboxamide–1-β-D-ribofuranoside
Albiglutide, 58–59
Alkylglucosides, 113
Allosteric activator, 365, 366, 368, 372, 373, 378–381
Alogliptin, 63
5-Aminoimidazole–4-carboxamide–1-β-D-ribofuranoside (AICAR), 309, 312, 314–318, 320–321
AMP-activated protein kinase (AMPK), 303–321
AMPK. *See* Adenosine monophosphate kinase; AMP-activated protein kinase
AMP mimetic, 286–288, 296
Amyloid polypeptide, 11
Anandamide, 76

B

Animal models, 117
Antidiabetic drugs, 359, 360, 374, 388, 389
Apoptosis, 156, 157
Appetite, 78, 193
Arachidonic acid, 76
2-Arachidonoylglycerol (2-AG), 76
Arterial hypertension, 10
Atherosclerosis, 156
Atomic force microscopy (AFM) studies, 112
ATP-sensitive potassium channel, 10, 12

B

Bezafibrate, 40–41
11β-Hydroxysteroid dehydrogenase–1 (11β-HSD1), 128, 215
Binding pocket, 111, 113
Binding site, 108–111, 113
Binding studies, 112
Blood glucose, 115, 117, 119, 120
Blood sugar, 106
Body weight, 118–120
Brain-derived neurotrophic factor, 220
Bromocriptin, 14
Brown adipose tissue, 190
Brush border, 107, 108, 112, 116

C

Calorie restriction, 306, 311, 321
Canagliflozin, 115, 118, 120
Cannabinoid CB1 receptor, 76
Cannabinoid CB2 receptor, 76
Cannabinoids, 96
Cardiac muscle, 339
Cardiovascular disease (CVD), 10, 14–18
Carnitine palmitoyltransferase–1 (CPT–1), 10–13
Cav1. *See* Caveolin–1
Caveolae, 166, 168–170, 173–175

M. Schwanstecher (ed.), *Diabetes - Perspectives in Drug Therapy*, Handbook of Experimental Pharmacology 203, DOI 10.1007/978-3-642-17214-4, © Springer-Verlag Berlin Heidelberg 2011